T0320645

TRANSACTIONS OF THE
INTERNATIONAL ASTRONOMICAL UNION
VOLUME XXVIIIA

REPORTS ON
ASTRONOMY

INTERNATIONAL ASTRONOMICAL UNION

UNION ASTRONOMIQUE INTERNATIONALE

International Astronomical Union

REPORTS

ON

ASTRONOMY

TRANSACTIONS
OF THE
INTERNATIONAL ASTRONOMICAL UNION
VOLUME XXVIIIA

Edited by

IAN F. CORBETT
General Secretary

CAMBRIDGE
UNIVERSITY PRESS

Shaftesbury Road, Cambridge CB2 8EA, United Kingdom

One Liberty Plaza, 20th Floor, New York, NY 10006, USA

477 Williamstown Road, Port Melbourne, VIC 3207, Australia

314–321, 3rd Floor, Plot 3, Splendor Forum, Jasola District Centre, New Delhi – 110025, India

103 Penang Road, #05–06/07, Visioncrest Commercial, Singapore 238467

Cambridge University Press is part of Cambridge University Press & Assessment, a department of the University of Cambridge.

We share the University's mission to contribute to society through the pursuit of education, learning and research at the highest international levels of excellence.

www.cambridge.org
Information on this title: www.cambridge.org/9781107019874

First published 2012

A catalogue record for this publication is available from the British Library

ISBN   978-1-107-01987-4   Hardback

Table of Contents

Preface

As background information to the IAU XXVIII General Assembly in Beijing 20–31 August 2012, these Transactions provide an account of what has been happening within the IAU through the words of the Presidents and Chairs of the Divisions, Commssions and Working Groups which make up the working levels of the IAU. These were completed towards the end of 2011 and provide a narrative record of activities over the triennium 2009–2011. For a variety of reasons, not all the Commissions and Working Groups submitted reports.

It is my pleasure to thank those IAU Division and Commission Presidents and Working & Program Group chairpersons who have provided these reports and the staff of the IAU Secretariat, Mme Vivien Reuter and Mme Jana Žilová, for their continued invaluable assistance in making the IAU function. We look forward to a most memorable XXVIII General Assembly in the historic city of Beijing.

Ian F. Corbett
IAU General Secretary
Paris, January 2012

Transactions IAU, Volume XXVIIIA
Reports on Astronomy 2009–2012   © International Astronomical Union 2012
Ian F. Corbett, ed.   doi:10.1017/S1743921312002566

DIVISION I   FUNDAMENTAL ASTRONOMY
ASTRONOMIE FONDAMENTALE

Division I provides a focus for astronomers studying a wide range of problems related to fundamental physical phenomena such as time, the inertial reference frame, positions and proper motions of celestial objects, and precise dynamical computation of the motions of bodies in stellar or planetary systems in the Universe.

PRESIDENT	Dennis D. McCarthy
VICE-PRESIDENT	Sergei A. Klioner
PAST PRESIDENT	Jan Vondrák
BOARD	Dafydd Wyn Evans, Catherine Y. Hohenkerk,
	Mizuhiko Hosokawa, Cheng-Li Huang,
	George H. Kaplan, Zoran Knežević,
	Richard N. Manchester, Alessandro Morbidelli,
	Gérard Petit, Harald Schuh,
	Michael H. Soffel, Norbert Zacharias

DIVISION I COMMISSIONS

Commission 4	Ephemerides
Commission 7	Celestial Mechanics and Dynamical Astronomy
Commission 8	Astrometry
Commission 19	Rotation of the Earth
Commission 31	Time
Commission 52	Relativity in Fundamental Astronomy

DIVISION I WORKING GROUPS

Division I WG	Nomenclature for Fundamental Astronomy,
Division I WG	Astrometry by Small Ground-Based Telescopes
Division I WG	Natural Satellites
Division I WG	Standards of Fundamental Astronomy (SOFA)

INTER-DIVISION WORKING GROUPS

Division I-III WG	Cartographic Coordinates and Rotational Elements
Division I-III WG	Natural Satellites

TRIENNIAL REPORT 2009–2012

1. Introduction

The goal of the division is to address the scientific issues that were developed at the 2009 IAU General Assembly in Rio de Janeiro. These are:

- Astronomical constants
    —Gaussian gravitational constant, Astronomical Unit, GM_{Sun}, geodesic precession-nutation

1

- •Astronomical software
- •Solar System Ephemerides
- —Pulsar research
- —Comparison of dynamical reference frames
- •Future Optical Reference Frame
- •Future Radio Reference Frame
- •Exoplanets
- —Detection
- —Dynamics
- •Predictions of Earth orientation
- •Units of measurements for astronomical quantities in relativistic context
- •Astronomical units in the relativistic framework
- •Time-dependent ecliptic in the GCRS
- •Asteroid masses
- •Review of space missions
- •Detection of gravitational waves
- •VLBI on the Moon
- •Real time electronic access to UT1-UTC

In pursuit of these goals Division I members have made significant scientific and organizational progress, and are organizing a Joint Discussion on *Space-Time Reference Systems for Future Research* at the 2012 IAU General Assembly. The details of Division activities and references are provided in the individual Commission and Working Group reports in this volume. A comprehensive list of references related to the work of the Division is available at the IAU Division I website at http://maia.usno.navy.mil/iaudiv1/.

2. Developments within the past triennium

The Journées 2010 meeting "New challenges for reference systems and numerical standards in astronomy" (http://syrte.obspm.fr/journees2010/)organized by the Paris Observatory and co-sponsored by the IAU was held September 2010, in Meudon, France. Among the topics was new information on the latest fundamental solar system ephemerides from JPL (Pasadena), IMCCE (Paris), and IAA (St. Petersburg). Links to all three ephemerides can be found at the Commission 4 web site.

The Journées 2011 meeting "Earth rotation, reference systems and celestial mechanics: Synergies of geodesy and astronomy" (http://info.tuwien.ac.at/hg/meetings/journees11/), organized by the Technical University of Vienna and co-sponsored by the IAU was held in September, 2011 in Vienna, Austria. It focused on issues related to recent developments in fundamental astronomy, time and relativity, plans for the next generation of space-time reference systems, astronomical space and time reference systems, Earth rotation and global geodynamics, celestial mechanics of solar system bodies, space observations and dedicated missions for geodesy and astronomy.

Thanks to a joint initiative by members of the Working Group on Astrometry by Small Ground Based Telescopes and astronomers from France (IMCCE), Turkey (Tubitak Obs.), Ukraine (Nikolaev Obs.) and China (Shanghai Obs. and Jinan Univ. of Guangzhou), a Summer School on Astrometry was held in Antalya (Turkey), on September 5-9, 2011. The purpose was to encourage students to do research in astrometry and to study new astrometric methods. Nineteen students and young astronomers attended lectures and observing sessions. A workshop "Astrometry Now and in the Future" was organized at the end of the school.

2.1. *Science*

Members of Commission 4 continued their work in the computation of fundamental solar system ephemerides and practical astronomical data, including geocentric and topocentric coordinates of stars and solar system objects; the prediction of astronomical phenomena, parameters describing the apparent orientation and illumination of solar system objects; and various quantities used to transform directions among standard reference systems. The Commission established a Working Group on Standardizing Access to Ephemerides. It is to provide guidance on a consistent format for ephemerides of solar system bodies for the astronomical community. The working group has tentatively decided to recommend the Spacecraft and Planetary Kernal (SPK) file format used in the SPICE Library written and maintained by the Navigation and Ancillary Information Facility (NAIF) of the Jet Propulsion Laboratory, but this depends on NAIF providing a detailed specification of those portions of the format needed for the ephemerides and a standalone routine for reading SPK files.

There is some uncertainty in how almanac-producing institutions will respond to a possible change in the definition of Coordinated Universal Time (UTC)if approved by the International Telecommunications Union in 2012. While this would not represent a major obstacle to Commission 4 affiliated institutions, it would require decisions on providing data to users in the future. Also, institutions computing fundamental solar system ephemerides will have to respond to the potential change in the definition of the astronomical unit.

Members of Commission 7 produced papers considering methods of determining orbital class and degree of chaos. Most are based on the calculation of the Lyapunov characteristic exponents and their derivatives. The role of chaotic orbits in the evolution and dynamics of spheroids was considered in several papers. Work was also done to establish the influence of chaos, and in particular of 'sticky' or 'confined' orbits, on galactic structures. Manifolds, linked to the unstable periodic orbits around the L1 (L2) Lagrange points are proposed to explain both the spirals and the rings in barred galaxies.

The chaotic behavior of planetary orbits in the Solar System was discovered using numerical integration of averaged equations of motion. This chaotic behavior manifests itself in an exponential divergence of nearby trajectories, which limits the validity of the solutions to less than 100 Myr. A numerical integration of the Solar System motion over 5 Gyr can only be considered as a random sample of its possible evolution and statistical studies are required. The search for the best solution for the Earth's orbit over millions of years is motivated by the possibility of calibrating recent geological timescales by correlating the geological stratigraphic data to the computed variation of the insolation on Earth.

Around 1% of solar-type stars host Jupiter-mass planets with the semimajor axis a less than 0.1 AU. It is thought that these so-called *hot Jupiters*, did not form *in situ* because the region within 0.1 AU was too hot and rarefied for a Jupiter-mass planet to form. Instead, they probably formed several AU from their host stars, and later migrated inward due to their interaction with the protoplanetary gas disk from which they formed.

The ICRF2 containing accurate positions of 3414 compact, extragalactic radio sources on the system of the original ICRF was adopted by the IAU in 2009 as the new fundamental celestial reference frame. It has a noise floor of $\sim$40 μas, some 5-6 times better than ICRF, and an axis stability of ~10 μas, nearly twice as stable as ICRF. Alignment of ICRF2 with the International Celestial Reference System (ICRS) was made using 138 stable sources common to both ICRF2 and ICRF-Ext2. The UCAC project concluded with the publication of its 4th data release (all-sky, over 100 million stars). IAU Commission 8 (Astrometry) also reports that the astrometric satellite Gaia which is expected to achieve parallax accuracies of about 10 to 300 μas for 6 to 20 mag. is scheduled to be launched in 2013. The JASMINE (Japan Astrometry Satellite Mission for Infrared Exploration) project proposes to launch three satellites [Nano-JASMINE, Small-JASMINE and (Medium-sized) JASMINE. It willl launch the Nano-JASMINE satellite in 2012 and continue development of its other satellites which will complement the Gaia project. The Joint Institute for VLBI in Europe (JIVE), in cooperation with the European VLBI Network (EVN), EC FP7 EuroPlaNet and ESPaCE and other partners, continued developing VLBI applications for ultra-precise tracking of planetary and other space science missions. Its high accuracy enables multi-disciplinary applications including celestial mechanics and astrometry of the Solar System, fundamental physics and studies of interiors and atmospheres of planets. In the reporting period, the activities focused on the Planetary Radio Interferometry and Doppler Experiment (PRIDE)

The WG on Natural Planetary Satellites encouraged the making of astrometric observations of all planetary satellites. A campaign of observations of the mutual events of the main satellites of Jupiter and Saturn has been made in 2009. These photometric observations provide highly accurate astrometric positions. The WG encouraged also progress in astrometric reductions in order to prepare for the arrival of the Gaia star catalogue. The International Occultation Timing Association (IOTA) reports that over the reporting period some 400 high-precision relative astrometric positions of asteroids derived from occultation observations were reported to the Minor Planet Centre. Over the reporting period, profiles with a resolution of a few km were obtained for about 35 asteroids. This included the binary asteroid 90 Antiope, where the profiles of both components were well-resolved. Occultation results have been combined with light curve inversion models to obtain scaled 3-D models of 44 asteroids. All observations are archived with NASA's Planetary Data System (http://sbn.psi.edu/pds/resource/occ.html).

During the past triennium a new Commission 19 website was established providing the Commission's terms of reference, membership directory,information on upcoming scientific meeting and the history of the Commission. Three workshops about Earth orientation variations were held during the past triennium: (1) a joint Upper Mantle Dynamics and Quaternary Climate in Cratonic Areas (DynaQlim)/Global Geodetic Observing System (GGOS) workshop on "Understanding Glacial Isostatic Adjustment" was held in Espoo, Finland in 2009; (2) an IERS workshop on "EOP Combination and Prediction' was held in Warsaw, Poland in 2009; and (3) a joint GGOS / IAU Commission 19 workshop on "Observing and Understanding Earth Rotation' was held in Shanghai, China in 2010. The proceedings of all of these meetings are to be published.

The proposal to eliminate leap seconds from Coordinated Universal Time (UTC) continues under discussion. Commission 31 members contributed to the discussion by the International Telecommunications Union Radiocommunications Section (ITU-R). Working Party 7A (Time Signals and Frequency Standard Emissions) of ITU-R met in Geneva on 5-11 October, 2010. During this meeting the possible revision of ITU-R Recommendation TF 460-6 to abolish leap seconds was discussed. Prior to this meeting, the IAU was requested to report on whether or not it supported this proposal. A poll of Commission 31 members especially requesting input from those opposed to the recommendation resulted in three opposing responses. Responses supporting the recommendation were also received. Based on this and other input from IAU members, the IAU submitted a document to the ITU-R supporting the proposed change. The proposed revision of Recommendation ITU-R.460-6 is to be sent to the Radio Assembly which will meet in Geneva in January 2012. A vote will take place during that meeting, and at least 70% of the votes are required for the approval. The IAU is a recognized international organization in the ITU and is not a Member State. Only Member States have the right to vote.

Commission 52 considered units of measurements for astronomical quantities in a relativistic context leading to the proposed redefinition of the astronomical unit in meters and abandoning the system of astronomical constants based on the Gaussian gravitational constant k. Members also reviewed or developed models for space applications such as Gaia and Lunar Laser Ranging and worked on improving the relativistic formulations and semantics of the 2010 edition of the IERS Conventions.

The Standards of Fundamental Astronomy (SOFA) Center made available its 8th release of software. It includes 18 new routines in the Time Scales section of the Astronomy Library and three new routines in the Operations on Angles section of the Vector Matrix Library for both the Fortran 77 and ANSI C releases. A new cookbook, Time Scale and Calendar Tools, has also been made available in separate versions for Fortran and C users, with common text and appropriately tailored examples. Over 500 people have registered for e-mail updates. An article about SOFA, may be found at http://www.scholarpedia.org/article/Standards_of_Fundamental_Astronomy. SOFA, which stands for Standards of Fundamental Astronomy, is a service operated by IAU Division I to provide authoritative fundamental-astronomy algorithms. To do this SOFA is made up of the SOFA Board, an international panel that reports through Commission 19 (Rotation of the Earth), the software (algorithms) the SOFA Collection and lastly the SOFA Centre, the website that is the public interface to SOFA that makes the software freely available. The SOFA Collection now contains 186 routines provided in Fortran 77 and ANSI C, while the latest Cookbook is tailored so that the content is identical but the examples are pertinent to the particular programming language.The SOFA Centre makes the software and documentation (including all previous releases) available for viewing and downloading. A major upgrade to this website occurred in January 2010, when a new URL (www.iausofa.org) was acquired that is independent of location.

The Working Group (WG) on Numerical Standards for Fundamental Astronomy has expanded its web pages to document ongoing efforts. A report on that work was published in Celestial Mechanics and Dynamical Astronomy summarizing the numerical values for the constants and providing the justification for the values chosen. Members are drafting procedures to adopt new Current Best Estimates and is beginning to test these procedures on a suggested new value for the mass of Mercury. The Working Group members participated in the drafting of a proposed IAU Recommendation on the definition of the astronomical unit.

The Working Group on Astrometry by Small Ground Based Telescopes continues to update and maintain information on astrometric programs and activities carried out by small telescopes through web pages and e-mails and to facilitate the coordination of activities from ground-based telescopes. Its scientific goals are to foster the follow-up of small bodies detected by the large surveys; to set up a dedicated observation network for the follow-up of objects which will be detected by Gaia; to contribute to the observation campaigns of the mutual events of natural satellites, stellar occultations, and binary asteroids and to contribute to some large astrometric studies of stars or other astrophysical objects. The fundamental purpose is to identify and coordinate the astrometric activities well adapted to telescopes with diameter less than 2 m. In addition to these activities, the WG has the important role to encourage teaching of astrometry and to prepare the next generation for the new astrometry challenges.

The WG on Natural Satellites continued to maintain ephemerides of all planetary satellites at www.imcce.fr/sat and at lnfm1.sai.msu.ru/neb/nss/nssephme.htm. Through an astrometric database and these ephemerides, observed positions of the giant planets may be deduced from satellites observations. Satellites ephemerides are also provided by JPL at ssd.jpl.nasa.gov and by the MPC for the irregular satellites of the giant planets at cfa-www.harvard.edu/iau/NatSats/NaturalSatellites.html The WG on Natural Planetary Satellites encouraged studies on the dynamics of the natural satellites systems. Progress has been made for the theoretical modeling of the motions of the main satellites of Jupiter and Saturn by including tidal effects, introducing constraints on the internal structures. A campaign of observations of the mutual events of the main satellites of Jupiter and Saturn was made in 2009. These photometric observations provide highly accurate astrometric positions.

The IAU Working Group on Cartographic Coordinates and Rotational Elements published its triennial (2009) report containing current recommendations for models for solar system bodies. It introduced improved values for the pole and rotation rate of Mercury, returned the rotation rate of Jupiter to a previous value, introduced improved values for the rotation of five satellites of Saturn, and added the equatorial radius of the Sun for comparison purposes. It also adds or updates size and shape information for the Earth, Mars satellites, the four Galilean satellites of Jupiter, and 22 satellites of Saturn. Pole, rotation, and size information has been added for various solar system objects. The high precision realization for the pole and rotation rate of the Moon is also updated. The WG adopted the IAU Working Group for Planetary System Nomenclature (WG-PSN) and the IAU Committee on Small Body Nomenclature (CSBN) definition of dwarf planets. As a result, Pluto and Charon now use the positive right handed coordinate system adopted for dwarf planets, minor planets, their satellites, and comets. The Working Group is considering providing limited updates to its recommendations on its web site(http://astrogeology.usgs.gov/Projects/WGCCRE). This will not remove the need for triennial reports, and in the next report the usefulness of these interim procedures will be considered.

The WG also provided general recommendations regarding urgent needs relating to the development of planetary cartographic products. These include planning and funding geodetically controlled cartographic products, updating the Mars orientation model, and resolving various determinations for the rotation of Jupiter and Saturn. In its next report, to be completed at the Working Group meeting at the IAU General Assembly in Beijing in 2012, the WG anticipates using an improved lunar ephemeris to define the Moon's orientation, and updates due to new results from on-going space Missions and Earth-based observations. The Working Group is also looking into establishing or re-establishing links to organizations, such as the International Association of Geodesy and the International Society for Photogrammetry and Remote Sensing.

The International VLBI Service for Geodesy and Astrometry (IVS) continued to provide products for the densification and maintenance of the celestial reference frame as well as for monitoring Earth orientation parameters (EOP). It held two General Meetings, one in Hobart, Tasmania, Australia in February 2010 and the other in Madrid, Spain in March 2012. Further, two Technical Operations Workshops were held at MIT Haystack Observatory in Westford, MA in April 2009 and May 2011, respectively. Another important meeting was the VLBI2010 Workshop on Technical Specifications in Bad Ktzting, Germany in March 2012.

As an activity for the International Year of Astronomy 2009, the IVS organized a very large astrometry session. On 18/19 November 2009, thirty-four VLBI antennas observed the largest astrometry session ever scheduled. The previous record was 23 stations in a single session. The scientific goals of this session were to strengthen the ICRF2 by observing as many ICRF2 defining sources as possible in one single session and to provide the arc lengths between all sources without relying on source overlaps. The session was accompanied by press releases through the IYA09 (IAU), IVS, and other organizations and open doors at the participating stations. It resulted in news coverage in regional and national media. Further, the Bordeaux group created a dynamic Web page which allowed interested people to watch the progress of the session via the Internet in real time. The response to the Web page was very positive with about 1000 users during the session; hence, it was decided to extend this service to the entire IVS observing program. More information on IYA09 is available at http://ivscc.gsfc.nasa.gov/program/iya09/.

In September 2011, a 15-day continuous VLBI observation campaign called CONT11 was observed. The network consisted of thirteen IVS stations, nine in the northern hemisphere and four in the southern hemisphere, giving the best geographical distribution and coverage in the series of CONT campaigns.

IVS Live is a generalized version of the IYA09 dynamic Web site, developed to provide easy access to the entire IVS observing plan. It has grown into a new tool that can be used to follow the observing sessions organized by the IVS, navigate through past or coming sessions, or search and display specific information related to sessions, sources (especially the most recent VLBI images) and stations. The IVS Live user interface and all its functionalities are accessible at the URL: http://ivslive.obs.u-bordeaux1.fr/.

The current VLBI system (S/X system, legacy system) was conceived and constructed in the 1960s and 1970s. Aging antennas, increasing radio frequency interference (RFI) problems, obsolete electronics, and high operating costs make it increasingly difficult to sustain the current level of accuracy, reliability, and timeliness. Recognizing these shortcomings, the IVS has been developing the next generation VLBI system, commonly known as the VLBI2010 system. It is envisaged that the VLBI2010 system will replace the current S/X system in the next several years. In 2009, a progress report outlined recommendations for the next generation system in terms of systems, analysis, operations, and network configuration. Currently two complete VLBI2010 signal paths have been completed and data are being produced. A VLBI2010 Project Executive Group (V2PEG) has been created to provide strategic leadership. A number of VLBI2010 projects are underway; several antennas have been erected and construction of about ten antennas is at various stages of completion. The next generation IVS network is growing, with an operational core of stations becoming available within the next few years, plus further growth continuing into the foreseeable future.

The International Earth Rotation and Reference Systems Service continued to provide Earth orientation data, terrestrial and celestial references frames, as well as geophysical fluids data to the scientific and other communities. Work on new realizations of the International Terrestrial Reference System (ITRF2008) and the International Celestial Reference System (ICRF2) was finished. In 2009, Bulletin B was revised following a survey made among the community. In order to be consistent with ITRF2008, the IERS EOP C04 was revised again in 2011. The new solution 08 C04 is the reference solution which started on 1 February 2011. The system of the Bulletin A was changed to match the system of the new 08 C 04 series. The IERS Conventions (i.e. standards etc.) have been updated regularly,a new revised edition was published at the end of 2010. The Global Geophysical Fluids Centre (GGFC) restructured to allow for the establishment of operational products. A new Working Group on Combination at the Observation Level was established in October 2009.

The IERS web site www.iers.org and about 15 individual web sites of IERS components were updated. The following workshops were held, partially co-organized with GGOS: IERS Workshop on EOP Combination and Prediction, Warsaw, Poland, October 2009; Second GGOS Unified Analysis Workshop, San Francisco, CA, USA, December 2009; Third GGOS Unified Analysis Workshop, Zurich, Switzerland, September 2011. Abstracts and presentations of all these workshops are available at the IERS web site.

2.2. *Organization*

All Division I Commissions have reviewed and updated their terms of reference.

Patrick Wallace (RAL, UK), the Chair of SOFA for over 15 years, has stepped down, but remains a member of the Board. Division I acknowledges his leadership over SOFA's first fifteen years, and the huge contribution that he has made and is still making to SOFA and to the wider astronomical community. Catherine Hohenkerk (HMNAO, UK) was elected by the Board as Chair in 2010.

The subcommittee on Division I structure discussed the possible creation of Working Groups on Extrasolar Planets and Near-Earth Objects. As a proposal is currently in circulation regarding possible changes in divisional structure including the formation of a Commission on Extrasolar Planets and since it appears that this will be discussed at length in the near future the subcommittee made no recommendations pending the outcome of the IAU Executive Committee meeting. Potential changes in the names of commissions 19, 31 and 52 were also discussed but no action was recommended by the subcommittee.

The co-affiliation of Commission 7 with Division III while retaining its affiliation with Division I was proposed and officially supported by the Division I Organizing Committee. Consent by Division III is expected.

IAU Division I is proposing that the IAU consider modifying existing statutes, bye-laws and working rules to allow Divisions to create two new organizational elements within the current Divisional structure. These would be designated "Services" and "Standing Committees." A Service would be a quasi-permanent organization created to provide well-defined products, such as data or software, to the astronomical community. A Standing Committee would be organized to provide specific specialized information, such as adopted values of astronomical quantities, to the astronomical community. It would be composed of a limited number of subject-matter experts and instituted by the Division Organizing Committee. Specifically, IAU Division I proposes that the IAU Executive Committee create a committee to draft the appropriate changes to the existing IAU statutes, bye-laws and working rules to permit the existence of these organizational elements within Divisions for consideration at the next General Assembly. If the Division I proposal to change the IAU rules is not accepted by the IAU Executive Committee. IAU Division I would propose to change the name of Commission 4 to "Ephemerides and Astronomical Constants." Also, an initiative for the Asteroid Dynamic Site (AstDyS) to become a permanent IAU service was proposed in response to this broader initiative.

Dennis D. McCarthy
president of the Division

Transactions IAU, Volume XXVIIIA
Reports on Astronomy 2009–2012
Ian Corbett, ed.

© International Astronomical Union 2012
doi:10.1017/S1743921312002578

COMMISSION 4

EPHEMERIDES
ÉPHÉMÉRIDES

PRESIDENT
VICE-PRESIDENT
PAST PRESIDENT
ORGANIZING COMMITTEE

George H. Kaplan
Catherine Y. Hohenkerk
Toshio Fukushima
Jean-Eudes Arlot, John A. Bangert,
Steven A. Bell, William M. Folkner,
Martin Lara, Elena V. Pitjeva,
Sean E. Urban, Jan Vondrák

COMMISSION 4 WORKING GROUP

Div. I / Commission 4 WG   Standardizing Access to Ephemerides

TRIENNIAL REPORT 2009–2012

1. Introduction

The Commission 4 Organizing Committee began its work for the 2009-2012 triennium by revising the commission's terms of reference, which serve as our "mission statement." The new terms of reference are:

(*a*) Maintain cooperation and collaboration between the national offices providing ephemerides, prediction of phenomena, astronomical reference data, and navigational almanacs.

(*b*) Encourage agreement on the bases (reference systems, time scales, models, and constants) of astronomical ephemerides and reference data in the various countries. Promote improvements to the usability and accuracy of astronomical ephemerides, and provide information comparing computational methods, models, and results to ensure the accuracy of data provided.

(*c*) Maintain databases, available on the Internet to the national ephemeris offices and qualified researchers, containing observations of all types on which the ephemerides are based. Promote the continued importance of observations needed to improve the ephemerides, and encourage prompt availability of these observations, especially those from space missions, to the science community.

(*d*) Encourage the development of software and web sites that provide astronomical ephemerides, prediction of phenomena, and astronomical reference data to the scientific community and public.

(*e*) Promote the development of explanatory material that fosters better understanding of the use and bases of ephemerides and related data.

There are two broad kinds of work that the commission supports. The first is the computation of fundamental solar system ephemerides, that is, using gravitational theory along with observations of many types to determine the orbits of bodies in the solar system. The second kind of work uses these fundamental ephemerides to compute practical astronomical data, such as the geocentric or topocentric coordinates of the Sun, Moon, planets and stars for any given time; the prediction of times of astronomical phenomena, such as the times of rise, set, and transit, and eclipse phenomena; the parameters that describe the apparent orientation and illumination of solar system objects at specific times; and various quantities that allow knowledgeable users to transform coordinates or vectors between standard reference systems.

Fundamental solar system ephemerides are now being produced by three groups, at JPL, IAA, and IMCCE (see institutional reports below). Session 2 of the 2010 *Journées* conference in Paris was devoted to a discussion and comparison of these ephemerides. In 2010, the commission Organizing Committee, with the concurrence of Division I, established a Working Group on Standardizing Access to Ephemerides, to facilitate the use of these products. A short report on its work is given immediately below.

2. Working Groups

The Working Group on Standardizing Access to Ephemerides consists of James Hilton (USNO, USA), chair; Jean-Eudes Arlot (IMCCE, France); Steve Bell (HMNAO, UK); Olga Bratseva (IAA, Russia); Nicole Capitaine (Paris Observatory, France); Agnès Fienga (Besançon Observatory, France); William Folkner (JPL, USA); Mickael Gastineau (IM-CCE, France); Elena Pitjeva (IAA, Russia); and Vladimir Skripnichenko (IAA, Russia). The purpose of this working group is to to provide a common software interface to all of the fundamental solar system ephemerides, based on a consistent distribution format. It is hoped that this will facilitate the use of the three sets of ephemerides now being regularly produced, as well as others in the future, allowing users to easily switch among them. The working group has tentatively decided to support the Spacecraft and Planetary Kernel (SPK) file format used in the SPICE Library written and maintained by the Navigation and Ancillary Information Facility (NAIF) of the Jet Propulsion Laboratory. This decision is dependent on NAIF providing a detailed specification of those portions of the SPK file format needed for the ephemerides and producing a standalone routine for reading SPK files.

The Division I – Division III Working Group on Satellites continues to maintain ephemerides of all planetary satellites at http://www.imcce.fr/sat and at http://lnfm1.sai.msu.ru/neb/nss/nssephme.htm. When used with an astrometric database, these ephemerides allow observed positions of the giant planets to be determined from observations of their satellites. Satellite ephemerides are also provided by JPL and by the Minor Planet Center for the irregular satellites of the giant planets.

Commission 4 members are also active in the Division I Working Group on Numerical Standards for Fundamental Astronomy, the Division I Board of the Standards of Fundamental Astronomy (SOFA), and the Division I – Division III Working Group on Cartographic Coordinates and Rotational Elements. Details can be found in the Division I report.

3. Reports of the Institutions

3.1. *Real Instituto y Observatorio de la Armada (ROA), Spain*

The Ephemerides Department of the Spanish Naval Observatory publishes the following printed publications:

- *Efemérides Astronómicas*: a yearly publication containing: data for eclipses, transits and other astronomical phenomena; ephemerides of solar system bodies; and apparent places of 194 stars. The ephemerides are computed from JPL DE405/LE405 fundamental ephemerides and USNO/AE98.
- *Fenómenos Astronómicos*: a biennial publication containing data for the general public. Most of the data is extracted from *Efemérides Astronómicas*.
- *Almanaque Náutico*: the Spanish nautical almanac, containing necessary data for celestial navigation.

In addition, ROA regularly issues a nautical almanac in electronic format, valid for a ten year period, and maintains a web site with interactive computation of astronomical phenomena and other astronomical information (http://www.armada.mde.es/roa/).

— submitted by Teodoro López Moratalla

3.2. *Institut de Mécanique Céleste et de Calcul des Éphémérides (IMCCE), France*

The ephemerides service of IMCCE has three main missions:

- performing research activities on the motions of the solar system objects;
- making the French official ephemerides on behalf of Bureau des longitudes; and
- providing calculations on request for professionals, space agencies, and the public.

During the reporting period, J. Berthier was assigned responsibility for the Service des éphémérides, and D. Hestroffer was appointed head of the IMCCE to succeed W. Thuillot. Various planetary ephemerides solutions (planets, Moon and the Sun) are used at the operational service des éphémérides; the ones developed at IMCCE are the VSOP model (Secular Variations of Planetary Orbits) and the INPOP model (Planetary Numerical Integration of Paris Observatory). The latter is an original 4-D theory based upon a high precision model for planets and the Moon, fitted to space observations, and developed in support of the Gaia mission. INPOP06 and INPOP08 have been published in the past (Fienga *et al.* 2008, A&A, 477, 315; Fienga *et al.* 2009, A&A, 507, 1675). Publication of INPOP10 with TCB–TDB is pending, based on LLR data, new radio-science data, and a new model for asteroid perturbations (Fienga *et al.* 2011, CMDA, in press).

For the natural satellites, the NOE model (Numerical Orbit and Ephemerides) is used for the Martian satellites, the Galileans and the main Uranians (Lainey *et al.*, A&A 2006, 456, 783; Arlot *et al.* A&A 2006, 456, 1173). An estimation of the propagation of the ephemerides error is provided (Desmars *et al.* A&A 2009, 499, 321). The ephemerides web server (Multi-sat ephemerides) is made in collaboration with the Sternberg Astronomical Institute in Moscow.

Distribution of the ephemerides is done by yearly publications, dedicated CD-ROM, and electronic ephemerides on the web. The yearly printed publications are:

- *Connaissance des temps*, since 1679, for high precision ephemerides;
- *Annuaire du Bureau des longitudes*, since 1795, for the general public; and
- *Ephemerides nautiques*, the French nautical almanac for the Navy.

An electronic version is available for the *Connaissance des temps*.

Electronic ephemerides are provided through Internet at http://www.imcce.fr (approximately 5 Mhits per month). In addition to these positional ephemerides, ephemerides for the physical observations of the planets and small bodies are also provided. Publication of occultations of planets and stars by the Moon has been suspended, because the ILOC (Japan) has stopped computing such data and predictions provided by IOTA are not compatible with our editorial calendar constraints. Stellar occultations by asteroids are provided. Specific web services are provided in the Virtual Observatory framework using VO standard protocols, metadata and VOTable for exchange of self-defined information. The Sky Body Tracker facility identifies any solar system object in any field of view. The asteroid search application, based upon pre-calculated ephemerides for $> 500{,}000$ asteroids (taken from the Astorb database, Lowell Observatory), has been extended to cover the time span 1889–2060; it is updated daily and interfaced with Aladin Sky Atlas V7. Implementation to include all natural satellites and comets is in progress.

Several developments are foreseen over the next triennium for the ephemerides service, including upgrade of the software accompanying *Connaissance des temps*, improving access to the Miriade system (http://vo.imcce.fr/webservices/miriade) for distributing VO compliant data, and distribution of physical ephemerides. A major revision of the *Annuaire du Bureau des longitudes* content was undertaken in 2011.

— submitted by Daniel Hestroffer and J. Berthier

3.3. *United States Naval Observatory (USNO), U.S.A.*

This report covers activity in the Astronomical Applications (AA) Department since the XXVIIth General Assembly in Rio de Janeiro. The AA Department employs 13 scientists in three divisions: the Nautical Almanac Office (NAO), the Software Products Division (SPD), and the Science Support Division (SSD). J. Bartlett was appointed chief of the SPD in late 2010.

Publication of *The Astronomical Almanac* and *The Astronomical Almanac Online*, *The Nautical Almanac*, *The* (U.S.) *Air Almanac*, and *Astronomical Phenomena* continued as a joint activity of Her Majesty's Nautical Almanac Office (HMNAO) of the United Kingdom and the NAO. This activity is governed by a formal memorandum of understanding between the parent organizations. Major changes in *The Astronomical Almanac* include updated and improved computations of the lunar librations (2011 edition) and revisions of Sections E and G required to accommodate Pluto's new status as a "dwarf planet" (2013 edition).

A major revision of the *Explanatory Supplement to the Astronomical Almanac*, produced in collaboration with P. K. Seidelmann (University of Virginia) and numerous contributors, was completed and should be available for purchase in 2012.

Version 3.0 of the Naval Observatory Vector Astrometry Software (NOVAS) in C and Fortran editions was released in December 2009, followed by version 3.1 in March 2011 (http://aa.usno.navy.mil/software/novas/novas_info.php). The first Python edition of NOVAS, which is based on version 3.1 of the C edition, was released in June 2011. NOVAS fully implements recent IAU resolutions in positional astronomy, including the latest reference system definitions and models for precession and nutation.

Version 2.2.1 of the *Multiyear Interactive Computer Almanac* (MICA) was released in April 2010, to be followed by version 2.2.2 in early 2012. MICA is available for computers running Microsoft Windows and Apple Mac OS operating systems.

The department operated two public Web sites during the reporting period. The main site (http://aa.usno.navy.mil/) underwent a full content review and update in 2011. *The Astronomical Almanac Online* (http://asa.usno.navy.mil/), updated annually, is maintained jointly with HMNAO and mirrored at both organizations.

A modest research program in positional astronomy, dynamical astronomy, and navigation continued within the department. Research topics included the spin theories of Mercury and Iapetus, the theory of bodily tides, relativistic celestial mechanics, and new methods of celestial navigation.

Other projects underway at USNO and of interest to Commission 4 include the USNO CCD Astrograph Catalog (UCAC), and observations of solar system bodies made with the Flagstaff Astrometric Scanning Transit Telescope (FASTT). Additional information on these projects can be found at http://www.usno.navy.mil/USNO/astrometry/.

— submitted by John A. Bangert

3.4. *Institute of Applied Astronomy (IAA), Russia*

Fundamental ephemerides: During the period 2009–2012, the regular publication of *The Russian Astronomical Yearbook* has been continued. Planetary and lunar ephemerides are based on the numerical model EPM2004. Ephemerides for planetary configurations, eclipses and occultations, as well as the ephemerides of the Moon (as Tchebyshov polynomials), the mutual phenomena of the Galilean satellites of Jupiter, and the configurations of eight satellites of Saturn are updated and located at http://quasar.ipa.nw.ru/PAGE/EDITION/ENG/newe.htm. The P03 precession and IAU 2000A nutation theories have been introduced. Fundamental catalogues FK6 and HIPPARCOS have been used for the calculation of star positions. The matrix for conversion from the ICRS to the CIRS is also given.

Special ephemerides: *The Naval Astronomical Yearbook* (NAY) (annual issues for 2010–2013) and biennial *The Nautical Astronomical Almanac* (NAA-2) (issues 2011–2012, 2013–2014) were published. The basic purpose of producing the Almanac is to increase the applicability of the NAY data, at the same accuracy, but in a smaller physical volume. The explanation is given in both Russian and English.

Software: Constructing numerical dynamical models, fitting the ephemerides to observations, as well as preparation of the ephemerides for publication are all carried out within the framework of the universal program package ERA — Ephemerides for Research in Astronomy (ftp://quasar.ipa.nw.ru/incoming/era/). The electronic version of *The Personal Astronomical Yearbook* (PersAY) for 2010–2015 was constructed. It is intended for calculation of the ephemerides published in *The Astronomical Yearbook*, including the topocentric ephemerides for any observer. A demonstration version of PersAY for interval 2010–2011, based on the fundamental ephemerides DE405/LE405 and EPM2004, is available at ftp://quasar.ipa.nw.ru/pub/PERSAY/persay.zip. The electronic system *Navigator* for the solution of basic naval astronavigating tasks is available at http://shturman.ipa.nw.ru/.

Research work: The updated Ephemerides of Planets and the Moon — EPM2010 (Pitjeva E., Bratseva O., Panfilov V., 2011, Proc. Journées 2010, 49) differ from the previous versions by the addition of perturbations from the ring of Trans-Neptunian objects (TNO), a corrected model of the dissipative effect of the lunar rotation, new values of the planetary masses adopted at the last IAU GA, an improved mass for Mercury from the Messenger mission, improvements in the reductions of observations, and an extended database of observations. Ephemerides were constructed by the simultaneous numerical integration of the equations of motion for all the major planets, the Sun, the Moon, the largest 301 asteroids, 21 TNOs, and the lunar libration, taking into account the perturbations from the solar oblateness, the asteroid belt (consisting of the remaining smaller asteroids), as well as the ring of the TNOs at a mean distance of 43 au. The 400-year integration (1800-2200) was performed in the barycentric system of coordinates for the epoch J2000.0. The free parameters of EPM2010 were determined from 16131 lunar laser ranging measurements (1970–2010), as well as more than 635000 planetary and spacecraft observations (1913–2010) of different types. A direct estimate of $GM_\odot$ and its annual rate of change have been obtained: $GM_\odot = (132712440032.7 \pm 0.7)$ km^3sec^{-2}, $G\dot{M}_\odot/GM_\odot = (-5.0 \pm 4.1) \cdot 10^{-14}$ (Pitjeva E.V., Pitjev, N.P., 2012, Solar System Research, n.1).

The differences between the time scales TT and TDB were also constructed for the EPM2004 and EPM2008 ephemerides. Access to these ephemerides, as well as those of Ceres, Pallas, Vesta, Eris, Haumea, Makemake, and Sedna, using the new program package Calc_Eph, is available via ftp://quasar.ipa.nw.ru/incoming/EPM.

Numerical ephemerides of the satellites of Mars and the main satellites of the outer planets have also been developed, fit to modern observations.

— submitted by Elena V. Pitjeva and Marina V. Lukashova

3.5. *Her Majesty's Nautical Almanac Office (HMNAO), U.K.*

For the last triennium, HM Nautical Almanac Office (HMNAO) has remained at the UK Hydrographic Office (UKHO) in Taunton. Initially, HMNAO was part of the UKHO Operations Division but now resides within that of the National Hydrographer, improving our links with our military and international customers. During this time, the office has taken on a new staff member, Dr. Julia Weratschnig, welcome evidence of succession planning for the office, bringing the total number of staff to four. She has taken on the preparation of The UK *Air Almanac*, now a free PDF download, *Astronomical Phenomena*, and is currently taking a leading role in the updating of Section G of *The Astronomical Almanac*, "Dwarf Planets and Small Solar System Bodies," in accordance with the resolutions reclassifying Pluto and Ceres at the 2006 GA.

In line with the requirements of the UK government's Central Office of Information directives on streamlining web sites and improving their accessibility, five web sites operated by HMNAO have been amalgamated into one, which can now be accessed at http://astro.ukho.gov.uk/. A significant amount of work on improving the accessibility of these web sites has also been undertaken. A new edition of *NavPac and Compact Data 2011–2015* has been released in tandem with web-based registration and downloading of updates.

This year, 2011, marks the centenary of our collaboration with the Astronomical Applications Department of USNO. Schedules have been maintained for all the joint publications of both offices as well as those produced solely by HMNAO. Indeed, Volume 1 of *Rapid Sight Reduction Tables*, AP3270, is being added to the body of joint publications (the corresponding U.S. publication is *Sight Reduction Tables for Air Navigation*, Pub. 249). To mark the strong ties between the USNO and HMNAO, an exchange scheme for staff from both offices to gain experience of each others' methods of working and to exchange expertise has been inaugurated. Eric Barron from USNO spent part of August at the UKHO looking at future directions for providing new services and products over the Internet. Further evidence of the fruitful collaboration is a comprehensive technical note describing the new process for calculating lunar librations in *The Astronomical Almanac*, using JPL rotational ephemerides, prepared by members of both offices and by Dr. Andrew Sinclair, formerly head of HMNAO.

Catherine Hohenkerk is now chair of the SOFA board and members of the HMNAO support a number of IAU activities. Don Taylor continues to work from Rutherford Appleton Laboratory working on technical notes on lunar librations and map projections.

— submitted by Steven A. Bell

4. Closing remarks

The commission appears to have a bright and active future. There is some uncertainty in how the almanac-producing institutions will respond to a change in the definition of Coordinated Universal Time (UTC), if that is approved by the International Telecommunications Union in 2012. Such a change would not represent a major obstacle to Commission 4 affiliated institutions, but would rather require some decisions on how best to provide data to users in the future. I expect that if the change is made, this will be a continuing point of discussion in the next triennium. Similarly, the institutions that are computing the fundamental solar system ephemerides will have to respond to a potential change in the definition of the astronomical unit, although this has been well discussed over the past several years and the path forward seems clear.

George H. Kaplan   *President of the Commission*

Transactions IAU, Volume XXVIIIA
Reports on Astronomy 2009–2012
Ian Corbett, ed.

© International Astronomical Union 2012
doi:10.1017/S174392131200258X

COMMISSION 7	**CELESTIAL MECHANICS AND DYNAMICAL ASTRONOMY** *MÉCANIQUE CÉLESTE ET ASTRONOMIE DYNAMIQUE*

PRESIDENT	Zoran Knežević
VICE-PRESIDENT	Alessandro Morbidelli
PAST PRESIDENT	Joseph A. Burns
ORGANIZING COMMITTEE	Evangelia Athanassoula, Jacques Laskar, Renu Malhotra, Seppo Mikkola, Stanton J. Peale, Fernando Roig

TRIENNIAL REPORT 2009-2012

1. Introduction

At the time this report was written, Commission 7 according to the IAU database had 304 registered members. The activities of Commission 7 in the past triennium closely followed the plan outlined at the business meeting of the Commission held on August 5, 2009 in Rio de Janeiro (Transactions IAU, Volume XXVIIB, 119-120). They can be summarized as follows:

• Preparation of the Terms of Reference of Commission 7. The document contains the scientific rationale and lists a number of the most important objectives of Commission 7, thus justifying its continued existence and activity. Several colleagues contributed to the preparation of this important document, its final form being done by Z. Knežević and A. Morbidelli.

• A complete remake and transfer of the Commission 7 web site to a new address: http://staff.on.br/iaucom7/. The web page has a new design, authored by F. Roig; it contains a number of new features and useful information, and is regularly updated.

• Taking advantage of the possibility offered by the current Statutes and By-Laws of the IAU, the initiative for the co-affiliation of Commission 7 with Division III while keeping its original affiliation with Division I, has been proposed by A. Morbidelli, who also prepared the rationale for the proposal. Commission 7 had a vote on the proposal which showed that a vast majority of the Commission members favor the co-affiliation. The initiative has also been officially supported by the Division I Organizing Committee. We are expecting only the consent by Division III to submit the proposal to the IAU.

• An initiative for the Asteroid Dynamic Site (AstDyS) to become a permanent IAU service has been proposed by A. Milani, Z. Knežević and M. E. Sansaturio. The proposal is in response to a wider initiative of Division I for establishment of permanent services under the auspices of the IAU. The motion is currently awaiting the completion of the ongoing restructuring of the IAU Divisions to be submitted to the relevant IAU bodies.

• Commission 7 officers actively participated in the activities of Division I Organizing Committee. A. Morbidelli served as a representative of Commission 7 in the Subcommittee for restructuring, created by the Division I.

• A significant effort has been put in an attempt to win organization of an IAU symposium in 2012. A. Milani, S. Mikkola and Z. Knežević prepared the two-stage proposal. Unfortunately, the proposal was not accepted by the IAU, thus our field remains without an IAU Symposium for an extended period of time.

• The President, vice-president and secretary of the Commission took care of various administrative and current businesses which were brought to their attention by members of the commission, officers of Division I and of the IAU.

• Commission 7 was involved with the interdivision Working group on Natural Planetary Satellites , which made progress in the theoretical modeling of motions of the main satellites of Jupiter and Saturn by including tidal effects and introducing constraints on the internal structures. The detailed report is included in the IAU Transactions.

• Finally, an important task of the Commission 7 was the preparation of this report.

2. Developments within the past triennium

2.1. *Chaos in galaxies*

E. Athanassoula, M. Romero-Gómez & E. Vasiliev

A number of papers considered various methods of determination of orbital class and degree of chaos. Most of them are based on the calculation of the Lyapunov characteristic exponents (LCE) and their derivatives. Skokos (2010) presents a comprehensive review of various applications of LCE. Manos *et al.* (2011) applied the Generalized ALignment Index (GALI) to study dynamical properties of orbits in the vicinity of a periodic orbit. Maffione *et al.* (2011a) focused on the MEGNO indicator (Mean Exponential Growth of Nearby Orbits) and applied it to study orbits in a triaxial galactic potential, while Maffione *et al.* (2011b) presented a detailed comparison of several chaos indicators. They find that FLI/RLI (Fast and Relative Lyapunov Indicator) are most suitable for analyzing global dynamics of the system, while MEGNO and GALI perform better at the level of individual orbits. Among the methods not related to LCE, 4D surfaces of section were used by Katsanikas *et al.* (2011) in application to multiply-periodic orbits. Bountis *et al.* (2011) used probability distributions of sum of coordinates to distinguish weakly chaotic from strongly chaotic orbits in a barred galaxy potential.

The role of chaotic orbits in the evolution and dynamics of spheroids was considered in several papers. Muzzio *et al.* (2009) find that triaxial cuspy models can have a large fraction of chaotic orbits and still be reasonably stable over many dynamical times. Valluri *et al.* (2009) studied the effect of baryonic contraction on the orbital structure of triaxial DM halos and the change of shape induced by the evolution of chaotic orbits, while Valluri *et al.* (2011) applied also frequency analysis to study the Galactic halo. Deibel *et al.* (2011) surveyed the orbital structure of triaxial cuspy galaxies with slow figure rotation. Vandervoort (2011) explores chaotic behavior in a model of homologous oscillations of an axisymmetric galaxy (whose size but not shape changes in time).

Work was also done to establish the influence of chaos, and in particular of 'sticky' or 'confined' orbits, on galactic structures. Manifolds, linked to the unstable periodic orbits around the L1 (L2), are proposed to explain both the spirals and the rings in barred galaxies. Athanassoula *et al.* (2009a) find that the bar strength determines whether the morphology is that of a ring or a spiral. In the former case the R1, R2 and R1R2 morphologies are explained, while in the latter, a clear link between the bar strength and the spiral pitch angle is found (Athanassoula *et al.* 2009b) and confirmed by Martínez-García (2011). Finally, Athanassoula *et al.* (2010) present comparisons to observations, while Romero-Gómez et al. (2011) apply this theory to our Galaxy. A different approach, still based on manifolds, but relying on the apsidal sections (pericenters and apocentres) of the manifolds was elaborated by Tsoutsis et al. (2009), Contopoulos & Harsoula (2010), Efthymiopoulos (2010) and Harsoula et al. (2011). This second approach necessitates a bar that either does not evolve, or evolves at a much slower rate than what is necessary for the first approach, i.e. no or little secular evolution. Patsis et al. (2009, 2010) analysed the orbital structure in NGC 3359 and 1300, using response models and making morphological comparisons. Harsoula & Kalapotharakos (2009) used frequency analysis to distinguish between regular and chaotic orbits in an N-body system. Manos & Athanassoula (2011) find a strong correlation between the bar strength and the fraction of orbits in the bar region that is chaotic. Finally, Shevchenko (2011) gave estimates of the Lyapunov and diffusion timescales in the solar neighborhood.

References

Athanassoula, E., Romero-Gómez, M., & Masdemont, J. J. 2009, *MNRAS*, 394, 67
Athanassoula, E., Romero-Gómez, M., Bosma, A., & Masdemont, J. J. 2009, *MNRAS*, 400, 1706
Athanassoula, E., Romero-Gómez, M., Bosma, A., & Masdemont, J. J. 2010, *MNRAS*, 407, 1433
Bountis, T., Manos, T., & Antonopoulos, Ch., 2011, arXiv:1108.5059
Contopoulos, G. & Harsoula, M., 2010, *CeMDA*, 107, 77
Deibel, A., Valluri, M., & Merritt, D., 2011, *ApJ*, 728, 128
Efthymiopoulos, C., 2010, *The European Physical Journal Special Topics*, 186, 91
Harsoula, M. & Kalapotharakos, C., 2009, *MNRAS*, 394, 1605
Harsoula, M., Kalapotharakos, C., & Contopoulos, G., 2011, *MNRAS*, 411, 1111
Katsanikas, M., Patsis, P., & Pinotsis, A., 2011, arXiv:1103.3981
Maffione, N., Giordano, C., & Cincotta, P., 2011a, arXiv:1108.5481
Maffione, N., Darriba, L., Cincotta, P., & Giordano, C., 2011b, arXiv:1108.2196
Manos, T. & Athanassoula, E., 2011, *MNRAS*, 415, 629
Manos, T., Skokos, Ch., & Antonopoulos, Ch., 2011, arXiv:1103.0700
Muzzio, J., Navone, H., & Zorzi, A., 2009, *CeMDA*, 105, 379
Patsis, P. A., Kaufmann, D. E., Gottersman, S. T., & Boonyasait, V. 2009, *MNRAS*, 394, 142
Patsis, P. A., Kalapotharakos, C., & Grosbol, P. 2010, *MNRAS*, 408, 22
Romero-Gómez, M., Athanassoula, E., Antoja, T., & Figueras, F., 2011, arXiv:1108.0660
Shevchenko, I. I., 2011, *ApJ*, 733, 39
Skokos, Ch., 2010, *Lect. Notes Phys.* 790, 63
Tsoutsis, P., Kalapotharakos, C., Efthymiopoulos, C., & Contopoulos, G. 2009, *A& A*, 495, 743
Valluri, M., Debattista, V., Quinn, T., & Moore, B., 2010, *MNRAS*, 403, 525
Valluri, M., Debattista, V., Quinn, T., Roskar, R., & Wadsley, J., 2011, arXiv:1109.3193
Vandervoort, P., 2011, *MNRAS*, 411, 37

2.2. *New results on the chaotic behavior of the Solar System*

J. Laskar

The discovery of the chaotic behavior of the planetary orbits in the Solar System (Laskar 1989) was obtained using numerical integration of averaged equations of motion. This chaotic behavior of the Solar System manifests itself by an exponential divergence of nearby trajectories, the distance between two orbital solutions being multiplied by ten every ten million years which limits the validity of the solutions to less than 100 Myr. A numerical integration of the Solar System motion over 5 Gyr can only be considered as a random sample of its possible evolution and statistical studies are required.

The first study over several Gyr made use of the numerical integration of the averaged equations (Laskar 1994) revealing the possibility for the eccentricity of Mercury to reach very high values, beyond 0.7, allowing for collisions with Venus in less than 5 Gyr. The drawback of this early study is that the averaged equations are no longer justified in the vicinity of a collision. This result has been unchallenged for 15 years.

Large progress has been achieved recently. A first statistical study using the secular equations for 1000 solutions over 5 Gyr has shown that the probability of a large increase of the eccentricity of Mercury over 5 Gyr is about 1% for the full model, but raises to more than 50% when general relativity is not taken into account in the model, which was confirmed by a test over 10 solutions with the non averaged Newtonian equations (Laskar 2008). At the same time, Batyugin & Laughlin (2008) have reproduced the experiment of Laskar (1994) with a Newtonian model and have confirmed that high values of the eccentricity of Mercury can occur, leading to possible collision with Venus. The large increase of the eccentricity of Mercury is obtained through a secular resonance between the perihelion motion of Mercury and the perihelion motion of Jupiter (Laskar 2008; Batyugin & Laughlin 2008). When general relativity (GR) is removed, the perihelion speed of Mercury gets closer to the one of Jupiter, which explains why the non relativistic system is much more unstable than the full model (Laskar 2008).

Finally, a full scale statistical simulation over 5 Gyr, with the full Solar System with GR and non-averaged equations, has been performed, gathering 2501 solutions with very close initial conditions (Laskar & Gastineau 2009).This study confirmed that the probability for a destabilization of the system, with a very large increase of the eccentricity of Mercury is about 1%. It was also shown that once the eccentricity of Mercury raise to very high level, actual collisions of Mercury with Venus or with the Sun are the most probable outcome. In some cases, the large increase of Mercury's eccentricity was followed by a total destabilization of the inner Solar System, with possible collision of any two planets of the inner Solar System (Mercury, Venus, Earth, Mars) in less than 5 Gyr. A joined test with pure Newtonian equations over 200 orbits shows that in this case the probability for high increase of Mercury leading to possible collision increases to about 60%.

The search for the best solution for the Earth's orbit over millions of years is motivated by the possibility to calibrate the recent geological timescales by correlation of the geological stratigraphic data to the computed variation of the insolation on Earth. Indeed, in the latest geological timescale adopted by the International Union of Geological Sciences (IUGS), the Neogene period (0-23.03 Ma) is now calibrated astronomically (Lourens et al. 2004). Since then, there has been a continuous effort to extend this astronomical calibration to the entire Cenozoic era, over about 65 Ma. The latest improvement in the long-term astronomical solution has been obtained through a complete revision of the numerical algorithm and the construction of the high-precision short-term planetary ephemerides INPOP (Fienga et al. 2008; 2009; 2011) that were extended over 1 Myr as a reference for the long-term solutions. Despite this effort, this new solution is valid over 50 Myr only (Laskar et al. 2011a).

The reasons of these difficulties in the construction of a long-term solution for the planetary orbits are now identified and are due to the highly chaotic behavior of Ceres and Vesta resulting from close encounters (Laskar et al. 2011b). As a result, their positions will be totally lost in less than 400 kyr. Despite their small mass, Ceres and Vesta perturb the planetary orbits. Due to these interactions, the eccentricity of the Earth becomes unpredictable after 60 Myr, somewhat less than the duration of the Cenozoic era.

As the NASA/Dawn spacecraft will be orbiting Vesta for several months before continuing its route towards Ceres, one can expect that the precision on the positions of these minor bodies will be improved. But this will be of no use for paleoclimate studies. Indeed, even if the initial error in these positions is reduced to 1.5 mm, their positions will still be in total error after less than 500 kyr, and there will be no significant change in the time of validity of the orbital solution of the Earth. This limit of 60 Myr thus appears as an absolute limit for a precise prediction of the Earth's eccentricity, that will not be beaten easily in the future.

References

Batygin, K. & Laughlin, G. 2008, *ApJ*, 683, 1207
Fienga, A., Manche, H., Laskar, J., & Gastineau, M. 2008, *A& A*, 477, 315
Fienga, A., Laskar, J., Morley, T., *et al.* 2009, *A& A*, 507, 1675
Fienga, A., Laskar, J., Kuchynka, P., *et al.* 2011, *CeMDA*, 111, 363
Laskar, J. 1989, *Nature*, 338, 237
Laskar, J. 1994, *A& A*, 287, L9
Laskar, J. 2008, *Icarus*, 196, 1
Laskar, J. & Gastineau, M. 2009, *Nature*, 459, 817
Laskar, J., Robutel, P., Joutel, F., *et al.* 2004, *A& A*, 428, 261
Laskar, J., Fienga, A., Gastineau, M., & Manche, H. 2011a, *A& A*, 532, 89
Laskar, J., Gastineau, M., Delisle, J., Farrés, A., & Fienga, A. 2011b, *A& A*, 532, L4
Lourens, L., Hilgen, F., Laskar, J., Shackleton, N., & Wilson, D. 2004, in: F. Gradstein, J. Ogg, & A. Smith (eds.) ed. F. Gradstein, J. Ogg, & A. Smith *A Geological Timescale 2004*, 409–440

2.3. *Dynamical mechanisms for the origin of retrograde extrasolar planets*

D. Nesvorny & R. Malhotra

Around 1% of solar-type stars host Jupiter-mass planets with the semimajor axis a less than 0.1 AU. It is thought that these planets, the so-called *hot Jupiters*, did not form in situ because the region within 0.1 AU was too hot and rarefied for a Jupiter-mass planet to form. Instead, the hot Jupiters probably formed several AU from their host stars, and later migrated inward.

Planets are thought to have radially migrated due to their interaction with the protoplanetary gas disk from which they formed. Yet, several orbital properties of hot Jupiters suggest that at least some of the observed hot Jupiters, if not most, arrived to their current orbits *after* the gas disk dispersal. For example, many hot Jupiters with $0.05 < a < 0.1$ AU have unexpectedly large orbital eccentricities ($e > 0.2$), whereas the the gas disk should efficiently damp e and produce nearly circular orbits, $e \approx 0$.

To date, the Rossiter-McLaughlin effect has been measured for 37 planets (mostly hot-Jupiters; Moutou *et al.* 2011). In about half of these cases the planet orbit normal is probably aligned with the stellar spin vector ($|\lambda| < 30°$, where λ is the projected spin-orbit misalignment angle), however, in 8 cases ($\sim 20\%$) the planets are inferred to have retrograde orbits. The origin of retrograde hot Jupiters is an intriguing problem.

To produce large values of $|\lambda|$, it is either necessary to tilt the spin axis of the star so that it ends up being misaligned with the original protoplanetary disk in which the planets formed, or to tilt the planetary orbit. A tilt of the star's axis can be produced by a number of effects including the late non-isotropic (Bondi-Hoyle) accretion on the star (Throop & Bally 2008; see also Thies *et al.* 2011), and interaction between the stellar magnetic field and protoplanetary disk (Lai *et al.* 2011).

An orbital tilt can be produced by: (i) Planetary scattering followed by Kozai migration (Nagasawa *et al.* 2008); (ii) Kozai migration due to a distant perturber in an inclined orbit (Fabrycky & Tremaine 2007, Wu *et al.* 2007, Naoz *et al.* 2011, Correia *et al.* 2011, Katz *et al.* 2011); and (iii) Secular migration in well-spaced, eccentric, and inclined planetary systems (Wu & Lithwick 2011).

The spin- and orbit-tilt theories have different implications. For example, the spin-tilt theory implies that co-planar planetary systems with large values of $|\lambda|$ should be relatively common, while the mutual inclinations of planets should be generally high according to the orbit-tilt theory. Interestingly, at least some planetary systems have large orbital inclinations (McArthur *et al.* 2010).

In addition, the observed large eccentricities of exoplanets can be best explained if the original, closely-packed planetary systems underwent a dynamical instability followed by planet scattering (Weidenschilling & Marzari 1996, Rasio & Ford 1996), and tidal circularization of close-in planets (Jackson *et al.* 2008). As planet scattering naturally leads to large orbital inclinations as well, the orbital tilt theories are therefore a logical extension of the planet scattering model.

Beaugé & Nesvorný (2011) performed numerical simulations of planet scattering followed by tidal circularization and migration of planets that evolved into highly eccentric orbits. They found that orbits typically acquire high eccentricities and high inclinations due to close encounters and subsequent slow secular interactions, rather than due to the sole effect of the Kozai resonance. Their final results provide a good match to the period and eccentricity distribution of hot Jupiters.

The inclination distribution of hot Jupiters appears to be sensitive to the number of planets in the initial systems, N. While very few hot Jupiters form in retrograde orbits for $N = 3$, the case with $N = 4$ shows a larger proportion ($\sim 10\%$), and a wider spread in inclination values. As the latter result better agrees with observations, this may suggest that the planetary systems with observed hot Jupiters were originally rich in the number of planets, some of which were ejected. In a broad perspective, this hints on an unexpected link between the hot Jupiters and recently discovered free floating planets (Sumi *et al.* 2011).

References

Beauge, C. & Nesvorny, D. 2011, arXiv:1110.4392
Correia, A. C. M., Laskar, J., Farago, F., & Boué, G. 2011, *CeMDA*, 111, 105
Fabrycky, D. & Tremaine, S. 2007, *ApJ*, 669, 1298
Jackson, B., Greenberg, R., & Barnes, R. 2008, *ApJ*, 678, 1396
Katz, B., Dong, S., & Malhotra, R. 2011, arXiv:1106.3340
Lai, D., Foucart, F., & Lin, D. N. C. 2011, *MNRAS*, 412, 2790
McArthur, B. E., Benedict, G. F., Barnes, R., *et al.* 2010, *ApJ*, 715, 1203
Moutou, C., Díaz, R. F., Udry, S., *et al.* 2011, *A& A*, 533, A113
Nagasawa, M., Ida, S., & Bessho, T. 2008, *ApJ*, 678, 498
Naoz, S., Farr, W. M., Lithwick, Y., Rasio, F. A., & Teyssandier, J. 2011, *Nature*, 473, 187
Rasio, F. A. & Ford, E. B. 1996, *Science*, 274, 954
Sumi, T., Kamiya, K., Bennett, D. P., *et al.* 2011, *Nature*, 473, 349
Thies, I., Kroupa, P., Goodwin, S. P., Stamatellos, D., & Whitworth, A. P. 2011, *MNRAS*, 417, 1817
Throop, H. B. & Bally, J. 2008, *AJ*, 135, 2380
Weidenschilling, S. J. & Marzari, F. 1996, *Nature*, 384, 619
Wu, Y. & Lithwick, Y. 2011, *ApJ*, 735, 109
Wu, Y., Murray, N. W., & Ramsahai, J. M. 2007, *ApJ*, 670, 820

Zoran Knežević
president of the Commission

Transactions IAU, Volume XXVIIIA
Reports on Astronomy 2009–2012   © International Astronomical Union 2012
Ian Corbett, ed.   doi:10.1017/S1743921312002591

COMMISSION 8   ASTROMETRY

PRESIDENT	Dafydd Wyn Evans
VICE-PRESIDENT	Norbert Zacharias
PAST PRESIDENT	Irina Kumkova
ORGANIZING COMMITTEE	Alexandre Andrei, Anthony Brown, Naoteru Gouda, Petre Popescu, Jean Souchay, Stephen Unwin, Zi Zhu

TRIENNIAL REPORT 2009–2012

1. Scientific Highlights

Gaia is continuing well with its development and construction and is expected to achieve parallax accuracies of about 10 to 300 μas for 6 to 20 mag. It is scheduled to launch in 2013.

The ICRF2 was adopted by the IAU in 2009 as the new fundamental celestial reference frame.

The UCAC project concluded with the publication of its 4th data release (all-sky, over 100 million stars).

The JASMINE project will launch the Nano-JASMINE satellite in 2012 and continues with the development of its other satellites which will complement the Gaia project.

The PPMXL (Heidelberg) and XPM (Ukraine) catalogues provide improved astrometric accuracies from new reductions of USNO-B (USNO-A) and 2MASS data.

It is disappointing that the SIM project was cancelled during this triennium. It would have further complemented the above two satellite projects and would have gone even beyond Gaia in astrometric accuracy.

2. Instrumentation and Reduction Methods

ARGENTINA: The instruments at the Complejo Astronómico El Leoncito continue to operate regularly and represent interesting resources for astrometric observations. Also the Estacion Astronomica Rio Grande continued the work on polar motion. The Facultad de Ciencias Astronómicas y Geofísicas of the Universidad Nacional de La Plat kept its research group on reference systems for astronomy and geodesy. Also the Observatorio Astronomico de Cordoba, worked on Galactic astrometry.

BELGIUM: At the Royal Observatory of Belgium a high-precision scanner, called Damian, has been built capable of digitizing astrophotographic plates up to 35×35 cm. It is described in De Cuyper *et al.* (2011). A report by Robert *et al.* (unpublished) shows a positional accuracy of better than 0.1 μm.

CHINA PR: Yanben Han (Beijing) reports that the San Juan SLR station, working in cooperation with National Astronomical Observatories of the Chinese Academy of Sciences (NAOC) and Universidad Nacional de San Juan (UNSJ) of Argentina, has operated for five years since the end of February 2006; the observations of the SLR station have made contributions to the ILRS and IERS. In June 2009, San Juan SLR station began to upgrade the system for daylight and kHz tracking. The San Juan SLR system upgrading will be completed in 2012. The 40-metre radio telescope project (VLBI) comes into operation in 2011. The San Juan Station will become an integrated observational station with SLR, GPS and VLBI in coming years (Han *et al.* 2008).

Zhenghong Tang (Shanghai) reports that a prototype of a rotating-drift-scan CCD has been developed and installed on a 300 mm (F = 250 mm, FoV=8.3° × 8.3°) telescope to observe faint space debris (Tang *et al.*, 2010). With 5s observations, it is possible to observe space debris as faint as 13 mag. The astrometric performance of this telescope has been analyzed in Yu *et al.* (2010).

A new method to determine real-time atmospheric refraction was presented by Yu *et al.* (2009). The method does not depend on strict local parameters and precise instruments. Observations were carried out with a simple telescope at Xinglong station of the National Astronomical Observatories, The atmospheric refraction from the zenith distance of 44.8° to 87.5° can be obtained by this method.

Zhenghong Tang (Shanghai) reports that the components included in the Time of Arrival (ToA) of pulses from millisecond pulsars are parsed by Zhao & Huang (2009). From this work, it is not necessary to calculate the differential ToA as well as differential gravitational delay between the observer and the Solar System Barycentre.

Zhenghong Tang (Shanghai) reports that the first successful CVN (Chinese VLBI Network) phase-referencing observation of the pulsar B0329+54 was carried out on October 16, 2008 in the S band (Guo *et al.*, 2010). The fitted position of the pulsar relative to the calibrator is accurate to the level of tens of μas, which is comparable with the present differential VLBI astrometric accuracy, proving that CVN will be a potential powerful tool for astrometry and astrophysics.

Zhenghong Tang (Shanghai) reports that a new station, 'Jiang-Nan-Tian-Chi station', of the Shanghai Astronomical Observatory (λ = 119.5978°, ϕ = 30.4694°, height = 958.409 m) has been established. The night sky brightness and atmospheric extinction at the station were determined by Yao & Tang (2009).

RUSSIA: Malkin (Pulkovo) reports about the investigation of the impact of the Galactic aberration on the precession-nutation model. It is estimated to be small, but not negligible: up to 20 mas per century. Thus, this effect should be modelled during processing of high-accuracy observations (Malkin 2011).

UKRAINE: Pakuliak (Kiev) reports that the digitization of observational archives has resulted in about 3,000 digitized images of the MAO NASU collection and about 2,000 digitized images of the AO Lviv NU collection, which are available from the search pages of the Ukrainian VO (`http://gua.db.ukr-vo.org`). On the basis of these two digitized archives, and in cooperation with Mykolaiv AO, the prototype of the Joint Digitized Archive as a core of the astrometric component of the Ukrainian VO has been created and is intended to present the tools for access to the digitized observational archives of all Ukrainian observatories.

UNITED STATES: The USNO Robotic Astrometric Telescope (URAT, Zacharias *et al.* 2012) has become operational. A single exposure with the "red lens" onto 4 large CCDs covers 28 sq.deg. with an astrometric precision below 20 mas. A new all-sky, astrometric survey has begun for the northern hemisphere at NOFS in 2011 to provide positions and proper motions for all stars in the R = 9 to 17.5 mag range and parallaxes of nearby stars without selection bias.

An astrometric pipeline has been developed for LSST (Monet 2011, LSST science book 2009).

3. Space Astrometry

Prusti (ESA) reports on the Gaia astrometric mission in the science programme of the European Space Agency. The satellite is currently (September 2011) in the qualification and assembly phase since successfully passing the mission critical design review. Gaia is scheduled for launch in 2013.

The main scientific aim of Gaia is to reveal the structure and kinematics of our Galaxy. The science requirements deduced from the main goal and amended by many other science cases have resulted in a mission conducting an astrometric, photometric and spectroscopic survey of the full sky. Gaia is anticipated to detect and measure more than 1 billion objects astrometrically and photometrically. In addition, spectroscopy will provide radial velocities for an estimated 150 million stars.

All spacecraft sub-systems have been delivered to the prime industrial contractor, EADS Astrium in Toulouse, France. The scientific performance estimates have been calculated based on tests and measurements at sub-system level. In late 2011, the sub-systems will be integrated and the performances will be verified at the system level. Scientific performance estimates are for the bright stars (6–12 mag), sky-averaged parallax accuracy between 5 and 14 μas. Stars brighter than about 6 mag will not be detected by Gaia as they saturate the detectors. For stars fainter than 12 mag, the parallax accuracy degrades to about 25 μas at 15 mag and to about 300 μas at the limiting magnitude of 20 for Gaia. The accuracies of positions and annual proper motions are roughly at similar μas accuracy levels.

In addition to astrometry, Gaia will provide photometry for *e.g.* stellar classification and spectroscopy mainly for radial velocities. The scientific performances are compliant to the original science requirements. The current photometric and spectroscopic performance estimates, as well as those for astrometry, can be found on the Gaia web-pages (http://www.rssd.esa.int/Gaia).

The ground segment is also preparing for the operational phase of the mission. In addition to the ESA provided elements for the operations, there is a large consortium of scientists and engineers entrusted with the scientific data processing task. More than 400 individuals are providing on the average about half of their time to Gaia data processing and analysis tasks. Early preparations are needed to cope with the huge amount of astronomical measurements from Gaia. On average, Gaia will produce some 400 million measurements daily which need processing to catalogue values for use by the astronomical community. Although the final catalogue will appear only in the next decade, Gaia will early on provide science alerts and intermediate catalogues that are planned some two years after the start of operations.

FRANCE: Astrometric tests using the Planck satellite and asteroids, together with the development of an optical image database, are currently under progress to prepare the Ground Based Optical Tracking of the Gaia mission. Several 1m class telescopes are used to determine the best way to obtain the precise position (20 mas) of the satellite during its mission among the reference stars in the same field of view. Once launched, the celestial coordinates of Gaia will be used to improve the coordinates that will be given in the final catalogue of the mission.

The Bordeaux Group (Krone-Martins & Ducourant), together with the Sao Paulo group (Teixeira), have developed specific tools to analyze the morphology of the tiny galaxies (few arcseconds of extension) that will be observed by Gaia. The algorithms of morphological classification and profile fitting developed and deployed at the Data Processing Centre in CNES are based on support vector machines, genetic algorithms, cross-entropy methods and the Radon transform. About five million of these objects, that cannot be observed from the ground, are expected (Krone-Martins, PhD thesis).

NETHERLANDS: The Joint Institute for VLBI in Europe (JIVE), in cooperation with the European VLBI Network (EVN), EC FP7 EuroPlaNet and ESPaCE and other partners, continued developing VLBI applications for ultra-precise tracking of planetary and other space science missions. The technique enables determination of positional components of spacecraft state-vectors at distances of several astronomical units with metre-level accuracy. Such a high accuracy enables multi-disciplinary applications of this VLBI tracking technique, including celestial mechanics and astrometry of the Solar System, fundamental physics and studies of interiors and atmospheres of planets. In the reporting period, the activities focused at the Planetary Radio Interferometry and Doppler Experiment (PRIDE) which exploits the technique of near-field VLBI tracking and has been demonstrated for ESA's Venus Express (VEX) and MarsExpress (MEX) missions (Molera Calves et al. 2011, Duev et al. 2011).

JAPAN: JASMINE is an abbreviation of Japan Astrometry Satellite Mission for Infrared Exploration. Three satellites are planned in the JASMINE series as a step-by-step approach to overcome technical issues and promote scientific results (Gouda et al. 2010, Gouda et al. 2011). These are Nano-JASMINE, Small-JASMINE and (Medium-sized) JASMINE.

Nano-JASMINE is a small satellite of which the size and weight are $(50cm)^3$ and about 35 kg respectively (Hatsutori et al. 2011). The diameter of the primary mirror is 5 cm. A fully depleted CCD is at the focal plane of the telescope (Kobayashi et al. 2010). The flight model of Nano-JASMINE has been fabricated and it will operate in the zw-band (0.6–1.0 μm). The target accuracy of parallaxes is about 3 mas at zw=7.5 mag (Kobayashi et al. 2011). Moreover, high-accuracy proper motions ($\sim$0.1 mas/year) can be obtained by combining the Nano-JASMINE catalogue with that of Hipparcos, as the decrease in the error of the proper motions is proportional to the inverse of the epoch difference between the two catalogues, which for the Hipparcos and Nano-JASMINE catalogues will exceed 20 years.

The observing strategy and methods used in the data analysis for Nano-JASMINE will be similar to what is planned for Gaia. Thus the use of Nano-JASMINE data is useful to check algorithms that are to be used in the Gaia data analysis (Yamada et al. 2011). Nano-JASMINE is scheduled to be launched in 2012 from the Alcantara space centre in Brazil by a Cyclone-4 rocket developed in Ukraine and will be put in a Sun-synchronized orbit with an altitude of about 800 km.

Small-JASMINE will determine positions and parallaxes accurate to 10–50 μas towards a $3° \times 3°$ region around the Galactic centre and other small regions which include scientifically interesting objects brighter than Hw=11.5 mag (Hw-band: 1.1–1.7 μm). Proper motion accuracies of between 10 and 50 μas/year are expected (Yano et al. 2011). The target launch date is around 2017. The JASMINE group aims at a proposal for the Small-JASMINE mission to JAXA, to get launch approval and the required budget from the Japanese government in the near future.

The main science objective of Small-JASMINE is to clarify the formation model of the Galactic bulge structure, star formation histories around the Galactic centre (Tsujimoto 2011) and the evolution of the super-massive black hole located at the centre of the Galaxy.

(Medium-sized) JASMINE is an extended mission of Small-JASMINE, which will observe towards almost the whole region of the Galactic bulge with accuracies of 10 μas in the Kw-band ($\sim$2.0 μm). The target launch date is the first half of the 2020s.

UNITED STATES: During the development of the Space Interferometry Mission (SIM), key projects conducted numerous investigations relating to astrometry (Unwin 2008).

Makarov (2005) developed a theoretical framework for the astrometric grid that may be relevant also to Gaia. Geisler (2006) described spectroscopic and photometric monitoring of candidate grid stars suitable as astrometric references at the μas level. Shao (2010) showed the importance of astrometry as a major technique for low-mass exoplanet detection and Traub (2009) demonstrated SIM's ability to detect multi-planet systems astrometrically. Gould (2005) showed how the microlens parallax allows precision astrometric measurements by stellar microlensing events.

Henry (2006) and Benedict (2007) explored the many contributions to precision stellar astrophysics. Kenneth Johnston (2009), Zacharias (2009) and Unwin (2009) discussed astrometric measurements of quasar nuclei as science targets and as anchors for a fundamental reference frame. Majewski (2007), Shaya (2009) and Kathryn Johnston (2009) showed how SIM could contribute to understanding the formation of the Galaxy and the Local Group and the dynamical role of dark matter. See also the NExScI website: http://nexsci.caltech.edu/missions/SIMPQ/SIMSciStudies/accepted.shtml.

NASA cancelled SIM in December 2010.

The JMAPS programme (USNO) continued throughout this reporting period.

4. Reference Frames

CHINA PR: Zi Zhu (Nanjing) reports that Xie and Kopeikin overviewed a set of post-Newtonian reference frames for a comprehensive study of the orbital dynamics and rotational motion of Moon and Earth by means of lunar laser ranging (LLR). They employ a scalar-tensor theory of gravity depending on two post-Newtonian parameters and utilize the relativistic resolutions on reference frames adopted by the IAU in 2000. The theoretical advantage of this work is in a simpler mathematical description. They also derive the post-Newtonian coordinate transformations between the frames investigated and analyze the residual gauge freedom (Xie & Kopeikin, 2010).

Zi Zhu (Nanjing) reports that extensive analyses of two large catalogues (PPMX and UCAC3) have been made in order to determine the local and overall systematic biases. The regional and magnitude dependent differences in stellar position and proper motion are comparable to random errors. The global orientation bias vector between the two systems is also significant (up to 17 mas). However, the term for the global rotation vector ω is small (tenths of mas per year): it is reasonable to believe that the PPMX and UCAC3 reference frames do not rotate with respect to each other. Because of plate dependent and field-to-field errors in the UCAC3 catalogue, they suggest that positions and proper motions of UCAC3 stars in the northern hemisphere ($\delta > -20°$) should be used with caution (Liu & Zhu, 2011).

Zi Zhu (Nanjing) reports that the Galactic coordinate system, initially defined by the IAU in 1958, was thereafter transformed in 1984 from the B1950.0 FK4-based system to the J2000.0 FK5-based system. In 1994, although the IAU recommended that the dynamical reference system FK5 be replaced by the ICRS, the definition of the Galactic coordinate system was not updated. They consider that the present Galactic coordinates may be problematic because of the unrigorous transformation method from the FK4 to the FK5 and of the non-inertiality of the FK5 system with respect to the ICRS. They suggest reconsidering the definition of the Galactic coordinate system which should be directly connected to the ICRS for precise observations at the μas level (Liu et al., 2011).

FRANCE: A new version of the LQAC (Large Quasar Astrometric Catalogue), called LQAC-2, has been produced (Souchay et al., 2011). It contains about 180,000 quasars compared to 110,000 in the first version (Souchay *et al.*, 2009) and contains more items such as a specific identification number, more precise coordinates from the LQRF (Andrei *et al.*, 2009) and morphological indices.

Souchay (Paris) reports on the use of extragalactic sources to materialize the current ICRF. In the optical domain, these sources exhibit variations of their luminosity. These variations, due to astrophysical phenomenon, could be correlated to astrometric variations of the photocentre. Different investigations are in progress to find such a correlation in a set of targets that could ensure the link between the ICRF and the future Gaia Celestial Reference Frame. These activities are carried out as part of the ICRS-PC activities.

RUSSIA: The Allan variance (AVAR) technique has been improved to allow the processing of unevenly weighted and multidimensional data. The proposed AVAR modifications are used to analyze radio source position catalogues and time series at the Pulkovo observatory (Malkin 2011).

UNITED STATES: The ICRF2 was released (IERS, Fey *et al.* 2009) containing accurate positions of 3414 compact, extragalactic radio sources on the system of the original ICRF.

The Navy Prototype Optical Interferometer continued with absolute observations of a sample of bright stars (Benson *et al.* 2010).

5. Positions and Proper Motions

ARGENTINA: The Yale/San Juan Southern Proper Motion Catalogue, SPM4, was published containing absolute proper motions, celestial coordinates and photometry for over 103 million objects between the south celestial pole and -20° declination.

CHINA PR: Qingyu Peng (Guangzhou) reports that following CCD observations of the open cluster NGC2168 using the newly-fitted 2k by 2k CCD attached to the 1-m telescope at the Yunnan Observatory with two different orientations, significant distortions in the observations were found using UCAC2 as reference. After correction, the positional measurement accuracy of a bright star is about 7 mas in each direction (60 s exposure and near the zenith) (Peng & Fan, 2010). Observations have also been taken of M67 (NGC2682) and a technique developed using many overlapping CCD frames to remove the distortions (Peng & Tu, 2011).

DENMARK: Høg reports that the Brorfelde Schmidt CCD Catalogue (BSCC) has been published in collaboration with N. Zacharias, USNO (arXiv:1006.4602). It contains about 13.7 million stars, north of +49° declination with precise positions and V and R photometry. The catalogue has been constructed from the reductions of 18,667 CCD frames observed with the Brorfelde Schmidt Telescope between 2000 and 2007.

GERMANY: Röser (Heidelberg) reports on the PPMXL catalogue of about 900 million objects with a new determination of mean positions and proper motions on the ICRS system by combining USNO-B1.0 and 2MASS astrometry. Typical individual mean errors of the proper motions range from 4 mas/yr to more than 10 mas/yr depending on the observational history. The mean errors of positions at epoch 2000.0 are 80–120 mas, if 2MASS astrometry could be used, 150–300 mas otherwise (Röser, Demleitner & Schilbach 2010).

UKRAINE: Pinigin and Maigurova (Nikolaev) report ecliptic zone observations with the Axial Meridian Circle of RSI MAO have been completed. The resulting catalogue of 141,927 stars has accuracies of 20–90 mas for the magnitude range 9–15.5. Observations of high proper motion stars with the AMC and Mobitel telescopes have been carried out since 2009.

Also the reprocessing of observations of 171 extragalactic radio sources using UCAC3 as the reference catalogue was carried out. The average values of the differences (optical minus radio) are (-6 ± 4) mas and (11 ± 4) mas for RA and declination, respectively.

Pinigin (Nikolaev) and Ryl'kov (Pulkovo) presented a compiled catalogue (Pul-ERS) of reference stars down to 17 mag around 240 extragalactic radio sources. Comparisons with UCAC3 show average differences of 12 mas in RA and 7 mas in Dec.

Fedorov (Kharkiv) reports that the XPM catalogue of positions and absolute proper motions of 314 million objects in the magnitude range $10 < B < 22$ was derived from the 2MASS and the USNO A2.0 positions. The catalogue is available from the CDS (Fedorov+ 2010). The proper motion accuracies for the Northern and Southern hemispheres are 3–8 mas/yr and 5–10 mas/yr, respectively. The mean formal error of absolute calibration is less than 1 mas/yr (P. Fedorov et al. 2009, 2010).

Comparisons of the proper motions between XPM and HCRF showed that the coordinate axes defined by the XPM catalogue are non-rotating with respect to distant extragalactic objects to within ±0.20 mas/yr. It was concluded that the derived system of the XPM proper motions is an independent realization of the ICRS in the optical and near-infrared wavelength range (P. Fedorov et al. 2010, 2011).

Comparisons of the XPM proper motions with those of other catalogues show appreciable systematic errors of proper motions with magnitude in these catalogues (Yatskiv et al. 2010, 2011; P. Fedorov et al. 2011). Comparing the absolute proper motions from the XPM catalogue with Tycho-2, UCAC-3, XC1 and PPMXL, showed that the coordinate axes of these catalogues have a rotation: ω_x and ω_y are very small, but the ω_z differences are ∼2 mas/yr and are present up to 14 mag.

UNITED STATES: Yale University reports that the SPM catalogue was released in early November 2009 and covers $\delta < -20°$. The catalogue lists absolute proper motions, positions, and B,V photometry for over 103 million stars and galaxies. It is complete to roughly V=17.5. The final precision of the SPM positions and absolute proper motions is 30–150 mas, and 2–10 mas/yr respectively. Systematic errors in the proper motions are estimated to be of the order of 1 mas/yr (Girard et al. 2011).

A separate proper-motion catalogue (based on the same SPM material) for a 450-square degree area, that encloses the Magellanic Clouds, was released in 2009 for 1.4 million objects (Vieira et al. 2010). The proper motions of the Clouds are derived with an estimated error of about 0.27 mas/yr in the inertial system of the Hipparcos catalogue. This study also provides the best-ever estimate of the relative proper motion of one Cloud with respect to the other (accuracy of 0.15 mas/yr) thanks to the well-defined reference frame encompassing both Clouds in this large-area study.

UCAC4 (Zacharias et al. 2012) is the final release of the USNO CCD Astrograph Catalogue. Positions, proper motions, and photometry are given for over 100 million stars, all-sky to about magnitude R = 16.

The 7th and 8th data release of the SDSS project occurred in 2009 and 2011, respectively. Proper motion results were published by Munn (2010).

6. Trigonometric Parallaxes (nearby and high proper-motion objects)

CHILE: The exceptional quality of Chilean skies has led to several astrometric investigations in cooperation with astronomers from ESO, USNO and Latin American institutes. An example of this is the IPERCOOL project on the parallaxes of dwarf stars.

CHINA PR: Shulin Ren (Nanjing) reports that the kinematical parameters of dozens of spectroscopic binaries are determined, or improved, based on the revised Hipparcos

Intermediate Astrometric Data by using an efficient fitting method which reduces the number of non-linear model parameters as far as possible. This method can also be used in the data processing of the Gaia binaries (Ren & Fu, 2010 & 2011). The statistical analysis shows that the classical double two-body model cannot be used to describe the kinematics of most hierarchical triple star systems with the precision of present day observations (1 mas). A new description of the kinematics of six hierarchical triple star systems with sufficient observations are discussed (Liu, *et al.*, 2010). A V-band Mass-Luminosity Relation is improved, based on the dynamical masses and luminosities of 203 main sequence stars by using a fitting method which can reasonably assign weights to the observational data including two quantities with different dimensions (Xia & Fu, 2010).

FRANCE: Parallax projects on targets of astrophysical interest are developed at Bordeaux (Ducourant & Teixeira). Brown dwarfs and very young associations are targeted to solve the questions of membership and kinematics of groups (Teixeira *et al.* 2009). The TW Hydrae association is the centre of a project developed at ESO/NTT aiming at the determination of parallaxes of 15 members. The idea is to determine the dynamical age of this association by back-tracing its members until they define a minimum volume.

GERMANY: Scholz (Potsdam) reports on a high proper motion ($\mu > 0.14$ arcsec/yr) survey for extremely faint ($i > 21$) objects in SDSS stripe 82 covering about 275 square degrees. The newly discovered objects were classified spectroscopically as L dwarfs (13), late-M dwarfs (8), and cool white dwarfs (4), all showing thick disk and halo kinematics (Scholz, Storm, Knapp, & Zinnecker 2009). A faint common proper motion companion of a nearby L dwarf was found serendipitously in UKIDSS and SDSS data. The companion's colour is typical of a late-T spectral type, the astrometric measurements are consistent with a physical pair (separation 75 AU) at a distance of 8 pc (Scholz 2010a). Eleven new late-T dwarf candidates with high proper motions ($0.1 < \mu < 0.8$ arcsec/yr), including two wide companions of Hipparcos stars, were discovered in a systematic search with UKIDSS and SDSS (Scholz 2010b). Two ultracool (T8–T10) brown dwarfs at spectroscopic/photometric distances of only 5 pc were discovered from their large proper motions of 2.5 and 1.5 arcsec/yr by combining the preliminary WISE data release with 2MASS and SDSS (Scholz, Bihain, Schnurr, & Storm 2011).

UKRAINE: Ivanov (Kiev) reports that a catalogue of more than 2,500,000 stars with high proper motions ($>0.04''$/year) was compiled from the FONAK1.1 catalogue (itself taken from 790 other published catalogues and sources). This catalogue will be placed at CDS.

7. Double Star Astrometry

UNITED STATES: Mason and Hartkopf report that considerable effort has been spent identifying known pairs as optical or physical. Of the 102,000 pairs in the last major release of the Washington Double Star Catalogue (2006.5), 1.3% were optical and 1.8% physical. For the current catalogue (n=115,000), 2.0% are optical and 7.8% physical. While the majority remain unknown, the statistics are improving.

The parallax programmes at the 61 inch NOFS reflector (Vrba *et al.* 2011) and the Allegheny Observatory (Gatewood, 2009) continued.

8. Solar System

The WG on Natural Planetary Satellites encouraged the making of astrometric observations of all planetary satellites. A campaign of observations of the mutual events of the main satellites of Jupiter and Saturn have been made in 2009. These photometric observations provide highly accurate astrometric positions. The WG encouraged also progress in astrometric reductions in order to prepare for the arrival of the Gaia star catalogue. Through a collaboration between USNO, IMCCE and the Royal Observatory of Belgium, the scanning of old photographic plates have been made using the Damian scanning machine, showing that new accurate astrometric information may be extracted from these plates.

BELGIUM: Damian has been used to digitise USNO plates spanning more than 30 years to determine very accurate ephemerides of the four Galilean satellites of Jupiter (results in De Cuyper et al. (2009) and Robert et al. (2011)), the natural satellites of Saturn and the major planets.

BOLIVIA: The Observatorio Astronomico Nacional Santa Ana is making developments on NEAs and asteroids including astrometric aspects for rapid orbit determination.

CHINA PR: Rongchuan Qiao (Xian) reports that they present 112 new CCD astrometric positions of Nereid, the second satellite of Neptune. They observed Nereid in 2006–2007 with the 1m and 2.16m telescopes of the Xinglong Station near Beijing, both equipped with large CCD detectors of 1340×1300 and 2080×2048 pixels, respectively. UCAC2 was used in the reduction so that a classical astrometric calibration could be used. They have shown that these observations appear to be of equal or higher precision (RMS=0.2″) than most of the previous CCD ones (Qiao, 2008).

Astrometric observations of Phoebe, the ninth satellite of Saturn (V=16.5 mag), were performed during the four successive 2005, 2006, 2007 and 2008 oppositions. A total of 1250 new observed positions of Phoebe were obtained in 30 nights involving six missions, by using three different telescopes. A comparison of these new observed positions with the latest JPL Phoebe ephemerides shows that the RMS of the residuals is better than 0.1 arcsec (Qiao, 2011). These observations have been made available to update Phoebe's orbit and represent 2994 positions, spread over a large time interval of 105 years, from 1904 to 2009. The accuracy of the updated orbit of Phoebe presented here is about 0.1 arcsec and has been significantly improved. The new orbit is in good agreement with reliable JPL ephemeris within less than 20 mas (Shen, 2011). Also, during this triennium, they have obtained more than 7,500 observations, which include the main satellites of Saturn and Uranus: Phoebe, Triton and Nereid.

FRANCE: The automatic Bordeaux meridian circle is still active and continues programmes of astrometry of solar system satellites and planets and photometric variability of blazars.

RUSSIA: L'vov (Pulkovo) reports that the tables of approaches of planets to extragalactic radio sources and occultations of radio sources by planets have been computed for the period till 2050 and are available at http://www.gao.spb.ru/english/as/ac_vlbi. These tables can be used for planning observations aimed at testing gravity theories and Solar system studies. (L'vov, Malkin, Tsekmeister, 2010, 2011)

UKRAINE: Filonenko (Kharkov) reports on DSLR observations of comets C/2006 W3 Christensen, 103P/Hartley 2 and C/2009 P1 (Garradd) using the 20 cm reflector AZT-7 and the 40 cm Baker-Schmidt telescope.

Ivantsov (Nikolaev) reports that for the collaboration "Observations and research of small Solar system bodies before Gaia" between institutions in Ukraine, Russia, Turkey and France, there were made 1834 astrometric observations of 61 asteroids at the RTT150 telescope (Tubitak National Observatory, Turkey) and 285 astrometric positions of 17 asteroids at the AZT-8 telescope (National Centre of Control and Test of Space facilities, Evpatoriya). The reductions were made using the UCAC3 catalogue. Standard deviations of the astrometric measurements for the asteroids were 0.19″ for the RTT150 telescope and 0.29″ for the AZT-8 telescope.

Shulga (Nikolaev) reports that the 0.5/3.0m Mobile Telescope (MBTL) equipped with a 3k×3k 12μ Apogee CCD and field rotator was installed and began test observations in 2010. Due to an original Combined Observation Method (electronic tracking technique) the MBTL can observe NEOs with apparent motion from 3 to 300″/min with a 90s exposure, as well as space debris on all orbit types with a 1s exposure for low Earth orbits. Following the tests, observations of asteroids and comets were taken in 2010–2011: 382 topocentric positions of 15 asteroids (11 of them with apparent motion from 3 to 140″/min) and 203 positions of 8 comets have been obtained; 304 positions of 10 asteroids reported to the MPC (under code 089). CCD observations of selected geosynchronous space debris were taken using the 0.3/1.5m FRT in 2009–2010 in order to maintain a catalogue of positions and orbital elements. The ephemerides accuracy is better than 0.15° on the prediction period of 200 days.

UNITED STATES: The International Occultation Timing Association (IOTA) reports that over the reporting period some 400 high-precision relative astrometric positions of asteroids derived from occultation observations were reported to the Minor Planet Centre.

Over the reporting period, profiles with a resolution of a few km were obtained for about 35 asteroids. This included the binary asteroid 90 Antiope, where the profiles of both components were well-resolved (www.asteroidoccultation.com/observations/ Results/Data2011/20110719_AntiopeProfile2.gif).

Occultation results have been combined with light curve inversion models to obtain scaled 3-D models of 44 asteroids (Durech *et al.* 2011). All observations are archived with NASA's Planetary Data System (http://sbn.psi.edu/pds/resource/occ.html). The observers are located mainly in Australasia, Europe, Japan and North America.

Minor planet observations continued with the 8 inch automated transit circle at NOFS (Harris, Murison 2011).

URUGUAY: There is strong astrometry of solar system bodies at the Observatorio Astronomico Los Molinos with work focusing on NEAS, asteroids, and comets. The department of Astronomia of the Universidad de la Republica also works on satellites and asteroids.

9. Open and Globular Clusters and the Galaxy

CHINA PR: Li Chen (Shanghai) reports on investigations on the kinematics and chemical properties of star clusters based on the SDSS DR7 spectroscopic data (Gao & Chen, 2010). Further studies have been carried out on NGC7380 which included CFHT/MegaCam observations of its proper motion (Chen *et al.*, 2011).

Investigations of young star clusters, confirm that subvirial and fractal-structured clusters will dynamically mass segregate on a short timescale (within 0.5 Myr) (Yu *et al.*, 2011).

Zi Zhu (Nanjing) reports that the astrometric data and radial velocities of carbon stars near the Galactic plane were used to investigate the kinematics of the Milky Way. The intention is to study the Galactic rotation curve up to 15 kpc of carbon stars and carbon-rich Mira variables identified toward the anti-centre direction. For these tracers, a flat rotation curve of 210 ± 12 km/s was found (Liu & Zhu, 2010).

FRANCE: The Bordeaux PM2000 proper motion catalogue has been used by the Bordeaux group to analyze the kinematics of known or suspected open clusters (Krone-Martins *et al.*, 2010). A catalogue was produced comprising all open clusters in the Bordeaux PM2000 region including their kinematical parameters and associated membership probability lists. For five open clusters, this is the first determination of their proper motions. Also confirmed were the non-existence of two kinematical populations in 15 previously suspected non-existent NGC objects and 2 additional open clusters.

GERMANY: Röser (Heidelberg) reports on a new deep ($r' < 17$) all-sky census of the Hyades based on the PPMXL catalogue and the application of the convergent point method to determine probable kinematic members. 724 stellar systems co-moving with the bulk Hyades space velocity are identified, representing a total mass of 435 solar masses. The tidal radius is about 9 pc, and 364 systems (with 275 solar masses) are found to be gravitationally bound (Röser *et al.* 2011).

JAPAN: Since 2004 the Mizusawa VLBI observatory of NAOJ has been operating VERA (VLBI Exploration of Radio Astrometry) for conducting VLBI astrometry of Galactic maser sources associated with star-forming regions and AGB stars. It has monitored about 100 sources in the Galaxy, and parallax measurements are already available for more than 20 sources with distances ranging from a few 100 pc to 5 kpc. The most recent results are summarized in the PASJ special issue for VERA (2011), where 9 papers report parallax measurements. Among them, Honma *et al.* (2011) reported a parallax for IRAS 05137+3919 to be 0.086±0.027 mas, locating this source in the far outer region beyond a Galacto-centeric distance of 15 kpc. Also, Hirota *et al.* (2011) reported a distance of the Perseus molecular cloud to be 232±18 pc based on observations of L1448C. Niinuma *et al.* (2011) measured parallax and proper motions of IRAS 06061+2151 and located it in the Perseus arm. This source shows a considerable non-circular motion with respect to Galactic rotation, confirming previous findings for other Perseus-arm sources. In addition to papers in the PASJ special issue, Nagayama *et al.* (2011) determined the distance of G48.61+0.02 to be 5.03±0.19 kpc, indicating that this source is associated with the supernova remnant W51C and star-forming region W51M, probably as a consequence of sequential star-formation triggered by a supernova whose remnant is W51C. The observations with VERA will be continued in 2012, and accurate determinations of Galactic parameters (such as distance to the Galaxy centre and rotation speed of LSR) are expected to be conducted in the coming year.

UKRAINE: Kharchenko (Kiev) reports on an investigation of 650 open clusters using the ASCC-2.5 catalogue of 2.5 million stars. From this, a catalogue was formed of the astrometric and photometric data and probabilities of the membership of stars in the 650 sky areas with Galactic open clusters. Also in this catalogue are the characteristics of the open clusters: coordinates, core and corona radii, shape parameters, proper motions, radial velocities, distances, reddening, ages, tidal radii and masses, integrated luminosities and colours, which for many clusters were determined for the first time. A revision of kinematic, structural and evolutionary parameters of the Galaxy disk was carried out on this set of open clusters, as basic representatives of the population of the Galaxy.

Rybka and Yatsenko (Kiev) present the RCGP catalogue of more than 0.5 million candidate Red Clump giants brighter than Ks=9.5 mag. These stars were selected from the PPMX catalogue as the most probable Red Clump members on colour-reduced proper motion diagrams, constructed from PPMX proper motions and 2MASS J and Ks-photometry. Based on the reddening of the extracted stars, Ks-band extinction was determined and taken into account. Using a two-dimensional Galactic rotation model, generalized by Ogorodnikov, the tangential velocity field of selected Red Clump members (mostly thin disk) was investigated within 1.5 kpc of the Sun. The values of kinematic parameters and solar components were determined as a function of height above the Galactic plane and heliocentric distance.

Fedorov (Kharkiv) reports that using the three-dimensional Ogorodnikov-Milne model, the kinematic parameters of the Galaxy were obtained by means of the absolute proper motions of faint stars in the XPM catalogue. The Oort constants were found to be A=8.2±0.44 km/s/kpc, B=11.26±0.34 km/s/kpc and the angular velocity of the Galaxy $\omega = 4.11 \pm 0.29$ mas/yr (Akhmetov *et al.* 2010, 2011).

UNITED STATES: Absolute proper motions for nine globular clusters located in the inner regions of the Milky Way were determined by Yale University with errors ranging between 0.4 and 0.9 mas/yr. The velocity structure of the Thick Disk was analyzed (Casetti-Dinescu *et al.* 2011).

The Kapteyn Selected Area Proper Motion Programme (Yale University) provides absolute proper motions in some 50 40×40 arcmin fields located in three declination zones: 0° and ±15°. The programme has a limiting magnitude of V~20 for a typical field, and V~22 for a few specific deep fields. This is based on Du Pont photographic plates, POSSI plates, 60-inch Mount Wilson plates and a few 4-m Mayall plates, with an overall baseline of ~80 years and typical errors of 1 to 2 mas/yr. Recent publications include results in the tidal streams and overdensities detected in SDSS.

The PanSTARRS programme has been observing with PS1 for over a year and first astrometric results have been obtained (Monet *et al.* 2011).

First astrometric results were obtained from the Kepler mission (Monet *et al.* 2010).

10. Education in Astrometry

The peer reviewed Scholarpedia project has established a section about "Astrometry" (www.scholarpedia.org/article/Category:Astrometry).

DENMARK: The history of astrometry during the past 2000 years has been studied by Erik Høg in a dozen reports which are available at arXiv:1104.4554.

Høg has written about the early years of Gaia development: "Astrometry history: Roemer and Gaia", placed at arXiv:1105.0879. The evolution of optics and detection in this period is the main subject of the report.

11. Symposia, Colloquia, Conferences

Summer School on Astrometry, 5–9 September, 2011, Antalya, Turkey.

Unresolved Galaxies, QSOs, Reference Frames, June 14–16, 2010, IAP, Paris, France (http://www.oca.eu/rousset/EGSG)

Journées 2010: New challenges for reference systems and numerical standards in astronomy, September 20–22, 2010, Meudon, France

Journées 2011: Earth rotation, reference systems and celestial mechanics: Synergies of geodesy and astronomy, September 19–21, 2011, Vienna, Austria.

Acknowledgements

The designated national representatives are thanked for their contributions to this report.

Dafydd Wyn Evans
president of the Commission

Transactions IAU, Volume XXVIIIA
Reports on Astronomy 2009–2012
Ian Corbett, ed.

© International Astronomical Union 2012
doi:10.1017/S1743921312002608

COMMISSION 19

ROTATION OF THE EARTH

ROTATION DE LA TERRE

PRESIDENT	Harald Schuh
VICE-PRESIDENT	Chengli Huang
SECRETARY	Florian Seitz
PAST PRESIDENT	Aleksander Brzezinski
ORGANIZING COMMITTEE	Christian Bizouard, Ben Chao,
	Richard Gross, Wieslaw Kosek,
	David Salstein
	IVS representative: Oleg Titov
	IERS representative: Bernd Richter
	IAG representative: Zinovy Malkin

TRIENNIAL REPORT 2009–2012

1. Introduction

The Commission supports and coordinates scientific investigations about Earth rotation and related reference frames. Above all C19 encourages and develops cooperation and collaboration in observation and theoretical studies of Earth orientation (the motions of the pole in the terrestrial and celestial reference systems and the rotation about the pole). The Commission serves the astronomical community by linking it to the official organizations providing the International Terrestrial and Celestial Reference Systems/Frames (ITRS/ITRF and ICRS/ICRF) and Earth orientation parameters (EOP): International Association of Geodesy (IAG), International Earth Rotation and Reference System Service (IERS), International VLBI Service for Geodesy and Astrometry (IVS), International GNSS Service (IGS), International Laser Ranging Service (ILRS), International DORIS Service (IDS). Among the most important activities are the development of methods for improving the accuracy and understanding of Earth orientation and related reference systems/frames. Further, C19 ensures the agreement and continuity of the reference frames used for Earth orientation with other astronomical reference frames and their densifications and provides means of comparing observational and analysis methods and results to ensure accuracy of data and models.

During the 27th IAU General Assembly in Rio de Janeiro the Organizing Committee (OC) of C19 has been re-structured and filled with new members. It consists of three ex-officio members (Commission President, Vice-President, Past President), three representatives from international services related to Earth rotation (International Association of Geodesy (IAG), International Earth Rotation and Reference Systems Service (IERS), International VLBI Service for Geodesy and Astrometry (IVS)) and five elected members of which all are in their first of two possible terms. Furthermore the position of a Commission Secretary has been established.

During the past triennium a new commission website has been established (www.iaucomm19.org). It lists the Commission's terms of reference as well as information on upcoming scientific meeting and the history of C19. In addition it provides the members' list of the Commission. During the past triennium the members' directory has been updated. It is now on an up-to-date status after former members were removed who are not active and/or interested in Earth rotation research any more.

A brief description of the most important developments in the fields related to C19 is given below. The following sections contain the reports of cooperating services/institutions. The list of references comprise only the most important papers which have been published in the past years; an extended list of references provided by the members of C19 will be posted at the Commission website. Two review works on Earth rotation shall be mentioned here that were published in the past triennium: A book chapter by Seitz and Schuh (2010) and a section in an encyclopedia by Schuh and Böhm (2011).

2. Scientific sessions, workshops and special publications within the past triennium

Three special workshops about Earth orientation variations were held during the past triennium: (1) a joint Upper Mantle Dynamics and Quaternary Climate in Cratonic Areas (DynaQlim)/Global Geodetic Observing System (GGOS) workshop on 'Understanding Glacial Isostatic Adjustment' was held in Espoo, Finland during June 23-26, 2009; (2) an IERS workshop on 'EOP Combination and Prediction' was held in Warsaw, Poland during October 19-21, 2009; and (3) a joint GGOS / IAU Commission 19 workshop on 'Observing and Understanding Earth Rotation' was held in Shanghai, China during October 25-28, 2010. The joint DynaQlim/GGOS workshop included discussions of the importance of Earth orientation measurements to both constrain GIA models and to verify and validate those models. The proceedings of the workshop will be published as a special issue of the journal *Physics and Chemistry of the Earth*. The IERS workshop included discussions of the determination, combination, and prediction of Earth orientation variations. The proceedings papers of the workshop have been published in volume 45 of the journal *Artificial Satellites*. The joint GGOS/IAU C19 workshop included discussions of all aspects of the Earth's rotation, including the observations and theory of the Earth's time varying rotation, the causes of the observed variations, the consistency of Earth rotation observations with global gravity and shape observations, and the combination of Earth rotation, gravity, and shape observations to gain greater understanding of the mass load acting on the surface of the solid Earth. The proceedings of the workshop will be published as a special issue of the *Journal of Geodynamics*. Further scientific sessions on Earth rotation which can be reported here are sessions at the annual General Assembly of the European Geosciences Union (EGU; Vienna, Austria), as well as at the Journées 'Systemes de reference spatio-temporels' in 2010 (Paris, France) and in 2011 (Vienna, Austria).

3. Report of national projects and individual institutions

3.1. *Report of activities in Australia*

By *O. Titov*. Activities in Australia the field of the Earth rotation during 2008-2011 focused on the following topics:

- New VLBI network included three AuScope radio telescopes (Hobart, Yarragadee, Katherine) started operation in 2011, particularly, in IVS-R1, R4 sessions.
- General IVS meeting held in Hobart, Tasmania, in February, 2010, hosted by the University of Tasmania (UTAS).

- Observational program for identification of the reference sources in radio/optics started in 2008 (PI: Oleg Titov). This program focuses on the future link between radio reference frame produced by VLBI and optical reference frame which will be produced by GAIA, and comprises several large optical facilities: The 3.5 meter New Technology Telescope (NTT, ESO) in Chile, the 6-meter Big Telescope Azimuthal (BTA, SAO) in Russia, and the two 8 meter Gemini Telescopes in Chile and Hawaii. Spectroscopic observations in optics of the reference radio sources to determine redshifts, and, thus, confirm their extragalactic nature are undertaken. Observational programs on all four telescopes are continuing at this stage. A proposal was submitted to reserve observing time at the 10 meter Gran Telescope Canarian (GTC) in Spain. The program is performed in collaboration with scientists from Russia, Australia, Germany, France, Spain and the USA. Redshifts of 50 reference radio sources have been measured by date.

- Indication of the Galactocentric aberration which is measured as 'secular aberration drift' from a global set of geodetic VLBI data may affect the future IAU Resolutions/IERS Recommendations.

- A significant investment in ground GNSS infrastructure over the last four years has included the expansion of the Australian GNSS network through federal government investment in geospatial infrastructure through AuScope. This funding has seen the construction of 48 GNSS sites with another 52 under construction. Significant research has also been undertaken by a number of Australian researchers investigating systematic error sources and mitigation strategies within analyses of global GNSS networks. This data from the Australian GNSS network is made available to the International GNSS Service and contributes to the IGS Earth Orientation parameter estimates.

- Satellite Laser Ranging (SLR) at the Mount Stromlo (Canberra) and Yarragadee (Western Australia) facilities have continued throughout the 2007-2011 period. In 2008, $80K of AuScope funding was used to upgrade the power of the laser in the Mt. Stromlo SLR system. This allows ranging to high Earth orbit satellites such as GNSS. The Australian systems were operated by Electro Optic Space Systems under contract to Geoscience Australia. The two SLR observatories contribute to the International Laser Ranging Service (ILRS) and subsequently their Earth Orientation parameter estimates. In 2007, Geoscience Australia became accredited as an associate analysis centre of the ILRS.

3.2. Report of activities at the Institute of Geodesy and Geophysics (IGG) of the Vienna University of Technology, Austria

By *T. Nilsson, M. Schindelegger and S. Böhm.* The main research area at IGG is the study of short-period and episodic variations in Earth rotation and the estimation of these effects using the very long baseline interferometry (VLBI) and other space geodetic techniques. Hourly Earth rotation parameters (ERP) estimated from the two-week long VLBI campaigns CONT02, CONT05, and CONT08, have been compared to estimates of GPS as well as to the IERS model for high frequency ERP variations (Nilsson *et al.*, 2010). Furthermore, time series of polar motion and DUT1 with sub-daily resolution estimated from all available geodetic VLBI observations have been used to create empirical models of the diurnal and sub-diurnal ERP (Böhm *et al.*, 2011). The obtained parameters have been compared to those derived from ocean tidal models, e.g. to the high frequency ERP variations as predicted by the IERS Conventions. Significant differences between the VLBI results and the IERS Conventions model were found at several tidal frequencies. The reasons for these discrepancies need further investigation. The possibility to combine data from the ring laser gyroscope 'G' in Wettzell with VLBI observations in order to estimate accurate hourly ERP have also been studied (Nilsson *et al.*, 2011a). The results show that a combination on the normal equation level is possible, although the contribution from the ring laser observations is presently very small since the accuracy of the VLBI data is about one order of magnitude higher.

Atmospheric excitations of Earth rotation at diurnal and sub-diurnal frequencies have been studied by analysing atmospheric angular momentum (AAM) functions calculated from numerical weather models. AAM series from a special set of hourly ECMWF analysis fields during the period of CONT08 (Schindelegger *et al.*, 2011a), from 3-hourly ECMWF delayed cut-off analysis fields (Schindelegger *et al.*, 2011b), and from the standard 6-hourly ECMWF operational analysis data, have been used. An interesting discovery from this investigation is that the excitations caused by atmospheric mass variations and those caused by winds tend to counteract each other. The reason for this counteraction has been carefully investigated. Changes in Earth rotation caused by the large Earthquakes in Chile, February 2010, and Japan, March 2011, have been studied. The expected variations in the ERP and the corresponding excitation values were calculated from a model. The changes are however so small, that it has presently not been possible to validate the model results using observations (Nilsson *et al.*, 2011b).

3.3. *Report of activities at the Royal Observatory of Belgium*

By *V. Dehant.* The main goal of the team 'Earth rotation' at the Royal Observatory of Belgium (ROB) is to better understand and model the Earth rotation and orientation variations, to study physical properties of the Earth's interior as well as the interactions between the solid Earth and the geophysical fluids. The work is based on theoretical developments as well as on the analysis of data from Earth rotation monitoring and general circulation models of the atmosphere, ocean, and hydrosphere. The scientists involved in this project work on the improvement of the processing of Very Long Baseline Interferometry (VLBI) and GNSS observations, on the determination of geophysical parameters from these data, and on analytical and numerical Earth rotation models. They study the angular momentum budget of the complex system composed of the solid Earth, the core, the atmosphere, the ocean, the cryosphere, and the hydrosphere at all timescales. This allows them to better understand the dynamics of all the components of the Earth rotation, such as Length-of-day variations (LOD), polar motion (PM), and precession/nutation, as well as to improve their knowledge and understanding of the system, from the external fluid layers to the Earth deep interior. In particular for the last years (2009–2011):

- ROB has developed a strategy for combining VLBI- and GPS-based normal equations in order to achieve a better accuracy and a better consistency in the resulting nutation series.
- ROB has performed an inversion of different VLBI nutation data sets in order to estimate parameters characterizing physical properties of the deep Earth. This inversion was performed using longer data sets and a new inversion strategy.
- ROB has analyzed the Earth's interior parameters inferred from the inversion of nutation observations in terms of existing models of the mechanical coupling at the boundaries of the fluid core. Estimations of physical properties of the deep Earth have been obtained.
- ROB has computed the topographic torque at the core-mantle boundary and its effects on nutations, and ROB has shown that some harmonics of the topography are enhanced due to the coupling of the nutation forcing and the topography itself.

3.4. *Report of activities in the People's Republic of China*

By *C. Huang.* The dynamic coupling between magnetic field and nutation near core-mantle-boundary was discussed in a numerical integration approach. Results showed that the contribution from magnetic field is approximately one order of magnitude smaller than required to fill the differences of -1.0-year nutation and FCN period between theoretical value and observation (Huang *et al.* 2011). A generalized theory of the gravity potential and figure interior was developed and applied to study the global dynamic fattening (H) and nutation. It was shown that the traditional 1.1% difference between H_{PREM} and H_{obs} can be reduced by 2/3 (Huang and Liu, 2011). A triaxial Earth rotation theory to incorporate all relevant physical processes was studied. The dynamic equations are formulated and the normal modes for an Earth model with a triaxial anelastic mantle, a triaxial fluid core, and dissipative oceans are obtained (Chen *et al.*, 2009, 2010; Shen *et al.*, 2011).

Prediction of EOP by artificial neural networks (ANN) has been studied by several colleagues. The accuracy for prediction over 60-360 days is quite good, and its accuracy for prediction of UT1 by integrated LOD is significantly improved for shorter lead times (Liao *et al.* 2011). The axial sequence of AAM is introduced into the forecasting model of ANN (Wang *et al.* 2008a); the operational prediction series of axial AAM is incorporated into the ANN model as an additional input in the real-time rapid prediction of LOD variations with 1-5 days ahead (Wang *et al.* 2008b); A non-linear ANN, general regression neural network (GRNN) model to forecast the LOD change was also studied (Zhang *et al.*, 2011). Different combinations of least squares(LS), ANN, autoregressive (AR) and Kalman filter (LS+ANN, LS+AR, and LS+AR+Kalman) in prediction of the EOP were studied (Xu *et al.*, 2010). The individual tropospheric and stratospheric wind contributions to the Earth's variable rotation were investigated. For the axial component these two terms are essentially additive; for the equatorial components these two terms cancel significantly (Zhou *et al.*, 2008).

The evidence of El Niño-related signals (sea-surface temperature anomalies) in earth rotation variation in interannual band was investigated (Zhao and Han, 2008). Plumb line (vertical) variations of the order of 0.2-0.4" has been found in North China, which are caused by the underground matter changes before and after an earthquake, and can be determined by ground astrometric techniques. It hints that the new application of this classical technique is expectable (Li and Li, 2009). A GGOS/IAU joint Science workshop *Observing and Understanding Earth Rotation* was held during Oct. 25–28, 2010 at the Shanghai Astronomical Observatory, China. There were 70 participants and 53 presentations.

3.5. *Report of activities in the Czech Republic*

By *J. Vondrák.* Long-periodic variations of Earth's rotation and their correlations with different geophysical phenomena were studied, in cooperation with Ya. Chapanov, Bulgaria (Chapanov *et al.*, 2008, 2009, 2010a,b, 2011). New celestial reference frame, as defined by our new star catalogue EOC-4 (Vondrák and Štefka 2010, Vondrák *et al.* 2010a), was used to reduce anew the classical astrometric observations to derive the Earth orientation parameters in the 20th century (Vondrák *et al.* 2010b, 2011c). New model of precession, valid for very long time intervals (J2000.0 ± 200 millenia), was derived in cooperation with N. Capitaine (France) and P. Wallace (UK). Its accuracy is comparable to the one of the present IAU 2006 model in the interval ± several centuries around the central epoch, J2000.0, and it deteriorates to several arcminutes at both ends of the interval studied (Vondrák *et al.* 2009, 2011a, b). The influence of geophysical (namely atmospheric and oceanic) excitation on nutation, including the excitation of the

Free Core Nutation, was studied (Vondrák 2009, Vondrák and Ron 2009, 2010, Ron and Vondrák 2011, Ron *et al.* 2011). The non-rigorous method of combining observations of Earth orientation parameters by modern space techniques (VLBI, GNSS, SLR) was further developed and improved (Štefka 2010, 2011, Štefka *et al.* 2009, 2010), so that also station coordinates are estimated.

3.6. *Report of activities in France*

By *C. Bizouard.*

1. Paris Observatory/SYRTE department

EOP Determination: The Paris Observatory/SYRTE department is in charge of the IERS Earth Orientation Center, which collects solutions of Earth Orientation Parameters (EOP) of many institutes, delivers them after validation in IERS format through WEB/FTP, and produces the reference EOP solution IERS C04 by combination of EOP series. Important tasks within the last four years have been the development of web services for the distribution of the Earth orientation matrix and EOP (http://hpiers.obspm.fr:eop-pc, Bizouard and Becker 2008) and the implementation of the new combined EOP solution C04 consistent with the International Terrestrial Reference Frame 2008 (Bizouard and Gambis 2008, Gambis and Bizouard 2010). These service activities are complemented by the long term project of combination at observational level. In cooperation with laboratories of the GRGS (Groupe de Recherche de Géodésie Spatiale), the Paris Observatory produces an EOP multi-technique solution from VLBI, GPS, SLR and DORIS normal equations since 2010 (Biancale *et al.*, 2011). In order to validate this multi-technique solution, GRGS and other European organizations have initiated the international campaign 'COL' (Combination at the Observation Level). The IERS working group COL has been created, which regularly examines the achieved progress and problems to be solved during international workshops. The third COL workshop was organized by the Paris Observatory in November 2011. The determination of EOP is also performed from individual techniques. The IVS center managed by Sébastien Lambert (and Anne-Marie Gontier, deceased in 2010) delivers SINEX normal equations that are included in the IVS EOP combined solution. The department also gives attention to Lunar Laser Ranging observations, which may improve the knowledge of multi-year variations of the celestial pole offsets (Zerhouni and Capitaine 2009).

Fundamental Research: Astro-geodetic work is strongly reinforced by theoretical works and EOP analysis. They concern both long term astrometric modeling of the Earth rotation (Vondrák *et al.*, 2011b) and geophysical analysis of its irregularities: the interpretation of the polar motion in light of the hydro-meteorological excitation (Zotov and Bizouard 2011, Bizouard *et al.* 2011, Bizouard and Seoane 2009) also in link with gravimetric GRACE data (Seoane *et al.* 2009, Seoane *et al.* 2011) and the consideration of asymmetric effects in the pole tide excitation (Bizouard 2011).

2. Research activities in the field of Earth rotation at other French institutions

While Paris Observatory/SYRTE department is the French reference for Earth rotation studies, many other French institutes or organizations deal with this subject. Especially, many groups are involved in the treatment of space geodetic observations (VLBI, GPS, SLR, LLR, DORIS), in particular for determining ERP: Observatoire de la Cte d'Azur (OCA), Observatoire Midi-Pyrénées/CNES, French private firm CLS, Bordeaux Observatory, and Institut Géographique National (LAREG department). Most of these groups together with Paris Observatory are federated within GRGS. One also finds geophysical institutes or universities, where some colleagues develop theoretical aspects of the Earth rotation in the light of geophysical processes (e.g., M. Leftz and O. de Viron in the Institut de Physique du Globe de Paris, Y. Rogister in the Institut de Physique du Globe de Strasbourg, J. Laskar at Paris Observatory/IMCEE department).

3.7. Research activities in the framework of the DFG-Research Unit FOR584/2 in Germany

By *J. Müller*. For the integrated study of Earth rotation and related global dynamic processes, the joint research initiative *Earth Rotation and Global Dynamic Processes* with partners from Germany, Austria and Switzerland has now been working for more than 5 years. There, 10 inter-related sub-projects (with 12.5 co-workers from 11 universities and research institutions) are funded by the German Research Foundation (DFG). Goal was the consistent modelling, analysis and interpretation of all relevant features related to Earth rotation (observation techniques, data processing, the geophysical processes in the Earth system, etc.) covering all essential time scales. The inter-disciplinary research group comprises competences from geodesy, geophysics, meteorology and oceanography. The complete project with a funding line of 3 + 3 years started in 2006 and has now entered the last phase.

The scientific challenges (for each of them, one or more projects have been defined) have been:

- Relativistic modelling of rotation, better nutation theory
- Consistent modelling and interactions of geophysical fluids relevant for EOP research
- Un-explained signals in laser gyros + combination with other techniques
- Lunar Laser Ranging modelling/analysis insufficient
- Consistent combined processing and analysis (techniques and parameters)
- Close internal and external cooperation, sustainable data and EOP tools

Some highlights of the past research years were

- Extension of the post-Newtonian nutation/precession theory to a non-rigid earth
- Determination of long term EOP parameters (precession/nutation as well as trends in UT and polar motion) from the analysis of lunar laser ranging data
- Assimilation of Earth rotation and GRACE parameters into ocean models
- Improved forward modeling for core/mantle interaction (electromagnetic, topographic and gravitational coupling)
- Analysis of climate variability from fully coupled atmosphere-hydrosphere models and its influence on EOP, evaluation of NAO index values determined from various input quantities
- Inversion of a dynamic Earth system model for the estimation of physical Earth parameters from EOP, determination of Love number k2
- Analysis and interpretation of observations of ringlaser gyros (including sub-daily and episodic variations of Earth rotation, development of a 3-D FEM topographic model for surface deformation)
- Earth rotation parameters with hourly resolution estimated from the combination of VLBI and ring laser data
- Study on the impact of earthquakes on Earth rotation
- Simultaneous estimation of consistent high-quality time series of EOP and TRF parameters from integrated VLBI, SLR, and GNSS analysis

- Combination of EOP and 2nd-degree harmonic gravity field coefficients for the separation of single mass contributions and motion effects as well as for mutual cross-validation
- Determination of tidal terms in diurnal and semi-diurnal polar motion and UT1 using VLBI and GPS
- Studying a possible coupling of non-linear changes of station coordinates into EOP
- Further development of the Earth Rotation Information System (ERIS) for providing EOP data, excitation functions, interactive analysis tools for Earth rotation studies

More information about the work of the research unit FOR584 and related publications can be found at its website www.erdrotation.de.

3.8. *Report of activities in the Space Research Centre of the Polish Academy of Sciences*

By *A. Brzeziński, J. Nastula, B. Kołaczek, M. Paśnicka, W. Kosek, M. Kalarus, T. Niedzielski and W. Popiński* Modeling perturbations in Earth rotation with subdaily to seasonal periods: We studied perturbations in Earth rotation caused by the influence of external fluid layers, the atmosphere, the oceans and the land hydrology, by using the available time series of the global angular momentum of those fluids, AAM, OAM and HAM, respectively. An important part of this work concerned the seasonal balance of excitation (Brzeziński et al., 2009; Dobslaw et al., 2010). A separate study had been devoted to the geophysical excitation of the free Chandler wobble (Brzeziński et al., 2011). We also investigated the possibility of modeling and observation of the perturbations in Earth rotation having very short periods, daily and subdaily. The so-called complex demodulation technique (Brzeziński, 2011b) appeared to be very useful tool for studying such high frequency geophysical effects (Brzeziński, 2009; 2011a). In addition, we developed analytical model of the so-called libration in UT1/LOD, the semidiurnal variation due to the lunisolar torque on the triaxial figure the Earth (Brzeziński and Capitaine, 2010).

Research on the geophysical excitation function of polar motion: Contributions to polar motion excitation determined from HAM models and harmonic coefficients of the Earth gravity field obtained from GRACE mission were analyzed. These contributions are different for different HAM models. None of the HAM functions closes the excitation budget of polar motion (Brzeziński et al., 2009; Nastula et al., 2011a, 2011b; Seoane et al., 2009a, 2009b, 2011). Geodetic residuals of polar motion excitation computed by removing the atmospheric and oceanic effects from geodetic determinations of polar motion are different for different ocean models. These differences are of the order of hydrological excitation of polar motion (several mas) and prove deficiencies of ocean models (Kołaczek et al., 2012; Paśnicka et al., 2012a). In addition regional variations of AAM, OAM and HAM (computed either from hydrological models or from gravimetric GRACE data) were computed and widely analyzed (Nastula et al., 2009; Nastula and Salstein, 2011; Nastula et al., 2012). Our attention was focused on two cases of regional distribution of OAM: in seasonal spectral band and in the wide band around the Chandler period.

Modeling, statistical analysis and prediction of Earth rotation: The EOP Prediction Comparison Campaign showed that ensemble prediction of pole coordinates data errors are less than the errors of the individual prediction techniques and the Kalman filter which involves short term prediction of axial component of the AAM is the most accurate prediction technique of UT1-UTC (Kalarus et al. 2010). The combination of the least squares and the multivariate autoregressive method involving axial component of the AAM was proposed to predict UT1-UTC (Niedzielski and Kosek 2011). It was found that pole coordinates data prediction errors are caused by wide band short period oscillations in joint ocean and atmospheric excitation functions (Kosek *et al.* 2009, Kosek 2010, 2011). Using semblance filtering the common oscillations were found in the geodetic and joint atmospheric-ocean excitation functions of polar motion (Kosek *et al.* 2011). The semblance showed that addition of the hydrology excitation function to the joint atmospheric-ocean excitation of polar motion improves the agreement of these functions in the annual frequency band (Kosek *et al.* 2011). The probability distribution of the EOP has also been studied (Niedzielski *et al.* 2009, Sen *et al.* 2009).

3.9. *Report of activities in Russia*

By *Z. Malkin*. Five institutes in Russia have been working on processing of space geodesy observations: Institute of Applied Astronomy, Institute of Astronomy, Institute of Time and Space Metrology (VNIIFTRI), Pulkovo Observatory, Sobolev Astronomical Institute (St. Petersburg State University), Sternberg Astronomical Institute (Moscow State University). Derived results are EOP , terrestrial and celestial reference frames, troposphere delays and geocenter motion. Most of the solutions are regularly submitted to IERS, IVS and IDS. More than 20 permanent VLBI, GPS and SLR stations are included in the IVS, IGS, EPN and IDS networks and used for derivation of IERS, IVS, IGS, EUREF, and IDS products and TRF densification. The Institute of Applied Astronomy started in regular (currently weekly) EOP determination on the Russian Quasar VLBI network (Finkelstein *et al.* 2011). Both Russian VLBI and GPS/GLONASS networks are used for EOP determination in the framework of the Russian State EOP Service (Kaufman and Pasynok 2010).

Several groups are working on investigation of Earth rotation variations at different time scales from intra-day to decadal and their geophysical causes. Akulenko *et al.* (2010, 2011) improved an Earth rotation model and used it to investigate the interconnection between fluctuations in the Atmospheric Angular Momentum and LOD variations, and improve the accuracy of interpolation and prediction of the Earth's axial rotation. Malkin and Miller (2010) and Miller (2011) investigated Chandler wobble variations using a 165-year IERS Polar motion series and a 170-year series of the Pulkovo latitude variations respectively. Gorshkov (2010) analyzed several EOP series to investigate LOD variations with periods of 2-7 years and their connection with various geophysical phenomena. Various aspects of improvement of Polar motion and UT1 predictions are discussed in Malkin (2010a) and Tissen *et al.* (2010). The accuracy of the celestial pole offset prediction has been assessed in Malkin (2010b). The impact of the Celestial Pole Offset modelling on the VLBI UT1 Intensive results was investigated in Malkin (2011a). Gubanov (2010) analyzed the Free Core Nutation period and amplitude variations. Malkin (2011b) estimated the impact of the Galactic aberration on precession and long-term nutation parameters derived from VLBI observations.

3.10. *Report of activities in United States of America*

By *R. Gross*. During the past triennium, investigations in the U.S. of the Earth's time varying rotation followed a number of themes, including theoretical studies (Gross, 2011), tidal variations (Gross, 2009a, 2009b, 2009c; Dickman, 2010; Dickman and Gross, 2010; Gross and Dickman, 2011), glacial isostatic adjustment (Matsuyama *et al.*, 2010; Mitrovica and Wahr, 2011), effects of global geophysical fluids (Dey and Dickman, 2010; Dickey *et al.*, 2010; Landerer *et al.*, 2009; Marcus *et al.*, 2010; Nastula *et al.*, 2009; Schindelegger *et al.*, 2011;), effects of earthquakes (Gross and Chao, 2010), effects of the core (Buffett, 2010a, 2010b; Buffett *et al.*, 2009; Dickey and de Viron, 2009; Dickey *et al.*, 2011), comparisons with gravity measurements (Cheng *et al.*, 2011; Gross *et al.*, 2009; Jin *et al.*, 2010, 2011), and improving predictions (Chin *et al.*, 2009; Gambis and Luzum, 2011;

Gambis *et al.*, 2011; Kalarus *et al.*, 2010; Luzum, 2010; Luzum and Nothnagel, 2010). Attention has been given to the use of atmospheric models to investigate the changes in Earth rotation that might be expected due to climate variability and possible secular changes (Salstein *et al.*, 2011).

3.11. *Report of activities related to Earth rotation in the International Association of Geodesy (IAG)*

By *Z. Malkin.* IAG continued to develope the Global Geodetic Observing System (GGOS) to provide observations of the three fundamental geodetic observables and their variations, that is, the Earth's shape, the Earth's gravity field and the Earth's rotational motion integrating different techniques.

IAG Commission 1 *Reference Frames* coordinated, in particular, researches in several directions related to the investigation of the Earth rotation. Three sub-commissions participated in this activities are:

- SC 1.1 Coordination of Space Techniques (President M. Rothacher)
- SC 1.2 Global Reference Frames (President C. Boucher)
- SC 1.4 Interaction of Celestial and Terrestrial Reference Frames (President H. Schuh)

The most important IAG meetings held in 2009–2011 were:

- IAG General Assembly at the IUGG 2011, Melbourne, Australia, 28 June-7 July 2011. Three symposia at this meeting discussed various matters related to the Earth rotation and reference frames: JG05 Integrated Earth Observing Systems; G01 Reference Frames from Regional to Global Scales; G03 Monitoring and Modelling Earth Rotation.
- IAG Commission 1 Symposium 2010 on Reference Frames for Applications in Geosciences (REFAG2010), Marne-La-Vallee, France, 4–8 October 2010.

3.12. *Report of the International VLBI Service for Geodesy and Astrometry (IVS)*

By *H. Schuh.* The IVS continued to fulfill its role as a service within the IAU as well as within the IAG, International Association of Geodesy. A main task of the IVS is the provision of products for the Earth orientation parameters, in particular UT1 and precession/nutation, as well as for the realizations of the celestial reference system (by the ICRF-2, the International Celestial Reference Frame) and the terrestrial reference system (by the ITRF, the International Terrestrial Reference Frame). More details about the IVS and about VLBI2010, the next generation VLBI system, are given in the Report of IAU Division I and can also be found on http://ivscc.gsfc.nasa.gov.

3.13. *Report of the International Earth Rotation and Reference Systems Service (IERS)*

By *W. Dick and C. Ma.* The International Earth Rotation and Reference Systems Service continued to provide Earth orientation data, terrestrial and celestial references frames, as well as geophysical fluids data to the scientific and other communities. Work on new realizations of the International Terrestrial Reference System (ITRF2008) and the International Celestial Reference System (ICRF2) was finished. In 2009, Bulletin B was revised following a survey which was made among the community. In order to be consistent with ITRF2008, the IERS EOP C04 was revised again in 2011. The new solution 08 C04 is the reference solution which started on 1 February 2011. The system of the Bulletin A was changed to match the system of the new 08 C 04 series. The IERS Conventions (i.e. standards etc.) have been updated regularly; a new revised edition was published at the end of 2010. The Global Geophysical Fluids Centre (GGFC) restructured to allow for the establishment of operational products. A new Working Group on Combination at the Observation Level was established in October 2009.

The following IERS publications and newsletters appeared between 2008 and 2011: A.L. Fey, D. Gordon, and C.S. Jacobs (eds.): The Second Realization of the International Celestial Reference Frame by Very Long Baseline Interferometry, 2009 (IERS Technical Note No. 35); G. Petit and B. Luzum (eds.): IERS Conventions (2010), 2010 (IERS Technical Note No. 36); IERS Annual Report 2007; IERS Bulletin A, B, C, and D (weekly to half-yearly); IERS Messages Nos. 132 to 196. The central IERS web site www.iers.org and about 15 individual web sites of IERS components have been updated, improved and enlarged continually.

The following workshops were held, two partially co-organized with GGOS: IERS Workshop on EOP Combination and Prediction, Warsaw, Poland, 19–21 October 2009; Second GGOS Unified Analysis Workshop, San Francisco, CA, USA, 11–12 December 2009; Third GGOS Unified Analysis Workshop, Zurich, Switzerland, 16–17 September 2011. Abstracts and presentations of all these workshops are available at the IERS web site.

4. Closing remarks

The last three years have shown great progress in Earth rotation research in terms of quality of the observations (accuracy, time resolution, time from observation until results) as well as in modeling the causes of variations of Earth rotation and in prediction of the EOP. This interesting field of research interfacing astronomy and geodesy has provided extremely useful output for many related disciplines such as meteorology, climatology, oceanography and other Earth sciences. As it has also attracted many young scientists working on Earth rotation, an ongoing goal is to include this new generation of Earth rotation researchers in the activities of Commission 19.

Florian Seitz   Harald Schuh
Secretary of Commission 19   *President of Commission 19*

References

Akulenko L., Markov Yu., Perepelkin V., & Rykhlova L., 2010, *Astronomy Reports*, 54(3), 260
Akulenko L., Markov Yu., Perepelkin V., & Rykhlova L., 2011, *Astronomy Reports*, 55(9), 849
Biancale R., Gambis D., Richard J. Y., & Bizouard C., 2011, *IAG Symposia*, in pess.
Bizouard, C., 2011, *Proc. Journées 2011*, submitted
Bizouard C. & Becker O., 2008, *Proc. 5th IVS General Assembly*, 199
Bizouard, C. & Gambis, D., 2008, *IAG Symposia*, 134
Bizouard C., Rémus F., Lambert S., Seoane L., & Gambis D., 2011, *A&A*, 526, A106
Bizouard C. & Seoane L., 2010, *J. Geodesy*. 84, 19
Böhm S., Brzeziński, A., & Schuh, H., 2011, *J. Geodynamics*, in press
Brzeziński A., 2009, *Proc. Journées 2008*, 89
Brzeziński A., 2011a, *Proc. Journées 2010*, 131
Brzeziński A., 2011b, *J. Geodynamics*, in press
Brzeziński A., N. Capitaine, 2010, *Proc. IAU*, Vol. 5, H15, 217
Brzeziński A., H. Dobslaw, R. Dill, M., & Thomas M., 2011, *IAG Symposia*, 136, 497
Brzeziński A., J. Nastula, & B. Kolaczek, 2009, *J. Geodynamics*, 48, 235
Buffett, B., 2010a, *Nature*, 468, 952
Buffett, B., 2010b, *EPSL*, 296, 367
Buffett, B. A., J. Mound, & A. Jackson, 2009, *GJI*, 177, 878
Capitaine N., Mathews P. M., Dehant V., *et al.*, 2011, *Celest.Mech.Dyn.Astr.*, 103, 179

Chapanov Ya., Vondrák J., & Ron C., 2008, *AIP Conference Proceedings*, 1043, 197
Chapanov Ya., Vondrák J., & Ron C., 2009, *Proc. Journées 2008*, 178
Chapanov Ya., Vondrák J., & Ron C., 2010a, *IAU Symposia*, 264, 407
Chapanov Ya., Vondrák J., & Ron C., 2011, *Proc. Journées 2010*, 149
Chapanov Ya., Vondrák J., Ron C., Štefka V., *et al.*, 2010b, *Akademperiodika*, 180
Chen, W. & W. Shen, 2010, *JGR*, 115, B12419
Chen, W., W. Shen, & J. Han, 2009, *Surv. Geophys.*, 30, 39
Cheng, M.-K., J. C. Ries, & B. D. Tapley, 2011, *JGR*, 116, B01409
Chin, T., R. Gross, D. Boggs, & J. Ratcliff, 2009, *Interplanetary Network Progr. Rep., 42-176*, JPL
Correia A., Levrard B., & Laskar J., 2008 *A&A* 488, L63
Dey, N. & S. R. Dickman, 20, *JGR*, 115, C09016
Dickey, J. O. & O. de Viron, 2009, *GRL.*, 36, L15302
Dickey, J. O., S. L. Marcus, & O. de Viron, 2010, *GRL*, 37, L03307
Dickey, J. O., S. L. Marcus, & O. de Viron, 2011, *J. Climate*, 24, 569
Dickman, S. R., 2010, *JGR*, 115, B12407
Dickman, S. R., & R. S. Gross, 2010, *J. Geodesy*, 84, 457
Dobslaw H., R. Dill, A. Grötzsch, A. Brzeziński, & M. Thomas, 2010, *JGR*, 115, B10406
Finkelstein A., Salnikov A., Ipatov A., *et al.*, 2011, *Proc. 20th Meeting of the EVGA*, 82
Gambis D. & C. Bizouard, 2010, *Highlights of Astronomy*, 15, 207
Gambis, D. & B. Luzum, 2011, *Metrologia*, 48(4), 165
Gambis, D., D. A. Salstein, & S. Lambert, 2011, *J. Geod.*, 85, 435
Göttl, F. & Seitz, F., 2008, *IAG Symposia*, 133, 439
Gorshkov V. L., 2010, *Solar System Research*, 44(6), 487
Govind, R., F. Lemoine, J. Valette, D. Chinn, & N. Zelensky, 2010, *Adv. Space Res.*, 46, 1593
Gross, R. S., 2009a, *J. Geodyn.*, 48, 219
Gross, R. S., 2009b, *J. Geodesy*, 83(7), 635
Gross, R. S., 2009c, *Proc. Journées 2008*, 95
Gross, R. S., 2011, *IAG Symposia*, 137, in press
Gross, R. S. & B. F. Chao, 2010, *IAG Symposia*, 135, 643
Gross, R. S. & S. R. Dickman, 2011, *Proc. Journées 2010*, in press
Gross, R. S., D. A. Lavallee, G. Blewitt, & P. J. Clarke, 2008, *IAG Symposia*, 133, 463
Gubanov V. S., 2010, *Astronomy Letters*, 36(6), 444
Hense, A.; Sündermann, J.; Drewes, H. *et al.*, 2009, *German Geodetic Commission*, B317
Huang C. L., Dehant V., Liao X. H., Van Hoolst T., & Rochester M. G., 2011, *JGR*, 116, B03403
Huang C. L. & Liu Y., 2011, *Proc. IUGG GA*, submitted
Jin, S., D. P. Chambers, & B. D. Tapley, 2010, *JGR*, 115, B02403
Jin, S., L. J. Zhang, & B. D. Tapley, 2011, *GJI*, 184, 651
Kalarus, M., H. Schuh, W. Kosek, *et al.*, 2010, *J. Geodesy*, 84, 587
Karatekin Ö., de Viron O., Lambert S., *et al.*, 2011, *Planet. Space Sci.*, 59(10), 923
Kaufman M. & Pasynok S., 2010, *Artificial Satellites*, 45(2), 81
Kołaczek B., Paśnicka M., & Nastula J., 2012, *Proc. Journées 2011*, submitted
Koot, L. & Dumberry M., 2011 *EPSL*, 308(3-4), 343
Koot L., Dumberry M., Rivoldini A., de Viron O., & Dehant V., 2010, *GJI*, 182(3), 1279
Koot, L., Rivoldini A., de Viron O., & Dehant V., 2008, *JGR*, 113, B08414
Kosek, W., 2010, *Proc. 6th Orlov Conference*, 96
Kosek, W., 2011, *IAG Symposia*, 136, 511
Kosek W., Popiński W., & Niedzielski T., 2011, *Proc. Journées 2010* ,168
Kosek, W., Rzeszótko A., & Popiński W., 2009, *Proc. Journées 2008*, 168
Kudryashova, M., Lambert, S., Dehant, V., Bruyninx, C., *et al.*, 2011, *Proc. Journées 2010*, 202
Lambert, S., Dehant, V., & Gontier, A.-M., 2009, *A&A*, 481(2), 535
Landerer, F. W., J. H. Jungclaus, & J. Marotzke, 2009, *GRL*, 36, L17603
Li, Z. X. & Li, H. , 2009, *Studia Geophysica et Geodaetica*, 53(2), 185
Liao, D. C., Wang, Q. J., Zhou, Y. H., Liao, X. H., & Huang C. L., 2011, *J. Geodyn.*, submitted

Luzum, B., 2010, *Artificial Satellites*, 45(2), 107

Luzum, B. & A. Nothnagel,, 2010, *J. Geodesy*, 84, 399

Malkin Z., 2010a, *Artificial Satellites*, 45(2), 87

Malkin Z, 2010b, *Astronomy Reports*, 54(11), 1053

Malkin Z., 2011a, *J. Geodesy*, 85(9), 617

Malkin Z., 2011b, *Astronomy Reports*, 55(9), 810

Malkin Z. & Miller N., 2010, *Earth Planets Space*, 62(12), 943

Marcus, S. L., O. de Viron, & J. O. Dickey, 2010, *JGR*, 115, B12409

Maslennikov K., A. Boldycheva, Z. Malkin, & O. Titov, 2010, *Astrophysics*, 53, 147

Matsuyama, I., J. X. Mitrovica, A. Daradich, & N. Gomez, 2010, *JGR*, 115, B05401

Miller N., 2010, *Solar System Research*, 45(4), 342

Mitrovica, J. X., & J. Wahr, 2011, *Ann. Rev. Earth Planet. Sciences*, 39, 577

Mocquet, A., Rosenblatt, P., Dehant, V., & Verhoeven, O., 2011, *Planet. Space Sci.*, 59(10), 1048

Nastula, J., Gross, R., & Salstein, D., 2012, *J. Geodynamics*, submitted.

Nastula, J., Paśnicka, M., & Kołaczek B., 2011a, *Acta Geophysica*, 59(3), 561

Nastula, J., Paśnicka, M., & Kołaczek, B., 2011b, *Proc. Journées 2010*, 165

Nastula, J. & Salstein, D., 2011, *IAG Symposia*, 136

Nastula, J., D. Salstein, & B. Kołaczek, 2009, *JGR*, 114, B04407

Niedzielski, T. & Kosek, W., 2011, *IAG Symposia*, 137, 153

Niedzielski, T., Sen, A. K., & Kosek, W., 2009, *Artificial Satellites*, 44, 33

Nilsson, T., Böhm J., & Schuh, H., 2010, *Artificial Satellites*, 45(2), 49

Nilsson, T., Böhm J., & Schuh, H., 2011a, *J. Geodynamics*, submitted

Nilsson, T., Böhm J., & Schuh, H., 2011b, *Proc. Journées 2010*, 176

Paśnicka, M., Nastula, J., & Kołaczek, B., 2012, *J. Geodynamics*, submitted.

Ron, C. & Vondrák, J., 2011, *Acta Geodyn. Geomat.*, 8(3), 243

Ron, C., Vondrák, J., & Štefka, V., 2011, *Proc. Journées 2010*, 221

Salstein, D., K. Quinn, & R. Abarca del Rio, 2011, *EGU Research Abstracts*, EGU2011-5270-1

Schindelegger, M., Böhm, J., Salstein, D., & Schuh, H., 2011a, *J. Geodesy*, 85(7), 425

Schindelegger, M., Böhm, J., Schuh, H., & Salstein,, 2011b, *Proc. Journées 2010*, 180

Schuh, H. & Böhm, S., 2011, Earth Rotation, in: *Encyclopedia of Solid Earth Geophysics*, 1, 123

Seitz, F. & Drewes, H., 2009, *Proc. Journées 2008*, 123

Seitz, F., Kirschner, S., & Neubersch, D., 2012, *IAG Symposia*, Proc. QuGOMS'11, in press.

Seitz, F. & Schuh, H., 2010, Earth Rotation, in: *Sciences of Geodesy I*, 185

Seitz, F., Schuh, H., & Müller, J. et al., 2010, *IAU Transactions B*, XXVIIB, 137

Sen A. K., Niedzielski T., & Kosek, W., 2009, *Proc. Journées 2008*, 105

Seoane, L., Nastula, J., Bizouard, C., & Gambis, D., 2009a, *GJI*, 178, 614

Seoane, L., Nastula, J., Bizouard, C., & Gambis, D., 2009b, *Proc. Journées 2008*, 149

Seoane, L., Nastula, J., Bizouard, C., & Gambis, D., 2011, *Int. J. Geophys.*, doi:10.1155/2011/174396.

Shen, W. B., Sun, R., & Chen, W, et al., 2011, *Ann. Geophys.*, 54(4), 436

Štefka, V., 2010, *Acta Geodyn. Geomater.*, 7(3), 275

Štefka, V., 2011, *Acta Geodyn. Geomater.*, 8(3), 237

Štefka, V., Kostelecký, J., & Pešek, I., 2009, *Acta Geodyn. Geomater.*, 6(3), 239

Štefka V., Pešek I., Vondrák J., & Kostelecký J., 2010, *Acta Geodyn. Geomat.*, 7(1), 29

Tissen V. , Tolstikov A., Balahnenko A., & Malkin Z., 2010, *Artificial Satellites*, 45(2), 111

Titov, O., D. Jauncey, H. Johnston, R. Hunstead, & L. Christensen, 2011, *AJ*, 142, 165

Titov, O., S. Lambert, & A.-M. Gontier, 2011, *A&A*, 529, A91

Titov, O. & Z. Malkin, 2009, *A&A*, 506, 1477

Verhoeven, O., Mocquet, A., Vacher, P., Rivoldini, A., et al., 2009, *JGR*, 114(B3), B03302

Vondrák, J., 2009, *Proc. VI Serbian-Bulgarian Astronomical Conference*, 9, 143

Vondrák, J., Capitaine, N., & Wallace, P., 2009, *Proc. Journées 2008*, 23

Vondrák, J., Capitaine, N., & Wallace, P., 2011a, *Proc. Journées 2010*, 24

Vondrák, J., Capitaine, N., & Wallace P., 2011b, *A&A*, 534, A22

Vondrák J. & Ron C., 2009, *Acta Geodyn. Geomat.*, 6(3), 217

Vondrák, J. & Ron C., 2010, *Acta Geodyn. Geomat.*, 7(1), 19

Vondrák, J., Ron, C., & Štefka, V., 2010b, *Acta Geodyn. Geomat.*, 7(3), 245

Vondrák, J., Ron, C., Štefka, V., & Chapanov, Ya., 2011c, *Proc. VII Bulg.-Serb. Astr. Conf.*, subm.

Vondrák, J. & Štefka, V., 2010, *Astron. Astrophys.*, 509, A3

Vondrák, J., Štefka, V., & Ron, C., 2010a, *Proc. 6th Orlov Conference*, 64

Wang, Q. J., Liao, D. C., Zhou, Y. H., & Liao, X. H., 2008a, *Chin Astron. Astrophys.*, 32, 342

Wang, Q. J., Liao, D. C., & Zhou, Y. H., 2008b, *Chinese Science Bulletin*, 53(7), 969

Winkelnkemper, T., Seitz, F., Min, S.-K., & Hense, A., 2008, *IAG Symposia*, 133, 447

Xu, X. Q. & Zhou, Y. H., 2010, *J. Spacecraft TT&C Tech.*, 29, 70

Zerhouni, W. & Capitaine N., 2009, *A&A* 507, 1687

Zhang, X. H., Wang, Q. J., & Zhu, J. J., 2011, *Chin Astron. Astrophys.*, in press

Zhao, J. & Han, Y. B., 2008, *Pure & Applied Geophys.*, 165(7), 1435

Zhou, Y. H., Chen, J. L., & Salstein, D. A., 2008, *GJI*, 174, 453

Zotov, L. & Bizouard, C., 2011, *J. Geodymics*, in pess

Transactions IAU, Volume XXVIIIA
Reports on Astronomy 2009–2012
Ian Corbett, ed.

© International Astronomical Union 2012
doi:10.1017/S174392131200261X

COMMISSION 31

TIME

L'HEURE

PRESIDENT	**Richard Manchester**
VICE-PRESIDENT	**Mizuhiko Hosokawa**
PAST PRESIDENT	**Pascale Defraigne**
ORGANIZING COMMITTEE	**Felicitas Arias**
	ShouGang Zhang
	Philip Tuckey
	Vladimir Zharov

TRIENNIAL REPORT 2009–2012

IAU Commission 31 "Time" is part of Division I of the IAU. It currently has 103 members, a net increase of about 15% over the triennium, and 9 Consultants. The Commission web page is: www.atnf.csiro.au/iau-comm31/. It maintains a current membership list, with links to the IAU membership database, links to past and up-coming meet- ings related to Time and links to relevant websites and documents.

The Commission has participated in discussions on the reorganisation of the IAU Division/Commission structure, planning for the Joint Discussion "Space-time Reference Systems for Future Research" to be held at the 28th General Assembly in Beijing and various other small matters related to its Terms of Reference.

However, by far the most significant issue dealt with during the triennium was the Interna- tional Telecommunications Union proposal (ITU-R TF.460-6) to redefine UTC, eliminating the occasional insertion of leap seconds to make UTC a uniform timescale. Following the request from the Study Group 7 (SG7) of ITU-R to the IAU for an opinion on this issue, Commission 31 members were polled. Although there were some dissenting views, the ma- jority of Commission 31 members either supported the ITU proposal or had no opinion. This was reported to the General Secretary and became the basis of the IAU response in favour of the proposal. SG7 was unable to reach a consensus and the matter was referred to the Radiocommunications Assembly which takes place in Geneva 16–20 January, 2012. Member states of the ITU will vote at the Assembly and a 70% majority is needed to accept the proposal. The matter is now largely out of the hands of the IAU, although individual IAU members may seek to influence their country's vote at the Assembly. Commission 31 has served as a useful focus for IAU discussions related to time, in particular for the leap-second debate. Timescales form part of the fundamental reference systems of astronomy and it is important that the IAU maintains expertise in this area. Astronomy is regaining a position of importance in time-keeping with pulsar timescales now a reality. For these reasons we recommend that that the Commission be retained for the next Triennium.

R N Manchester (on behalf of the OC)
31 October, 2011

Transactions IAU, Volume XXVIIIA
Reports on Astronomy 2009–2012   © International Astronomical Union 2012
Ian Corbett, ed.   doi:10.1017/S1743921312002621

COMMISSION 52

RELATIVITY IN
FUNDAMENTAL ASTRONOMY
RELATIVITÉ DANS
ASTRONOMIE FONDAMENTALE

PRESIDENT   Gérard Petit
VICE-PRESIDENT   Michael Soffel
PAST PRESIDENT   Sergei A. Klioner
ORGANIZING COMMITTEE   Victor A. Brumberg, Nicole Capitaine
                       Agnès Fienga, Bernard Guinot
                       Cheng Huang, François Mignard
                       Ken Seidelmann, Patrick Wallace

COMMISSION 52 WORKING GROUPS

TRIENNIAL REPORT 2009–2012

1. Introduction

The IAU Commission 52 "Relativity in Fundamental Astronomy" (RIFA) has been established during the 26th General Assembly of the IAU (Prague, 2006). The general scientific goals of the Commission were identified as:

• clarify geometrical and dynamical concepts of Fundamental Astronomy within a relativistic framework,

• provide adequate mathematical and physical formulations to be used in Fundamental Astronomy,

• deepen the understanding of the above results among astronomers and students in astronomy,

• promote research needed to accomplish these tasks.

In September 2011 the Commission has 72 full members. The web page of the Commission is at http://astro.geo.tu-dresden.de/RIFA.

2. Developments 2009–2012

When starting its work in 2006, the Commission had identified three scientific topics to consider first:

1. Units of measurements for astronomical quantities in the relativistic context;

2. Astronomical units in the relativistic framework;

3. Time-dependent ecliptic in the GCRS.

Topic 1 has been mostly carried out over the previous triennium and led to publications; Topic 2 is being considered over this triennium, and resulted in a Resolution proposed by the Division 1 working group Numerical Standards in Fundamental Astronomy; Topic 3 has not been formalized as a task group at this time, reflexion is ongoing considering the implications of IAU Resolution B1 (2006).

Although discussions on these scientific issues took place within the Organizing Committee, the Commission did not undertake new projects in a formalized manner over the period covered by this report and the President takes responsibility for this situation. Nevertheless, many members of the Commission were active in relation to these topics and to other RIFA-related topics, some examples follow:

• The discussion on astronomical units led to avoid considering redefinition in a relativistic framework, but to propose a new conventional definition of the astronomical unit of length as a multiple of the meter. This leads to the abandon of the "System of astronomical constants" based on the Gaussian gravitational constant k.

• The review of models or development of new models for space geodesy and other space applications, e.g. GAIA and Lunar Laser Ranging.

• The work on the improvement of the relativistic formulations (e.g. coordinate time transformations) and of the semantics of the IERS Conventions has been continued. This appears in the 2010 edition of the Conventions (Petit & Luzum (2010)).

The Commission has also taken part to the Division 1 subcommittee, which was set-up to discuss the future structure of the Division.

Gérard Petit
President of the Commission

Reference

Petit, G. & Luzum, B., (eds.), 2010, IERS Conventions (2010), IERS Technical Note 36, Verlag des BKG, Frankfurt/Main.

Transactions IAU, Volume XXVIIIA
Reports on Astronomy 2009–2012   © International Astronomical Union 2012
Ian Corbett, ed.   doi:10.1017/S1743921312002633

DIVISION I/WORKING GROUP
NUMERICAL STANDARDS OF FUNDAMENTAL ASTRONOMY

CHAIR	Brian Luzum
BOARD	Nicole Capitaine
	Agnès Fienga
	William Folkner
	Toshio Fukushima
	James Hilton
	Catherine Hohenkerk
	Gérard Petit
	Elena Pitjeva
	Michael Soffel
	Patrick Wallace

TRIENNIAL REPORT 2009–2012

1. Introduction

At the 2006 General Assembly of the International Astronomical Union (IAU), a proposal was adopted to form the Working Group (WG) for Numerical Standards of Fundamental Astronomy (NSFA). The goal of the WG is to update the "IAU Current Best Estimates" conforming with IAU Resolutions, the International Earth Rotation and Reference Systems Service (IERS) Conventions, and the Système International d'Unités (SI). The need for changes to the numerical standards are due mainly to the adoption of a new precession model, the redefinition of Barycentric Dynamical Time (TDB), and the significant improvement of the accuracy of recent estimates for a number of constants. The work from the first triennium culminated in the acceptance at the 2009 IAU General Assembly of Resolution B2 which adopted the NSFA list of Current Best Estimates as the IAU (2009) System of Astronomical Constants.

2. Developments within the past triennium

The NSFA WG has achieved several significant milestones within the last triennium.

• The WG published its report on the IAU (2009) System of Astronomical Constants (Luzum *et al.*, 2011). The report summarizes the numerical values for the constants and also provides the justification for the values chosen. This report has been published as an open access document to facilitate the availability of the information to users.

• Primarily through the efforts of WG member C. Hohenkerk, the NSFA WG web pages have been redesigned. In addition to being more visually appealing, the new web pages provide a better organizational structure and provide more complete information regarding the efforts and the results of the WG.

• The WG has drafted procedures for adopting new CBEs and is beginning to test these procedures on a suggested new value for the mass of Mercury. Based on this experience, refinements to the procedure are possible.

• WG member N. Capitaine is the primary author of a proposed IAU Resolution regarding the re-definition of the astronomical unit for consideration at the 2012 IAU General Assembly. The WG has provided feedback on the proposal in preparation for a wider discussion with the astronomical community.

3. Closing remarks

The NSFA WG continues to provide a service to the IAU and the broader astronomical community by using its knowledge and expertise to assemble a consistent set of Current Best Estimates. We are hoping to be able to continue this work within the IAU, but to continue this work with the current WG would require a modification to the structure of the IAU to allow for either a standing WG or a Service. We are expecting this issue will be resolved at the upcoming 2012 IAU General Assembly.

Brian Luzum
chair of Working Group

Reference

Luzum, B., Capitaine, N., Fienga, A., Folkner, W., Fukushima, T., Hilton, J., Hohenkerk, C., Krasinsky, G., Petit, G., Pitjeva, E., Soffel, M., & Wallace, P. 2011, *Celest. Mech. Dyn. Astr.* 110(4), 293, DOI:10.1007/s10569-011-9352-4.

Transactions IAU, Volume XXVIIIA
Reports on Astronomy 2009–2012
Ian Corbett, ed.

© International Astronomical Union 2012
doi:10.1017/S1743921312002645

DIVISIONS I and III / WORKING GROUP CARTOGRAPHIC COORDINATES AND ROTATIONAL ELEMENTS

CHAIR	Brent A. Archinal
VICE-CHAIR	vacant
PAST CHAIR	P. Kenneth Seidelmann
MEMBERS	Michael F. A'Hearn
	Albert R. Conrad
	Guy J. Consolmagno
	Régis Courtin
	Toshio Fukushima
	Daniel Hestroffer
	James L. Hilton
	Gregory A. Neumann
	Jürgen Oberst
	Philip J. Stooke
	David J. Tholen
	Peter C. Thomas
	Iwan P. Williams

TRIENNIAL REPORT 2009–2012

1. Introduction

As in the past, the primary activity of the IAU Working Group on Cartographic Coordinates and Rotational Elements has been to prepare and publish a triennial ("2009") report containing current recommendations for models for Solar System bodies (Archinal *et al.* (2011a)). The authors are B. A. Archinal, M. F. A'Hearn, E. Bowell, A. Conrad, G. J. Consolmagno, R. Courtin, T. Fukushima, D. Hestroffer, J. L. Hilton, G. A. Krasinsky, G. Neumann, J. Oberst, P. K. Seidelmann, P. Stooke, D. J. Tholen, P. C. Thomas, and I. P. Williams. An erratum to the "2006" and "2009" reports has also been published (Archinal *et al.* (2011b)). Below we briefly summarize the contents of the 2009 report, a plan to consider requests for new recommendations more often than every three years, three general recommendations by the WG to the planetary community, other WG activities, and plans for our next report.

2. 2009 Report Contents

The 2009 WG report introduces improved values for the pole and rotation rate of Mercury (based on, but not precisely as recommended in Margot (2009)), returns the rotation rate of Jupiter to a previous value, introduces improved values for the rotation of five satellites of Saturn, and adds the equatorial radius of the Sun for comparison purposes. It also adds or updates size and shape information for the Earth, Mars' satellites Deimos and Phobos, the four Galilean satellites of Jupiter, and 22 satellites of Saturn.

Pole, rotation, and size information has been added for the asteroids (21) Lutetia, (511) Davida, and (2867) Šteins. Pole and rotation information has been added for (2) Pallas. Pole and rotation and mean radius information has been added for (1) Ceres. Pole information has been updated for (4) Vesta. The high precision realization for the pole and rotation rate of the Moon is updated to use the JPL DE 421 lunar ephemeris, but rotated (by small fixed angles) to represent the mean Earth/polar axis system. The WG has adopted the IAU Working Group for Planetary System Nomenclature (WGPSN) and the IAU Committee on Small Body Nomenclature (CSBN) definition of dwarf planets. As a result, Pluto and Charon now use the positive right handed coordinate system adopted for dwarf planets, minor planets, their satellites, and comets.

Upon request and to provide information more often than every three years (e.g., for use by missions and for new cartographic products) the WG announced it will consider providing limited updates to its recommendations on its (soon to be updated) web site (http://astrogeology.usgs.gov/Projects/WGCCRE). The tentative plan is to determine every 6 months whether time-critical updates are necessary and, if so, announce them on the site. We will also offer newly published and (preferably) peer-reviewed determinations related to Solar System coordinate systems. These postings do not remove the need for our triennial reports, in which we will continue to publish the majority of our recommendations. In our next report, we will consider the usefulness of these interim procedures and whether they should continue. Input for such updates (whether for WG consideration or information only) and comment on these procedures from the community is welcome.

For the first time, the WG also provided some general recommendations regarding current urgent needs relative to the development of planetary cartographic products. These include the following (paraphrased) recommendations: 1) The advantages of geodetically controlled cartographic products are many and well known, yet the current trend seems to be that such products are often not planned for or funded. It is strongly recommended that this trend be reversed and that such products be planned for and made as part of the normal mission operations and data analysis process. 2) The WG recommends that the Konopliv *et al.* (2006) Mars orientation model be updated or a similar model be developed that takes advantage of the substantial additional Mars data available since the time of their work, so that it can be adopted by the WG and operational Mars missions in the 2012 time frame. 3) The WG urges the planetary community to jointly address resolving the various determinations for the rotation of Jupiter and Saturn and to develop consensus determinations, such as was done in the past for Jupiter (Riddle and Warwick (1976)). The WG would like to hear feedback from the planetary community regarding these recommendations, e.g., actions planned or taken, the appropriateness of such recommendations, or even suggestions for further such general recommendations.

3. Other Activities

As part of other WG activities, a meeting of this group took place at the IAU General Assembly in Rio de Janeiro in 2009. Archinal was re-elected as chairperson and plans were discussed for the 2009 report. During the triennium members of the WG have also provided information and advice on planetary coordinate systems to various individuals, instrument teams, and missions. In particular, substantial advice was provided to MESSENGER mission representatives as to how best to update Mercury's coordinate system, to Cassini mission personnel on the updating of the coordinate system of various Saturnian satellites, to the Rosetta mission regarding the coordinate systems for (2867)

Šteins and (21) Lutetia, and to the Dawn mission regarding the coordinate system for (4) Vesta.

In the latter case, the Dawn mission proposed to establish a new prime meridian for Vesta in a substantially different location from the existing one, using primarily the argument that it was important to make it clear that the new high resolution data from Dawn were different from previously acquired datasets and therefore most appropriately presented in a different system. At the request of the Dawn mission, the WG formally voted on that type of system and voted to not concur with the proposed system. This was on the basis that it did not meet the long standing WG recommendation that "... once an observable feature at a defined longitude is chosen, the longitude definition origin should not change except under unusual circumstances..." (Archinal et al. (2011a)). The Dawn mission at the present time has stated that it plans to proceed with the new definition in spite of a lack of concurrence with IAU recommendations. The WG indicated that it will attempt to obtain feedback from the planetary community to see if having more than one prime meridian definition on a given body was now considered useful. We will take such feedback into account and consider for our next report whether to modify the WG recommendation on maintaining existing longitude systems that are tied to fixed features on a body.

The WG is also actively writing abstracts and making presentations at meetings to increase awareness of our activities. Examples have been presentations at the Lunar and Planetary Science Conference (Archinal (2011c)), the International Primitive Body Exploration Working Group meeting (Archinal (2011d)), the European Planetary Science Congress – Division of Planetary Sciences Joint Meeting (Archinal et al. (2011e)), and the International Society for Photogrammetry and Remote Sensing (ISPRS) Workshop on Geospatial Data Infrastructure (Archinal and Kirk (2011f)).

4. Current WG Plans

The Working Group is currently reformulating its membership and is starting to address changes needed for the next report. The WG currently anticipates updates or new values in several areas including a) the use of an improved lunar ephemeris to define the Moon's orientation, either from JPL or others; b) updates for the orientation of Mars, Jupiter, and Saturn (as already noted); and c) updates due to new results from on-going missions (Mercury, Saturnian satellites, (4) Vesta, (21) Lutetia) and Earth-based observations (various asteroids). The WG also will look into establishing or re-establishing links to other organizations, such as the International Association of Geodesy and the ISPRS. On a "reasonable effort" basis, it will continue to provide assistance on coordinate system and mapping issues to members of the international planetary science community (space agencies, missions, instrument teams, product developers, etc.).

A draft of the next report will be presented for discussion and completion at the Working Group meeting at the IAU General Assembly in Beijing in 2012.

Brent A. Archinal
chair of Working Group

References

Archinal, B. A. et al. 2011a, Cel. M&DA 109, 101–135, doi 10.1007/s10569-010-9320-4
Archinal, B. A. et al. 2011b, Cel. M&DA 110, 401–403, doi 10.1007/s10569-011-9362-2
Archinal, B. A. 2011c, LPS XLII, abs. 2715

Archinal, B. 2011d, *IPEWG 2011*, http://ipewg.caltech.edu/
Archinal, B. *et al.* 2011e, *EPSC-DPS Joint Meeting 2011*, abs. 1553
Archinal, B. A. & Kirk, R. L. 2011f, *ISPRS Workshop*, http://isprs-wg41.nsdi.gov.cn/
Konopliv, A. S. *et al.* 2006, *Icarus* 182, 23–50
Margot, J.-L. 2009, *Cel. M&DA* 105, 329–336, doi 10.1007/s10569-009-9234-1
Riddle, A. C. & Warwick, J. W. 1976, *Icarus* 27, 457–459

Transactions IAU, Volume XXVIIIA
Reports on Astronomy 2009–2012
Ian Corbett, ed.

© International Astronomical Union 2012
doi:10.1017/S1743921312002657

DIVISIONS I, III / COMMISSIONS 4, 7, 8, 16, 20 / WORKING GROUP ON NATURAL PLANETARY SATELLITES

CHAIR   J.E. Arlot
BOARD   K. Aksnes, C. Blanco, N. Emelianov
               R.A. Jacobson, D. Pascu, Q.Y. Peng
               P.K. Seidelmann, M. Soma, D.B. Taylor
               R. Vieira-Martins, G.V. Williams

TRIENNIAL REPORT 2009–2011

1. Introduction

The Working Group on the Natural Planetary Satellites has been created to promote the development of high-quality ephemerides. The Working Group encourages theoretical studies, coordinated observations, and makes all data available to the users through the NSDC web site (http://www.imcce.fr/nsdc).

2. Activities of the Working Group on Natural Planetary Satellites

The Working Group had continued its main activities: maintaining and feeding the astrometric database of observations of the natural planetary satellites NSDB and providing ephemerides of all known satellites. These ephemerides named MULTI-SAT are available at www.imcce.fr/sat (IMCCE) or at lnfm1.sai.msu.ru/neb/nss/nssephme.htm. (trilingual version of SAI) described by Emelianov and Arlot in Astronomy and Astrophysics, vol. 487, p. 759. Ephemerides MULTI-SAT are continuously updated using the database. The accuracy of the ephemerides has been explored by Desmars *et al.* in Astronomy and Astrophysics, Volume 499, Issue 1, 2009, p. 321. The accuracy of the MULTI-SAT ephemerides has been estimated for all the 107 outer planetary satellites by Emelyanov (2010, Planetary and Space Science, 58, p. 411) over the 1975-2020 time interval. As it is shown in Emelyanov 2010 for a number of satellites new observations are of vital importance for maintaining a precision of the ephemerides allowing the identification of satellites during the reduction of observations. For some satellites the precision of their ephemerides is of the order of the sizes of their orbits and such satellites can be considered to have been lost.

NASA's JPL maintains ephemerides for all of the natural planetary satellites and makes them available electronically through JPL's On-Line Solar System Data Service known as Horizons (at http://ssd.jpl.nasa.gov/) or in the form of Spice Kernels (SPK files) through NASA's Navigation and Ancillary Information Facility (NAIF).

Original orbits and ephemerides for the outer irregular satellites of the giant planets are available from the Minor Planet Center (http://www.minorplanetcenter.net/iau/NatSats/NaturalSatellites.html). New astrometric observations and orbits are published monthly in the *Minor Planet Circulars* (new orbits appear nightly on the *DOU MPECs*).

An European contract had linked several European laboratories for improving the natural satellites ephemerides for space projects purpose. A standardisation of the NSDB database will be made through this project in a near future.

The NSDB database of astrometric observations maintained by the working group has been described by Arlot and Emelianov in Astronomy and Astrophysics, volume 503, p. 631 and is accessible on the Internet: http://www.imcce.fr/nsdc or http://lnfm1.sai. msu.ru /neb /nss /nssnsdcme.htm. Efforts have been made in order to help the observers for the reduction of data through a software PRAIA (Platform for Reduction of Astronomical Images Astrometrically) proposed by Assafin *et al.* and described in Astronomy and Astrophysics, volume 515, p. A32. Similarly, a reduction procedure for mutual events of the main satellites of the giant planets has been proposed by Emelianov and Gilbert in Astronomy and Astrophysics, volume 453, p. 1141.

A Summer School on astrometry has been organized in Antalya (Turkey) on September 5-10, 2011 and special lectures have been given on the reduction of observations of planetary satellites, either as by direct astrometric imaging or by photometric observation of phenomena (http://www.tug.tubitak. gov.tr/ aass/).

During the period 2008-2011, mutual events of the Galilean satellites and the main satellites of Saturn occur and observations were organized. Only 25 observations of the Saturnian satellites events were made because of the conjunction Saturn/Sun but more than 500 observations of the Galilean satellites events were made through a worldwide network including many amateur astronomers.

Studies have been made in order to know the interest of a new reduction of old observations of the natural planetary satellites. Plates made at USNO during the period 1967-1998 have been reduced using the UCAC2 catalogue after being scanned on the DAMIAN machine of the Royal Observatory of Belgium. Results have been published by Robert *et al.* in Monthly Notices of the Royal Astronomical Society, volume 415, p. 701.

In a workshop dedicated to the Gaia observations of solar system objects held in Pisa (Italy) in May 2011, it has been stated that the natural satellites will benefit of the new astrometric catalogue from Gaia for the astrometric reduction. Only 50 observations of each natural satellites until magnitude 20 will be provided by Gaia except the Galilean moons and Titan for which no observation is possible.

3. Selected works performed during the triennum

The accuracy of astrometric observations and the improvement of the dynamics of the satellites systems allow now to get information on the physical nature of the satellites. The comparison with physical data from space probes or from large ground based telescopes appears to be fruitful.

3.1. *all satellites: observations and theoretical studies*

A publication gathering astrometric observations of planetary satellites, close approaches and occultations of stars by asteroids and mutual events in the systems of planetary satellites with the 26-in. refractor of Pulkovo observatory in 1995–2006 was made by Kiseleva *et al.* in Planetary and Space Science, Volume 56, Issue 14, p. 1908. Another publication gathers many observations made with the transit circle of Bordeaux observatory during 1997–2007 by Arlot *et al.* in Astronomy and Astrophysics, Volume 484, Issue 3,2008, p. 869. A theoretical study on Tidal dynamics of extended bodies in planetary systems was published by Mathis and Le Poncin-Lafitte, in Astronomy and Astrophysics, Volume 497, Issue 3, 2009, p. 889. Astrometric corrections for geometric distorsion were proposed by Peng *et al.* (2010, Chinese Science Bulletin, 55, p. 791 and Sci Sin Phys Mech Astron, 2011, 41, p. 1126) and applied to Phoebe.

3.2. The Martian satellites

New ephemerides available at JPL were published by Jacobson in AJ 139, 2010, p. 668. A paper on the origin of the Martian moons has been published by Rosenblatt in Astronomy and Astrophysics Review, Volume 19, article id.44. Mars Express data have been used for improving the values of the masses of these moons and their ephemerides: results are published by Rosenblatt *et al.* in Planetary and Space Science, volume 56, p. 1043.

3.3. The Galilean satellites

Predictions of mutual events occurring in 2009 has been published by Arlot in Astronomy and Astrophysics, Volume 478, Issue 1, 2008, p. 285-298. Some results of the former campaigns of observations of the mutual events have been published:

- the 1997 campaign: by Emel'Yanov and Vashkov'yak in Solar System Research, Volume 43, Issue 3, p. 240
- the 2003 campaign: by Emelyanov in Monthly Notices of the Royal Astronomical Society, vol. 394, Issue 2, p. 1037; by Arlot *et al.* in Astronomy and Astrophysics, Volume 493, Issue 3,2009, p. 1171;
- the 2009 campaign: by Zhang *et al.* in Astronomy and Astrophysics, volume 532, p. A36; by Emelyanov *et al.*in Solar System Research, Volume 45, Issue 3, p. 264; by Marino *et al.* in Astronomia, n3, p. 18.

Tidal effects on Ganymede have been explored by Bland *et al.* in Icarus, Volume 200, Issue 1, p. 207 (2009). The rotation of Io with a liquid core was studied by Henrard and published in Celestial Mechanics and Dynamical Astronomy, Volume 101, Issue 1-2, p. 1 (2008). The analysis of a large set of astrometric data and comparison to a complete theory of the motion of Io including tides led to a thermal equilibrium of Io as published by Lainey *et al.* in Nature, Volume 459, Issue 7249, p. 957 (2009).

3.4. The satellites of Saturn

A new catalogue of improved astrometric observations was published by Desmars *et al.* in Astronomy and Astrophysics, Volume 493, Issue 3,2009, p. 1183. The predictions of the events of the satellites of Saturn during the 2009 equinox has been published by Arlot and Thuillot in Astronomy and Astrophysics, vol. 485, 2008, p. 293.

An updated Phoebe's orbit has been published by Shen *et al.* in MNRAS, Online Early and CCD astrometric observations of Phoebe made in 2005-2008 published by Qiao *et al.* in MNRAS, Volume 413, Issue 2, pp. 1079-1082. JPL continuously improves the ephemerides through Cassini data as presented at the 41st and 42nd annual meetings of the AAS DDA.

An Analytical description of physical librations of saturnian coorbital satellites Janus and Epimetheus has been published by Robutel *et al.* in Icarus, Vol.211, p. 758. Numerical exploration of resonant dynamics in the system of Saturnian major satellites has been made by Callegari and Yokoyama in Planetary and Space Science, Volume 58, p. 1906. A Theory of the rotation of Janus and Epimetheus was made by Noyelles in Icarus, Vol.207, p. 887 and by Tiscareno *et al.* in Icarus, Vol.204, p. 254.

The Rotational modeling of Hyperion was explored by Harbison *et al.* in Celestial Mechanics and Dynamical Astronomy, Vol.110, p. 1. Very Long Baseline Array Astrometric Observations of the Cassini Spacecraft at Saturn were published by Jones *et al.* in AJ, Vol.141, article id. 29 (2011) and may be useful for satellites studies.

3.5. The Outer satellites of Jupiter and Saturn

Discoveries occurred during the past triennium: new satellites of Jupiter S/2010 J 1 and S/2010 J 2 by Jacobson *et al.* in 2011 (Central Bureau Electronic Telegrams 2734, 1). A new analytical solution of the equations describing secular and long-period solar perturbations of mean orbits of outer satellites of giant planets was published (M.A.Vashkov'yak, N.M.Teslenko. Astronomy Letters, 2009, Vol. 35, pp. 850-865; M.A.Vashkov'yak. Solar System Research, 2010, Vol. 44, No. 6, pp. 527-540). A paper on the history of the Irregular Satellites of the Giant Planets was published by Nicholson *et al.* in The Solar System Beyond Neptune, M. A. Barucci, H. Boehnhardt, D. P. Cruikshank, and A. Morbidelli (eds.), University of Arizona Press, Tucson, p. 411. Another paper on the origin of the Irregular satellites of Jupiter (capture configurations of binary-asteroids) was published by Gaspar *et al.* in Monthly Notices of the Royal Astronomical Society, Volume 415, Issue 3, p. 1999.The photometric model parameters for the 97 new outer satellites of Jupiter, Saturn, Uranus, and Neptune have been determined by N. V. Emel'yanov and Ural'skaya 2011 (Solar System Research, 45, 377–385) from the magnitudes accompanying the results of astrometric observations published in Minor Planet Circulars (MPC).

A paper on the rotation of the outer irregular satellites was published by Melnikov and Shevchenko in Icarus, Volume 209, Issue 2, p. 786 (2010). A precise modeling of Phoebe's rotation was published by Cottereau *et al.* in Astronomy and Astrophysics, Volume 523, id.A87 (2010) and a paper on the Rotational Behavior of Nereid was published by Alexander *et al.* in The Astronomical Journal, Volume 142, Issue 1, article id. 1 (2011).

3.6. *The satellites of Uranus*

Mutual events have occurred in the Uranian system in 2007 and the list of events together with possibilities of observations were published by Arlot and Sicardy in Planetary and Space Science, Volume 56, Issue 14, p. 1778. The Observation of an eclipse of U-3 Titania by U-2 Umbriel on December 8, 2007 made with ESO-VLT was published by Arlot *et al.* in Astronomy and Astrophysics, vol. 492, 2008, p. 599 and several observations analyzed in Assafin *et al.* Astron. J. 137, 4046-4053 (2009). A Photometric and astrometric analysis of a mutual event between the Uranian satellites Miranda and Oberon was published by Birlan *et al.* in Astronomische Nachrichten, Vol.329, p. 567.

Updated ephemerides of the irregular Uranian satellites were published by Brozovic and Jacobson, 2009, AJ 137, p. 3834.

3.7. *The satellites of Neptune*

The orbit of Nereid based on astrometric observations was recalculated and published by Emelyanov and Arlot in Monthly Notices of the Royal Astronomical Society, Online Early. The origin of Triton was explored in a paper by Nogueira *et al.* published in Icarus, Volume 214, Issue 1, p. 113-130.

Observations of the faint Halimede, Psamathe, Sao, Laomedeia and Neso were used for new ephemerides (Brozovic, M., Jacobson, R. A., Sheppard, S. S., 2011, AJ 141, p. 135).

3.8. *The Pluto system*

New results on Pluto'system were presented at the meeting Nix and Hydra: Five Years after Discovery, at STScI, Baltimore, MD. A new satellite of Pluto named S/2011 (134340) was discovered and publications are in preparation.

3.9. *The satellites of asteroids*

These objects are in fact binary or triple objects, the center of mass of the system being not inside the largest object. Nowadays a lot of binary (or triple) systems have been discovered. No data base of astrometric observations is available yet and no ephemeris is published.

Transactions IAU, Volume XXVIIIA
Reports on Astronomy 2009–2012   © International Astronomical Union 2012
Ian Corbett, ed.   doi:10.1017/S1743921312002669

DIVISION II   SUN and HELIOSPHERE
SOLEIL et HELIOSPHERE

Division II of the IAU provides a forum for astronomers and astrophysicists studying a wide range of phenomena related to the structure, radiation and activity of the Sun, and its interaction with the Earth and the rest of the solar system. Division II encompasses three Commissions, 10, 12 and 49, and four Working Groups.

PRESIDENT	Valentín Martínez Pillet
VICE-PRESIDENT	James A. Klimchuk
PAST PRESIDENT	Donald B. Melrose
BOARD	Gianna Cauzzi, Lidia van Driel-Gesztelyi, Natchimuthuk Gopalswamy, Alexander Kosovichev, Ingrid Mann and Carolus J. Schrijver

DIVISION II COMMISSIONS

Commission 10	Solar Activity
Commission 12	Solar Radiation & Structure
Commission 49	Interplanetary Plasma & Heliosphere

DIVISION II WORKING GROUPS

Division II WG	Solar Eclipses
Division II WG	Comparative Solar Minima
Division II WG	International Collaboration on Space Weather
Division II WG	Communicating Heliophysics to the Public

TRIENNIAL REPORT 2009-2012

1. Introduction. The status of ground and space based solar and heliospheric astronomy

The solar activity cycle entered a prolonged quiet phase that started in 2008 and ended in 2010. This minimum lasted for a year longer than expected and all activity proxies, as measured from Earth and from Space, reached minimum values never observed before (de Toma, 2012). The number of spotless days from 2006 to 2009 totals 800, the largest ever recorded in modern times. Solar irradiance was at historic minimums. The interplanetary magnetic field was measured at values as low as 2.9 nT and the cosmic rays were observed at records-high. While rumors spread that the Sun could be entering a grand minimum quiet phase (such as the Maunder minimum of the XVII century), activity took over in 2010 and we are now well into Solar Cycle 24 (albeit, probably, a low intensity cycle), approaching towards a maximum due by mid 2013. In addition to bringing us the possibility to observe a quiet state of the Sun and of the Heliosphere that was previously not recorded with modern instruments, the Sun has also shown us how little we know about the dynamo mechanism that drives its activity as all solar cycle predictions failed to see this extended minimum coming.

While the most veteran solar missions are being slowly decommissioned (such as the ESA/NASA SOHO mission) or have been completely switched off (such as the NASA SMEX TRACE mission), others are continuing to observe the Sun and the Heliosphere and new ones have been launched. SOHO spacecraft is functioning at the moment with the irreplaceable LASCO coronagraphs that still provides unique images of Coronal Mass Ejections (CMEs) at distances of several solar radii. It is unclear for how long will ESA and NASA provide support to this landmark mission. The Japanese-led Hinode mission (launched in 2006) is still observing the Sun at a resolution, continuity and sensitivity never achieved before. The database built up by this mission will provide unparalleled insights about the solar magnetic field in years to come. The twin-spacecraft STEREO mission has finally benefited from an active Sun after its first cruise years where the Sun was sitting in a deep activity minimum with a strongly reduced CME production. We have now crucial 3D information about these events that represent the most energetic processes that shape the Heliosphere. Finally, in 2010 NASA launched the SDO mission that provides full Sun coverage in white-light and UV wavelengths together with high temporal cadences. While the outstanding images provided by this mission have already impressed scientists and laymen, their quantitative aspects (vector magnetic fields, perfect alignment between visible and UV frames, etc.) are of an importance that we are barely starting to fully appreciate. Last, but not least, the SUNRISE stratospheric balloon flight (launched during the activity minimum of June 2009, Solanki *et al.* 2010) has shown the enormous potential of these platforms half-way in between ground based telescopes and space observatories. The SUNRISE mission has provided half an hour long time series of Doppler velocities and vector magnetic fields with a sensitivity, cadence and spatial resolution (0.15 arcsec) that will be difficult to match any time soon. A second flight of this balloon-based platform with the Sun in maximum activity conditions promises a scientific return as large as that from the first one.

Clearly, the status of space based solar physics shows a wealth of missions providing state-of-the-art data that ensures substantial quantitative and qualitative progress in our understanding of the Sun and its surrounding Heliosphere. The longer-term future is also bright given the recent selection by ESA of the Solar Orbiter mission (launch 2017) with its unique suite of in-situ and remote sensing observations filling the gap left by the highly successful Ulysses mission. Solar Orbiter will be flown in combination with the NASA Solar Probe+ mission that, thanks to their proximity to the Sun, will provide a step forward in our predicting capability of solar storms and other Space Weather phenomena.

Ground-based solar physics is also moving forward. The Advanced Technology Solar Telescope is near its construction phase with a firm funding line and a very competitive set of first-light instruments (Keil *et al.* 2010). In Europe, a telescope with a slightly different scientific scope (European Solar Telescope, Collados *et al.* 2010), but also in the 4m aperture class, is being studied and its feasibility phase was successfully completed in 2011. As this 4m-class era of solar telescopes approaches, successful first-light images are being obtained from the 1.5m aperture telescopes that are starting operations both in the USA and Europe. The NST (Big Bear Observatory, Goode *et al.* 2010) has obtained astonishing images of the Sun in white-light that make the comparison between MHD simulations and these observations straightforward. The Gregor telescope (Volkmer *et al.* 2010) has received its 1.5m mirror and its near its commissioning phase. Both Gregor and NST represent a much needed intermediate step while the solar community approaches the exciting era of a less photon-starved telescopes of the 4m class later this decade.

2. Structure of the Division

Division II includes three Commissions, 10, 12 and 49, and four Working Groups. The current Commission presidents, elected at the Rio de Janiero General Assembly, are Lidia van Driel-Gesztelyi, Commission 10, Alexander Kosovichev, Commission 12, and Natchimuthuk Gopalswamy, Commission 49. Also at the IAU XXVII General Assembly, in Rio, 2009, Lidia van Driel-Gesztelyi was asked to continue as the Division's first secretary. Commission 10 Solar Activity, focuses on transient aspects of the Sun, including flares, prominence eruptions, CMEs, particle acceleration, magnetic reconnection and topology, coronal loop heating, and shocks in the corona. Commission 12 Solar Radiation and Structure, emphasizes steady-state aspects of the Sun, including long-term irradiance, helioseismology, magnetic field generation, active regions, photosphere, and chromosphere. Commission 49 Interplanetary Plasma and Heliosphere, studies the solar wind, shocks and particle acceleration, both transient and steady-state, e.g., corotating, structures within the Heliosphere, and the termination shock and boundary of the Heliosphere. There can be considerable overlap among the Commissions, such as in the areas of magnetic activity, solar evolution, particle acceleration, and space weather.

The four Division Working Groups are explained in the next section.

The organizing committee of Division II includes the president, vice-president, secretary and immediate past president of the Division, together with the presidents and vice-presidents of the three Commissions. The position of Divisional secretary had been discussed and agreed in principle at the Sydney General Assembly, but it was only at the Prague General Assembly that it was decided to elect a secretary from among the vice-presidents of the Commissions. This structure was agreed to be maintained in the Rio General Assembly.

Communication with the members of the Division has been relatively fluent by using the e-mail system announced in the Rio GA (<iaudivii@iac.es>). This e-mail system has been mostly used to remind Division II members about the various deadlines, in particular for Symposium proposals. This has indeed had a considerable impact in terms of the number of Division II proposals received in the various calls.

2.1. Working Groups

The four current Working Groups of the Division are, from the oldest to the newest, Solar Eclipses, International Collaboration on Space Weather, Comparative Solar Minima and Communicating Heliophysics to the Public.

It is worth mentioning that the Division II Organizing Committee (OC) has proposed to the IAU Executive that all members of the WG on Solar Eclipses that belong to the WG-OC but are non-professional astronomers should be nominated for IAU membership in the XXVIII GA in Beijing.

2.2. WG on Solar Eclipses (Jay M. Pasachoff)

The Working Group on Solar Eclipses is chaired by Jay M. Pasachoff (USA) and for the 2009-2012 triennium also included Iraida S. Kim (Russia), Hiroki Kurokawa (Japan), Jagdev Singh (India), Vojtech Rusin (Slovakia), Atila Ozguc (Turkey), Yihua Yan (China), Fred Espenak (USA), Jay Anderson (Canada, consultant on meteorology), Glenn Schneider (USA), and Michael Gill (UK, maintainer of Solar Eclipse Mailing List). For the 2012-2015 triennium, the Working Group will again include Jay M. Pasachoff (USA, chair), Iraida S. Kim (Russia), Hiroki Kurokawa (Japan), Jagdev Singh (India), Vojtech Rusin (Slovakia), Fred Espenak (USA), Minde Ding (China), Jay Anderson (Canada, consultant on meteorology), Glenn Schneider (USA), and Michael Gill (UK, maintainer of Solar Eclipse Mailing List). New members will include Nick Lomb (Australia), Haisheng

Ji (China), Bill Kramer (www.eclipse-chasers.com), Xavier Jubier (online maps), and Michael Zeiler (www.eclipse-maps.com, maps).

The WG has web sites www.eclipses.info and www.totalsolareclipse.net. The WG has as its task the coordination of solar eclipse efforts, particularly making liaisons with customs and other officials of countries through which the path of totality passes and providing educational information about the safe observation of eclipses for the wide areas of the Earth in which total or partial eclipses are visible. The work is coordinated with that of the Program Group on Public Information at the Times of Eclipses of IAU Commission 46 on Education and Development. Two members, Espenak and Anderson, have produced the widely used NASA Technical Publications with eclipse paths and detailed information, available as hard copies or online, linked through www.eclipses.info or via the NASA Eclipse Web Site at eclipse.gsfc.nasa.gov. The bulletins are directly at URL eclipse.gsfc.nasa.gov/SEpubs/bulletin.html; at this writing, bulletins are in preparation for ASE2012 and TSE2012.

Eclipse maps, with the paths marked in Google Maps or in Google Earth, and local circumstances presented for wherever the user clicks, are provided by Xavier Jubier as the first entries at www.eclipses.info, or directly, for the Google maps, to xjubier.free.fr/en/site_pages/SolarEclipsesGoogleMaps.html. See also xjubier.free.fr/en/site_pages/Solar_Eclipses.html. Fred Espenak's World Atlas of Solar Eclipse Paths is at eclipse.gsfc.nasa.gov/SEatlas/SEatlas.html.

Two review articles by the WG chair in Nature (freely available in Pasachoff, 2009a) and in Research in Astronomy and Astrophysics (Pasachoff, 2009b) describe the scientific value of eclipse observations.

During the past triennium, we had total solar eclipses of 11 July 2010, visible across the Pacific Ocean, including some islands in French Polynesia and, later, Easter Island.

A comparison of the observations from the two sites separated by 93 minutes appeared in Pasachoff et al. (2011). The thermodynamics of the corona and the observed evolution of the magnetic field during this same eclipse was described in Habbal et al. (2011).

There were no central eclipses in 2011. Partial eclipses were visible on 4 January 2011 over a wide area of Europe, Africa, and Asia; on 1 June 2011 in Iceland and northern Scandinavia, as well as northenmost Canada, Alaska, and Siberia; on 1 July 2011 in a small area of the ocean off the coast of Antarctica, barely touching the coast (so was, for all practical purposes, invisible; most unusually, probably nobody on Earth viewed this solar eclipse); and on 25 November 2011, in southern New Zealand, Tasmania, and southern South Africa. See maps at eclipse.gsfc.nasa.gov/OH/OH2011.html. 2012 boasts of two central solar eclipses. The 20 May 2012 annular eclipse will be visible from sites in Japan and then across the Pacific to the western United States. The 13/14 November 2012 total solar eclipse's path of totality will cross northerneasternmost Australia, principally Cairns and Port Douglas, before heading entirely over the ocean. Partial phases will be visible throughout Australia, New Zealand, eastern Indonesian islands, and Papua New Guinea. See xjubier.free.fr/en/site_pages/solar_eclipses/ TSE_2012_GoogleMapFull.html. Discussions of the 2012 eclipses and maps appeared online at the minus-first anniversary of the next total solar eclipse at www.365daysofastronomy.org for 13 and 14 November 2011.

Other total eclipses in the triennium are 3 November 2013 across Africa from Gabon to northern Kenya, xjubier.free.fr/en/site_pages/solar_eclipses/ HSE_2013_GoogleMapFull.html; and 20 March 2015 in Svalbard or the Faroe Islands, up to the north pole, xjubier.free.fr/en/site_pages/solar_eclipses/ TSE_2015_GoogleMapFull.html. Partial phases will be visible throughout Europe, northern Africa, and north-central Asia. Annular eclipses will be visible on 10 May 2013 in northern Australia, including the outback and the Cape York peninsula, and Kiribati `xjubier.free.fr/en/site_pages/solar_eclipses/ASE_2013_GoogleMapFull.html`; and, undoubtedly unobserved, on 29 April 2014 in Antarctica `xjubier.free.fr/en/site_pages/solar_eclipses/ASE_2014_GoogleMapFull.html`. See `xjubier.free.fr/en/site_pages/SolarEclipsesGoogleMaps.html`.

A new set of eclipse maps, from Michael Zeiler, is available at `www.eclipse-maps.com`. Jay Anderson has maps, cloud statistics, and weather discussions at `home.cc.umanitoba.ca/~jander/=http://eclipser.ca`.

Additional online resources that have become available in the past three years include an enhanced local circumstances calculator at `www.eclipse-chasers.com/tseCalculator.php` with Baily's Beads simulations using either the US Naval Observatory's charts by Watts showing the lunar profile or the JAXA Kaguya lunar data reduced by International Occultation Timing Association (IOTA) Australia member David Herald. Charts showing an exaggerated lunar profile with predicted bead placements can be created for one-tenth second intervals. Advanced local circumstance calculations incorporate the dip of the horizon and refraction estimates for more accurate timing predictions.

Eclipse chasers can log their eclipse experiences and upload pictures to Kramer's log at `www.eclipse-chasers.com/tseChaserLogSums.php`. Google maps showing the eclipse paths as pins or weather icons (based on submitted observations) can be displayed for those eclipses with log data. Over 100 observers with an average of six total solar eclipses each have logged their eclipse observations thus far. Eclipse-observer statistics are also logged at Sheridan Williams' site at `www.clock-tower.com`.

2.3. *WG on Communicating Heliophysics to the Public (C. Briand)*

The Working Group on Solar Eclipses is chaired by Carine Briand (France) and includes Francesco Berilli (Italy), Maria-Cristina Rabello-Soares (USA), Jean-Pierre Raulin (France), Deborah Scherrer (USA), Yihua Yan (China).

Communication to and with the public takes an increasing place in the researcher activities. If for a long time it was felt as a time consuming obligation, the scientists now consider it as part of their professional work and a way to increase the impact of scientific results and indeed recognize the benefit of such activity. Most of scientific disciplines develop outreach activities. Of course, astrophysicists are also deeply implicated in many public outreach activities. In fact, it results that the astronomical community is one the best organized community for Education/Public Outreach (E/PO).

Images of galaxies, deep space and planets have been largely spread in the public. But heliophysicists also have numerous original assets to communicate. The high spatial resolution of the telescopes provide amazing details of the Sun, unachievable on other stars. Many solar structures have lifetime from a few minutes to a few hours, allowing the production of fascinating movies. Last but not least, sonification of radio dynamic spectra offer an original way to "look" at our environment. These media provide exciting means to catch the public's attention and to help delivering more complex messages about physics.

Owing to the space age and the development of large international instruments, the heliophysical community is particularly well organized compare to other fields of physics. The International Heliophysical Year 2007 has demonstrated the capability of our community to mobilize the researchers all over the world to participate to E/PO activities (see IAU report 2006-2009 for details). Now, the link with the other fields of astrophysics needed to be reinforced in particular through a strongest participation to the activities of the Commission 55 of the IAU. The aim of this working group is twofold. We first aim to join the IAU-C55 efforts to provide information from the heliophysics community, and second to discuss the best practices to share experience and documents within our own community.

The working group was proposed during the General Assembly of IAU in 2009. It is composed of six persons representing most of the regions of the world: two from Europe, two from the United States, one from China and one from Brazil. When possible, we also invited other scientists to participate to our works. For example, Jean-Pierre Lebreton (ESA) joined a discussion of the WG during the AGU 2009 in San Francisco.

We took advantage of the presence of Jean-Louis Bougeret (former chair of the Commission 49) at the CAP2010 to take the first formal contacts with the Commission 55. Strongest bounds were created during the CAP2011 meeting in Beijing. The organizers of the meeting provided us a time slot to present our community and highlight some specific activities. Most of the activities fully correspond to the CAP conferences and journal but also to the "Best Practices" working group of C-55. We will thus encourage all heliophysicists to present their works and to participate to the discussions of C55 during the next IAU-GA.

If space agencies provide numerous resources, the researchers also developed their own activities and documents, in particular in their own language. The second goal of the WG is to promote these public outreach activities and to exchange experiences. The best way to share the documents and to highlight unusual or impressive actions is through a web page. It is dedicated to the heliophysics community but also to the other astronomical communities. At the time of writing those lines, the website is under development. However, it should be opened for the next IAU-GA. Different sections are already written like "Exhibits", "Movies", "Books" and specific E/PO meetings. All researchers will be invited to provide information to appear on this web site.

The working group should evolve by including more heliophysicists to discuss the best practice to share experience. The web we are developing is one option but other media can be used also (in particular social media). The implication of the people specialized in E/PO must increase and the link between them and the scientists should be reinforced. This may be a line of thought for the next period.

2.4. *WG on Comparative Solar Minima (S. Gibson)*

The Working Group on Comparative Solar Minima is chaired by Sarah Gibson (USA) and includes Ed Cliver (USA), Hebe Cremades (Argentina), Peter Fox (USA), Nat Gopalswamy (USA), Gustavo Guerrero (Sweden), Margit Haberreiter (Switzerland), Joanna Haigh (UK), Janet Kozyra (USA), Kanya Kusano (Japan), Cristina Mandrini (Argentina), Georgeta Maris (Romania), Valentin Martinez Pillet (Spain), Pete Riley (USA), Barbara Thompson (USA), Andrey Tlatov (Russia), Ilya Usoskin (Finland) and David Webb (USA).

The mission of the WG is to facilitate international and interdisciplinary research that focusses on the coupled Sun-Earth system during solar minimum periods. Such research seeks to characterize the system at its most basic, "ground state", but also to understand the degree and nature of variations within and between solar minima.

The WG has a web site ihy2007.org/IAUWG/WEBPAGES/IAUWG.shtml which includes a listing of meetings and workshops organized by working group members and also a bibliography of multi-disciplinary papers relevant to the WG. In addition, there is a project page summarizing research comparing the last two solar minimum.

One of the main activities of the WG pertained to the Whole Heliosphere Interval - an international coordinated observing and modeling effort to characterize the three-dimensional interconnected solar-heliospheric-planetary system during the last minimum ihy2007.org/WHI/WHI.shtml. A WHI workshop was held in Boulder, Colorado in November 2009, and an AGU special session in August 2010 in Iguazu, Brazil.

Moreover, at the WHI workshop a topical issue of Solar Physics was successfully proposed to gather papers regarding the Sun-Earth Connection near Solar Minimum, with emphasis on the WHI period. The Issue will be published in December 2011, in a double issue of Solar Physics (Vol. 274).

2.5. *WG on International Collaboration on Space Weather (D. Webb)*

The Working Group for International Collaboration on Space Weather has as its main goal to help coordinate the many activities related to space weather at an international level. It is chaired by David Webb and its website is at www.iac.es/proyecto/iau_divii/IAU-DivII/main/spaceweather.php. The site currently includes the international activities of the International Heliospheric Year (IHY), the International Living with a Star (ILWS) program, the CAWSES (Climate and Weather of the Sun-Earth System) Working Group on Sources of Geomagnetic Activity, and Space Weather studies in China.

The International Heliospheric Year was an international program of scientific collaboration during the time period 2007-2009. The IHY has officially closed, and has been superseded under UN auspices as the International Space Weather Initiative. The website is at: www.iswi-secretariat.org. David Webb is the IAU scientific representative for ISWI. ISWI is a program of international cooperation to advance the space weather science by a combination of instrument deployment, analysis and interpretation of space weather data from the deployed instruments in conjunction with space data, and communicate the results to the public and students. Although ISWI follows-on to the successful IHY, it focuses exclusively on space weather. The goal of the ISWI is to develop the scientific insight necessary to understand the science, and to reconstruct and forecast near-Earth space weather.

The goal of the ILWS program is to stimulate, strengthen, and coordinate space research to understand the governing processes of the connected Sun-Earth System as an integrated entity. The website is at: ilwsonline.org. The kickoff meeting was held in 2002 and since then several Steering Committee and science meetings have been held. The latest was the 2011 ILWS Science Workshop "Towards the Next Solar Maximum" held August 28-September 1, 2011 in Beijing, China.

The CAWSES Working Group has as its objective to understand how solar events impact geospace by investigating the underlying science and developing prediction models and tools. CAWSES has been extended as CAWSES II for 2009-2013. The website is at: www.cawses.org/CAWSES/Home.html. CAWSES-II supports a framework for sustaining international and interdisciplinary collaborations by leveraging the full potential for scientific discovery and learning inherent in past investments in instrumentation and facilities. CAWSES-II focuses on "grand challenge" questions that can only be addressed through interdisciplinary research and international collaboration. It supports development of a virtual community to produce the scientific breakthroughs enabled by advances in cyber infrastructure and to help revolutionize the way in which collaborative scientific research is done. CAWSES-II has 6 task groups that address science and integrative activities to focus its activities.

The working group on Space Weather Studies in China is chaired by Jingxiu Wang and is involved with many new initiatives on space weather. In the near future we plan to add information and website links to recent and active space weather studies in other countries, such as in India, Russia and the Americas.

3. IAU Meetings

The Division has played a leading role in the following IAU meetings, that have been held since the last triennial report by Melrose *et al.* (2009):

• IAUS 273 on the Physics of Sun and star spots (Debi Prasad Choudhary, Klaus G. Strassmeier, eds.), August 23 - 26 2010, Los Angeles, USA

• IAUS 286 on the Comparative magnetic minima: characterizing quiet times in the Sun and stars (Cristina Mandrini, David Webb, eds.), October 3 - 7, 2011, Mendoza, Argentina

At the XXVII IAU General Assembly in Rio de Janeiro, the following Division II leaded events took place

• IAUS 264 on Solar and Stellar Variability - Imapct on Earth and Planets (Alexander Kosovichev, ed.), August 3 - 7, 2009

• JD16 International Heliospheric Year Global Campaign - Whole Heliosphere Interval (Barbara J. Thompson), August 12 - 14, 2009

• JD11 New Advances in Helio- and Astero-Seismology (Junwei Zhao, ed.), August 10 - 11, 2009

Valentín Martínez Pillet
president of the Division

References

Collados, M., Bettonvil, F., Cavaller, L. *et al.* 2010, *AN* 331, 615

Goode, P. R., Coulter, R., Gorceix, N., *et al.* 2010, *AN* 331, 620

Habbal, S. R., Druckümuller, M., Morgan, H. *et al.* 2011, *ApJ*, 734, 88

Keil, S. L., Rimmele, T. R., & Wagner, J., ATST team 2010, *AN* 331, 609

Melrose, D. B., Martínez Pillet, V., Webb, D. F. *et al.* 2009, *Transactions IAU* 4, 27A, 73

Pasachoff, J. 2009a, *Nature*, 459, 789

Pasacoff, J. 2009ib, *Research in Astronomy and Astrophysics*, 9, 613

Pasachoff, J., Rušin, V., Druckmüllerová, H. *et al.* 2011, *ApJ*, 734, 114

Solanki, S. K., Barthol, P., Danilovic, S. *et al.* 2010, *ApJ* 723, L127

de Toma, G. 2012, in: C. H. Mandrini, J., & D. Webb (eds.), *Comparative Magnetic Minima: characterizing quiet times in the Sun and stars*, Proc. IAU Symp. No. 286, Mendoza, Argentina, 3-7 October 2011 (Cambridge: University Press), in press

Volkmer, R., von der Lühe, O., Denker, C. *et al.* 2010, *AN* 331, 624

Transactions IAU, Volume XXVIIIA
Reports on Astronomy 2009–2012
Ian Corbett, ed.

© International Astronomical Union 2012
doi:10.1017/S1743921312002670

COMMISSION 10

SOLAR ACTIVITY

ACTIVITE SOLAIRE

PRESIDENT	Lidia van Driel-Gesztelyi
VICE-PRESIDENT	Carolus J. Schrijver
PAST PRESIDENT	James A. Klimchuk
ORGANIZING COMMITTEE	Paul Charbonneau, Lyndsay Fletcher, S. Sirajul Hasan, Hugh S. Hudson, Kanya Kusano, Cristina H. Mandrini, Hardi Peter, Bojan Vršnak, Yihua Yan

TRIENNIAL REPORT 2009–2012

1. Introduction

Commission 10 of the International Astronomical Union has more than 650 members who study a wide range of activity phenomena produced by our nearest star, the Sun. Solar activity is intrinsically related to solar magnetic fields and encompasses events from the smallest energy releases (nano- or even picoflares) to the largest eruptions in the Solar System, coronal mass ejections (CMEs), which propagate into the Heliosphere reaching the Earth and beyond. Solar activity is manifested in the appearance of sunspot groups or active regions, which are the principal sources of activity phenomena from the emergence of their magnetic flux through their dispersion and decay. The period 2008–2009 saw an unanticipated extended solar cycle minimum and unprecedentedly weak polar-cap and heliospheric field. Associated with that was the 2009 historical maximum in galactic cosmic rays flux since measurements begun in the middle of the 20th Century. Since then Cycle 24 has re-started solar activity producing some spectacular eruptions observed with a fleet of spacecraft and ground-based facilities. In the last triennium major advances in our knowledge and understanding of solar activity were due to continuing success of space missions as SOHO, Hinode, RHESSI and the twin STEREO spacecraft, further enriched by the breathtaking images of the solar atmosphere produced by the Solar Dynamic Observatory (SDO) launched on 11 February 2010 in the framework of NASA's Living with a Star program. In August 2012, at the time of the IAU General Assembly in Beijing when the mandate of this Commission ends, we will be in the unique position to have for the first time a full 3-D view of the Sun and solar activity phenomena provided by the twin STEREO missions about 120 degrees behind and ahead of Earth and other spacecraft around the Earth and ground-based observatories. These new observational insights are continuously posing new questions, inspiring and advancing theoretical analysis and modelling, improving our understanding of the physics underlying magnetic activity phenomena. Commission 10 reports on a vigorously evolving field of research produced by a large community. The number of refereed publications containing 'Sun', 'heliosphere', or a synonym in their abstracts continued the steady growth seen over the preceding decades, reaching about 2000 in the years 2008–2010, with a total of close to 4000 unique authors. This report, however, has its limitations and it is inherently incomplete, as it was prepared jointly by the members of the Organising Committee of

Commission 10 (see the names of the primary contributors to the sections indicated in parentheses) reflecting their fields of expertise and interest. Nevertheless, we believe that it is a representative sample of significant new results obtained during the last triennium in the field of solar activity.

2. Solar-cycle activity (P. Charbonneau & H.S. Hudson)

This triennium witnessed the end of Hale Cycle 23 and the beginning of Cycle 24, and this seemingly routine evolution produced surprises. The Cycle 23/24 minimum itself, nominally at the very end of 2008, was exceptionally protracted, more so than any from the past century. This confounded almost all of the experts who dared to predict the onset time of the new cycle (e.g., Pesnell, 2008). At the same time various indices related to solar activity achieved record levels of – passivity. Finally, in fits and starts, activity resumed to the point where no less than seven GOES X-class flares have occurred in the first three quarters of 2011.

Several conferences have already been devoted to this anomalous minimum, including IAU Symposiums No. 273 (Ventura, California, 2010) and No. 286 (Mendoza, Argentina, 2011). A literature search for articles whose abstracts included the term "solar minimum" produced more than 300 hits in 2010, threefold of the average number of papers during Cycle 23. The minimum between Cycles 23 and 24 has caught our attention! One is reminded of Mark Twain's quote, "There is something fascinating about science. One gets such wholesale returns of conjecture out of such a trifling investment of fact." But the phenomena we are experiencing are not trifling at all, since life on Earth depends so intimately on the Sun, and we need to understand its behavior better than we presently do.

The indices of activity have behaved erratically. The sunspot number (SSN), the longest continuous record of sunspot activity, had correlated very well with the F10.7 radio index, introduced in 1947. But in this minimum, the correlation changed significantly (e.g. Svalgaard & Hudson 2010), suggesting either an altered Sun or else a new bias in the interpretations of these indices. Thanks to the excellent material now available for monitoring solar activity, we know that these signs relate to the behavior of the ever-fascinating solar magnetic field.

Numerous attempts have been made to understand the peculiar features of this extended minimum using various types of dynamo models. Attention has focused primarily on flux transport dynamos, in which the large-scale meridional flow in the convection zone regulates both the equatorward propagation of activity belts, as well as poleward transport of the surface magnetic field and reversal of the Sun's dipole moment. In such models, relatively modest, persistent variations of the meridional flow speed can lead to significant variations in overall cycle amplitudes (e.g. Lopes & Passos 2009). The challenge provided by the Cycle 23/24 minimum has led to more detailed and targeted analyses. The low polar field strength and long duration characterizing this minimum could be reproduced by a flux transport dynamo model, provided the meridional flow sped up in the rising phase of the cycle, and subsequently slowed down in the descending phase (Nandy et al. 2011). This proposed phasing is actually contrary to that inferred observationally (Hathaway & Rightmire 2011), but the discrepancy may hinge on the development of a high-latitude counterrotating flow cell, which in itself could yield weaker polar surface fields (Jiang et al. 2009) as well as a delayed start for Cycle 24 (Dikpati et al. 2010).

These dynamo-based "explanations" of the Cycle 23/24 minimum all rely on the kinematic approximation, where the magnetic back-reaction of the dynamo-generated magnetic field is altogether neglected, or at best subsumed in simple, *ad hoc* algebraic quenching nonlinearities. Significant progress has also taken place with fully dynamical magnetohydrodynamical simulation of dynamo action, particularly in the successful production of large-scale magnetic fields undergoing polarity reversals (Ghizaru *et al.* 2010; Käpylä *et al.* 2010; Brown *et al.* 2011). The Ghizaru *et al.* simulations are noteworthy in being solar-cycle-like in a number ways, including regular polarity reversals on a multi-decadal timescale, good cross-hemispheric coupling, torsional oscillations of a few nHz amplitude, and the buildup of a high-latitude counterrotating flow cell in the descending phase of the simulated cycles. Detailed analyses of such simulations may provide insight on what was dynamically peculiar –or not– in the transition from Cycle 23 to 24.

This was the third cycle for which stable total solar irradiance (TSI) measures were available, from a series of spacecraft. Within a precision expressed in tens of parts per million, the first two minima agreed well. During the Cycle 23/24 minimum, though, TSI dropped significantly below this level (Fröhlich 2009; Kopp & Lean 2011), confounding those experts who had expected solar minima to reflect a basal level of solar luminosity. Associated with this variation, which could be due either to a true luminosity variation or to superficial activity (Fröhlich 2009), apparently all measures of solar magnetic activity seem to have diminished. The EUV flux decreased relative to the previous minimum (Solomon *et al.* 2010). The solar polar fields, thought to supply the heliospheric field, decreased (Wang *et al.* 2009). A diminished interplanetary magnetic field suggested a lower "floor" than seen before (Owens *et al.* 2008). At the same time cosmic-ray fluxes achieved record high levels (Mewaldt *et al.* 2010), consistent with a magnetic deflation of the heliosphere resulting from the diminished magnetic activity.

During the maximum of Cycle 23, and continuing at present, Livingston and Penn have been tracking sunspot magnetic fields and brightnesses in a uniform and stable manner, and have surprisingly found a secular trend in field intensity: their magnetic fields are decreasing – and in umbral brightness: their darkness is also decreasing (Penn & Livingston 2010). Such long-term variations, centered now on the anomalous Cycle 23/24 minimum, have never before been observed.

On a separate subject, the absolute value of the solar luminosity may require revision. The new value of TSI at solar minimum, from improved measurements, turned out to be about 0.33% lower than previously thought (Kopp & Lean 2011).

3. The solar large-scale magnetic field (C.J. Schrijver)

Our understanding of the Sun's large-scale magnetic field was aided in the past few years by two unique events. First, there was the extended sunspot minimum which pushed the solar and heliospheric fields, and the associated galactic cosmic ray fluxes, to a state never before seen in the modern instrumental era. Second, the instrumentation available to study the Sun's magnetic field from the surface into the heliosphere was strengthened by a new capability: the achievement of the first full-sphere view of the Sun (in fact, of any star) from the STEREO spacecraft as they drifted past quadrature from the Sun-Earth line on 6 February 2011, in combination with unrivaled view of the front side offered by the Solar Dynamics Observatory (launched on 11 February 2010).

The unusually weak polar-cap and heliospheric fields appear to be a consequence of a mild but sustained increase in the poleward meridional advection, as demonstrated by surface flux transport modeling by Wang *et al.* (2009). That change in the meridional surface velocity was observationally confirmed by Hathaway & Rightmire (2011).

The full-sphere view brought the phenomenon of sympathetic activity back into focus. One set of events, studied by Schrijver & Title (2011) using SDO observations combined with STEREO coronal images, revealed that a series of flares and filament eruptions (evolving into CMEs) were all connected through a network of topological divides. This series of events was possibly initiated by the emergence of a new active region some 90 degrees away (visible only in the STEREO/B images). MHD modeling by Török *et al.* (2011) supports that hypothesis, by showing how the eruption of one flux rope can cause a neighboring one – until then marginally stable – to erupt, cascading from one eruption to another in the set of four model flux ropes. Jiang *et al.* (2011) discuss observations of such a nature, including the formation of transient coronal dimmings between the erupting structures.

Insights on the coupling of the Sun's surface to the heliospheric field also came from theory and modeling. Work by Yeates *et al.* (2010a), for example, shows that whereas the potential-field source-surface (PFSS) modeling works remarkably well in mapping the heliospheric field to its source regions, non-potential effects are significant: their magnetofrictional modeling reveals how the introduction of electrical currents by shearing of the field by differential rotation can almost double to Sun's open flux at maximum, adding in contrast only one quarter around cycle minimum relative to a purely potential PFSS model. Yeates *et al.* (2010b) discuss CME formation in the context of the shearing of large-scale field, stressing – as have other studies – that knowledge of the full-sphere magnetic field is important in both the study of the background heliospheric field as well as the eruptive phenomena that rock its foundations.

It appears that the pattern of overarching helmet structures is related to flaring activity: Svalgaard *et al.* (2011) report on a correlation of flaring with the Hale sector structure. But whether that reflects patterns in the sources of the flux or is related to the interaction of emerging flux with the pre-existing large-scale field remains to be established.

One of the outstanding problems in coronal-heliospheric studies is the origin of the slow solar wind, with a composition and ionization balance reminiscent of that of the closed-field active region environment. Antiochos *et al.* (2011) argue that the slow solar wind may in fact originate from the boundary region between open and closed fields which maps into a "web of separatrices" into the heliosphere; the nature of these narrow coronal channels appears to commonly be a singular line that is in fact a separatrix lying on the solar surface (Titov *et al.* 2011). Linker *et al.* (2011) discuss the implications of these studies for the evolution of the heliospheric field in an MHD model: they find that the open field evolves across the solar surface through reconnection with the closed field, remaining closed underneath the helmet streamers until field migrates out into open-field coronal holes again, thus arguing against the diffusive transport of open field through closed-field regions as proposed by, e.g., Fisk (2005).

4. Simulation studies (K. Kusano)

Numerical simulation is now an indispensable methodology for the study of various disciplines, and the rapid growth of supercomputers is further increasing its importance for the understanding of complex phenomena like solar dynamics. Radiative magnetohydrodynamic (MHD) simulations have improved rapidly over the past years. In particular, sunspot models were dramatically advanced by Rempel *et al.* (2009) and Rempel & Schlichenmaier (2011). These first-principles MHD models of a whole sunspot compare well with the high resolution observations. Our understanding of penumbral structure and Evershed flow were greatly improved recently by the comparison between simulations and observations, although the convective structure in penumbra is still under investigation.

Another new trend in simulation studies is the forward modelling of wave propagation through sunspots based on 3D MHD simulations (Cameron *et al.* 2008; Parchevsky &

Kosovichev 2009; Moradi *et al.* 2009). This type of study is important for the detailed assessments of the validity of helioseismic interpretations to constrain the subsurface models. Flux emergence is another central subject of the radiative 3D MHD simulations (Cheung *et al.* 2010), enabling calculations of the whole evolution from small-scale turbulence in the early stage of flux emergence to the formation of an active region. Simulations of flux emergence also help us understand the building-up process of free energy within an active region (Hood *et al.* 2011).

Data-driven simulation, in which the observed data are used for the initial and boundary conditions, is also a recent new challenge (Fan 2011; Wu *et al.* 2009), and it is promising as the new method of realistic simulations. However, it is still under development and how to evaluate its physical reality is also a future issue.

5. Transition region (H. Peter)

As the interface between the chromosphere and the corona, where the change from plasma-dominated to magnetic-field-dominated takes place, the transition region is of vital interest to understand the dynamics of the corona.

The redshifts in the transition region and the blueshifts in the low corona are well established. New evidence has been found that in the quiet Sun, for temperatures above 1 MK the line shifts tend to drop to zero (Dadashi *et al.* 2011). On the modeling side there are two new proposals to understand the transition-region line shifts. In their 3D MHD models Hansteen *et al.* (2010) find that plasma is heated low in the atmosphere to high temperatures and only then does it rise before it cools and falls down again. This process gives very small shifts at high temperatures (not quite the observed blueshifts). A similar model by Zacharias *et al.* (2011b) identifies up- and downflows in the legs of coronal loops, depending on the surface region they are connected to, which can change in time. They also find fingers of cool gas reaching high into the corona. It remains to be seen which of these processes (or both or different ones) govern the observed line shifts.

The interpretation through 3D MHD models of the transition region already hints at high spatial and temporal complexity. Based on spectroscopic observations McIntosh & De Pontieu (2009) analyzed SUMER data, showing the long-known transition-region line asymmetries re-interpreting them in terms of a background and a fast upflow component in the spectral profile. They show that the upflow component could carry enough energy and mass to power the solar corona. Later Tian *et al.* (2011) conducted a similar study for data from EIS where they also compared their interpretation to artificially created spectra, with similar results. Martínez-Sykora *et al.* (2011a) did a comprehensive study of the line asymmetries based on a 3D MHD model in order to better understand how to interpret the line asymmetries.

6. Reconstruction of the 3D solar corona (C.H. Mandrini)

Observational methods for reconstruction of the three dimensional (3D) structure of the solar corona can be divided in two broad categories: solar stereoscopy (SS), and solar rotational tomography (SRT). These complementary methods have different purposes, strengths, and limitations, and can provide great observational insight, as well as powerful constraints to MHD models. Both techniques have been boosted by the STEREO mission, and will greatly benefit from the expanded temperature coverage, and increased spatio-temporal resolution, provided by the SDO/AIA instrument. A very recent comprehensive review on coronal 3D reconstruction methods can be found in Aschwanden (2011).

Solar stereoscopy using images simultaneously taken from two different spacecraft has been applied to reconstruct (by triangulation) the geometry of coronal loops, and combined with multiple-band EUV DEM analysis to provide the electron density and temperature distribution along them (Aschwanden *et al.* 2008). Some recent works in which stereoscopic triangulations have been applied are Aschwanden (2009) to oscillating loops, Feng *et al.* (2009) to polar plumes, Liewer *et al.* (2009) and Bemporad *et al.* (2009) to an erupting filament/prominence, and Thompson (2011) to a strongly rotating, erupting, quiescent polar crown prominence.

White-light tomography allows determination of the 3D electron density distribution. Recent developments include a new approach called Qualitative SRT, which has been applied to LASCO/C2 data covering the complete Solar Cycle 23 (Morgan & Habbal 2010). This technique has been recently applied to STEREO/COR1 data (Kramar *et al.* 2009), and used with total brightness measurements for the first time (Frazin *et al.* 2010). Frazin *et al.* (2009a) discuss the possibility of using advanced image processing concepts to reconstruct CMEs from the view-points provided by SOHO and STEREO.

Tomography has been recently applied to EUV data for the first time, and combined with local DEM analysis to form a novel technique named differential emission measure tomography (DEMT; Frazin *et al.* 2009b). This allows construction of 3D electron density and temperature maps. DEMT has been applied to STEREO/EUVI data to analyze coronal cavities by Vásquez *et al.* (2009), and combined with magnetic extrapolations to study the large scale thermodynamic structure of the solar corona by Vásquez *et al.* (2011). Also, DEMT is currently being used to provide density and temperature constraints on the inner corona MHD component of the Space Weather Modeling Framework of the University of Michigan (Jin *et al.* 2011). The effects of dynamics in tomographic reconstructions can be mitigated by using time-dependent approaches, such as the Kalman-filtering method (Butala *et al.* 2010). Another recent effort concerning multi-instrumental tomography is the work by Barbey *et al.* (2011), who developed a parallelized open-source tomography software, able to handle different data sources such as STEREO/EUVI and STEREO/COR1, with the inclusion of both a static and a smoothly varying temporal evolution models.

7. Solar flares (L. Fletcher)

As is well known, solar flares are due to the release of magnetic energy. In the few days before a flare a distinct increase has recently been reported in both the subphotospheric kinetic helicity density (a measure of twist which can be imparted to the field) determined by local helioseismology and the photospheric helicity (Reinard *et al.* 2010; Komm *et al.* 2011; Magara & Tsuneta 2008; Park *et al.* 2008). This behavior reinforces our physical picture of active region energy loading. With vector fields now available from Hinode Solar Optical Telescope (SOT), and routinely from the Solar Dynamics Observatory (SDO), studies of magnetic free energy, and its evolution (Jing *et al.* 2009, 2010) are feasible, though still problematic. The active-region magnetic topology is also core to flare evolution. Combining MHD simulations and flare observations, Masson *et al.* (2009) suggest a coronal null topology supporting 'slip-running' flare reconnection. Topological studies by Des Jardins *et al.* (2009) indicate that the strongest flare energy release occurs near topological structures known as separators. As a flare is an essentially magnetic disturbance, it makes sense to try and understand how magnetic energy is transported through the corona and converted to other forms by magnetic perturbations (e.g. Birn *et al.* 2009; Longcope *et al.* 2009; Reeves *et al.* 2010).

Electron acceleration and the radiation of non-thermal X-rays are core to flare physics. New visibility-based approaches to RHESSI imaging (Kontar *et al.* 2008; Dennis & Pernak 2009) are being used to quantify the sizes and shapes of flare X-ray sources, and multi-wavelength observations constrain the number of non-thermal electrons required to produce the observed flare emission in the chromosphere (Watanabe *et al.* 2010a; Krucker *et al.* 2011) and the corona (Krucker *et al.* 2010). Recent work confirms that a large – some might say worryingly large – fraction of electrons needs to be accelerated, prompting investigations of 'reacceleration' models (Brown *et al.* 2009). Electron acceleration and propagation remain a focus of modelling, with recent investigations including wave-particle interactions in beam-generated turbulence (Hannah & Kontar 2011; Zharkova & Siversky 2011). Stochastic (Petrosian & Chen 2010; Bian *et al.* 2010), current-sheet (Gordovskyy *et al.* 2010; Karlický & Bárta 2011; Mann & Warmuth 2011) and shock acceleration models (Warmuth *et al.* 2009) all continue to compete.

Optical emission occurs in flares of all sizes (Jess *et al.* 2008; Kretzschmar 2011) but the emission mechanism is still unclear. Potts *et al.* (2010) find evidence for an optically thin chromospheric source, but other events look more photospheric in origin. The photosphere is clearly affected by flares: Fe I line profiles from the SDO Helioseismic Magnetic Imager (HMI) show blue-shifts at the time of the impulsive phase (Martínez Oliveros *et al.* 2011) simultaneous with a change in the line-of-sight photospheric magnetic field (see also Wang & Liu 2010). The excitation of flare-associated helioseismic disturbances may be related to this (Martínez Oliveros & Donea 2009; Matthews *et al.* 2011).

The Extreme ultraviolet Imaging Spectrometer (EIS) on Hinode has been used to probe flare chromospheric sources (Watanabe *et al.* 2010b; Graham *et al.* 2011; Milligan 2011) typically showing hot, dense, redshifted plasmas. However, the standard assumptions of EUV spectroscopy (ionization equilibrium, Maxwellian distribution of electron speeds) may be invalid during a flare, motivating investigations of out-of-equilibrium plasmas (e.g. Dzivčáková & Kulinová 2010; Kulinová *et al.* 2011). With the anticipated launch of the Interface Region Imaging Spectrometer (IRIS), large ground-based telescope projects such as the Atacama Large Millimeter Array (ALMA), the Advanced Technology Solar Telescope (ATST) and its possible European counterpart, and smaller ground-based instruments (e.g. ROSA, IBIS), chromospheric flare observations should prosper in the coming years. Given the rich physics of this layer, and its role as the source of the greater part of a flare's radiation, the flare community would do well to look to the chromosphere, and prepare to take advantage of these data.

8. Flux rope formation and CME initiation (L. van Driel-Gesztelyi)

Flux rope-like magnetic structures are commonly seen in coronal mass ejections (CMEs) as well as measured in situ in their related 'magnetic clouds' in the interplanetary medium. It has been an open question, however, whether a flux-rope is present prior to the CME eruption or forms from a sheared magnetic arcade through magnetic reconnection during the eruption. There has been significant progress in our understanding of the CME process from its initiation through its evolution and structure over the last three years. Here I show a few highlights.

We have known since long from vector magnetic field observations that magnetic flux emerges twisted. However, a sub-surface flux rope cannot directly cross the photosphere, but it must be destroyed and transformed by reconnection before it can enter the corona (Hood *et al.* 2009, and references therein). Nevertheless, signatures of a global twist are recognisable at a glance in emerging active regions (ARs) even in longitudinal magnetograms as large-scale yin yang pattern of the polarities due to contribution from the azimuthal vector component (Luoni *et al.* 2011). Flux rope formation continues as the active region decays and its field is getting dispersed due to its interaction with turbulent (super)granulation eddies. Flux dispersion towards the internal magnetic inversion line of an AR leads to magnetic cancellations there and gradually transforms a sheared arcade into a flux-rope, which may erupt in a CME. Key observational evidence was provided for this scenario by Green *et al.* (2011), in agreement with 3-D MHD simulations by Aulanier *et al.* (2010). With Hinode/EIS spectral measurements filament rotation ($v \leqslant 20$ km s^{-1}), consistent with the expansion of a twisted structure, was found *prior to* a CME eruption and X-class flare (Williams *et al.* 2009).

The twist is further increased during the eruption through a series of reconnections with the surrounding sheared arcade along a current sheet which forms *under* the departing flux rope (see Amari *et al.* 2010 for an MHD simulation and Liu *et al.* 2010 for observational evidence). The flux rope is very hot ($T \approx 10MK$), while at lower-T it appears as a void (a cavity), as evidenced by high-resolution multi-wavelength SDO observations (Cheng *et al.* 2011). Long, thin, straight, bright structures in the wake of CMEs have been interpreted as imaging evidence of current sheets (Ciaravella & Raymond 2008; Saint-Hilaire *et al.* 2009).

What is the cause of the eruption? Aulanier *et al.* (2010) has argued that the eruption is *not* caused by magnetic cancellations, nor coronal tether-cutting – these processes simply *build a flux-rope* and make it slowly rise to the critical height above the photosphere at which the torus instability can set in. The eruptive threshold is determined by the vertical gradient of the magnetic field in the low-β corona. Démoulin & Aulanier (2010) have shown that the *loss of equilibrium* is in fact equivalent with the *torus instability*, which is identified as the principal driver of CMEs. Kink instability, another ideal MHD instability, may simply raise the flux rope high enough for the torus instability to set in. On its own, kink instability leads to a confined eruption.

9. Coronal waves (B. Vršnak)

The past three years were very dynamic and fruitful in research on coronal large-scale large-amplitude propagating disturbances. About sixty peer-reviewed papers were published on EUV waves, coronal shocks and Moreton waves. In the majority of the papers the initiation, morphology, and kinematics of waves were analyzed, and the results were most often interpreted in terms of fast-mode MHD waves. However, in some events various non-wave or slow-mode-wave interpretations were proposed. Following this track, Warmuth & Mann (2011) provided evidence that there are physically different classes of propagating coronal disturbances, where a part of them are not really waves. Consequently, Long *et al.* (2011) proposed a new term, Coronal Bright Fronts, to avoid inconsistency in terminology. Considering the wave – driver relationship, all studies showed that initially it is difficult to resolve the wave and the eruption. Later on, the wave detaches from the eruption, and continues as a freely propagating wave. Furthermore, due to the high sensitivity of new instruments, it became possible to recognize the full EUV-wave dome around the eruption (e.g. Veronig *et al.* 2010), giving an insight into the 3-D kinematics and morphology of the disturbance. Studies of coronagraphic white-light signatures of the coronal shocks, as well as studies of the relationship between EUV waves, chromospheric Moreton waves, and radio type II bursts provided additional information about the shock formation and evolution. Several studies demonstrated that the wave is initiated during the temporary lateral acceleration of CME flanks, and that after the driven phase, it evolves as freely propagating blast.

Exceptionally important studies on EUV waves are those employing spectral analysis, since they provide detailed plasma diagnostics. The Hinode/EIS campaign HOP-180, targeted specifically to hunt EUV waves, provided a unique high-cadence spectroscopic data of an EUV wave, showing downflows of 20 km s^{-1} at the wavefront, analogous to that in Moreton waves (Harra *et al.* 2011). As a response to great advances in observations, several theoretical studies were published. The presented MHD simulations dealt with the relationship between the CME expansion and the resulting response of the ambient atmosphere, demonstrating very persuasively all the complexity of wave phenomena associated with coronal arcade eruption. It should be also noted that several review papers were published (Wills-Davey & Attrill 2009; Warmuth 2010; Gallagher & Long 2011; Zhukov 2011).

10. Solar radio bursts and particle acceleration (Y. Yan)

Solar radio emission provides important complementary diagnostics of activity events to optical, EUV, X-ray and γ-ray observations. Multi-wavelength analyses have advanced our knowledge on thermal and non-thermal phenomena in the solar atmosphere and interplanetary space (e.g. Cliver & Ling 2009; Kaufmann *et al.* 2009; Lee *et al.* 2009; Minoshima *et al.* 2009; Krucker *et al.* 2010; Klein *et al.* 2010; Reid *et al.* 2011).

Analysis of radio bursts with fine structures have attracted much attention so as to identify acceleration mechanisms of fast particles and their propagation in the flare area (e.g. Ning *et al.* 2009; Chernov *et al.* 2010; Kumar *et al.* 2010). For example, the coherent characteristics of the zebra pattern structures i.e. quasi-parallel narrow-band stripes in the microwave dynamic spectra can be applied to coronal magnetic field diagnostic in addition to that of other plasma parameters (e.g. Zlotnik *et al.* 2009; Chen *et al.* 2011).

Theoretical research on radio emission generated by electrons accelerated in solar flares/CMEs advanced as well. Li & Fleishman (2009) found that the radio emission produced by stochastic acceleration due to cascading MHD turbulence and regular acceleration in collapsing magnetic traps, are distinctly different. Li *et al.* (2011) presented simulations for decimetric type III radio bursts at twice the local electron plasma frequency, extending their previous model of metric-wave type III bursts to the lower corona.

Prototype studies on new-generation solar-dedicated imaging-spectroscopy facilities in decimetric and microwave ranges have started (Yan *et al.* 2009; Sawant *et al.* 2009; Wang *et al.* 2010). The analysis of the first interferometric observation of a zebra-pattern radio burst with simultaneous high spectral and temporal resolutions (Chen *et al.*2011) indicates that the different models can be examined with strict constraints. In the near future high-resolution true imaging-spectroscopy observations are expected to greatly advance our capability to measure and map coronal magnetic fields, to understand the physics of solar flares, and their influence on space weather.

11. Generation of hot plasma outflows from active regions - a potential source of the slow solar wind (L. van Driel-Gesztelyi)

One of the major discoveries in coronal physics made by Hinode/EIS was the presence of persistent high-temperature high-speed outflows from the periphery of ARs with line-of-sight velocities up to 50 km s^{-1} (Harra *et al.* 2008; Doschek *et al.* 2008) and spectral line asymmetries nearly 200 km s^{-1}, being suggestive of multiple components (Bryans *et al.* 2010; Peter 2010; Brooks & Warren 2011). Outflow locations are of low electron density and low radiance (Del Zanna 2008) and they originate over monopolar magnetic field concentrations (Doschek *et al.* 2008; Baker *et al.* 2009). Two main categories of mechanism have been proposed to be the driver of these outflows (i) magnetic reconnection-related (see Baker *et al.* 2009; Del Zanna *et al.* 2011 and references therein) and (ii) compression of surrounding fields by AR expansion (Murray *et al.* 2010). The blue-shifted plasma flows are believed to be a possible source of the slow solar wind

(Brooks & Warren 2011) when topological conditions in the large-scale solar magnetic fields are favourable (see the last paragraph in Section 3 and references therein).

12. Closing remarks

Studies of solar activity have produced far more exciting new results than we were able to cover and cite in this brief report – only the solar cycle was at minimum, solar activity studies have continued to increase both in volume and quality. The wealth of key findings indicates a vigorous, active community organised and supported by Commission 10.

Lidia van Driel-Gesztelyi
President of the Commission

References

Amari, T., Aly, J.-J., Mikic, Z., & Linker, J. 2010, *ApJL*, 717, L26

Antiochos, S. K., Mikić, Z., Titov, V. S., Lionello, R., & Linker, J. A. 2011, *ApJ*, 731, 112

Aschwanden, M. J., 2009, *Space Sci. Rev.*, 149, 31

Aschwanden, M. J. 2011, *Living Rev. Sol. Phys.* lsrp-11-5

Aschwanden, M. J., Nitta, N. V., Wülser, J.-P., & Lemen, J. R. 2008, *ApJ*, 680, 1477

Aulanier, G., Török, T., Démoulin, P., & DeLuca, E. E. 2010, *ApJ*, 708, 314

Baker, D., van Driel-Gesztelyi, L., Mandrini, C. H., Démoulin, P., & Murray, M. J. 2009, *ApJ*, 705, 926

Barbey, N., Guennou, C., & Auchére, F. 2011, *Sol. Phys.*, in press, DOI: 10.1007/s11207-011-9792-8

Bemporad, A., Del Zanna, G., & Andretta, V. *et al.* 2009, *Ann. Geophys.*, 27, 3841

Bian, N. H., Kontar, E. P., & Brown, J. C. 2010, *A&A*, 519, A114

Birn, J., Fletcher, L., Hesse, M., & Neukirch, T. 2009, *ApJ*, 695, 1151

Brooks, D. H. & Warren, H. P. 2011, *ApJL*, 727, L13

Brown, B. P., Miesch, M. S., Browning, M. K., Brun, A. S., & Toomre, J. 2011, *ApJ*. 731, 69

Brown, J. C., Turkmani, R., & Kontar, E. P. *et al.* 2009, *A&A*, 508, 993

Bryans, P., Young, P. R., & Doschek, G. A. 2010, *ApJ*, 715, 1012

Butala, M. D., Hewett, R. J., Frazin, R. A., & Kamalabadi, F. 2010, *Sol. Phys.*, 262, 495

Cameron, R., Gizon, L., & Duvall, T. L., Jr. 2008, *Sol. Phys.*, 251, 291

Ciaravella, A. & Raymond, J. C. 2008, *ApJ*, 686, 1372

Chen, B., Bastian, T. S., Gary, D. E., & Jing, J. 2011, *ApJ*, 736, 64

Cheng, X., Zhang, J., Liu, Y., & Ding, M. D. 2011, *ApJL*, 732, L25

Cheung, M. C. M., Rempel, M., Title, A. M., & Schüssler, M. 2010, *ApJ*, 720, 233

Chernov, G. P., Yan, Y. H., Tan, C. M., Chen, B., & Fu, Q. J. 2010, *Sol. Phys.*, 262, 149

Cliver, E. W. & Ling, A. G. 2009, *ApJ*, 690, 598

Dadashi, N., Teriaca, L., & Solanki, S. K. 2011, *A&A*, 534, A90

Del Zanna, G. 2008, *A&A*, 481, L49

Del Zanna, G., Aulanier, G., Klein, K.-L., & Török, T. 2011, *A&A*, 526, A137

Démoulin, P. & Aulanier, G. 2010, *ApJ*, 718, 1388

Dennis, B. R. & Pernak, R. L. 2009, *ApJ*, 698, 2131

Des Jardins, A., Canfield, R., Longcope, D., Fordyce, C., & Waitukaitis, S. 2009, *ApJ*, 693, 1628

Dikpati, M., Gilman, P. A., de Toma, G., & Ulrich, R. K. 2010, *GRL* 371, L14107

Doschek, G. A., Warren, H. P., & Mariska, J. T., *et al.* 2008, *ApJ*, 686, 1362

Dzifćáková, E. & Kulinová, A. 2010, *Sol. Phys.*, 263, 25

Fan, Y. 2011, *ApJ*, 740, 68

Feng, L., Inhester, B., & Solanki, S. K., *et al.* 2009, *ApJ*, 700, 292

Fisk, L. A. 2005, *ApJ*, 626, 563

Frazin, R. A., Jacob, M., & Manchester, W. B., IV *et al.* 2009a, *ApJ*, 695, 636

Frazin, R. A., Vásquez, A. M., & Kamalabadi, F. 2009b, *ApJ*, 701, 547

Frazin, R. A., Lamy, P., Llebaria, A., & Vásquez, A. M. 2010, *Sol. Phys.*, 265, 19
Fröhlich, C. 2009, *A&A* 501, L27
Gallagher, P. T. & Long, D. M. 2011, *Space Sci. Rev.* 158, 365
Ghizaru, M., Charbonneau, P., & Smolarkiewicz, P. K. 2010, *ApJ*, 715, L133
Gordovskyy, M., Browning, P. K., & Vekstein, G. E. 2010, *ApJ*, 720, 1603
Graham, D. R., Fletcher, L., & Hannah, I. G. 2011, *A&A*, 532, A27
Green, L. M., Kliem, B., & Wallace, A. J. 2011, *A&A*, 526, A2
Hannah, I. G. & Kontar, E. P. 2011, *A&A*, 529, A109
Hansteen, V. H., Hara, H., De Pontieu, B., & Carlsson, M. 2010, *ApJ*, 718, 1070
Harra, L. K., Sakao, T., & Mandrini, C. H., *et al.* 2008, *ApJL*, 676, L147
Harra, L. K., Sterling, A. C., Gömöry, P., & Veronig, A. 2011, *ApJL* 737, L4
Hathaway, D. H. & Rightmire, L. 2011, *ApJ*, 729, 80
Hood, A. W., Archontis, V., Galsgaard, K., & Moreno-Insertis, F. 2009, *A&A*, 503, 999
Hood, A. W., Archontis, V., & MacTaggart, D. 2011, *Sol. Phys.*, 157
Jess, D. B., Mathioudakis, M., Crockett, P. J., & Keenan, F. P. 2008, *ApJ*, 688, L119
Jiang, J., Cameron, R., Schmitt, D., & Schüssler, M. 2009, *Astrophys. J.* 693, L96
Jiang, Y., Yang, J., Hong, J., Bi, Y., & Zheng, R. 2011, *ApJ*, 738, 179
Jin, M., Manchester, W. B., & Van der Holst, B., *et al.* 2011, *ApJ*, in press
Jing, J., Chen, P. F., Wiegelmann, T., Xu, Y., Park, S.-H., & Wang, H. 2009, *ApJ*, 696, 84
Jing, J., Tan, C., & Yuan, Y., *et al.* 2010, *ApJ*, 713, 440
Käpylä, P. J., Korpi, M. J., & Brandenburg, A. *et al.* 2010, *Astron. Nachr.* 331, 73
Karlický, M. & Bárta, M. 2011, *ApJ*, 733, 107
Kaufmann, P., Trottet, G., & Giménez de Castro, C. G., *et al.* 2009, *Sol. Phys.*, 255, 131
Klein, K.-L., Trottet, G., & Klassen, A. 2010, *Sol. Phys.*, 263, 185
Komm, R., Ferguson, R., Hill, F., Barnes, G., & Leka, K. D. 2011, *Sol. Phys.*, 268, 389
Kontar, E. P., Hannah, I. G., & MacKinnon, A. L. 2008, *A&A*, 489, L57
Kopp, G. & Lean, J. L. 2011, *GRL* 380, L01706
Kramar, M., Jones, S., Davila, J., Inhester, B., & Mierla, M. 2009, *Sol. Phys.*, 259, 109
Kretzschmar, M. 2011, *A&A*, 530, A84
Krucker, S., Hudson, H. S., & Glesener, L. *et al.* 2010, *ApJ*, 714, 1108
Krucker, S., Hudson, H. S., & Jeffrey, N. L. S. *et al.* 2011, *ApJ*, 739, 96
Lee, J., Nita, G. M., & Gary, D. E. 2009, *ApJ*, 696, 274
Kulinová, A., Kašparová, J., & Dzifčáková, E., *et al.* 2011, *A&A*, 533, A81
Kumar, P., Srivastava, A. K., & Somov, B. V., *et al.* 2010, *ApJ*, 723, 1651
Li, B., Cairns, I. H., Yan, Y. H., & Robinson, P. A. 2011, *ApJL*, 738, L9
Li, Y. & Fleishman, G. D. 2009, *ApJL*, 701, L52
Liewer, P. C., de Jong, E. M., & Hall, J. R., *et al.* 2009, *Sol. Phys.*, 256, 57
Linker, J. A., Lionello, R., Mikić, Z., Titov, V. S., & Antiochos, S. K. 2011, *ApJ*, 731, 110
Liu, R., Liu, C., Wang, S., Deng, N., & Wang, H. 2010, *ApJL*, 725, L84
Long, D. M., Gallagher, P. T., McAteer, R. T. J., & Bloomfield, D. S. 2011, *A&A* 531, A42
Longcope, D. W., Guidoni, S. E., & Linton, M. G. 2009, *ApJ*, 690, L18
Lopes, I. & Passos, D. 2009, *Sol. Phys.* 257, 1
Luoni, M. L., Démoulin, P., Mandrini, C. H., & van Driel-Gesztelyi, L. 2011, *Sol. Phys.*, 270, 45
Magara, T. & Tsuneta, S. 2008, *PASJ*, 60, 1181
Martínez-Oliveros, J. C. & Donea, A.-C. 2009, *MNRAS*, 395, L39
Martínez Oliveros, J. C., Couvidat, S., & Schou, J., *et al.* 2011, *Sol. Phys.*, 269, 269
Martínez-Sykora, J., De Pontieu, B., Hansteen, V., & McIntosh, S. W. 2011, *ApJ*, 732, 84
Mann, G. & Warmuth, A. 2011, *A&A*, 528, A104
Masson, S., Pariat, E., Aulanier, G., & Schrijver, C. J. 2009, *ApJ*, 700, 559
Matthews, S. A., Zharkov, S., & Zharkova, V. V. 2011, *ApJ*, 739, 71
McIntosh, S. W. & De Pontieu, B. 2009, *ApJ*, 707, 524
Mewaldt, R. A., Davis, A. J., & Lave, K. A. *et al.* 2010, *ApJ*, 723, L1
Milligan, R. O. 2011, *ApJ*, 740, 70
Minoshima, T., Imada, S., & Morimoto, T., *et al.* 2009, *ApJ*, 697, 843

Moradi, H., Hanasoge, S. M., & Cally, P. S. 2009, *ApJL*, 690, L72

Morgan, H. & Habbal, S. R. 2010, *ApJ*, 710, 1

Murray, M. J., Baker, D., van Driel-Gesztelyi, L., & Sun, J. 2010, *Sol. Phys.*, 261, 253

Nandy, D., Munõz-Jaramillo, A., & Martens, P. C. H. 2011, *Nature* 471, 80

Ning, Z., Cao, W., & Huang, J., *et al.* 2009, *ApJ*, 699, 15

Owens, M. J., Crooker, N. U., & Schwadron, N. A. *et al.* 2008, *GRL* 352, L20108

Parchevsky, K. V. & Kosovichev, A. G. 2009, *ApJ*, 694, 573

Park, S.-H., Lee, J., & Choe, G. S. *et al.* 2008, *ApJ*, 686, 1397

Penn, M. & Livingston, W. 2010, *IAU Symp.* **273**; ArXiv 1009.0784

Pesnell, W. D. 2008, *Sol. Phys.* 252, 209

Peter, H. 2010, *A&A*, 521, A51

Petrosian, V. & Chen, Q. 2010, *ApJL*, 712, L131

Potts, H., Hudson, H., Fletcher, L., & Diver, D. 2010, *ApJ*, 722, 1514

Reeves, K. K., Linker, J. A., Mikić, Z., & Forbes, T. G. 2010, *ApJ*, 721, 1547

Reid, H. A. S., Vilmer, N., & Kontar, E. P. 2011, *A&A*, 529, A66

Reinard, A. A., Henthorn, J., Komm, R., & Hill, F. 2010, *ApJ* 710, L121

Rempel, M. & Schlichenmaier, R. 2011, *Living Rev. Sol. Phys.*, 8, 3

Rempel, M., Schüssler, M., Cameron, R. H., & Knölker, M. 2009, *Science*, 325, 171

Saint-Hilaire, P., Krucker, S., & Lin, R. P. 2009, *ApJ*, 699, 245

Sawant, H. S., Cecatto, J. R., & Mészárosová, H., *et al.* 2009, *Adv. Space Res.* 44, 54

Schrijver, C. J. & Title, A. M. 2011, *JGR* 116, 4108

Solomon, S. C., Woods, T. N., Didkovsky, L. V., Emmert, J. T., & Qian, L. 2010, *GRL* 371, L16103

Svalgaard, L. & Hudson, H. S. 2010, in *ASP Conf. Ser.* 428, SOHO-23: Understanding a Peculiar Solar Minimum, S.R. Cranmer, J.T. Hoeksema, & J.L. Kohl (eds.), 325

Svalgaard, L., Hannah, I. G., & Hudson, H. S. 2011, *ApJ*, 733, 49

Thompson, W. T. 2011, *JASTP*, 73, 1138

Tian, H., McIntosh, S. W., & De Pontieu, B., *et al.* 2011, *ApJ*, 738, 18

Titov, V. S., Mikić, Z., Linker, J. A., Lionello, R., & Antiochos, S. K. 2011, *ApJ*, 731, 111

Török, T., Panasenco, O., & Titov, V. S., *et al.* 2011, *ApJL*, 739, L63

Vásquez, A. M., Frazin, R. A., & Kamalabadi, F. 2009, *Sol. Phys.*, 256, 73

Vásquez, A. M., Huang, Z., Manchester IV, W. B., & Frazin, R. A. 2011, *Sol. Phys.*, in press, DOI: 10.1007/s11207-010-9706-1

Veronig, A. M., Muhr, N., Kienreich, I. W., Temmer, M., & Vršnak, B. 2010, *ApJL* 716, L57

Wang, H. & Liu, C. 2010, *ApJ*, 716, L195

Wang, X., Ge, H., Gary, D. E., & Nita, G. M. 2009, *PASP*, 121, 1139

Wang, Y.-M., Robbrecht, E., & Sheeley, Jr., N. R. 2009, *ApJ* 707, 1372

Warmuth, A. 2010, *Adv. Space Res.* 45, 527

Warmuth, A. & Mann, G. 2011, *A&A* 532, A151

Warmuth, A., Mann, G., & Aurass, H. 2009, *A&A*, 494, 677

Watanabe, K., Krucker, S., & Hudson, H. *et al.* 2010a, *ApJ*, 715, 651

Watanabe, T., Hara, H., Sterling, A. C., & Harra, L. K. 2010b, *ApJ*, 719, 213

Williams, D. R., Harra, L. K., Brooks, D. H., Imada, S., & Hansteen, V. H. 2009, *PASJ*, 61, 493

Wills-Davey, M. J. & Attrill, G. D. R. 2009, *Space Sci. Rev.* 149, 325

Wu, S. T., Wang, A. H., & Gary, G. A., *et al.* 2009, *Adv. Space Res.*, 44, 46

Yan, Y., Zhang, J., & Wang, W., *et al.* 2009, *Earth Moon and Planets*, 104, 97

Yeates, A. R., Attrill, G. D. R., & Nandy, D., *et al.* 2010a, *ApJ*, 709, 1238

Yeates, A. R., Mackay, D. H., van Ballegooijen, A. A., & Constable, J. A. 2010b, *JGR*, 115, 9112

Zacharias, P., Peter, H., & Bingert, S. 2011, *A&A*, 531, A97

Zharkova, V. V. & Siversky, T. V. 2011, *ApJ*, 733, 33

Zhukov, A. N. 2011, *JASTP*, 73, 1096

Zlotnik, E. Y., Zaitsev, V. V., Aurass, H., & Mann, G. 2009, *Sol. Phys.*, 255, 273

Transactions IAU, Volume XXVIIIA
Reports on Astronomy 2009–2012
Ian Corbett, ed.

© International Astronomical Union 2012
doi:10.1017/S1743921312002682

COMMISSION 12

SOLAR RADIATION AND STRUCTURE

RAYONNEMENT ET STRUCTURE SOLAIRE

PRESIDENT   Alexander Kosovichev
VICE-PRESIDENT   Gianna Cauzzi
PAST PRESIDENT   Valentin Martinez Pillet
ORGANIZING COMMITTEE   Martin Asplund, Axel Brandenburg,
                   Dean-Yi Chou,
                   Jorgen Christensen-Dalsgaard,
                   Weiqun Gan, Vladimir D. Kuznetsov,
                   Marta G. Rovira, Nataliya Shchukina,
                   P. Venkatakrishnan

TRIENNIAL REPORT 2009–20012

1. Introduction

Commission 12 of the International Astronomical Union encompasses investigations of the internal structure and dynamics of the Sun, mostly accessible through the techniques of local and global helioseismology, the quiet solar atmosphere, solar radiation and its variability, and the nature of relatively stable magnetic structures like sunspots, faculae and the magnetic network. The Commission sees participation of over 350 scientists worldwide.

A brief review of Commission activity and some important developments in the field in the framework 2009–2011 is reported below. Several of these developments came about during the deep, unusually long minimum of solar magnetic activity which manifested itself between 2008 and 2009, and which was observed by a number of new missions and advanced ground-based instrumentation, so providing for a clearer view of the "basic" quiet Sun structure. As always, the report is by no means exhaustive, merely reflecting the main interests of the Commission Organizing Committee.

2. Organizational activities

Commission 12 proposed and organized IAU Symposium 264 "Solar and Stellar Variability - Impact on Earth and Planets" (Rio de Janeiro, Brazil, August 3-7, 2009), IAU Symposium 274 "Advances in plasma astrophysics" (Catania, Italy, September 6-10, 2010), IAU Symposium 294 "Solar and astrophysical dynamos and magnetic activity" (Beijing, China, August 27-31, 2012). It also participated in the organization of IAUS 286 "Comparative magnetic minima: characterizing quiet times in the Sun and stars" (Mendosa, Argentina; Oct. 3-7, 2011) and IAUS 271 "Astrophysical dynamics - from stars to galaxies" (Nice, France, June 21-25, 2010).

3. New observational facilities

The period of 2009-12 was marked tremendous advances in new observational facilities from the ground and space. These include the NASA's Solar Dynamics Observatory mission, launched in February 2010, for understanding the basics mechanisms of solar dynamics and variability, the PICARD mission (France) launched in June 2010 and dedicated to the simultaneous measurement of the absolute total and spectral solar irradiance, and the diameter and solar shape; Russian Solar mission CORONAS-Photon (Complex Orbital Observations Near-Earth of Activity of the Sun) to investigate the processes of free energy accumulation in the Sun's atmosphere, accelerated particle phenomena and solar flares, and the correlation between solar activity and magnetic storms on Earth. At the same time the previously launched space solar observatories: SOHO, STEREO, RHESSI and Hinode continue to operate. Together, these solar space telescopes have provided unprecedented amount of multi-wavelength data for complex investigations of the structure, dynamics and magnetism of the Sun from the deep interior to the outer corona.

In addition, first high-resolution observations of the Sun were obtained with two large telescopes: the 1-m optical telescope on balloon observatory SUNRISE which had the first 6-day flight in June 2009, and the 1.6-m New Solar Telescope (NST) at the Big Bear Solar Observatory. The telescope has a unique of-axis configuration that helps to solve the heat problem. This telescope in the first of the new generation of large solar telescopes. Two very large solar telescopes, Advanced Technology Solar Telescope (ATST) and European Solar Telescope (EST) are being developed.

The new telescopes combined with precise spectro-polarimetric instrumentation (del Toro Iniesta & Martinez Pillet 2010) provide important insight into the small-scale phenomena. This includes discovery of fine structure of magnetic elements in sunspot penumbrae, detection of small-scale vortices in granulation, overturning convection in penumbra filaments, superdiffusion of small magnetic elements etc.

Future observational projects include space missions: The Interface Region Imaging Spectrograph (NASA), Solar Orbiter (ESA), and InterHelioProbe and Polar Ecliptic Patrol (Russia).

The Interface Region Imaging Spectrograph (IRIS) (Lemen et al. 2011) addresses critical questions in order to understand the flow of energy and mass through the chromosphere and transition region, namely: (1) Which types of non-thermal energy dominate in the chromosphere and beyond? (2) How does the chromosphere regulate mass and energy supply to the corona and heliosphere? (3) How do magnetic flux and matter rise through the lower atmosphere, and what roles does flux emergence play in flares and mass ejections? These questions are addressed with a high-resolution imaging spectrometer that observes Near- and Far-VU emissions that are formed at temperatures between 5,000K and 1.5×10^6 K. IRIS has a field-of-view of 120 arcsec, a spatial resolution of 0.4 arcsec, and velocity resolution of 0.5 km/s. Members of the IRIS investigation team are developing advanced radiative MHD codes to facilitate comparison with and interpretation of observations.

The Solar Orbiter (Woch et al. 2008) orbit comprises initially a nearly Sun-synchronous phase at a distance of only 0.22 AU from Sun center. In a later stage, the orbital inclination will be raised, thus allowing Solar Orbiter to reach solar latitudes of about 35 degrees, and making it the first mission after Ulysses to study the Sun from a high-latitude vantage point. In contrast to Ulysses, however, Solar Orbiter will carry a complementary suite of both, in-situ and remote- sensing instruments, which will allow the study of the solar atmosphere to be extended to the largely unexplored polar regions of the Sun. The polar magnetic fields are responsible for the polar coronal holes driving the fast solar wind, but are poorly known. From its vantage point outside the ecliptic, Solar Orbiter will uncover the surface and sub-surface flows at the poles, the polar magnetic field structure and its evolution. It will provide new insights into the formation of the polar coronal holes, the nature of their boundaries and the acceleration of the fast solar wind emanating from the holes.

Phase B of the Interhelioprobe Mission (Zelenyi *et al.* 2004) started in 2009-2010. The mission is aimed at the study of the inner heliosphere and the Sun at short distances by using a spacecraft (SC) at heliocentric orbit formed by multiple gravity-assisted maneuvers at Venus. Interhelioprobe observations in the immediate proximity to the Sun combined with in-situ plasma measurements will contribute significantly to the solution of the problems of heating of the solar corona, solar wind acceleration, and the origin of major solar active events such as solar flares and coronal mass ejections. At the end of the mission, the gravity-assisted maneuvers at Venus can be used for inclining the SC orbit to the ecliptic plane and conducting out-of-ecliptic observations of the Sun, including its polar regions and ecliptic corona. The composition of the mission scientific payload has been determined. It will comprise the instruments for remote observations of the Sun (X-ray telescope-spectrograph, coronagraph, magnetograph, and photometer) and in-situ measurements in the heliosphere (magnetometer, solar-wind electron analyzer, plasma analyzer, analyzer of solar neutrons, detector of charged particles, gamma-ray spectrometer, X-ray spectrometer, and wave complex). The development of sketch design of the mission will be completed in July 2012.

Phase A of the Polar Ecliptic Patrol mission (PEP) started in 2009. The mission is aimed at the study of the global pattern of solar activity, including its manifestations in the heliosphere and near-Earth space. The mission will comprise two small satellites. By gravity-assisted maneuvers at Venus, the satellites will be placed on heliocentric orbits inclined to the ecliptic plane at an angle to each other at distances about 0.5 AU from the Sun. The satellites on the orbits will be shifted about one another by a quarter of a period (one period is about 130 days). Such a scheme will ensure continuous monitoring of the Sun-Earth line from one (and in some periods, from both) SC. When one SC is in the ecliptic plane, another is over one of the solar poles; as the first SC goes away from the ecliptic plane, the second one approaches it. Thus, simultaneous monitoring of the near-ecliptic and polar regions is carried. This will enable a continuous study of the slow- and high-speed solar wind and will provide a 3D pattern of the solar corona and ejections. Observing solar ejections from two spaced SC and from out-of-ecliptic position will allow us to determine their exact direction relative to the Sun-Earth line and their heliolatitude and heliolongitude extension. Stage A will be devoted to developing the details of the ballistic characteristics of the mission, its scientific tasks and instruments, and the tentative outward appearance of the spacecraft.

4. Research highlights

A revision of the progress made in these fields is presented. For some specific topics, the review has counted with the help of experts outside the Commission Organizing Committee that are leading and/or have recently presented relevant works in the respective fields.

4.1. *Solar irradiance and its variability*

Solar irradiance and variability has become a new hot topic during the past 3 years because of the unusual long minimum of solar activity in 2007-2009. During the recent minimum with an unusually long periods with no sunspots, TSI was also extremely low, namely 25% of a typical cycle amplitude lower than in 1996. Together with the values during the previous minima this points to a long-term change related to the strength of solar activity. On the other hand, activity indices as the 10.7 cm radio flux (F10.7), the CaII and MgII indices and also the Ly-α irradiance, showed a much smaller decrease. This means that proxy models for TSI based on the photometric sunspot index (PSI), and on e.g. MgII index to represent faculae and network have to be complemented by a further component for the long-term change (Fröhlich 2011). This problem was investigated in detail by Ball $et\ al.$ (2011) and also by Steinhilber (2010) who studied how well modeled solar irradiances agree with measurements from the SORCE satellite, both for total solar irradiance and broken down into spectral regions on timescales of several years. It was found that a model that assumes that all variation in solar irradiance is the result of changes in the distribution of magnetic features on the solar surface captures 97% of the observed TSI variation. However, the modeled spectral irradiance (SSI) showed significant disagreement with the SIM instrument on SORCE. If the data are correct this disagreement implies that some mechanism other than surface magnetism is causing SSI variations.

The solar X-ray continuum emission at five wavelengths between 3.495 $Å$ and 4.220 $Å$ for 19 flares in a 7-month period in 2002-2003 was observed by the RESIK (REntgenovsky Spektrometr s Izognutymi Kristalami) crystal spectrometer on CORONAS-F (Phillips $et\ al.$ 2010). In this wavelength region, free-free and free-bound emissions have comparable fluxes. With a pulse-height analyzer having settings close to optimal, the fluorescence background was removed so that RESIK measured true solar continuum in these bands with an uncertainty in the absolute calibration of $\pm20\%$. With an isothermal assumption, and temperature and emission measure derived from the ratio of the two GOES channels, the observed continuum emission normalized to an emission measure of 1048 cm^{-3} was compared with theoretical continua using the CHIANTI atomic code. The accuracy of the RESIK measurements allows photospheric and coronal abundance sets, important for the free-bound continuum, to be discriminated. It is found that there is agreement to about 25% of the measured continua with those calculated from CHIANTI assuming coronal abundances in which Mg, Si, and Fe abundances are four times photospheric.

The variable Sun is the most likely candidate for the natural forcing of past climate changes on time scales of 50 to 1000 years. Evidence for this understanding is that the terrestrial climate correlates positively with the solar activity. During the past 10,000 years, the Sun has experienced the substantial variations in activity and there have been numerous attempts to reconstruct solar irradiance. The recent deep solar minimum provided an opportunity to investigate the Sun's properties in its "minimum activity" state, and then use this knowledge for reconstructing the past solar irradiance using activity proxies. Shapiro $et\ al.$ (2011) assumed that the minimum state of the quiet Sun in time corresponds to the observed quietest area on the present Sun. Then they used available long-term proxies of the solar activity, which are ^{10}Be isotope concentrations in ice cores and 22-year smoothed neutron monitor data, to interpolate between the present quiet Sun and the minimum state of the quiet Sun. This determines the long-term trend in the solar variability, which is then superposed with the 11-year activity cycle calculated from the sunspot number. The time-dependent solar spectral irradiance from about 7000 BC to the present is then derived using a radiation code. It was found that the total and spectral solar irradiance that was substantially lower during the Maunder minimum than today. This leads to large historical solar forcing and indicates that the solar forcing probably played significant role in climate changes. However, a coupled climate model to explore the effect of a 21st-century grand minimum on future global temperatures, finding a moderate temperature offset of no more than -0.3 deg C in the year 2100 (Feulner & Rahmstorf 2010). This temperature decrease is much smaller than the warming expected from anthropogenic greenhouse gas emissions by the end of the century.

Currently, the solar irradiance is monitored by several satellites. The spectral irradiance is measured with high spectral and temporal resolution by the EVE instrument on SDO.

4.2. *Solar composition*

The chemical composition of the Sun is an important ingredient in our understanding of the formation, structure, and evolution of the Sun and the Solar System. Also, this is a reference standard for the composition of other astronomical objects (Asplund *et al.* 2009). Since about 2004, there have been debates among stellar physicists about the value of solar metallicity. Prior to 2004, the value of Z/X for the Sun was assumed to be 0.0231 (Grevesse & Sauval 1998)). In 2005, however, a series of papers were published with lower values for the abundance of O, C and N in the solar photosphere. Asplund *et al.* (2005) using a time-dependent, 3D hydrodynamical model of the solar atmosphere instead of 1D hydrostatic models determined that Z/X for the Sun is only 0.0165. This lowering of abundances has serious consequences for solar and stellar models. However, this new spectroscopically determined metallicity value disagreed with the standard solar models supported by helioseismology (Basu 2009). The problem of solar metallicity was studied in detail by the CIFIST team at the Paris Observatory Caffau *et al.* (2011). They found that a part of this effect can be attributed to an improvement of atomic data and the inclusion of NLTE computations, but also the use of hydrodynamical model atmospheres may also play a role as originally suggested by Asplund *et al.* (2005). The photospheric solar abundances of several elements, among them C, N, and O, was determined using a 3D simulation model of the solar atmosphere obtained with the CO^5BOLD code. The spectroscopic abundances were obtained by fitting the equivalent width and/or the profile of observed spectral lines with synthetic spectra computed from the 3D model atmosphere. It was found that, in fact, 3D effects are not responsible for the systematically low estimates of the solar abundances. The solar metallicity resulting from the new analysis is $Z = 0.0154$, and $Z/X = 0.0211$. This new result is in better agreement with the constraints of helioseismology than the previous 3D abundance results, but the discrepancy is still significant. This problem will certainly remain among the most important problems of solar physics and astrophysics.

Chlorine is an odd-Z element with low abundance in the solar photosphere and in meteorites. Chlorine has no photospheric lines in the visible spectrum available for abundance analysis. Cl XVI lines were observed with the RESIK crystal spectrometer on the CORONAS-F spacecraft during 20 solar flares, from which it was possible to determine much more definitively the Cl abundance for flare plasmas (Sylwester *et al.* 2011). The abundance of chlorine was determined from X-ray spectra obtained with the RESIK instrument on CORONAS-F during solar flares between 2002 and 2003. Using weak lines of He-like Cl, Cl XVI, between 4.44 and 4.50 A, and with temperatures and emission measures from GOES on an isothermal assumption, we obtained A(Cl) = 5.75 ± 0.26 on a scale A(H) = 12. The uncertainty reflects an approximately a factor of two scatter in measured line fluxes. Nevertheless, this value represents what is probably the best solar determination yet obtained. It is higher by factors of 1.8 and 2.7 than Cl abundance estimates from an infrared sunspot spectrum and nearby H II regions. The constancy of the RESIK abundance values over a large range of flares (GOES class from below C1 to X1) argues for any fractionation that may be present in the low solar atmosphere to be independent of the degree of solar activity.

4.3. *Structure of the solar interior*

Helioseismology received significant boost with the Helioseismic and Magnetic Imager on SDO. This instruments provides uninterrupted 4096×4096-pixel Doppler-shift oscillation data with high spatial (0.5 arcsec/pixel) and high temporal (45 sec) resolutions. These data resolve the whole solar oscillation spectrum from global low-degree modes to modes of very high angular degree and frequency and include all modes captured in the Sun's resonant cavity. To process such large amount of data special helioseismology tools ("pipelines") were developed and implemented in the Joint Science Operation Center (JSOC) at Stanford. In particular, a time-distance helioseismology pipeline provides full-disk maps of subsurface flows and wave-speed variations in the range of depths 0–20 Mm every 8 hours (Couvidat *et al.* 2010; Zhao *et al.* 2011a,b). In addition, pipeline processing is developed for global helioseismology, far-side imaging and ring-diagram analysis. The analysis data are available on-line on the Stanford JSOC web site.

New helioseismology experiment experiment SOKOL on board of CORONAS-PHOTON (2009) was directed to the study of characteristics and inner structure of the Sun on the basis of the spectra of global oscillations of the Sun (Lebedev *et al.* 2011). Such spectra were obtained by method of measurement of solar radiation intensity variations. This experiment was a continuation of the study of solar global fluctuations started on CORONAS-I and CORONAS-F satellites. Solar photometer SOKOL designed by IZMI-RAN observed the variations of intensity of solar radiation in seven optical ranges from near ultra-violet up to infra-red range of the spectrum. The spectra of p-modes fluctuations of the Sun and the dependence of amplitude of oscillations from the wavelength are obtained.

Local helioseismology provides 3D maps of subsurface flows and wave-speed anomalies below the solar surface. This is a relatively new discipline which is rapidly developing. It shows great potential for understanding the physical processes inside the Sun that lead to generation and transport of solar magnetic fields, formation of sunspot and active regions, and also initiation of flares and CMEs. There is no doubt that subsurface flows in the subsurface turbulent boundary layer play a key role in the mechanisms of solar activity. Local helioseismology studies greatly benefit from the new Helioseismic and Magnetic Imager data from SDO and continuing operation of the Global Oscillation Network (GONG).

Among highlights of local helioseismology was a statistical study of the GONG team that established a relationship between vorticity and helicity of subsurface flows in a 7 Mm deep layer and flaring activity of active regions. Komm *et al.* (2011) applied a discriminant analysis to 1023 active regions and their subsurface-flow parameters, such as vorticity and kinetic helicity density, with the goal of distinguishing between flaring and non-flaring active regions. Synoptic subsurface flows were obtained by analyzing GONG high-resolution Doppler data using the ring-diagram analysis. It was found that the subsurface-flow characteristics improve the ability to distinguish between flaring and non-flaring active regions. For the C- and M-class flare category, the most important subsurface parameter is the so-called structure vorticity, which estimates the horizontal gradient of the horizontal-vorticity components. The no-event skill score, which measures the improvement over predicting that no events occur, reaches 0.48 for C-class flares and 0.32 for M-class flares, when the structure vorticity at three depths combined with total magnetic flux are used. This analysis provides a basis for developing new physics-based forecasts of flaring activity of active regions.

Another very promising result was recently obtained by Ilonidis *et al.* (2011), who by using a specially designed deep-focus time-distance helioseismology scheme detected subsurface signatures of emerging sunspot regions before they appeared on the solar disc. Strong acoustic travel-time anomalies of an order of 12 to 16 seconds were detected as deep as 65,000 kilometers. These anomalies were associated with magnetic structures that emerged with an average speed of 0.3 to 0.6 km/s and caused high peaks in the photospheric magnetic flux rate 1 to 2 days after the detection of the anomalies. Thus, synoptic imaging of subsurface magnetic activity may allow anticipation of large sunspot regions before they become visible, improving space weather forecast. This result is causing debates because the emerging flux has not been observed at such depth by other helioseismology techniques, and also because the detected travel-time anomaly is stronger than this was anticipated from theoretical models. We expect a significant progress in developing such new local helioseismology schemes that allow to extract weak signals of deep perturbations insight the Sun by observing acoustic oscillations on the surface.

A substantial progress has been made in solving very difficult problems of imaging the structure of the solar tachocline and measuring meridional flows. Both of these problems are critical for understanding the mechanism of solar dynamo. However, the helioseismic signal in both cases is very weak compared to the "realization" noise of stochastically excited solar oscillations, and thus robust measurements require averaging over long periods. This, in turn, requires very high stability of helioseismology instruments and taking account various systematic effects, for instance, effects caused by seasonal variations and instrumental distortions. Very important role in these studies is played by numerical simulations which provide artificial data oscillation data that model the internal perturbations and flows and solar oscillations simulating stochastic acoustic sources close to the surface. Such 3D wave simulations and helioseismology testing have been performed for the global spherical Sun models of subsurface sound speed variations (Hartlep et al. 2011a,b), for global-scale solar convection (Hanasoge et al. 2010), for the meridional circulation (Hartlep et al. 2011b), and also for local scale MHD models of sunspots (Cameron et al. 2011; Parchevsky & Kosovichev 2009; Parchevsky et al. 2010a,b). The helioseismology simulations not only provided means for testing helioseismic inferences (Birch et al. 2011; Hartlep et al. 2011b; Moradi et al. 2010), but also help to detect new helioseismology effect and develop new diagnostics. For instance, Hartlep et al. (2011a) showed that under certain conditions, subsurface structures in the solar interior can alter the average acoustic power observed at the photosphere above them. By using numerical simulations of wave propagation, it was found that this effect is large enough for it to be potentially used for detecting emerging active regions before they appear on the surface. In the simulations, simplified subsurface structures are modeled as regions with enhanced or reduced acoustic wave speed. Observations from the SOHO/MDI prior and during the emergence of NOAA active region 10488 were used to test the use of acoustic power as a potential precursor of the emergence of magnetic flux.

The subsurface structure and dynamics of sunspots continues to be a central topic of helioseismology. The main difficulties in the diagnostics of sunspots are due to the strong inhomogeneity, magnetic field effects that can cause systematic phase of acoustic waves, which are not accounted for by the wave propagation models, absorption and transformation of acoustic wave, and nonuniform distribution of acoustic sources and their spectral characteristics (for a recent review see Kosovichev (2010) and references therein). Local helioseismic diagnostics of sunspots still have many uncertainties. However, there have been significant achievements in resolving these uncertainties, verifying the basic results by new high-resolution observations, testing the helioseismic techniques by numerical simulations, and comparing results obtained by different methods.

For instance, a recent analysis of helioseismology data from the Hinode space mission (Zhao et al. 2010) has successfully resolved several uncertainties and concerns (such as the inclined-field and phase-speed filtering effects) that might affect the inferences of the subsurface wave-speed structure of sunspots and the flow pattern. Zhao et al. (2010) analyzed a solar active region observed by the Hinode Ca II H line using the time-distance helioseismology technique, and inferred wave-speed perturbation structures and flow fields beneath the active region with a high spatial resolution. The general subsurface wave-speed structure is similar to the previous results obtained from SOHO/MDI observations. The general subsurface flow structure is also similar, and the downward flows beneath the sunspot and the mass circulations around the sunspot are clearly resolved. Below the sunspot, some organized divergent flow cells are observed, and these structures may indicate the existence of mesoscale convective motions. Near the light bridge inside the sunspot, hotter plasma is found beneath, and flows divergent from this area are observed. The initial acoustic tomography results from Hinode show a great potential of using high-resolution observations for probing the internal structure and dynamics of sunspots.

Initial steps to developing waveform tomography based on measurements of travel-time delays and amplitude variations for cross-correlations representing effective point wave sources have been made (Chou et al. 2009; Cameron et al. 2011; Ilonidis & Zhao 2011; Zhao et al. 2011c).

4.4. Structure of the chromosphere

Chromospheric research has witnessed a strong forward momentum in the last few years, owing to several important theoretical and observational developments. Ground-based imaging spectro-polarimetry performed with instruments such as IBIS (Cavallini 2006; Reardon & Cavallini 2008) and CRISP (Scharmer 2006) is providing novel observations of chromospheric signatures, in particular using the IR CaII 854.2 nm line. Long overlooked, this spectral line has come strongly forward as one of the best diagnostics to study both the dynamics (Cauzzi et al. 2008) and the magnetic structure of the chromosphere (Pietarila et al. 2007; de la Cruz Rodríguez & Socas-Navarro 2011). High resolution, highly stable limb and on-disk observations with the broad-band CaII H filter on SOT/Hinode have been exploited for investigating the role of chromospheric structures, such as spicules, in coronal heating and mass replenishment (De Pontieu et al. 2009; Sterling et al. 2010). In a parallel development, realistic 3-D MHD modeling of the entire regime between the upper convection zone and the corona is now within reach (Abbett 2007; Fang et al. 2010; Gudiksen et al. 2011) albeit still with important limitations such as the small size of the computational domain or the approximate treatment of radiative transfer. These developments have clarified at least some of the small-scale, dynamic chromospheric phenomena, while others are still very much debated.

A coherent picture has emerged that identifies the propagation and dissipation of photospheric acoustic waves at their dominant frequencies (from ~ 2 to $\sim 7\text{-}8$ mHz) into the chromospheric layers as important shapers of the chromospheric structure. These waves, uniformly excited at the solar surface by turbulent motions, are selected and guided to propagate in the upper atmosphere by the local and highly variable magnetic topology. Thus, a crucial role is played by the magnetic field, at once "passive" (brought about by convection) and "active" (waveguide) agent (Wedemeyer-Böhm et al. 2009; Carlsson et al. 2010). In the quietest portions of the chromosphere, as in the center of supergranular cells or in coronal holes, small scale acoustic shocks are ubiquitous (Vecchio et al. 2009) but lead to limited spatio–temporal heating. The recent 2D radiation-MHD modelling of Leenaarts et al. (2011), that takes into account non-equilibrium ionization of hydrogen and other effects, shows that such chromospheric areas can be as cold as 2000 K or less, provided that no significant magnetic heating is present. Chromospheric acoustic shocks are present also in or near magnetic structures such as network points. The shocks are stronger than those occurring in quiet Sun, but at they appear at a lower temporal frequency as the presence of an inclined field lowers the atmospheric cutoff frequency (Jefferies *et al.* 2006; Stangalini *et al.* 2011). An alternate explanation invokes convective downdrafts in the immediate surroundings of magnetic elements as responsible for the excitation of the slow modes eventually leading to chromospheric shocks (Kato *et al.* 2011), but in any case it is now established that such shocks are the cause of the dynamic phenomenon known as "dynamic fibrils" present in plage and strong network (De Pontieu *et al.* 2007; Langangen *et al.* 2008a). While the shocks produce a tangible heating in and around the magnetic elements, as Cauzzi *et al.* (2009) deduced from temporal evolution of chromospheric line widths, it remains unclear whether other effects such as Alfvénic turbulence might be the primary heating agent of these magnetic structures (Vigeesh *et al.* 2009; van Ballegooijen *et al.* 2011). Finally, the role of neutrals within the magnetic chromosphere is starting to be addressed in a variety of studies (Fontenla *et al.* 2008; Arber *et al.* 2009; Krasnoselskikh *et al.* 2010; Zaqarashvili *et al.* 2011; Song & Vasyliūnas 2011) but results are still very preliminary at this moment.

Spicules are another recurrent chromospheric feature occurring at the borders of network and plage, but of a very different nature. Indeed, much attention has been given in the last few years to the so-called "type II spicules", thin and very dynamic features, almost vertical, reaching up to 10 Mm and rapidly swaying, probably from Alfvén waves. They have been observed in detail off the solar limb with the Hinode Ca II H filter (de Pontieu *et al.* 2007; De Pontieu *et al.* 2007), and identified on the disk in the blue wings of Ca II 854.2 nm and Hα (Langangen *et al.* 2008b; Rouppe van der Voort *et al.* 2009). The formation of these ubiquitous, jet-like, type II spicules has been attributed to magnetic reconnection (de Pontieu *et al.* 2007). This hypothesis might be validated by the occurrence of a spicule-II-like event in the MHD simulation of Martínez-Sykora *et al.* (2011), deriving from the creation of a tangential field discontinuity in the chromosphere after small-scale flux emergence. Observations of dynamical features with similar characteristics in coronal EUV lines, presenting spatio-temporal correlation with the chromospheric signatures (De Pontieu *et al.* 2009; Tian *et al.* 2011), have made type II spicules a prime candidate for establishing a link between corona and chromosphere. Recently, De Pontieu *et al.* (2011) traced type II spicules directly from chromospheric jets to coronal heating events as shown in UV SDO images; estimates of the mass and energy flux density into the corona are compatible with the amount required to sustain the energy lost from the active-region corona. An alternate view on these thin structures has been recently proposed by Judge *et al.* (2011), which suggest that at least some of them correspond not to flux-tube aligned flows, but to warps in two-dimensional sheet-like structures, perhaps related to the magnetic tangential discontinuities naturally arising in low-β plasma conditions. An important consequences of this suggestion is the drastic reduction of the mass flux into the corona, as some of the signatures can be interpreted in terms of phase speeds of warps and Alfvén speeds rather than in terms of real flows.

The direct measurement of magnetic fields in the chromosphere is a long-standing goal of solar physics, as it could provide a direct constraint on the magnetic free energy available in the corona, and in general more reliable boundary conditions for force-free extrapolations. However, it is still a difficult task, as fields are weak and their signature embedded in broad and very dynamic spectral lines. A number of studies on this subject have been conducted in the last few years, with important results. A study on the magnetic field of spicules at the limb, conducted by Centeno *et al.* (2010) exploiting the remarkable polarization properties of the HeI 1083 nm triplet, retrieved field values as high as 50 G. Exploiting imaging spectro-polarimetry in the CaII 854.2 nm line with CRISP, de la Cruz Rodríguez & Socas-Navarro (2011) investigated for the first time whether the chromospheric fibrils, as seen in line core intensity images, always trace the direction of the magnetic field, as is normally assumed. Their answer is mostly positive, with however a few dubious cases that warrant further studies on the subject. Using IBIS, Wöger *et al.* (2009) were able to derive the three-dimensional topology of a strong magnetic structure extending from photosphere to chromosphere. While the magnetic structure became weaker with height, it developed a filament-like structure at larger heights, hinting at the actual chromospheric canopy. These authors remark that currently available data still prevents, in most cases, the detection of weak field strengths, especially at the temporal resolution implied by the intensity variations seen in chromospheric diagnostics. However, as remarked by Judge *et al.* (2010), the use of high cadence, high angular resolution images of fibrils in Ca II and Hα can much improve the chromospheric magnetic field constraints, under conditions of high electrical conductivity and hence field-aligned flows. Such work is possible only with time series data sets from two-dimensional spectroscopic instruments, under conditions of good seeing. Finally, new instruments like the 4-slits Facility Infrared Spectropolarimeter (FIRS, Jaeggli *et al.* 2010) are also now available to achieve diffraction-limited precision spectropolarimetry in the chromosphere by performing fast spectrographic scans of active regions in the HeI 1083 nm triplet (e.g. Schad & Penn 2011).

5. Activities for the 2012 General Assembly

Commission 12 actively promoted and endorsed submission of several proposals for solar-related meetings to be held during the upcoming IAU General Assembly of August 2012. Two such proposals were selected: a full Symposium on Astrophysical Dynamos, and a Special Session on Large Solar Telescopes (see below). Together with a Joint Discussion on 3-D views of Solar and Stellar Activity, and a Special Session on Star-Planet Relation & Public Outreach (chairs L. van Driel-Gesztelyi and J.L. Bougeret, respectively) these will make 2012 a record year for solar events at a General Assembly, a very important venue to present our activities to the astronomical community at large.

5.1. *IAU Symposium 294 "Solar and astrophysical dynamos and magnetic activity"; chair A. Kosovichev*

The goal of this symposium is to discuss the most important results of recent studies of the cosmic dynamo processes: the origin and evolution of magnetic fields in various astrophysical objects from planets, to stars and galaxies, solar and stellar activity cycles, advances in dynamo theories and numerical simulations, similarities and differences between the solar and stellar activity of different scales, driving mechanisms and triggers of solar and stellar magnetic relaxation phenomena, connections between the dynamo mechanisms in various objects, and other hot topics related to the solar and astrophysical dynamos.

The symposium will overview the state of our understanding of dynamo mechanisms in different astrophysical conditions, discuss new observational results, theoretical models, similarities and differences of the physical processes leading to magnetic field generation and formation of magnetic structures. It will focus on the link between theory and observation, and identify critical problems for future observations and modeling.

The symposium will bring together observers and theorists, and encourage discussions and co-operations among solar, stellar, planetary and galactic astronomers. It will help in the development of new ideas regarding the fundamental dynamo processes, and in understanding links between these processes and magnetic activity on various cosmic scales.

5.2. *IAU Special Session 6 "Science with large solar telescopes"; chair G. Cauzzi*

Efficient high order adaptive optics systems (Rimmele 2004), and technological developments proving the feasibility of air-cooled, open solar telescopes have allowed for the planning of facilities with apertures sensibly larger than the existing evacuated solar telescopes. Indeed, several of these innovative projects have become operational in the last few years: the balloon-borne 1-m telescope SUNRISE had a successful first flight in June 2009 (Solanki *et al.* 2010); the 1.6 m off-axis New Solar Telescope at Big Bear has been commissioned in 2010 and is producing first scientific results (Goode *et al.* 2010, e.g.); the German 1.5 m on-axis telescope GREGOR on Tenerife has recently achieved first light. Their large collecting area and imaging stability provide for the increased spatio-temporal resolution and spectro-polarimetric sensistivity necessary to address critical scientific questions such as the magneto-convective nature of sunspots' penumbrae and umbrae (Scharmer *et al.* 2011; Ortiz *et al.* 2010); the distribution of weak magnetic fields in the quiet photosphere (Danilovic *et al.* 2010; Abramenko *et al.* 2010) or the role of chromospheric heating events in replenishing the corona (Langangen *et al.* 2008b; Rouppe van der Voort *et al.* 2009).

The next generation 4-m Advanced Technology Solar Telescope has been funded in 2010 by the USA National Science Foundation for construction on Haleakala, and is currently scheduled for first light in 2017 (Rimmele *et al.* 2011). It will become the most powerful solar telescope in operation and a leading facility for studying the dynamics and magnetism of the solar atmosphere, with a unique emphasis on coronal conditions. On this wake, many other projects for large optical telescopes are also in various stages of their design, e.g. the 1.5 m coronograph COSMO optimized for measurements of the coronal magnetic field; the 2-m Indian National Large Solar Telescope; the Chinese 1-m Space Solar Telescope; the 4-m European Solar Telescope; the Japanese-led project for a space-based 1.5 m optical telescope aboard Solar-C; the Chinese 8-m Giant Solar Telescope.

New challenges will accompany operations of these advanced facilities, including innovative solutions to efficiently extract accurate scientific results from the expected enormous data volume (of the order of hundreds of TB/day); the development of robust data reduction pipelines to provide science-ready data to a user base sensibly larger than the current one; the adoption of more efficient modes of operation, i.e. scheduling observations on a flexible basis in order to best match science programs to observing conditions. Discussion on these issues within the community is just starting, but in the next few years we expect that the operation and scientific results of the new facilities, as well as instrumental upgrades in existing telescopes (including developments of Multi-Conjugate AO or coronal polarimetry) will yield much novel insight into the peculiarities and possibilities of observations with large solar telescopes.

The Special Session 6 during the next IAU General Assembly in August 2012 in Beijing has thus been proposed as a timely and focused opportunity for discussing such topics. The solar physics community, and members of Commission 12 in particular, will be able to voice and address critical issues for the development of future facilities that will be at the forefront of solar astrophysics in the next decades.

A. Kosovichev
president of the Commission

References

Abbett, W. P. 2007, *ApJ*, 665, 1469

Abramenko, V., Yurchyshyn, V., Goode, P., & Kilcik, A. 2010, *ApJ*, 725, L101

Arber, T. D., Botha, G. J. J., & Brady, C. S. 2009, *ApJ*, 705, 1183

Asplund, M., Grevesse, N., & Sauval, A. J. 2005, in Astronomical Society of the Pacific Conference Series, Vol. 336, Cosmic Abundances as Records of Stellar Evolution and Nucleosynthesis, ed. T. G. Barnes III & F. N. Bash, 25

Asplund, M., Grevesse, N., Sauval, A. J., & Scott, P. 2009, *ARA&A*, 47, 481

Ball, W. T., Unruh, Y. C., Krivova, N. A., Solanki, S., & Harder, J. W. 2011, *A&A*, 530, A71

Basu, S. 2009, in Astronomical Society of the Pacific Conference Series, Vol. 416, Solar-Stellar Dynamos as Revealed by Helio- and Asteroseismology: GONG 2008/SOHO 21, ed. M. Dikpati, T. Arentoft, I. González Hernández, C. Lindsey, & F. Hill, 193

Birch, A. C., Parchevsky, K. V., Braun, D. C., & Kosovichev, A. G. 2011, *Sol. Phys.*, 272, 11

Caffau, E., Ludwig, H.-G., Steffen, M., Freytag, B., & Bonifacio, P. 2011, *Sol. Phys.*, 268, 255

Cameron, R. H., Gizon, L., Schunker, H., & Pietarila, A. 2011, *Sol. Phys.*, 268, 293

Carlsson, M., Hansteen, V. H., & Gudiksen, B. V. 2010, *Mem. Soc. Astron. Italiana*, 81, 582

Cauzzi, G., Reardon, K., Rutten, R. J., Tritschler, A., & Uitenbroek, H. 2009, *A&A*, 503, 577

Cauzzi, G., Reardon, K. P., Uitenbroek, H., Cavallini, F., Falchi, A., Falciani, R., Janssen, K., Rimmele, T., Vecchio, A., & Wöger, F. 2008, *A&A*, 480, 515

Cavallini, F. 2006, *Sol. Phys.*, 236, 415

Centeno, R., Trujillo Bueno, J., & Asensio Ramos, A. 2010, *ApJ*, 708, 1579

Chou, D.-Y., Yang, M.-H., Zhao, H., Liang, Z.-C., & Sun, M.-T. 2009, *ApJ*, 706, 909

Couvidat, S., Zhao, J., Birch, A. C., Kosovichev, A. G., Duvall, T. L., Parchevsky, K., & Scherrer, P. H. 2010, *Sol. Phys.*, 260

Danilovic, S., Beeck, B., Pietarila, A., Schüssler, M., Solanki, S. K., Martínez Pillet, V., Bonet, J. A., del Toro Iniesta, J. C., Domingo, V., Barthol, P., Berkefeld, T., Gandorfer, A., Knölker, M., Schmidt, W., & Title, A. M. 2010, *ApJ*, 723, L149

de la Cruz Rodríguez, J. & Socas-Navarro, H. 2011, *A&A*, 527, L8

De Pontieu, B., Hansteen, V. H., Rouppe van der Voort, L., van Noort, M., & Carlsson, M. 2007, *ApJ*, 655, 624

de Pontieu, B., McIntosh, S., Hansteen, V. H., Carlsson, M., Schrijver, C. J., Tarbell, T. D., Title, A. M., Shine, R. A., Suematsu, Y., Tsuneta, S., Katsukawa, Y., Ichimoto, K., Shimizu, T., & Nagata, S. 2007, *PASJ*, 59, 655

De Pontieu, B., McIntosh, S. W., Carlsson, M., Hansteen, V. H., Tarbell, T. D., Boerner, P., Martinez-Sykora, J., Schrijver, C. J., & Title, A. M. 2011, *Science*, 331, 55

De Pontieu, B., McIntosh, S. W., Carlsson, M., Hansteen, V. H., Tarbell, T. D., Schrijver, C. J., Title, A. M., Shine, R. A., Tsuneta, S., Katsukawa, Y., Ichimoto, K., Suematsu, Y., Shimizu, T., & Nagata, S. 2007, *Science*, 318, 1574

De Pontieu, B., McIntosh, S. W., Hansteen, V. H., & Schrijver, C. J. 2009, *ApJ*, 701, L1

del Toro Iniesta, J. C. & Martinez Pillet, V. 2010, ArXiv e-prints, 1010.0504

Fang, F., Manchester, W., Abbett, W. P., & van der Holst, B. 2010, *ApJ*, 714, 1649

Feulner, G. & Rahmstorf, S. 2010, *Geophys. Res. Lett*, 370, 5707

Fontenla, J. M., Peterson, W. K., & Harder, J. 2008, *A&A*, 480, 839

Fröhlich, C. 2011, *Space Sci. Rev.*, 133

Goode, P. R., Yurchyshyn, V., Cao, W., Abramenko, V., Andic, A., Ahn, K., & Chae, J. 2010, *ApJ*, 714, L31

Grevesse, N. & Sauval, A. J. 1998, *Space Sci. Rev.*, 85, 161

Gudiksen, B. V., Carlsson, M., Hansteen, V. H., Hayek, W., Leenaarts, J., & Martínez-Sykora, J. 2011, *A&A*, 531, A154

Hanasoge, S. M., Duvall, Jr., T. L., & DeRosa, M. L. 2010, *ApJ*, 712, L98

Hartlep, T., Kosovichev, A. G., Zhao, J., & Mansour, N. N. 2011a, *Sol. Phys.*, 268, 321

Hartlep, T., Roth, M., Doerr, H., Zhao, J., & Kosovichev, A. G. 2011b, in AAS/Solar Physics Division Abstracts #42, 1611

Ilonidis, S. & Zhao, J. 2011, *Sol. Phys.*, 268, 377

Ilonidis, S., Zhao, J., & Kosovichev, A. 2011, *Science*, 333, 993

Jaeggli, S. A., Lin, H., Mickey, D. L., Kuhn, J. R., Hegwer, S. L., Rimmele, T. R., & Penn, M. J. 2010, *Mem. Soc. Astron. Italiana*, 81, 763

Jefferies, S. M., McIntosh, S. W., Armstrong, J. D., Bogdan, T. J., Cacciani, A., & Fleck, B. 2006, *ApJ*, 648, L151

Judge, P. G., Tritschler, A., & Chye Low, B. 2011, *ApJ*, 730, L4

Judge, P. G., Tritschler, A., Uitenbroek, H., Reardon, K., Cauzzi, G., & de Wijn, A. 2010, *ApJ*, 710, 1486

Kato, Y., Steiner, O., Steffen, M., & Suematsu, Y. 2011, *ApJ*, 730, L24

Komm, R., Ferguson, R., Hill, F., Barnes, G., & Leka, K. D. 2011, *Sol. Phys.*, 268, 389

Kosovichev, A. G. 2010, ArXiv e-prints, 1010.4927

Krasnoselskikh, V., Vekstein, G., Hudson, H. S., Bale, S. D., & Abbett, W. P. 2010, *ApJ*, 724, 1542

Langangen, Ø., Carlsson, M., Rouppe van der Voort, L., Hansteen, V., & De Pontieu, B. 2008a, *ApJ*, 673, 1194

Langangen, Ø., De Pontieu, B., Carlsson, M., Hansteen, V. H., Cauzzi, G., & Reardon, K. 2008b, *ApJ*, 679, L167

Lebedev, N. I., Kuznetsov, V. D., Zhugzhda, Y. D., & Boldyrev, S. I. 2011, Solar System Research, 45, 200

Leenaarts, J., Carlsson, M., Hansteen, V., & Gudiksen, B. V. 2011, *A&A*, 530, A124

Lemen, J., Title, A., De Pontieu, B., Schrijver, C., Tarbell, T., Wuelser, J., Golub, L., & Kankelborg, C. 2011, in AAS/Solar Physics Division Abstracts #42, 1512

Martínez-Sykora, J., Hansteen, V., & Moreno-Insertis, F. 2011, *ApJ*, 736, 9

Moradi, H., Baldner, C., Birch, A. C., Braun, D. C., Cameron, R. H., Duvall, T. L., Gizon, L., Haber, D., Hanasoge, S. M., Hindman, B. W., Jackiewicz, J., Khomenko, E., Komm, R., Rajaguru, P., Rempel, M., Roth, M., Schlichenmaier, R., Schunker, H., Spruit, H. C., Strassmeier, K. G., Thompson, M. J., & Zharkov, S. 2010, *Sol. Phys.*, 267, 1

Ortiz, A., Bellot Rubio, L. R., & Rouppe van der Voort, L. 2010, *ApJ*, 713, 1282

Parchevsky, K., Kosovichev, A., Khomenko, E., Olshevsky, V., & Collados, M. 2010a, ArXiv e-prints, 1002.1117

—. 2010b, Highlights of Astronomy, 15, 354

Parchevsky, K. V. & Kosovichev, A. G. 2009, in Astronomical Society of the Pacific Conference Series, Vol. 416, Solar-Stellar Dynamos as Revealed by Helio- and Asteroseismology: GONG 2008/SOHO 21, ed. M. Dikpati, T. Arentoft, I. González Hernández, C. Lindsey, & F. Hill, 61

Phillips, K. J. H., Sylwester, J., Sylwester, B., & Kuznetsov, V. D. 2010, *ApJ*, 711, 179

Pietarila, A., Socas-Navarro, H., & Bogdan, T. 2007, *ApJ*, 670, 885

Reardon, K. P. & Cavallini, F. 2008, *A&A*, 481, 897

Rimmele, T. R. 2004, in Society of Photo-Optical Instrumentation Engineers (SPIE) Conference Series, Vol. 5490, Society of Photo-Optical Instrumentation Engineers (SPIE) Conference Series, ed. D. Bonaccini Calia, B. L. Ellerbroek, & R. Ragazzoni, 34–46

Rimmele, T. R., Keil, S., & Wagner, J., ATST Team. 2011, in AAS/Solar Physics Division Abstracts #42, 801

Rouppe van der Voort, L., Leenaarts, J., de Pontieu, B., Carlsson, M., & Vissers, G. 2009, *ApJ*, 705, 272

Schad, T. A. & Penn, M. J. 2011, in AAS/Solar Physics Division Abstracts #42, 305

Scharmer, G. B. 2006, *A&A*, 447, 1111

Scharmer, G. B., Henriques, V. M. J., Kiselman, D., & de la Cruz Rodríguez, J. 2011, *Science*, 333, 316

Shapiro, A. I., Schmutz, W., Rozanov, E., Schoell, M., Haberreiter, M., Shapiro, A. V., & Nyeki, S. 2011, *A&A*, 529, A67

Solanki, S. K., Barthol, P., Danilovic, S., Feller, A., Gandorfer, A., Hirzberger, J., Riethmüller, T. L., Schüssler, M., Bonet, J. A., Martínez Pillet, V., del Toro Iniesta, J. C., Domingo, V., Palacios, J., Knölker, M., Bello González, N., Berkefeld, T., Franz, M., Schmidt, W., & Title, A. M. 2010, *ApJ*, 723, L127

Song, P. & Vasyliūnas, V. M. 2011, Journal of Geophysical Research (Space Physics), 116, 9104

Stangalini, M., Del Moro, D., Berrilli, F., & Jefferies, S. M. 2011, *A&A*, 534, A65

Steinhilber, F. 2010, *A&A*, 523, A39

Sterling, A. C., Moore, R. L., & DeForest, C. E. 2010, *ApJ*, 714, L1

Sylwester, B., Phillips, K. J. H., Sylwester, J., & Kuznetsov, V. D. 2011, *ApJ*, 738, 49

Tian, H., McIntosh, S. W., & De Pontieu, B. 2011, *ApJ*, 727, L37

van Ballegooijen, A. A., Asgari-Targhi, M., Cranmer, S. R., & DeLuca, E. E. 2011, *ApJ*, 736, 3

Vecchio, A., Cauzzi, G., & Reardon, K. P. 2009, *A&A*, 494, 269

Vigeesh, G., Hasan, S. S., & Steiner, O. 2009, *A&A*, 508, 951

Wedemeyer-Böhm, S., Lagg, A., & Nordlund, Å. 2009, *Space Sci. Rev.*, 144, 317

Woch, J., Solanki, S. K., & Marsch, E. 2008, AGU Fall Meeting Abstracts, A8

Wöger, F., Wedemeyer-Böhm, S., Uitenbroek, H., & Rimmele, T. R. 2009, *ApJ*, 706, 148

Zaqarashvili, T. V., Khodachenko, M. L., & Rucker, H. O. 2011, *A&A*, 529, A82

Zelenyi, L. M., Kuznetsov, V. D., Kotov, Y. D., Petrukovich, A. A., Mogilevsky, M. M., Boyarchuk, K. A., Zastenker, G. N., & Yermolaev, Y. I. 2004, in IAU Symposium, Vol. 223, Multi-Wavelength Investigations of Solar Activity, ed. A. V. Stepanov, E. E. Benevolen-skaya, & A. G. Kosovichev, 573–580

Zhao, J., Couvidat, S., Bogart, R. S., Duvall, Jr., T. L., Kosovichev, A. G., Beck, J. G., & Birch, A. C. 2011a, Journal of Physics Conference Series, 271, 012063

Zhao, J., Couvidat, S., Bogart, R. S., Parchevsky, K. V., Birch, A. C., Duvall, T. L., Beck, J. G., Kosovichev, A. G., & Scherrer, P. H. 2011b, *Sol. Phys.*, 163

Zhao, J., Kosovichev, A. G., & Ilonidis, S. 2011c, *Sol. Phys.*, 268, 429

Zhao, J., Kosovichev, A. G., & Sekii, T. 2010, *ApJ*, 708, 304

Transactions IAU, Volume XXVIIIA
Reports on Astronomy 2009–2012
Ian Corbett, ed.

© International Astronomical Union 2012
doi:10.1017/S1743921312002694

COMMISSION 49

INTERPLANETARY PLASMA AND HELIOSPHERE

PLASMA INTERPLANÉTAIRE ET HÉLIOSPHÈRE

PRESIDENT	Natchimuthuk Gopalswamy
VICE-PRESIDENT	Ingrid Mann
PAST PRESIDENT	Jean-Louise Bougeret
ORGANIZING COMMITTEE	Carine Briand, Rosine Lallement, David Lario, P. K. Manoharan, Kazunari Shibata, David F. Webb

TRIENNIAL REPORT 2009-2012

1. Introduction

Commission 49 (Interplanetary Plasma and Heliosphere) is part of IAU Division II (Sun and Heliosphere). The research topics include large-scale solar disturbances such as coronal mass ejections (CMEs), shocks, and corotating interaction regions (CIRs) propagating into the heliosphere. The disturbances propagate through the solar wind, which essentially defines the heliosphere. The solar disturbances provide large-scale laboratory to study plasma processes over various time and spatial scales, the highest spatial scale being the size of the heliosphere itself ($\sim$ 100 AU). These solar disturbances are related to solar activity in the form of active regions and coronal holes. Solar eruptions are accompanied by particle acceleration and the particles can be hazardous to life on earth in various ways from modifying the ionosphere to damaging space technology and increasing lifetime radiation dosage to astronauts and airplane crew. Particle acceleration in solar eruptions poses fundamental physics questions because the underlying mechanisms are not fully understood. One of important processes is the particle acceleration by shocks, which occurs throughout the heliosphere. The heliosphere has both neutral and ionized material, with interesting interaction between the two components.

The present triennial report covers a very interesting period marked by rapid explosion of research in the heliophysical processes, thanks to the wide range of observing facilities from space and ground. This period witnessed the two important milestones in the effort of sending mission to the Sun: the approval of the Solar Orbiter and Solar Probe Plus. When these missions come online, they will provide ground truth for a number of theories and greatly enhance our understanding of the near-Sun plasma and magnetic field. The triennial period also witnessed the formal conclusion of the International Heliophysical Year, although the efforts are continuing in the form of the International Space Weather Initiative (ISWI) program (www.iswi-secretariat.org). One of the IHY Coordinated Investigations Programs on the Whole Heliospheric Interval evolved into the IAU Symposium 286 on "Comparative Magnetic Minima: Characterizing quiet times in the Sun and stars" coordinated through Divisions II and IV and some of their associated commissions. The meeting was held during 3–7 October 2011 in Mendoza, Argentina. The ISWI program has been deploying instruments to study solar shocks, particles from the Sun and our galaxy, geospace processes, and others in collaboration with the United Nation's Office of Outer Space Affairs. ISWI recently crossed a milestone: ISWI instruments have been deployed in 101 countries. Two Space science summer schools were organized, one in Bahir Dar, Ethiopia (2010) and the other in Tatranska-Lomnica, Slovakia (2011). The Scientific Committee on Solar Terrestrial Physics (SCOSTEP, http://www.yorku.ca/scostep/) launched the second phase of the Climate and Weather of the Sun-Earth System (CAWSES -II). The CAWSES-II program is in full swing attempting to answer four basic questions: What is the solar influence on climate? How will geospace respond to an altered climate? How does short-term solar variability affect the geospace environment? What is the geospace response to variable inputs from the lower atmosphere?

One of the highlights of this present period is the launch of the Solar Dynamics Observatory (SDO) mission in February 2010. The SDO mission provides data of unprecedented quality to make progress in understanding the irradiance, magnetism, and atmospheric dynamics of the Sun, complementing the information provided by other missions such as SOHO, STEREO, and Hinode. Wind, ACE and SOHO continue to provide information on solar disturbances at 1 AU, which can be combined with the near-Sun observations obtained by remote-sensing for understanding the heliospheric propagation of solar disturbances. The STEREO mission was in a unique quadrature configuration during 2010 – 2011 with spacecraft along the Sun-Earth line providing unprecedented 3-D view of coronal mass ejections, which are the most energetic phenomenon in the heliosphere.

Section 2 provides a discussion on the small-scale density structures in the heliosphere. Section 3 describes recent results on dust and dust interactions in the interplanetary medium. Section 4 highlights recent results on solar eruptions and their heliospheric consequences. Information on recent reviews on CMEs and the heliospheric aspects of CMEs are presented in section 5. Use of the interplanetary scintillation technique to study various aspects of the three-dimensional heliosphere is presented in section 6. Section 7 highlights results on the energetic particles from the Sun and galactic cosmic rays. Section 8 presents a summary of recent results from studies on the outer heliosphere.

2. A View of the Heliosphere at Small Scale

Carine Briand
Observatoire de Paris, LESIA/CNRS, Meudon, France
carine.briand@obspm.fr

Ingrid Mann
EISCAT Scientific Association, Kiruna, Sweden
ingrid.mann@eiscat.se

The large scale structures of the interplanetary medium and spatial environment of the planets are deeply constrained by numerous microphysical processes. Langmuir waves are at the origin of the most intense radio emissions observed in the heliosphere. They also efficiently couple with the electron dynamics. For these crucial roles, they have deserved a lot of attention for many years, and in particular in the last years owing to the new measurements allowed by STEREO/WAVES. We present here the most outstanding results in this field obtained during the last four years.

2.1. Role of Density Fluctuations in the Dynamics of Langmuir Waves

The density fluctuations are thought to play an important role in the beam-Langmuir waves dynamics. Density wells result from the turbulent cascade of energy, from large to small scales. For many years, *in situ* electric field measurements in the solar wind and planetary environment (like the terrestrial foreshock) have revealed the presence of many intense, localized Langmuir wavepackets. In the last years new progresses in the understanding of the role of the density fluctuations on the dynamics of the waves have been achieved.

The Intense Langmuir Solitons (ILS) are localized Langmuir wavepackets of a few mV/m. They have been interpreted as eigenmodes of electron density cavities (Ergun *et al.* 2008; Malaspina & Ergun 2008). The cavity length have been estimated to peak at about 5 to 10 km (Malaspina *et al.* 2010a). Two models have been proposed to explain the formation of such wavepackets.

In a first approach, the growth of the Langmuir wave is sufficiently moderate so that the density fluctuations constrain the waves to directly develop as an eigenmode of the cavity. As described by Hess *et al.* (2010), larger and deeper cavities favor the Langmuir wave generation. Following a more recent study (Hess *et al.* 2011), both quasi-planar Langmuir waves and eigenmodes should be present in the solar wind, the latest showing a higher amplitude. The authors proposed an analytic formulation of the size and amplitude distribution of the eigenmode wavepackets over a large range of distance to the Sun, in particular close to it. Future missions like the Solar Orbiter or the Solar Probe Plus should be able to test this model.

A second approach to explain the presence of cavities eigenmodes was developed by Zaslavsky *et al.* (2009). In their model, the growth rate of Langmuir waves is supposed to be strong. The waves would be constrained by the presence of density fluctuations not in the growing phase (as the previous model) but during the coalescence phase, when the waves reach saturation.

Density fluctuations also play a key role in the Earth environment, in particular to explain the Langmuir waves strength distribution in the terrestrial foreshock. Indeed, Malaspina *et al.* (2009) deduced that the maximum foreshock Langmuir field strength falls with distance to the foreshock via a power law. Comparing STEREO/WAVES measurements with theoretical models, they concluded that scattering of Langmuir waves by density fluctuations is mandatory to reproduce such a behavior. Finally, LaBelle *et al.* (2010) also invoke the linear growth of Langmuir waves in an homogeneous medium to explain the highly modulated electric waveforms observed in the Earth polar cusp.

2.2. Langmuir Waves: from Linear to Nonlinear Plasma Physics

For the electromagnetic radiation to serve as a tool for the diagnostic of the interplanetary medium, the local processes at the origin of such radiation must be known with great details. Two lines of thought are currently debated. The first relies on the classical model of Ginzburg & Zheleznyakov where nonlinear waves coupling generated EM emission at twice the local plasma frequency. Observational evidences of such wave coupling have been presented by Henri *et al.* (2009) in the frame of interplanetary Type III radio bursts. Numerical Vlasov simulations have reinforced their observations. Henri *et al.* (2010) have indeed shown that the energy levels of the observed waves are in good agreement with the threshold of the parametric decay for *non monochromatic* waves.

In a recent work Malaspina *et al.* (2010b) proposed an alternative process to the classical model to explain the $2f_p$ radiation observed during Type III radio bursts. Intense localized Langmuir wavepackets trapped in density wells would drive secondary order currents that oscillate at twice the plasma frequency, producing an electromagnetic radiation at this frequency. Such antenna-like mechanism produces electric field amplitude compatible with STEREO/WAVES observations within an order of magnitude.

The nonlinear evolution of waves is usually classified in terms of "weak" and "strong" turbulence. The typical structures of the strong turbulence like the Langmuir solitons are usually thought to be absent of the interplanetary medium where conditions typical of the weak turbulence prevailed. However, Henri et al. (2011b) have demonstrated that the transition toward strong turbulence can also be a consequence of an initial weak turbulence inverse cascade. This result may encourage space physicists to revisit the waveform data in space plasma environments.

Langmuir waves are also observed inside small magnetic depression called magnetic holes. In a recent study Briand et al. (2010) have shown that the polarization of the waves is compatible with Langmuir/z-mode waves. Through combined observations of STEREO and CLUSTER they have also demonstrated that the presence of a pronounced electron strahl is required for the generation of Langmuir waves and that the electron distribution function inside the magnetic hole is more isotropic compare to outside the hole. Thus, important wave-particles interactions must take place in such environment. To go further in the understanding of such processes, kinetic numerical simulations are required.

2.3. Langmuir Waves: an Efficient Tool for Particle Measurements

In the last years, several studies have shown how Langmuir wave analyses can advantageously complete particles instruments measurements to locally infer density fluctuations, electron beam speed and/or electron temperature at high frequencies.

While the power spectrum describing the turbulent cascade is well known along the inertial range, it is still a source of large debate in the dissipation range. In particular, at electron scale the measurements are very difficult. Langmuir waves can give some information about the density fluctuations observed at "high" frequencies, if not in standard quiet solar wind conditions, at least in several other environments like the terrestrial foreshock.

Langmuir waves are very bursty: fluctuation of the order of tens of milliseconds are perfectly observed. In particular the Time Domain Sampler mode of STEREO/WAVES provides for the first time long time duration waveforms (about 130 msec compared to the 17 msec of the WIND/WAVES instrument). Malaspina et al. (2010a) interpret these fluctuations as a signature of local, small scale density variations of the plasma. They thus extend the power spectrum of density fluctuations in the solar wind by an order of magnitude toward the high frequencies compare to former studies.

When the Langmuir wave energy is large enough (typically $\geqslant 10^{-2}$), the pondemorotrice force can explain the formation of electron density fluctuations. Owing to its specific antenna mounting, the TDS mode of STEREO/WAVES allows for the measurement of both strong Langmuir electric field and density fluctuations on short time scale (a few milliseconds). Using STEREO as a density probe, Henri et al. (2011a) have provided the first observational evidence for ponderomotive effects in the solar wind. They have estimated the density fluctuations in the foreshock to be in the range 50 to 500 Debye lengths.

The determination of the beam speed is often difficult due to the reduced time resolution of the instruments. Malaspina et al. (2011) proposed to use the polarization of the Langmuir waves to deduce this velocity. They showed that the polarization of the waves perpendicular to the local magnetic field is more pronounced as the electron beam speed increases. Langmuir/z-mode propagating in a fluctuating density medium may explain such behavior: the small wavenumber associated with a high beam speed can be reduced to even lower values by density fluctuations, increasing the possibility for strong polarization. Thus, the measurement of the Langmuir waves polarization could provide a new method to deduce electron beam speed and density fluctuations.

Among the many questions regarding the interplanetary shocks, one is particularly well addressed by Langmuir wave measurements. It concerns the localization of the Type II emission generation: is this emission generated at the nose of the shock or on the trailing edge? Since Type II radio emissions take their free energy from accelerated electron beams, the detection and localization of Langmuir waves is a good indicator to answer this question. Pulupa et al. (2010) studied the physical conditions for the development of Langmuir waves activity upstream of a shock, providing thus new keys to study the dynamics of the shocks both some an observational point of view than a numerical one.

3. Cosmic Dust in the Heliosphere

The two major cosmic dust components in the heliosphere are the interstellar and interplanetary dust particles. The trajectories of interstellar dust particles that enter the heliosphere are influenced by the heliospheric magnetic field (see Sect. 3.1). Interplanetary dust particles form by fragmentation of comets, asteroids and meteroids, the latter being fragments of the two former parent objects. Heliospheric interactions of the interplanetary dust are recently studied especially for the nanodust among the interplanetary dust (see Sect. 3.2). Cosmic dust in the heliosphere also interacts with electrons, ions and neutrals in the interplanetary medium, but observational results are rare (see Sect. 3.3).

3.1. Interstellar Dust in the Heliosphere

The motion of the Sun and the heliosphere relative to the surrounding interstellar medium and its containing Local Interstellar Cloud (LIC) causes a flux of interstellar dust into the solar system. This LIC dust is the only dust component measurable in the Solar System that was not previously incorporated in larger Solar System objects.

3.1.1. Observations

Interstellar dust in the heliosphere was measured in detail during the Ulysses mission that orbited the Sun in high inclination orbit during almost three revolutions between about 1.3 and 5.4 AU. Krüger et al. (2010) published the final Ulysses dust data covering measurements of the last 3 years and comparing them to previous observations. The dust experiment provided 6719 dust data sets recorded of particles with masses 10^{-19} kg $\leqslant$ m $\leqslant 10^{-10}$ kg, a large fraction being classified as interstellar dust. During the mission Ulysses passed the same latitudes during different phases of the solar cycle and by comparing specific measurement intervals where a large faction of interstellar dust was observed Krüger et al. (2010) found that the impact rate of interstellar grains varied by more than a factor of two. This variation is commonly assumed to be primarily due to the change of the solar cycle. They also find a change in flux direction that was already noted in earlier observations. Krüger & Grün (2009) report, that while until 2004, the measured interstellar dust flow direction was close to the mean apex of the motion of the Sun through the LIC, this seems to be shifted later by approximately 30 degree away from the ecliptic. Detailed model calculations are needed to explain to what extent the shift is due to a shift of the initial flux into the solar system and to what extent it is due to the influence of Lorentz force onto the ISD within the heliosphere. Very recently the plasma wave instruments onboard the two STEREO spacecraft observe impacts of interstellar dust (Zaslavsky *et al.* 2011). The STEREO measurements provide a data set of interstellar dust flux near 1 AU during more than 3 years.

3.1.2. *Model Calculations*

New calculations are made to follow the trajectories of interstellar dust into and around the heliosphere. Slavin *et al.* (2010) calculate the entry of tiny interstellar dust into the heliosphere and find that the inclination of the interstellar magnetic field relative to the inflow direction generates an asymmetric distribution of the larger interstellar dust that crosses the heliopause (with sizes of the order of 0.1 micrometer). Sterken *et al.* (2011) present simulations of the interstellar dust trajectories within the heliosphere. Their model accounts for the influence of solar radiation pressure force and Lorentz force in the interplanetary magnetic field and they present the resulting dust densities, dust fluxes and flux directions. The developed tools can now be used for predicting the fluxes of interstellar dust at spacecraft locations.

3.1.3. *Review Papers*

Recent review papers address several different aspects of interstellar dust in the heliosphere. Krüger & Grün (2009) review the different measurements of interstellar dust with dust experiments on board spacecraft. Mann (2010) reviews the recent studies of interstellar dust in the solar system, discusses the entry of dust into the solar system and the comparison of derived properties to dust models: the current measurements suggests similarities in the composition of dust in the local interstellar cloud and pristine cometary dust; comparing different published meteor studies suggests that as to date there is no clear identification of interstellar meteors. Two other reviews consider the physics of the local interstellar cloud surrounding the heliosphere and the medium at the boundaries of the heliosphere — (Frisch *et al.* 2009, 2011). These two latter works take into account absorption line data that allow constraining the gas distribution surrounding the heliosphere, as well as pick-up ion observations that provide information on the neutral gas entering the heliosphere.

3.2. *Nanodust in the Interplanetary Medium*

Observations by the STEREO mission show for the first time that nanodust also exists widely distributed in the interplanetary medium (Meyer-Vernet *et al.* 2009). The detection of the nanodust is possible, because of its high impact speed onto the spacecraft (Meyer-Vernet *et al.* 2010). The nanodust most likely forms during collisions of larger dust in the inner solar system, is accelerated in the solar wind and observed near Earth orbit when moving outward. Czechowski & Mann (2010) have suggested a possibly scenario to generate the nanodust fluxes that are observed near 1 AU: the nanodust forms by collisional fragmentation of larger dust particles inside 1 AU. Initial velocities are close to that of Keplerian motion. Dust particles with a charge to mass ratio Q/m of the order of 10^{-4}–10^{-5} e/m_p (e=elementary charge, m_p=the proton mass) are either trapped in orbits with perihelia very close to the Sun and destroyed by sublimation and sputtering. Or they are ejected outward and accelerated to high velocities, of the order of 300 km/s. The charge to mass ratio of 10^{-4}–10^{-5} e/m_p for which acceleration is effective corresponds to dust with radii 3–10 nm (if the charging of larger dust can be extrapolated to the nanometric size). The trajectory calculations can explain the acceleration of the nanodust, while it is still open what causes the flux variations. While an error occurred in the published flux estimate (Czechowski & Mann 2011) the observed and estimated fluxes are within an acceptable range given the large uncertainties of the parameters that enter the problem. STEREO has now observed nanodust during 3 years and at the same time has also observed larger dust particles (Zaslavsky *et al.* 2011).

Interactions with the solar wind are also observed for the streams of nanodust that are ejected from the magnetospheres of Jupiter and Saturn. Hsu *et al.* (2010) studied the Jupiter and Saturn streams based on Cassini measurements and found that the detection patterns of the stream particles are correlated with the interplanetary magnetic field changing the stream direction and the strength of the stream. Similarly Flandes *et al.* (2011) argue based on a comparison of dust, magnetic field and solar wind measurements onboard Ulysses that the dust streams are affected by variations in the interplanetary magnetic field.

The dust formation by collisional fragmentation is a common process in the interplanetary dust cloud as well as in the interplanetary medium. The nanodust has different physical properties compared to larger particles. Extrapolating the collisional fragmentation laws to small-sized fragments therefore has a lower limit. This small size limit is not determined yet. The formation and observation of nanodust in the solar system is the content of a book with 9 different contributions, to be published by Springer in 2012 (Mann *et al.* 2012).

3.3. *Dust Interactions in the Interplanetary Medium*

While the dust is clearly influenced by the presence of the solar wind, current observations show no evidence for the influence of the dust particles on the solar wind on large scales. An exception are possibly the inner source pick-up ions that are observed in the solar wind with Ulysses (see Gloeckler *et al.* 2010 for the most recent results). Dust interactions with the solar wind produce neutral gas or ions in low charge states and dust destruction by mutual collisions, sublimation and sputtering provides a source of electrons and ions. Quantitative discussion of these processes (Mann *et al.* 2010b, 2010a) has shown that the inner source pick-up ions are possibly generated by dust, while other interactions are so far not clearly confirmed by observations (Mann *et al.* 2011). Recent studies have considered the influence that the presence of dust has on the solar wind. Russell *et al.* (2010) claims, for instance that the presence of dust particles causes field enhancements in the solar wind, however without providing a detailed description of the mechanism.

References

Briand, C., Soucek, J., Henri, P., & Mangeney, A. 2010, *J. Geophys. Res.*, 115, A12113
Czechowski, A. & Mann, I. 2010, *ApJ*, 714, 89
Czechowski, A., & Mann, I. 2011 *ApJ* 732, 127
Ergun, R. E., Malaspina, D. M., & Cairns, I. H., *et al.* 2008, *Phys. Rev. Lett.*, 101(5), 051101
Flandes, A., Krüger, H., & Hamilton, D. P., *et al.* 2011, *Planet. Space Sci.*, 59, 1455
Frisch, P. C., Bzowski, M., & Grün, E., *et al.* 2009, *Space Sci. Rev.*, 146, 235
Frisch, P. C., Redfield, S., & Slavin, J. D. 2011 *Annu. Rev. Astron. Astr.*, 49, 237
Gloeckler, G., Fisk, L. A., & Geiss, J. 2010 *Twelfth Int. Solar Wind Conf.*, 1216, 514
Henri, P., Meyer-Vernet, N., & Briand, C., Donato,S. 2011a, *Phys. Plasmas*, 18, 082308
Henri, P., Califano, F., Briand, C., & Mangeney, A. 2011b, *European Physics Letter*, in press
Henri, P., Califano, F., Briand, C., & Mangeney, A. 2010, *J. Geophys. Res.*, 115, A06106
Henri, P., Briand, C., & Mangeney, A., *et al.* 2010, *J. Geophys. Res.*, 114, A03103
Hess, S. L. G., Malaspina, D. M., & Ergun, R. E. 2011, *J. Geophys. Res.*, 116, A07104
Hess, S. L. G., Malaspina, D. M., & Ergun, R. E. 2010, *J. Geophys. Res.*, 115, A10103
Hsu, H.-W., Kempf, S., & Postberg, F., *et al.* 2010, *Twelfth Int. Solar Wind Conf.*, 1216, 510
Krüger, H., Dikarev, V., & Anweiler, B., *et al.* 2010, *Planet. Space Sci.*, 58, 951
Krüger, H. & Grün, E. 2009, *Space Sci. Rev.*, 143, 347

LaBelle, J., Cairns, I. H., & Kletzing, C. A. 2010, *J. Geophys. Res.*, 115, A10317

Malaspina, D. M. & Ergun, R. E. 2008, *J. Geophys. Res.*, 113, A12108

Malaspina, D. M., Li, B., Cairns, I. H., Robinson, P. A., Kuncic, Z., & Ergun, R. E. 2009, *J. Geophys. Res.*, 114, A12101

Malaspina, D. M., Kellogg, P. J., Bale, S. D., & Ergun, R. E. 2010, *ApJ*, 711, 322

Malaspina, D. M., Cairns, I. H., & Ergun, R. E. 2010, *J. Geophys. Res.*, 115, A01101

Malaspina, D. M., Cairns, I. H., & Ergun, R. E. 2011, *Geophys. Res. Lett.*, 38, L13101

Mann, I. 2010, *Annu. Rev. Astron. Astr.*, 48, 173

Mann, I., Czechowski, A., & Meyer-Vernet, N. 2010a *Twelfth Int. Solar Wind Conf.*, 1216, 491

Mann, I., Czechowski, A., Meyer-Vernet, N., Zaslavsky, A., & Lamy, H. 2010b *Plasma Phys. Contr. F.*, 52, 124012

Mann, I., Pellinen-Wannberg, A., & Murad, E., *et al.* 2011, *Space Sci. Rev.*, doi:10.1007/s11214-011-9762-3 published online

Mann, I., Meyer-Vernet, N., & Czechowski, A. 2012, (Heidelberg: Springer Verlag), submitted

Meyer-Vernet, N., Maksimovic, M., & Czechowski, A., *et al.* 2009, *Sol. Phys.*, 256, 463

Meyer-Vernet, N., Czechowski, A., & Mann, I., *et al.* 2010, *Twelfth Int. Solar Wind Conf.*, 1216, 502

Pulupa, M. P., Bale, S. D., & Kasper, J. C. 2010, *J. Geophys. Res.*, 115, A04106

Russell, C. T., Jian, L. K., Lai, H. R., Zhang, T. L., Wennmacher, A., & Luhmann, J. G. 2010, *Twelfth Int. Solar Wind Conf.*, 1216, 522

Slavin, J. D., Frisch, P. C., & Heerikhuisen, J., *et al.* 2010, *Twelfth Int. Solar Wind Conf.*, 1216, 497

Sterken, V. J., Altobelli, N., Kempf, S., Krüger, H., Grün, E., Srama, R., & Schwehm, G. 2010, *A&A*, in press

Zaslavsky, A., Volokitin, A. S., Krasnoselskikh, V. V., Maksimovic, M., & Bale, S. D. 2009, *J. Geophys. Res.*, 115, 108103

Zaslavsky, A., Meyer-Vernet, N., & Mann, I., *et al.* 2011, *Planet. Space Sci.*, submitted

4. Solar Eruptions and Their Interplanetary Manifestation

Nat Gopalswamy

NASA/GSFC, Greenbelt, MD, USA

nat.gopalswamy@nasa.gov

4.1. *Introduction*

Solar eruptions provide important tools to probe the interplanetary medium. Of particular importance are the type III and type II bursts because these bursts are produced by electron beams and shock waves, respectively propagating through the interplanetary medium. Type III bursts can occur with and without an associated CME. The radio dynamic spectra provide information on the speed of the electron beam and the shock if we have independent information on the density variation in the corona and IP medium. Analyzing the radio dynamic spectrum in conjunction with the eruption information from coronagraphic observations, one can derive the large-scale structure of the interplanetary magnetic field as well as its strength.

4.2. *Type III Bursts*

Cairns *et al.* (2009) presented a method for extracting the density profile of the corona from the time-varying frequencies of type III radio bursts in the frequency range 40–180 MHz. They found that wind-like regions (density falling off as the square of the heliocentric distance) occur quite often below ~ 2 Rs. This is different from the typical behavior where a much steeper index derived from eclipse observations. These authors provide a simple physical interpretation involving conical flow from a localized source (e.g., UV funnels observed near the photosphere). Cairns *et al.* (2009) were able to demonstrate this using the linear relationship found between the inverse frequency and time known in at much lower frequencies.

Reiner *et al.* (2009) made use of the stereoscopic view provided by the twin STEREO and the Wind spacecraft to make multipoint measurements of type III radio burst sources by three-spacecraft triangulation measurements. Using three-point measurements of the beaming characteristics for two type III radio bursts, these authors found that individual type III bursts exhibit a wide beaming pattern that is approximately beamed along the direction tangent to the Parker spiral magnetic field line at the source location. In another work involving ray tracing calculations, Thejappa & MacDowall (2010) showed that the radio emission from a localized source escapes as direct and reflected waves along different paths and that the reflected waves experience higher attenuation and group delay because they travel longer path lengths in regions of reduced refractive index. These authors were able to discern the direct and reflected components of a type III event observed by the STEREO spacecraft and found that the sources are located between the turning point of the ray and the harmonic plasma layer.

4.3. *Type II Bursts*

While the source of type III bursts has been more or less accepted to be the flare reconnection, the source of type II producing shocks has been controversial. Flare blast waves and CME-driven shocks have been the competing processes, but recent observations seem to indicate a CME origin. One of the major arguments in this direction has been the universal drift rate spectrum of the type II bursts, which has an index around 2 over the entire wavelength range of type II bursts (Gopalswamy *et al.* 2009a). If individual wavelength domains are considered, it was found that the power law index is < 2 at metric wavelengths (corresponding to the inner corona) and > 2 at kilometric wavelengths (far away from the Sun). These deviations can be explained by the increasing in shock speed near the Sun as a part of the eruption process and the declining speed in the interplanetary medium due to the drag force. However, there have been reports of occasional type II bursts associated with slow CMEs (Magdalenić *et al.* 2010; Nindos *et al.* 2011). One explanation is to attribute the type II shock to the impulsive increase of the pressure in the flare region (Magdalenić *et al.* 2010; Nindos *et al.* 2011). The other explanation is to attribute the slow CME association to the variability of Alfven speed in the corona that can vary over a factor of 4 (Gopalswamy *et al.* 2008) because the Alfven speed in the inner corona can be as low as 200 km/s, which is low enough for the slow CMEs to drive a shock. Furthermore, direct observation of shock formation in SDO/AIA images precisely coinciding with the appearance of a type II burst clearly shows the shock overlying the CME at a heliocentric distance of 1.2 solar radii (Gopalswamy *et al.* 2011). Multiple type II bursts observed in some events have also been interpreted as the flare and CME driving two different shocks causing two different type II bursts. However, the two type II bursts can be explained by a single CME (Gopalswamy *et al.* 2009a; Cho *et al.* 2011).

4.4. *EUV Waves*

EUV wave transients associated with CMEs have been studied extensively over the past decades, thanks to the excellent data from SOHO's Extreme-ultraviolet Imaging telescope (EIT), STEREO's EUV Imager (EUVI), and recently SDO/AIA. The physical nature of these so-called "EUV waves" has not been fully understood. It is natural to expect them to be MHD waves and/or shocks depending on the speed of the driving CME. When the EUV waves are associated with type II bursts, the EUV waves are expected to be shocks because the type II bursts are produced by CME-driven shocks. When there is no type II burst, the EUV wave may be a weak shock or simply fast-mode wave. EUV waves have also been interpreted as signature of magnetic reconfiguration due to the expansion of the associated CME into the ambient magnetized plasma. In addition to the association with type II bursts, wave reflection from a coronal hole (Gopalswamy *et al.* 2009b) also supports the wave interpretation. Veronig *et al.* (2010) presented clear evidence from STEREO/EUVI that the wave is a dome-shaped spherical wave surrounding the CME (see also Temmer *et al.* 2011). Chen & Wu (2011) interpret an EUV event using SDO/AIA data as consisting of a fast mode wave followed by a slower disturbance. They identify the slower wave with the EIT wave, while the leading one as the fast mode wave (Moreton wave). Warmuth & Mann (2011) analyzed a set of 176 EUV waves observed by SOHO/ EIT and STEREO/EUVI and found that the waves fall into three classes: 1) initially fast waves ($v \geqslant 320$ km s^{-1}) that show pronounced deceleration, 2) waves with moderate ($v \approx 170-320$ km s^{-1}) and nearly constant speeds, and 3) slow waves ($v \leqslant 130$ km s^{-1}) showing a rather erratic behavior. They concluded that class 1 waves are nonlinear large-amplitude waves or shocks that propagate faster than the ambient fast-mode speed and subsequently slow down due to decreasing amplitude. Class 2 waves are linear waves moving at the local fast-mode speed. Class 3 waves may be disturbances that could be attributed to magnetic reconfiguration. They suggest that a single model cannot explain all the three classes of EUV waves, a conclusion shared by others (Zhukov 2011; Gallagher *et al.* 2011).

4.5. *White-light Signatures of CME-driven Shocks: Four-part CME Structure*

Although there have been attempts to search for white-light features of the shocks in the past (Sheeley *et al.* 2000; Vourlidas *et al.* 2003), the diffuse feature has been recognized as a shock manifestation only recently (Gopalswamy 2009; Gopalswamy *et al.* 2009a; Ontiveros & Vourlidas, 2009; Gopalswamy, 2010; Eselevich & Eselevich, 2011). In coronagraphic difference images, one observes a diffuse feature surrounding the bright feature identified as the CME flux rope. The bright and diffuse structures are sometimes referred to as the main body and whole CME (Michalek *et al.* 2007, Gopalswamy *et al.* 2008; Yashiro *et al.* 2008). The diffuse feature surrounding the flux rope is identified with the compressed plasma known as shock sheath. The shock itself is too thin to be observed in coronagraphic images. The diffuse structure is only observed in relatively fast CMEs. Thus, fast CMEs have an additional shock sheath structure and hence should be referred to as CMEs with a four-part structure. More recently, a shock was identified very close to the Sun, using EUV images obtained by the Atmospheric Imaging Assembly (AIA) on board the Solar Dynamics Observatory (SDO). The eruptive prominence, the flux rope, and the shock sheath can all be tracked within the SDO/AIA field of view (Kozarev *et al.* 2011; Ma *et al.* 2011; Gopalswamy *et al.* 2011).

Identification of the shock sheath surrounding the CME flux rope has opened up new opportunities to derive the shock and ambient medium properties. The shock strength can be measured in terms of the density compression ratio downstream to upstream using brightness jump in the diffuse feature (Ontiveros & Vourlidas 2009; Eselevich & Eselevich 2011; Bemporad & Mancuso 2011). Bemporad & Mancuso (2011) used the compression ratio to infer the Alfvén Mach number at various locations of the shock front. They conclude that the CME-driven shocks could be efficient particle accelerators at the initiation phases of the event, while at later times they progressively loose energy, also losing their capability to accelerate high-energy particles. Gopalswamy & Yashiro (2011) introduced a new technique to measure the coronal magnetic field strength in the heliocentric distance range $6-23$ solar radii using the shock standoff distance and the CME radius of curvature in the SOHO and STEREO coronagraphic images. Assuming the adiabatic index, they determined the Alfvén Mach number, and hence the Alfvén speed in the ambient medium using the measured shock speed. By measuring the upstream plasma density using polarization brightness images, they obtained the magnetic field strength upstream of the shock. The estimated magnetic field decreased from $\sim$ 48 mG around 6 Rs to 8 mG at 23 Rs. The radial profile of the magnetic field can be described by a power law in agreement with other estimates at similar heliocentric distances. This is a new technique, which provides a means of estimating the magnetic field in the heliospheric region that will be probed by the Solar probe Plus. This technique was also applied to the SDO/AIA data for shock-driving CMEs associated with metric type II radio bursts. Obtaining the upstream parameters from the type II band-splitting, Gopalswamy *et al.* (2011) derived the coronal magnetic field very close to the Sun (1.5 Rs) to be $\sim$ 1.3 G. Recently, Maloney & Gallagher (2011) extended the stand-off distance measurement to 120 Rs, using observations from the STEREO Heliospheric Imager. This, coupled with the HELIOS observations from the past, provides an opportunity to estimate the magnetic field strength throughout the inner heliosphere.

4.6. *Variability in CME-driven Shocks at 1 AU*

One way to find out whether a CME drives a shock or not is to look at its radio emission properties (Gopalswamy *et al.* 2010). CME-driven shocks emitting type II radio bursts are known as radio-loud (RL) events as opposed to radio-quiet (RQ) events in which the shocks do not produce the bursts. Starting with the 200+ shocks detected at L1 by one or more of Wind, SOHO, and ACE spacecraft, the associated CMEs were grouped into RL and RQ events. When the CME properties were compared for the two groups, there were significant differences. The CME speeds were very different for the RL and RQ CMEs, while the corresponding IP shock speeds were similar for the two groups. The difference between RL and RQ events seems to be erased as the CMEs propagate into the interplanetary medium because of the momentum exchange between the CMEs and the ambient solar wind. Another significant difference was that the RQ CMEs accelerated on the average, while the RL CMEs decelerated near the Sun (within the coronagraphic field of view). In general, the RL CMEs drove shocks near the Sun, which weakened as they propagated into the IP medium. Among the RL CMEs, some were radio loud only near the Sun, some only near Earth, and others throughout the IP medium. The heliocentric distance range over which the radio emission occurred essentially depended on the CME speed.

Pulupa *et al.* (2010) attempted to identify shock parameters that favor production of upstream Langmuir waves, and hence type II radio bursts. Among the 178 interplanetary shocks observed by the Wind spacecraft, only 43 (or 24%) were found to produce upstream Langmuir waves, as measured by the enhancements in wave power near the plasma frequency. They found that the de Hoffmann-Teller speed is the best indicator of the Langmuir wave production, consistent with the fast Fermi model of electron acceleration. Several other parameters, including the magnetic field strength and the level of solar activity (but not the Mach number) were also found to be correlated with Langmuir wave production. They suggest that additional parameters may be associated with an increased level of shock front curvature or upstream structure, leading to the formation of upstream foreshock regions, or with the generation of an upstream electron population favorable for shock reflection.

Langmuir waves require electron acceleration in the shock, but the energetic storm particle (ESP) events indicate ion acceleration at the shock front. Looking at the properties of the driving CMEs near the Sun, Mäkelä *et al.* (2011) found that CMEs with an ESP-producing shock are faster than those driving shocks without an ESP event and have a larger fraction of halo CMEs (67% versus 38%). The fraction of halo CMEs in a population is indicative of how energetic the population is. It was also found that the Alfvénic Mach numbers of shocks with an ESP event are on average 1.6 times higher than those of shocks without. The ESP events occur more often in radio-loud shocks (RL, shocks producing type II bursts) than in radio-quiet shocks: 52% of RL shocks and only ~33% of RQ shocks produced an ESP event at proton energies above 1.8 MeV; in the keV energy range the ESP frequencies are 80% and 65%, respectively. The interplanetary shocks seem to be organized into a decreasing sequence by the energy content of the CMEs: radio-loud (RL) shocks with an ESP, followed by RL shocks without an ESP event, then radio-quiet (RQ) shocks with and without an ESP event.

4.7. Large-scale CME Deflections

CMEs can interact with other CMEs leading to deflection and/or merging. CMEs can also interact with neighboring streamers and coronal holes. Different processes dominate during different phases of the solar cycle. For example, deflection towards the equator in the solar minimum phase is expected because of the strong global solar field. During the solar maximum phase, CME-CME interaction is expected to be common. During the declining phase, the equatorial coronal holes appear in great abundance, so CME-coronal hole interaction is most common (Gopalswamy et al. 2009b). The CME deflection by coronal holes has also important consequences for space weather: a CME originating close to the disk center may not arrive at 1 AU. For example, shocks are observed at 1 AU without any discernible driver at 1 AU, although the associated CME is clearly observed near the Sun. These "driverless shocks" represent extreme deflection by coronal holes (Gopalswamy et al. 2009b, 2010). The CME deflection essentially makes disk center CMEs behave like limb CMEs. The amount of CME deflection needed for driverless shocks has been estimated to be in the range 20 – 60 degrees, depending on the orientation of the CME axis with respect to the ecliptic plane. For example, a high-inclination flux rope needs to be deflected less in the longitudinal direction compared to a low inclination one because of the smaller east-west extent of the CME. Gui et al. (2011) reported on the change in propagation directions for a set of eight CMEs. They interpreted the deflections as a consequence of the gradient of the magnetic energy density of the background medium. Lugaz et al. (2011) performed numerical simulations to understand the CME propagation in the heliosphere. In addition to the coronal hole deflection discussed above, they also suggested additional deflection caused by a second faster CME.

References

Bemporad, A. & Mancuso, S. 2011, *ApJ*, 739, L64

Cairns, I. H., Lobzin, V. V., Warmuth, A., Li, B., Robinson, P. A., & Mann, G. 2009, *ApJ*, 706, L265

Chen, P. F. & Wu, Y. 2011, *ApJ*, 732, L20

Cho, K.-S., Bong, S.-C., Moon, Y.-J., Shanmugaraju, A., Kwon, R.-Y., & Park, Y. D. 2011, *A&A*, 530, A16

Eselevich , M. V. & Eselevich, V. G. 2011, *Astron. Rep.*, 55(4), 359

Gallagher, P. T. & Long, D. M. 2011, *Space Sci. Rev.*, 158, 365

Gopalswamy, N. 2009, in: T. Tsuda, R. Fujii, K. Shibata, & M. A. Geller (eds.), *Climate and Weather of the Sun-Earth System (CAWSES): Selected Papers from the 2007 Kyoto Symposium*, (Tokyo: Terrapub), p. 77

Gopalswamy, N., Yashiro, S., & Xie, H., et al. 2008, *ApJ*, 674, 560

Gopalswamy, N., Thompson, W. T., & Davila, J. M., et al. 2009a, *Sol. Phys.*, 259, 227

Gopalswamy, N., Mäkelä, P., Xie, H., Akiyama, S., & Yashiro, S. 2009b, *J. Geophys. Res.*, 114, A00A22, doi:10.1029/2008JA013686

Gopalswamy, N. & Yashiro, S. 2011, *ApJ*, 736, L17

Gopalswamy, N., Xie, H., Mäkelä, P., Akiyama, S., Yashiro, S., Kaiser, M. L., Howard, R. A., & Bougeret, J.-L. 2010, *ApJ*, 710, 1111

Gopalswamy, N., Nitta, N., Akiyama, S., Mäkelä, P., & Yashiro, S. 2011, *ApJ*, in press

Gui, B., Shen, C., Wang, Y., Ye, P., Liu, J., Wang, S., & Zhao, X. 2011, *Sol. Phys.*, 271, 111

Kozarev, K. A., Korreck, K. E., Lobzin, V. V., Weber, M. A., & Schwadron, N. A. 2011, *ApJ*, 733, L25

Lugaz, N., Downs, C., Shibata, K., Roussev, I. I., Asai, A., & Gombosi, T. I. 2011, *ApJ*, 738, 127

Ma, S., Raymond, J. C., Golub, L., Lin, J., Chen, H., Grigis, P., Testa, P., & Long, D. 2011, *ApJ*, 738, 160

Magdalenić, J., Marqué, C., Zhukov, A. N., Vršnak, B., & Žic, T. 2010, *ApJ*, 718, 266

Mäkelä, P., Gopalswamy, N., Akiyama, S., Xie, H., & Yashiro, S. 2011, *J. Geophys. Res.*, 116, A08101

Maloney, S. A. & Gallagher, P. T. 2011, *ApJ*, 736, L5

Michalek, G., Gopalswamy, N., & Xie, H. 2007, *Sol. Phys.*, 246, 409

Nindos, A., Alissandrakis, C. E., Hillaris, A., & Preka-Papadema, P. 2011, *A&A*, 531, A31

Ontiveros, V. & Vourlidas, A. 2009, *ApJ*, 693, 267

Pulupa, M. P., Bale, S. D., & Kasper, J. C. 2010, *J. Geophys. Res.*, 115, A04106, doi:10.1029/2009JA014680

Reiner, M. J., Goetz, K., Fainberg, J., Kaiser, M. L., Maksimovic, M., Cecconi, B., Hoang, S., Bale, S. D., & Bougeret, J.-L. 2009, *Sol. Phys.*, 259, 255

Sheeley, N. R., Jr., Hakala, W. N., & Wang, Y.-M. 2000, *J. Geophys. Res.*, 105, 5081

Temmer, M., Veronig, A. M., Gopalswamy, N., & Yashiro, S. 2011, *Sol. Phys.*, Online First, doi:10.1007/s11207-011-9746-1

Thejappa, G. & MacDowall, R. J. 2010, *ApJ*, 720, 1395

Veronig, A. M., Muhr, N., Kienreich, I. W., Temmer, M., & Vršnak, B. 2010, *ApJ*, 716, L57

Vourlidas, A., Wu, S. T., Wang, A. H., Subramanian, P., & Howard, R. A. 2003, *ApJ*, 598, 1392

Warmuth, A. & Mann, G. 2011, *A&A*, 532, A151

Yashiro, S., Michalek, G., Akiyama, S., Gopalswamy, N., & Howard, R. A. 2008, *ApJ*, 673, 1174

Zhukov, A. N. 2011, *J. Atmos. Sol.-Terr. Phy.*, 73, 1096

5. Coronal Mass Ejections and Their Heliospheric Aspects

David Webb

Institute for Scientific Research, Boston College, Newton, MA, USA

david.webb@bc.edu

5.1. Recent Reviews of CMEs

Coronal mass ejections (CMEs) consist of large structures containing plasma and magnetic fields that are expelled from the Sun into the heliosphere. They are of interest for both scientific and technological reasons. Scientifically they are of interest because they are responsible for the removal of built-up magnetic energy and plasma from the solar corona, and technologically they are of interest because they are responsible for major space weather effects at Earth (Baker *et al.* 2009). Most of the ejected material comes from the low corona, although cooler, denser material probably of chromospheric or photospheric origin can be involved. The CME plasma is entrained on an expanding magnetic field, commonly of the form of helical field lines with changing pitch angles, i.e., a flux rope. Observations of Earth-directed CMEs, often observed as halos surrounding occulting coronagraphs, are important for space weather studies. In this section we emphasize results on the heliospheric aspects of CMEs, especially their imaging, during the last triennium.

Until the last decade, images of CMEs had been made near the Sun primarily by coronagraphs on board spacecraft. Coronagraphs view the flow of density structures outward from the Sun in broadband white light by observing Thomson-scattered sunlight from the free electrons in coronal and heliospheric plasma. This emission has an angular dependence which must be accounted for in the measured brightness (e.g., Vourlidas & Howard 2006; Howard & Tappin 2009). CMEs are faint relative to the background corona, but more transient, so some form of background subtraction is typically needed to identify them. The first spacecraft coronagraph observations of CMEs were made by the OSO-7 coronagraph in the early 1970s, followed by better quality and longer periods of CME observations using Skylab (1973 – 1974), P78-1 (Solwind; 1979 – 1985), and SMM (1980, 1984 – 1989). In late 1995, SOHO was launched and two of its three LASCO coronagraphs still operate today (Brueckner et al. 1995; Gopalswamy et al. 2009, 2010). Finally late in 2006, LASCO was joined by observations from the STEREO CORs (Howard et al. 2008; Russell, ed. 2008). These satellite observations have been complemented by white light data from the ground-based Mauna Loa Solar Observatory (MLSO) K-coronameters, currently the MK4 version viewing from $1.2 - 2.9 R_\odot$ (Fisher et al. 1981).

There were several excellent reviews of solar eruptive phenomena and CMEs published just before this triennium, including Gopalswamy et al. (2006), Kahler (2006), Aschwanden (2006). In addition, there are several Living Reviews of Solar Physics articles (http://solarphysics.livingreviews.org/), including "Space Weather: The Solar Perspective" (Schwenn 2006), and two, "Solar Eruptive Phenomena" (Webb & Howard 2011), and "Coronal Mass Ejections: Models and Their Observational Basis" (Chen 2011), published very recently. In addition, other published or planned LRSP articles include on prominences, flares, space weather and other related phenomena. An introductory text on CMEs (Howard 2011) has also recently been published. Finally, analyses of the CMEs observed during the Whole Heliosphere Interval (WHI) international campaign in 2008 were recently published: Cremades et al. (2011) and Webb et al. (2011).

5.2. Heliospheric Aspects of CMEs

Several decades ago, interplanetary transients were observed at larger distances from the Sun than viewed by coronagraphs using interplanetary radio scintillation (1964 – present; Hewish et al. 1964; Vlasov 1981) and from the zodiacal light photometers on the twin Helios spacecraft (1975 – 1983; Jackson 1985). The Helios photometers observed regions in the inner heliosphere from 0.3 – 1.0 AU but with a very limited field of view (FoV). The new millennium saw the arrival of a new class of detector, the heliospheric imager, with the Solar Mass Ejection Imager (SMEI) launched on board the Coriolis spacecraft early in 2003 and the Heliospheric Imagers (HIs) launched on the twin STEREO spacecraft in late 2006. LASCO has detected well over 10^4 CMEs during its lifetime (Gopalswamy et al. 2009; http://cdaw.gsfc.nasa.gov/CME_list/), SMEI has observed over 360 transients (Webb et al. 2006; Howard & Simnett 2008), and the number of "events" reported using the HIs is well over 500 (http://www.sstd.rl.ac.uk/stereo/Events Page.html), despite their operation during the least active Sun during the space age.

CMEs carry into the heliosphere large amounts of coronal magnetic fields and plasma, which can be detected by remote sensing and in-situ spacecraft observations. Here they have been called interplanetary CMEs or ICMEs. The term ICME was originally devised as a means to separate the phenomena observed far from the Sun (e.g., by in-situ spacecraft) and those near the Sun (e.g., by coronagraphs). However, in the STEREO era, where CMEs can now be tracked continuously from the Sun to 1 AU and beyond, the term has become largely redundant. Consequently, in a recent workshop on remote sensing of the heliosphere in Wales (http://heliosphere2011.dph.aber.ac.uk/) it was decided to no longer use the term ICME.

The passage of CME material past a single spacecraft is marked by distinctive signatures, but with a great degree of variation from event to event (e.g., Zurbuchen & Richardson 2006). These signatures include transient interplanetary shocks, depressed proton temperatures, cosmic ray depressions, flows with enhanced helium abundances, and unusual compositions of ions and elements. Often observed in in-situ data are highly structured magnetic field configurations corresponding to the arrival of a CME. The field assumes the structure of a helix and is accompanied by strong magnetic fields with low field variance, low plasma beta and low temperature. Such structures were called magnetic clouds by Burlaga et al. (1981). Such a structure is often modeled as a flux rope, having a series of helical field lines like the coils of a spring with pitch angles increasing toward the outer edge. Since many if not all CMEs are now considered to contain flux ropes, it is logical to expect magnetic clouds to form the core of CMEs. Models have been developed for the force free and non-force free states, the latter also known as the Grad–Shafranov technique. See recent Living Reviews of Solar Physics articles by Chen (2011) and Webb & Howard (2011) for detailed discussions of CME models. Around 30% to 50% of CMEs observed in-situ show a clear signature of a magnetic cloud. It remains unknown whether the remainder does not show the signature because the imbedded flux rope is less structured, is absent, or whether the spacecraft did not pass through the flux rope component (i.e., skirted its flank; e.g., Möstl 2010). The in-situ signatures of CMEs are well described in several recent reviews (Schwenn 2006; Zurbuchen & Richardson 2006; Richardson & Cane 2010).

Several techniques have been developed to remotely detect and track disturbances related to CMEs in the interplanetary medium (Jackson 1992; Jackson et al. 2011) and see http://heliosphere2011.dph.aber.ac.uk/. These have utilized radio and white light wavelengths to detect and image these structures. The techniques are kilometric radio observations from space and interplanetary scintillation (IPS) observations from the ground. The kilometric observations can track the emission typically from strong shocks traveling ahead of fast CMEs. Such instruments have been flown on the ISEE-3 and Ulysses spacecraft and are currently on board Wind and STEREO (Bougeret et al. 2008).

The IPS technique relies on measurements of the fluctuating intensity level of a large number of point-like distant meter-wavelength radio sources. They are observed with one or more ground arrays operating in the MHz range. IPS arrays detect changes to density in the (local) interplanetary medium moving across the line of sight to the source. Disturbances are detected by either an enhancement of the scintillation level and/or an increase in velocity. When built up over a large number of radio sources a map of the density enhancement across the sky can be produced. The technique suffers from relatively poor temporal (24 hour) resolution and has a spatial resolution limited to the field of view of the radio telescope. For example, high-latitude arrays such as the long-deactivated 3.5 ha array near Cambridge in the UK could not observe sources in the mid-high-latitude southern hemisphere. Scattering efficiency also poses a limitation on IPS measurements as increasing the frequency at which to measure the sources allows an observer to detect disturbances closer to the Sun. Higher frequencies means fewer sources, however, so the spatial resolution is effectively decreased. Finally ionospheric noise limits viewing near the Sun and near the horizon, and a model-dependence for interpreting the signal as density or mass. Workers have, however, been working with these difficulties for almost 50 years and a number of techniques have evolved to extract reliable CME measurements using IPS. Recent papers involving such measurements include Bisi *et al.* (2008), Jackson *et al.* (2011), Tappin & Howard (2010) and Manoharan (2010).

Today's heliospheric imagers are the successors to the zodiacal-light photometers (Leinert *et al.* 1975) on the twin Helios spacecraft flown in solar orbits in the 1970s and early 1980s. SMEI, in particular, was designed to exploit the heliospheric remote sensing capability demonstrated by that instrument (Jackson 1985; Webb & Jackson 1990). Unlike Helios, which could only observe a few narrow strips across the sky, this new generation of imager could observe large areas simultaneously. SMEI was the first, developed as a proof-of-concept U.S. Air Force experiment for operational forecasting. Launched in January 2003 on the Coriolis spacecraft, SMEI images nearly the entire sky in white light once per 102 minute spacecraft orbit, using three baffled camera systems with CCD detectors. Individual frames are mapped into ecliptic coordinates to produce a nearly complete sky map. SMEI has observed over 360 CMEs, many of which were Earth-directed allowing the comparison with in-situ spacecraft and prediction of arrival times and speeds. Unlike with in-situ spacecraft, however, SMEI enables the comparison with coronagraph events in any direction, enabling large-scale tracking and 3D reconstruction.

SMEI has been used for CME tracking, (e.g., Howard *et al.* 2007), space weather forecasting (Webb *et al.* 2009; Howard & Tappin 2010) and 3D reconstruction (Jackson *et al.* 2011; Tappin & Howard 2009). SMEI observations have been compared with coronagraph and in-situ spacecraft measurements (Howard *et al.* 2007; Howard & Simnett 2008; Webb *et al.* 2009) and with IPS observations (Jackson *et al.* 2008; Bisi *et al.* 2008). While it observes the entire sky beyond 20° elongation, its field of view is often obscured by energetic particle saturation during its passage through the magnetospheric polar caps and the South Atlantic Anomaly, and by hot pixel degradation.

In late 2006 the twin STEREO spacecraft were launched (Russell, ed. 2008) carrying the Heliospheric Imagers (HIs) (Howard *et al.* 2008; Eyles *et al.* 2009). The HIs view the inner heliosphere starting at an elongation of 4° from the Sun. HI-1 has a FoV of 20°, from 4–24° elongation ($\sim 12-85\,R_\odot$), and HI-2 of 70°, from ~ 19–89° elongation ($\sim 68-216\,R_\odot$). There is a 5.3° overlap between the outer HI-1 and inner HI-2 FoVs. The HIs do not cover the entire position angle (PA) range around the Sun, but observe up to a 90° range in PA, usually centered on the ecliptic and viewing either east (HI-A) or west (HI-B) of the Sun. They do not suffer the same problems with particle saturation as SMEI, but are constrained by their fields of view about the ecliptic plane. Combined with the coronagraphs, the HIs do provide for the first time a continuous view from the Sun to around 1 AU and the stereoscopic viewpoints enable the possibility for 3D reconstruction using the coronagraphs and HI-1.

The STEREO spacecraft share similar ~ 1 AU orbits about the Sun as the Earth but separate from the Sun-Earth line by 22.5° per year. STEREO-A (Ahead) leads the Earth in its orbit, while STEREO-B (Behind) lags. Most of the work involving the STEREO-HIs and CMEs to date have focused on their detection and tracking, and comparison with in-situ spacecraft. Publications include Harrison *et al.* (2008), Davies *et al.* (2009), Mierla *et al.* (2010), Möstl (2010), Davis *et al.* (2011), Kilpua *et al.* (2011), Liewer *et al.* (2011) and DeForest *et al.* (2011).

The important difference between heliospheric imagers and coronagraphs is that 3D information is available in heliospheric imagers that is not available in coronagraphs. This is because the assumptions imposed on coronagraphs (Thomson scattering assumptions, low angles) break down at large elongations and across large distances. This increases the difficulty of the analysis, but makes available additional information on the structure and kinematics of the CME. This thereby removes the need for auxiliary data to provide this information. The theory describing this ability is developed by Howard & Tappin

(2009). Recently papers have been published that consider the 3D structure of the CME, including Wood & Howard (2009), Lugaz *et al.* (2009, 2010) and Howard & Tappin (2009, 2010). Techniques involving the extraction of 3D properties from heliospheric image data are reviewed by Howard (2011).

References

Aschwanden, M. J. 2006, *Physics of the Solar Corona*, (Chichester: Springer-Verlag and Praxis Publ.), Chap. 17

Baker, D. N., Balstad, R., & Bodeau, J. M., *et al.* 2009, *Severe Space Weather Events — Understanding Societal and Economic Impacts: A Workshop Report*, (Washington DC: The National Academies Press)

Bisi, M. M., Jackson, B. V., & Hick, P. P., *et al.* 2008, *J. Geophys. Res.*, 113, A00A11, doi:10.1029/2008JA013222

Bougeret, J.-L. *et al.* 2008, *Space Sci. Rev.*, 136, 487

Brueckner, G. E., Howard, R. A., & Koomen, M. J., *et al.* 1995, *Sol. Phys.*, 162, 357

Burlaga, L., Sittler, E., Mariani, F., & Schwenn, R. 1981, *J. Geophys. Res.*, 86, 6673

Chen, P. F. 2011, *Living Rev. Solar Phys.*, 8, 1

Cremades, H., Mandrini, C. H., & Dasso, S. 2011, *Sol. Phys.*, doi:10.1007/s11207-011-9769-7

Davies, J. A., Harrison, R. A., & Rouillard, A. P., *et al.* 2009, *Geophys. Res. Lett.*, 36, L02102, doi:10.1029/2008GL036182

Davis, C. J., de Koning, C. A., & Davies, J. A., *et al.* 2011, *Space Weather*, 9, S01005, doi:10.1029/2010SW000620

DeForest, C. E., Howard, T. A., & Tappin, S. J. 2011, *ApJ*, 738, 103

Eyles, C. J., Harrison, R. A., & Davis, C. J., *et al.* 2009, *Sol. Phys.*, 254, 387

Fisher, R. R., Lee, R. H., MacQueen, R. M., & Poland, A. I. 1981, *Appl. Op.*, 20, 1094

Gopalswamy, N., Mikić, Z., & Maia, D., *et al.* 2006, *Space Sci. Rev.*, 123, 303

Gopalswamy, N., Yashiro, S., & Michalek, G., *et al.* 2009, *Earth Moon Planets*, 104, 295

Gopalswamy, N., Yashiro, S., & Michalek, G., *et al.* 2010, *Sun and Geosphere*, 5, 7

Harrison, R. A., Davis, C. J., & Eyles, C. J., *et al.* 2008, *Sol. Phys.*, 247, 171

Hewish, A., Scott, P. F., & Wills, D. 1964, *Nature*, 203, 1214

Howard, R. A., et al. 2008, *Space Sci. Rev.*, 136, 67

Howard, T. 2011, *Coronal Mass Ejections: An Introduction*, (New York: Springer)

Howard, T. A., Fry, C. D., Johnston, J. C., & Webb, D. F. 2007, *ApJ*, 667, 610

Howard, T. A. & Simnett, G. M. 2008, *J. Geophys. Res.*, 113, A08102, doi:10.1029/2007JA012920

Howard, T. A. & Tappin, S. J. 2009, *Space Sci. Rev.*, 147, 31

Howard, T. A. & Tappin, S. J. 2010, *Space Weather*, 8, S07004, doi:10.1029/2009SW000531

Jackson, B. V. 1985, *Sol. Phys.*, 100, 563

Jackson, B. V. 1992, in: Z. Svestka, B. V. Jackson, & M. E. Machado (eds.), *Eruptive Solar Flares, Lecture Notes in Physics*, 399 (Berlin: Springer-Verlag), p. 248

Jackson, B. V., Hick, P. P., & Buffington, A., *et al.* 2008, *Adv. Geosci.*, 21, 339

Jackson, B. V., Hick, P. P., & Buffington, A., *et al.* 2011, *J. Atmos. Sol.-Terr. Phy.*, 73, 1214

Kahler, S. W 2006, in: N. Gopalswamy, R. Mewaldt, and J. Torsti (eds.), *Solar Eruptions and Energetic Particles*, Geoph. Monog. Series, 165 (Washington D.C.: AGU), p. 21

Kilpua, E. K. J., Jian, L. K., Li, Y., Luhmann, J. G., & Russell, C. T. 2011, *J. Atmos. Sol.-Terr. Phy.*, 73, 1228

Leinert, C., Link, H., Pitz, E., Salm, N., & Knueppelberg, D. 1975, *Raumfahrtforschung*, 19, 264

Liewer, P. C., Hall, J. R., Howard, R. A., DeJong, E. M., Thompson, W. T., & Thernisien, A. 2011, *J. Atmos. Sol.-Terr. Phy.*, 73, 1173

Lugaz, N., Vourlidas, A., & Roussev, I. I. 2009, *Ann. Geophys.*, 27, 3479

Lugaz, N., Hernandez-Charpak, J. N., & Roussev, I. I., *et al.* 2010, *ApJ*, 715, 493

Manoharan, P. K. 2010, *Sol. Phys.*, 265, 137

Mierla, M., Inhester, B., & Antunes, A., *et al.* 2010, *Ann. Geophys.*, 28, 203

Möstl, C., Temmer, M., & Rollett, T., et al. 2010, Geophys. Res. Lett., 37, L24103, doi:10.1029/2010GL045175

Richardson, I. G. & Cane, H. V. 2010, Sol. Phys., 264, 189

Russell, C. T., (ed.) 2008, Space Sci. Rev., 136, issues 1–4

Schwenn, R. 2006, Living Rev. Solar Phys., 3, 2

Tappin, S. J. & Howard, T. A. 2009, Space Sci. Rev., 147, 55

Tappin, S. J. & Howard, T. A. 2010, Sol. Phys., 265, 159

Vlasov, V. I. 1981, Geomag. Aeron., 21, 441

Vourlidas, A. & Howard, R. A. 2006, Astrophys. J., 642, 1216

Webb, D. F. & T. A. Howard 2011, Living Rev. Solar Phys., in press

Webb, D. F., Howard, T. A., & Fry, C. D., et al. 2009, Space Weather, 7, S05002, doi:10.1029/2008SW000409

Webb, D. F. & Jackson, B. V. 1990, J. Geophys. Res., 95, 20641

Webb, D. F., Mizuno, D. R., & Buffington, A., et al. 2006, J. Geophys. Res., 111, A12101, doi:10.1029/2006JA011655

Webb, D. F., Cremades, H., Sterling, A. C., et al., 2011, Sol. Phys., doi:10.1007/s11207-011-9787-5

Wood, B. E. & Howard, R. A. 2009, ApJ, 702, 901

Zurbuchen, T. H. & Richardson, I. G. 2006, Space Sci. Rev., 123, 31

6. Interplanetary Scintillation and 3-D Heliosphere

P. K. Manoharan

Radio Astronomy Centre, National Centre for Radio Astrophysics,
Tata Institute of Fundamental Research, Udhagamandalam (Ooty), India.
mano@wm.ncra.tifr.res.in

6.1. Solar Wind Structures at Peculiar Solar Minimum

During the period 2009–2011, at the end of solar cycle 23 and the beginning of cycle 24, the Sun remained at a quiet level. Several studies have been made based on interplanetary scintillation (IPS) measurements taken from radio telescopes, located at Ooty, India (327 MHz), STELab, Japan (327 MHz), and Pushchino, Russia (111 MHz). During the deep minimum of activity, the distribution of low-speed solar wind occupied a wider latitude range of $\pm 60°$, whereas the high-speed regions were confined close to the poles. The above latitudinal structures were however different from that of the corresponding phase of the previous cycle 22, in which the low-speed wind was distributed within the equatorial latitude range of $\pm 30°$ (Manoharan 2010a; Tokumaru et al. 2010; Janardhan et al. 2011).

As the solar activity declined, the density turbulence around the Sun decreased rapidly and suggested an excessive reduction in solar wind mass flux of $\sim 30–40\%$ in comparison with previous minima. Moreover, the radial dependence of IPS during 2006–07 was weaker than that of an expected spherically symmetric solar wind model. It was consistent with the quiet levels of IPS and ionospheric scintillation as well as unusual weak radial dependence of scintillation index during 2008, at the deep solar minimum phase. Such a weak dependence can only be explained by the considerable concentration of absolute solar wind turbulence confined to the solar equatorial plane that was connected to the strong influence of pronounced heliospheric current sheet (Glubokova et al. 2010; Manoharan 2011; Shishov et al. 2010).

A co-ordinated study of solar wind structure in the inner heliosphere, combining IPS measurements from EISCAT, imaging of heliospheric structures from STEREO HI and in-situ measurements from the ASPERA instrument on Venus Express displayed well-defined dominant co-rotating structures throughout the declining phase of solar cycle 23 (Bisi *et al.* 2010a). In a multi-wavelength study, in-situ data and IPS measurements identified that the high-speed magnetic clouds associated with a magnetically complex solar source caused the severe geomagnetic storm of the cycle 23 (Dst = -457) (Schmieder *et al.* 2011).

6.2. *IPS Measurements and 3-D Reconstruction of Heliospheric Structure*

Several studies have been made using the time-dependent computer-assisted tomography technique (CAT) developed at the University of California, San Diego, by B. V. Jackson and his team to reconstruct the solar wind density and velocity in the inner heliosphere (see, e.g., Jackson *et al.* 2010, 2011a and references there in). The basic data sets required for the 3-D reconstruction are the time series of velocity and g-value for a number of lines of sight of the heliosphere. Particularly, when a large number of scintillators were employed, the 3-D reconstruction provided a better understanding of the evolution of coronal mass ejections (CMEs) and showed that the CMEs associated with flux-rope systems were magnetically driven. Such a magnetically energetic CME caused an intense geomagnetic storm, even if the trailing part of the CME passed through the Earth's magnetosphere (Manoharan 2010b).

The solar wind reconstructed structures have been compared with in-situ measurements from the Wind spacecraft orbiting the Sun-Earth L1-Point and the high correlation validated the 3-D tomographic reconstruction results of transients and quiet solar wind (Bisi *et al.* 2009b).

Regular IPS observations at 327 MHz using the four-station system of the Solar-Terrestrial Environment Laboratory (STEL) revealed that solar origin, dynamical behavior and 3-D feature of the solar wind are crucial to develop the space weather prediction model. The accuracy of prediction also critically depends on the number of IPS lines of sight (Fujiki *et al.* 2010).

6.3. *Coronal Mass Ejection Studies Using IPS Technique*

The CME study using the IPS picket-fence method provided evidence that the density turbulence (also density) embedded within the CME was large at small solar offsets and decreased at large solar distances. In the case of a fast CME, the shock was formed within $\sim$100 R$_\odot$ and the compression ahead of the shock increased with distance from the Sun (Manoharan 2010b).

The link between the travel time of the CME and the effective acceleration in the Sun-Earth distance (85 out of 91 events) showed the effects of aero-dynamical drag between the CME and the solar wind. In consequence, the speed of the CME equalized to that of the background solar wind. However, for a large fraction of CMEs (for $\sim$50% of the events), the inferred effective acceleration in the Sun-Earth line prevailed over the above drag force, suggesting an average dissipation of energy $\sim$10^{31} ergs per event, which was likely provided by the Lorentz force associated with the internal magnetic energy carried by the CME (Manoharan & Mujiber Rahman 2011).

In a multi-instrument, multi-technique, coordinated study of the solar eruptive event of 13 May 2005, it became evident that the 3-D structure of the CME event was complex, which was determined by asymmetries in the initial eruption as well as by interaction between the ICME/MC and the background solar wind during the interplanetary transit. The 3-D structure of the ICME also played an important role in governing the way in which it coupled into the magnetosphere and ionosphere of the Earth (Bisi *et al.* 2010b).

The UCSD time-dependent 3-D reconstruction of SMEI and IPS have been compared with measurements at the SOHO, Wind, ACE, and STEREO spacecraft. The analyses of these shocks from hour-averaged in-situ data showed that the enhanced density column associated with the shock response varied considerably between different instruments, even for in-situ instruments located at L1 near Earth. The relatively-low-resolution SMEI 3-D reconstructions generally showed density enhancements, and within errors, the column excesses match those observed in situ. In these SMEI 3-D reconstructions from remotely-sensed data, the shock density enhancements appeared not as continuous broad fronts, but as segmented structures. This may provide part of the explanation for the observed discrepancies between the various in-situ measurements at Earth and STEREO, but not between individual instruments near L1 (Jackson et al. 2011b).

6.4. The Whole Heliosphere Interval

The Whole Heliosphere Interval (WHI), an international campaign to study the 3D solar-heliospheric-planetary connected system near solar minimum. The data and models correspond to solar Carrington Rotation 2068 (20 March – 16 April 2008) extending from below the solar photosphere, through interplanetary space, and down to Earth's mesosphere. Nearly 200 people participated in aspects of WHI studies, analyzing and interpreting data from nearly 100 instruments and models in order to elucidate the physics of fundamental heliophysical processes. The solar and inner heliospheric data showed structure consistent with the declining phase of the solar cycle. A closely spaced cluster of low-latitude active regions was responsible for an increased level of magnetic activity, while a highly warped current sheet dominated heliospheric structure. The geospace data revealed an unusually high level of activity, driven primarily by the periodic impingement of high-speed streams. The WHI studies traced the solar activity and structure into the heliosphere and geospace, and provided new insight into the nature of the interconnected heliophysical system near solar minimum (Thompson et al. 2011).

In another study, the low-resolution 3-D reconstructed heliosphere with STELab IPS velocity data, from a central part of the WHI period (around 4 April 2008), showed what appears to be a co-rotating region passing across the Sun-Earth L1 point (and crossing STEREO-B first and STEREO-A later). The global reconstructed density and radial velocity compared well with multi-point in situ spacecraft measurements in the ecliptic, namely STEREO and Wind data, as the interplanetary medium passes over the spacecraft locations (Bisi et al. 2009a).

6.5. Turbulence in the Fast/Slow Solar Wind Flows

Measurements of the magnetic field are necessary to validate competing theories and models of solar wind dynamics. In the Potential Field Source Surface (PFSS) model, the photospheric magnetic field is mapped out to a boundary condition, such as a radial configuration set to a few solar radii and then associated with observed structure in the corona. The coronal field analysis of the 1983 Alfven wave event, based on Faraday rotation measurements, was utilized to estimate the background magnetic field strength, which was nicely predicated by the PFSS model (Jensen & Russell 2009).

Turbulence studies of the solar wind showed that MHD wave efflux energy contributes to heating. Alfven waves theoretically carry 10^{22} W in the chromosphere, 10^{20} W in the corona, and are measured to carry 3×10^{17} W at 1 AU. Analysis of the magnetic field perturbation of the 1983 Alfven perturbations inferred that the efflux energy at 4 solar radii was at least 6×10^{19} W. It approximately accounts for 20% of the wave energy required to accelerate the solar wind (Jensen & Russell 2009). Further, the Faraday rotation observations of CME/ICMEs have demonstrated the usefulness of determining the critical parameters, such as the orientation and radius of a flux cylinder in the interplanetary space (Jensen *et al.* 2010).

The radio sounding studies of pulsars at 111 MHz by the Large Phased Array of the Lebedev Physical Institute, during 2005 and 2007 indicated that the acceleration of fast, high-latitude solar wind continued to heliocentric distances of $5-10$ $R_\odot$. The mean plasma density at about 5 $R_\odot$ was $1.4\times10^4 \text{cm}^3$, which was substantially lower than that of the solar maximum period. A comparison of these results with Stanford coronal magnetic field data, STEREO/SECCHI, and SOHO/EIT synoptic maps showed that the solar wind from the polar coronal holes was associated with the above reduced density. Further, the estimated Faraday rotation at heliocentric distances of $6-7$ $R_\odot$ showed a modest deviation of magnetic field from the spherically symmetric distribution (Chashei *et al.* 2010).

A study on the turbulence spectrum of the solar wind in the near-Sun region $R<50$ $R_\odot$, obtained from IPS measurements with the Ooty Radio Telescope at 327 MHz showed that the scintillation was dominated by density irregularities of size about $100-500$ km. The scintillation at the small-scale side of the spectrum, although significantly less in magnitude, has a flatter spectrum than the larger-scale dominant part. Furthermore, the spectral power contained in the flatter portion rapidly increased closer to the Sun. These results on the turbulence spectrum for $R < 50$ $R_\odot$ quantified the evidence for radial evolution of the small-scale fluctuations ($\leqslant 50$ km) generated by Alfven waves (Manoharan 2010c).

6.6. *New IPS Facilities*

The Murchison Widefield Array (MWA) is a new radio interferometer, currently under construction in the radio-quiet Western Australian outback, which exploits the recent advances in digital signal processing to rise to this challenge. The MWA expected to play a very useful role in improving our understanding of both the quiet and the dynamic Sun, and of space weather phenomena (Oberoi *et al.* 2010, 2011b). The first spectroscopic images of solar radio transients from the prototype of MWA, observed on March 27, 2010, over a frequency band of $170.9-201.6$ MHz, showed broadband emission features and numerous short-lived, narrowband, non-thermal emission (Oberoi *et al.* 2011a).

A new antenna, Solar Wind Imaging Facility (SWIFT), dedicated for IPS observations, has been developed at the STELab, Japan. The SWIFT has an aperture size of 108m (N-S) by 38m (E-W), and allows IPS observation on more number of sources fainter than those observed in the existing system. The IPS systems at Fuji and Kiso observatories have also been updated in 2010. These updated systems enable to make cross correlation measurements at the solar wind speed between Fuji, Kiso, and Toyokawa (Tokumaru *et al.* 2011).

A new IPS observing system has been established at Urumqi Astronomical Observatory (UAO), China, and a series of experimental observations have been successfully carried out from May to December 2008, at bands of 327/611 MHz and 2.3/8.4 GHz (Liu *et al.* 2010).

The Ooty Radio Telescope (one of the radio observing facilities of TIFR, India) front-end is being upgraded. It involves installation of digital acquisition system at each output of 4-dipole combiner ($\sim$2m section of the telescope). The new system allows wider band-width and will also provide a flexible system to correlate signals between pair of above 2-m-section outputs. The daily IPS observation is expected to increase from $\sim$1000$-$1400 radio sources to $\sim$3 to 4 times (Prasad & Subrahmanya 2011).

References

Bisi, M. M., Jackson, B. V., Buffington, A., Clover, J. M., Hick, P. P., & Tokumaru, M. 2009a, *Sol. Phys.*, 256, 201

Bisi, M. M., Jackson, B. V., Clover, J. M., Manoharan, P. K., Tokumaru, M., Hick, P. P., & Buffington, A. 2009b, *Ann. Geophys.*, 27, 4479

Bisi, M. M., Jackson, B. V., Breen, A. R., Dorrian, G. D., Fallows, R. A., Clover, J. M., & Hick, P. P. 2010a, *Sol. Phys.*, 265, 233

Bisi, M. M., Breen, A. R., & Jackson, B. V., *et al.* 2010b, *Sol. Phys.*, 265, 49

Chashei, I. V., Shishov, V. I., & Smirnova, T. V. 2010, *Sol. Phys.*, 265, 129

Fujiki, K. M. & Ito, H., Tokumaru, M. 2010, *12th Solar Wind Conference, AIP Conference Proceedings*, 1216, 663

Glubokova, S. K., Tyul'bashev, S. A., & Chashei, I. V., Shishov V. I. 2010, *Geomagn. Aeronomy*, 51, 1

Jackson, B. V., Hick, P. P., Buffington, A., Bisi, M. M., Clover, J. M., & Tokumaru, M. 2010, *Advances in Geosciences* (World Scientific Publishing Co., USA), Vol. 21: Solar-Terrestrial (ST), 339

Jackson, B. V., Hick, P. P., Buffington, A., Bisi, M. M., Clover, J. M., Tokumaru, M., Kojima, M., & Fujiki, K. 2011a, *J. Atmos. Sol.-Terr. Phy.*, 73, 1214

Jackson, B. V., Hamilton, M. S., Hick, P. P., Buffington, A., Bisi, M. M., Clover, J. M., Tokumaru, M., & Fujiki, K. 2011b, *J. Atmos. Sol.-Terr. Phy.*, 73, 1317

Janardhan, P., Bisoi, S. K., Ananthakrishnan, S., Tokumaru, M., & Fujiki, K. 2011, *Geophys. Res. Lett.*, 38, L20108

Jensen, E. A. & Russell, C. T. 2009, *Geophys. Res. Lett.*, 36(5), L05104

Jensen, E. A., Hick, P. P., Bisi, M. M., Jackson, B. V., Clover, J., & Mulligan, T. 2010, *Sol. Phys.*, 265, 31

Liu, L.-J., Zhang, X.-Z., Li, J.-B., Manoharan, P. K., Liu, Z.-Y., & Peng, B. 2010, *Res. Astron. Astrophys.*, 10, 577

Manoharan, P. K. 2010a, in: A. G. Kosovichev, A. H. Andrei, & J.-P. Roelot (eds.), *Proc. IAU Symposium, Solar and Stellar Variability: Impact on Earth and Planets*, 264, 356

Manoharan, P. K. 2010b, *Sol. Phys.*, 265, 137

Manoharan, P. K. 2010c, in: S. S. Hasan & R. J. Rutten (eds.), *Astrophysics and Space Science Proceedings* (Berlin Heidelberg: Sringer), p. 324

Manoharan, P. K. 2011, in: I. F. Corbett (ed.), *Highlights of Astronomy*, 15, 484

Manoharan, P. K. & Mujiber Rahman, A. 2011, *J. Atmos. Sol.-Terr. Phy.*, 73, 671

Oberoi, D., Benkevitch, L., Cappallo, R. J., Lonsdale, R. J., Matthews, L. D., & Whitney, A. R., the MWA Collaboration 2010, White paper submitted to Decadal Strategy for Solar and Space Physics

Oberoi, D., *et al.* 2011a, *ApJ*, 728, L27

Oberoi, D., Lonsdale, C. J., & Benkevitch, L., *et al.* 2011b, *General Assembly and Scientific Symposium, 2011 XXXth URSI*, 1

Prasad, P. & Subrahmanya, C. R. 2011, *Exp. Astron.*, 31, 1

Shishov, V. I., Tyul'bashev, S. A., Chashei, I. V., Subaev, I. A., & Lapaev, K. A. 2010, *Sol. Phys.*, 265, 277

Schmieder, B., *et al.* 2011, *Adv. Space Res.*, 47, 2081

Thompson, B. J., *et al.* 2011, *Sol. Phys.*, in press

Tokumaru, M., Kojima, M., Fujiki, K. 2009, *Transactions of Space Technology Japan*, 7, 21

Tokumaru, M., Kojima, M., & Fujiki, K. 2010, *J. Geophys. Res.*, 115, A04102

Tokumaru, M., Kojima, M., Fujiki, K., Maruyama, M., Maruyama, Y., Ito, H., & Iju, T. 2011, *Radio Science*, 46, RS0F02, doi:10.1029/2011RS004694

7. Solar Energetic Particles and Galactic Cosmic Rays

David Lario

The Johns Hopkins University, Applied Physics Laboratory, Laurel, USA

david.lario@jhuapl.edu

With the exception of a few solar energetic particle (SEP) events that marked the beginning of solar cycle 24 in mid-August 2010, in early March, June, August 2011 and September 2011, the main characteristic of the period between September 2009 and September 2011 was the absence of large SEP events to analyze. Therefore, most of the research efforts in this time interval were focused on the study of the energetic particle enhancements associated with corotating interaction regions (CIRs), small ^{3}He-rich events observed by multiple spacecraft spread through the interplanetary medium, the record-setting intensity of galactic cosmic rays (GCRs) and long-term recompilations of SEP events observed during the last solar cycle.

Because CIR-associated particles are very prominent during solar minimum, the unusually long solar minimum period provided the opportunity to examine the overall organization of CIR energetic particles for a much longer period than ever before. Recurrent low-energy (<1 MeV) proton enhancements associated with CIRs were observed near 1 AU for many solar rotations (up to 30) due to several persistent high-speed solar wind streams (Lee *et al.* 2010). The multipoint observations (by near-Earth space observatories and the twin STEREO spacecraft) of CIR events provided evidence that CIR-associated energetic ions frequently show significant differences from spacecraft to spacecraft, particularly at sub-MeV energies. Discrepancies in the observed structures are due to the latitudinal separation between spacecraft, changes in the coronal hole generating the high-speed solar wind streams, and/or the presence of interplanetary coronal mass ejections (ICMEs) or small-scale interplanetary (IP) transients in the vicinity of or embedded within the CIRs (Gomez-Herrero *et al.* 2011). Temporal variations in the CIR-associated ion increases may also be due to concomitant SEP events that produce a mixing of SEP and CIR particle populations (Lee *et al.* 2010).

Energetic particles observed in association with CIRs are thought to be accelerated by distant shocks formed by the compression between fast and slow solar wind streams. However, unshocked compression regions associated with CIRs near 1 AU have been hypothesized as candidate to energize particles (Giacalone *et al.* 2002). Bucik *et al.* (2009) compared the predictions of compression acceleration with measurements of $\sim$0.1 to $\sim$1 MeV/n ion intensities. Observations show that the ion intensity in CIR events with in-situ reverse shocks is well organized by the parameters that characterize the compression region itself, like compression width, solar wind speed gradients and total pressure. In turn, for CIR events with the absence of shocks the model predictions are not fulfilled.

During this protracted solar minimum, small ^{3}He-rich SEP events were observed by both the ACE spacecraft and by the two STEREO spacecraft as they separated in longitude from ACE at a rate of $\sim$22 deg per year. In a widely-held view of impulsive solar energetic particle (ISEP) events, electrons and ions are accelerated at the site of a solar flare when magnetic energy is released by reconnection. When the reconnection involves some open field lines, those field lines provide a path for particles to escape into the heliosphere, leading to the observation of ISEP events. The very limited spatial and temporal extent of the acceleration and release was thought to imply that these particles should have a relatively narrow spread in heliolongitude. However, STEREO and ACE observations have shown that ISEP events may be simultaneously detected from well-separated longitudes (Wiedenbeck *et al.* 2010). To understand the spreading of particles in longitude from a flare site Wiedenbeck *et al.* (2010) studied the magnetic field configuration around active regions (ARs) via the potential field source surface (PFSS) model. Such model calculations frequently indicate that open field lines originating near an AR may spread by more than several 10's of degrees before reaching the source surface. Shocks driven by coronal mass ejections (CMEs) have also been suggested as responsible for the acceleration of SEPs as observed from distant longitudes. The presence of ubiquitous suprathermal tails with a solar cycle dependent composition (Dayeh *et al.* 2009) may also serve as seed population for the mechanisms of particle acceleration at traveling shocks.

Efforts to extract the properties of the CME-driven shocks from white-light coronagraph observations were performed (Ontiveros & Vourlidas 2009, see section 4.5 for more details). Extreme-ultraviolet observations from the Solar Dynamics Observatory (SDO) and Type II radio bursts observations were also used to characterize shocks forming low in the corona (Kozarev *et al.* 2011). By modeling the configuration of the overlying coronal magnetic fields (via PFSS) and considering the orientation of the shock fronts, it is possible to estimate the efficiency of the shocks in accelerating particles (Kozarev *et al.* 2011). The multipoint coronagraph observations from SOHO and the two STEREO allow the reconstruction of the 3D envelope of the shock. The spatial extent, radial coordinates and speed of the shocks can be used as input to numerically simulate the CME propagation. Comparison of both the SEP onset times as observed by three spacecraft separated in longitude with the times when the magnetic connection between each spacecraft and the shock is established, together with the evolution of the shock parameters at the region of the shock front connecting to each spacecraft, allowed Rouillard *et al.* (2011) to confirm the description of gradual SEP events established a decade earlier using combined simulations of shock and SEP transport (i.e., Lario *et al.* 1998).

Among the works recompiling properties of SEP events observed during the last solar cycle, Cane *et al.* (2010) examined the properties and associations of 280 proton events that extended above 25 MeV. The events were divided into five representative types based on the relative abundances and particle profiles to illustrate how particle characteristics vary with their associated solar parameters (i.e., CMEs, flares and radio emissions). A continuum of event properties with no indication of specific parameters that clearly separate the groups of events was found. There was, however, a reasonable separation of events based on the timing of the associated type III emissions relative to the Hα flare. Type III bursts indicate the presence of flare particles that escape to the IP medium. The least intense, relatively short-lived, proton events that are electron-rich have associated type III bursts that occur at the start of the flare, indicating rapid acceleration and escape of particles. In the largest events the type III emissions occur after the impulsive phase. Cane *et al.* (2010) suggested that this late acceleration and/or release of particles results in a composition different from that of impulsive acceleration and release; proposing a scenario in which concomitant flare processes contribute particles in the majority of SEP events. On the other hand, Gopalswamy & Mäkelä (2010) showed that the occurrence of a long-duration and low-frequency (<14 MHz) type III burst is not a good indicator for the occurrence of an SEP event, and neither the type III burst duration nor the burst intensity can distinguish between SEP and non-SEP events. The lack of solar energetic protons in association with a large complex type III burst that reached local plasma frequencies was explained by Gopalswamy & Mäkelä (2011) as a signature that the acceleration of low-energy electrons responsible for the type III burst at the flare site does not imply the acceleration of protons, which most likely occurs in the CME-driven shock.

Solar cycle 23 also showed a few SEP events that exceeded the previously determined streaming limit, even in their prompt component (Lario *et al.* 2009). The mechanisms leading to the exceeding of the streaming limit include, apart from an intense source of particles, the inhibition of amplification of waves by the streaming particles and/or the existence of large-scale IP structures able to modify the nominal conditions for SEP transport. Ng *et al.* (2010) presented new theoretical results on how the streaming limit depends on ion species and energy, ambient wave intensity spectrum, Alfvén speed, solar-wind speed, shock speed, and the presence of IP shocks and interaction regions. The potential relevance of the latitude of the observer in the SEP intensity-time profiles was investigated by Rodriguez-Gasen *et al.* (2011). The influence of IP structures on the SEP transport was also proven by using both SEP observations (Tan *et al.* 2009) and SEP transport simulations (Agueda *et al.* 2010).

During solar cycle 23 sixteen ground-level events (GLEs) were detected by neutron monitors. A study of their spectra in the energy range $\sim$0.1 to 700 MeV/n showed that the proton fluence for all 16 GLEs were well fit by the double power-law. Minimizing the difference between the spectral indices above and below the "break" energy minimizes also the energy requirements for accelerating enough 500 MeV protons for a detectable GLE (Mewaldt *et al.* 2009a).

STEREO observations were also used to discover the possible presence of energetic neutral hydrogen atoms (ENAs) emitted during the X9 solar flare event on 2006 December 5. Mewaldt *et al.* (2009b) concluded that the observed ENAs were most likely produced in the high corona and that charge-transfer reactions between accelerated protons and partially stripped coronal ions are, in general, an important source of ENAs in solar events. Taking into account ENA losses, the observed ENAs in the event were produced in the high corona at heliocentric distances $\geqslant$2 solar radii.

During this extended solar minimum, galactic cosmic rays (GCRs) achieved the highest intensities observed in the space age. In the energy interval from $\sim$70 to $\sim$450 MeV/n, the measured intensities of major species from C to Fe were each 20%-26% greater in late 2009 than in the 1997-1998 minimum and previous solar minima of the space age (1957-1997) (Mewaldt *et al.* 2010). The elevated intensities (also observed at neutron monitor energies; Ahluwalia & Ygbuhay 2010) were due to several unusual aspects of the solar cycle 23/24 minimum, including record-low interplanetary magnetic field (IMF) intensities, an extended period of reduced IMF turbulence, reduced solar-wind dynamic pressure, and extremely low solar activity. GCR intensity variations at 1 AU were found to lag IMF variations by 2-3 solar rotations, indicating that significant modulation occurs inside $\sim$20 AU. In 2010, the intensities suddenly decreased to 1997 levels following increases in solar activity and in the inclination of the heliospheric current sheet.

Hard X-ray and gamma-ray observations by RHESSI allowed the identification of the properties of the accelerated ions and electrons that interact in the solar atmosphere and photosphere during flares and relate them to the SEPs observed in space. These properties provide information on the acceleration processes and particle transport of particles in solar flares (e.g. Zharkova *et al.* 2011). It is generally agreed that magnetic reconnection is the energizing mechanism of solar flares and CMEs but the connection between these processes and observations involves intermediate processes of acceleration and heating of plasma particles, transport of particles and their radiative signatures that are still not fully understood. In most of the intense flares observed by RHESSI (and previously by SMM) the heavy interacting particles at the Sun have composition that is similar to gradual SEP events (i.e. coronal), but that in at least one flare it was found a composition close to that observed in impulsive SEP events (Murphy *et al.* 2011). On the other hand, comparison of the number of flare-accelerated 30 MeV protons that interact in the solar atmosphere (estimated using gamma-ray RHESSI data) with the deduced number of SEPs reaching 1 AU shows that the latter to be typically $\sim$10-100 times larger than those interacting in the solar atmosphere. This implies that the vast majority of the protons observed in situ are accelerated by a CME-driven shock or other coronal or interplanetary processes rather than by the flare; although flare-accelerated ions may still contribute to some SEP events, possibly for heavy elements (Mewaldt *et al.* 2008).

Studies of SEPs at 1 AU over the last few years have made very significant progress, but they also reveal that many key questions will remain unanswered until it is possible to fly fully instrumented spacecraft closer to the Sun where the bulk of SEP acceleration takes place. Solar cycle 24 will provide us with the opportunity to analyze SEP events from multiple points of view by using STEREO, near-Earth spacecraft and the MESSEN-GER spacecraft in orbit around Mercury. However, issues like the composition of seed particle populations, the nature of wave-particle interactions, the separation between the roles of SEP acceleration and transport processes, and the timing relation between SEP acceleration and solar eruptive events are best studied from distances close to the Sun. In this recent years, multiple efforts have been focused on the development of the future exciting new missions such as the ESA Solar Orbiter and the NASA Solar Probe Plus that will explore the inner heliosphere between 0.04 and 0.7 AU.

References

Agueda, N., *et al.* 2010, *A&A*, 519, A36
Ahluwalia, H. & Ygbuhay, R. 2010, *Proc. 12th Solar Wind Conf.*, AIP Conf. Proc., 1216, 699
Bucik, R., *et al.* 2009, *Ann. Geophys.*, 27, 3677
Cane, H. V., *et al.* 2010, *J. Geophys. Res.*, 115, A8, A08101
Dayeh, M. A., *et al.* 2009, *ApJ*, 693, 1588
Giacalone, J., *et al.* 2002, *ApJ*, 573, 845
Gopalswamy, N. & Mäkelä, P. 2010, *ApJ* (Letters), 721, L62
Gopalswamy, N. & Mäkelä, P. 2011, *Cent. Eur. Astrophys. Bull.*, 35, 71
Gomez-Herrero, R., *et al.* 2011, *J. Atm. Sol-Ter. Phys.*, 73, 551
Kozarev, K. A., *et al.* 2011, *ApJ* (Letters), 733, L25
Lario, D., *et al.* 1998, *ApJ*, 509, 415
Lario, D., *et al.* 2009, *Sol. Phys.*, 260, 407
Lee, C. O., *et al.* 2010, *Sol. Phys.*, 263, 239
Mewaldt, R. A., *et al.* 2008, *7th Ann. Int. Astrophys. Conf.*, AIP Conf. Proc., 1039, 111
Mewaldt, R. A., *et al.* 2009a, *Proc. 31st Int. Cosmic Ray Conf.*, Paper #0783
Mewaldt, R. A., *et al.* 2009b, *ApJ* (Letters), 693, L11
Mewaldt, R. A., *et al.* 2010, *ApJ* (Letters), 723, L1
Murphy, R., *et al.* 2011, *Bull. Amer. Astron. Soc.*, 43, Abstract #22.36
Ng, C. K., *et al.* 2010, *Amer. Geophys. Union, Fall Meeting 2010*, Abstract #SH41C-06
Ontiveros, V. & Vourlidas, A. 2009, *ApJ*, 693, 267
Rodriguez-Gasen, R., *et al.* 2011, *Adv. Space Res.*, 47, 2140
Rouillard, A. P., *et al.* 2011, *ApJ*, 735, 7
Tan, L. C., *et al.* 2009, *ApJ*, 701, 1753
Wiedenbeck, M. E., *et al.* 2010, *Amer. Geophys. Union, Fall Meeting 2010*, Abstract #SH42B-02
Zharkova, V. V., *et al.* 2011, *Space Sci. Rev.*, doi:10.1007/s11214-011-9803-y

8. The Outer Heliosphere

John D. Richardson
Kavli Institute for Astrophysics and Space Science, M.I.T., Cambridge, MA, USA
jdr@space.mit.edu

8.1. *Introduction*

Progress in understanding the outer heliosphere and its interaction with the local interstellar medium (LISM) continues its rapid advance. The previous two reports showcased the Voyager 1 (V1) and 2 (V2) crossings of the termination shock and first observations of the heliosheath. Highlights of the current report are the Interstellar Boundary Explorer (IBEX) observations of the the global heliosphere observed with energetic neutral atoms (ENAs) and the Voyager observations of the heliosheath, particularly recent results which suggest that Voyager 1 is close to the heliopause and the LISM. Theorists and modellers have been active in trying to explain these new observations.

8.2. *IBEX*

IBEX observes neutrals from 10 eV to 6 keV with the goal of mapping the 3D structure of the outer heliosphere. The first results were published in a series of Science papers in October 2009. The key result was a detection of ribbon of ENAs forming a 300° arc on the sky (Funsten *et al.* 2009) and passing between the Voyager spacecraft (McComas *et al.* 2009). The ribbon is observed at all energies above 200 eV with an average width of about 20° and has the largest intensities at about 1 keV (Fuselier *et al.* 2009), roughly the average solar wind energy. Fine-scale variations are observed within the ribbon, most notably one bright "knot" region and several weaker knots (McComas *et al.* 2009; Livadiotis *et al.* 2011). A similar but wider feature was observed at higher energies (6−16 keV) by the Cassini MIMI experiment (Krimigis *et al.* 2009). The ribbon seems to emanate from regions where the local interstellar magnetic field is perpendicular to the direction of Earth (Schwadron *et al.* 2009). These results suggest that the LISM magnetic and gas dynamic pressure both have a major influence on the LISM interaction with the heliosphere (McComas *et al.* 2009; Krimigis *et al.* 2009), placing the actual interaction between the two limiting cases discussed by Parker (1961).

The second IBEX map (a complete ENA map of the heliosphere is obtained every six months) showed the same basic ribbon structure, but some changes were observed (McComas *et al.* 2010). The bright knot spread out and became 25−35% less intense in the second map and the intensities from the polar regions decreased by 10−15%. Suggestions for the causes range from solar cycle variation to the smaller-scale solar wind variation driving waves which propagate back and forth through the heliosheath (Washimi *et al.* 2011).

Numerous possible explanations for the origin of the ribbon have been suggested with source regions ranging from the termination shock to the boundary of the local interstellar cloud (McComas *et al.* 2009, 2011; Grzedzielski *et al.* 2010; Heerikhuisen *et al.* 2010). One of the more developed hypotheses is that the ribbon is formed by secondary ENAs (McComas *et al.* 2009; Heerikhuisen *et al.* 2010). Solar wind protons charge exchange with LISM neutrals inside the heliosphere and move outward of the solar wind speed. These fast neutrals are re-ionized via charge exchange with interstellar ions outside the heliopause, then become secondary ENAs when they charge exchange with interstellar neutrals. These neutrals, which have undergone three charge exchange reactions, would then be observed by IBEX with energies near 1 keV. If the ions formed in the LISM charge exchange to form secondary ENAs before they scatter in pitch angle, then only neutrals from regions where the LISM magnetic field is perpendicular to Earth would be observed. If the pitch angle scattering time is assumed small, this mechanism matches the observations quite well (Heerikhuisen *et al.* 2010), although the validity of this assumption is still under discussion (Florinski *et al.* 2010; Gamayunov *et al.* 2010).

IBEX was recently shifted into a more stable orbit and should continue studying the variability of the heliosphere for many years.

8.3. Heliosheath

The Voyager spacecraft continue to explore the heliosheath on their way to the LISM. An exciting result suggests that V1 is approaching the heliopause (HP) and has entered a previously unknown heliopause boundary layer more than 4 AU thick where flow is parallel to the HP (Krimigis et al. 2011). The radial flow speed in the heliosheath at V1 derived from the LECP instrument using the Compton-Getting effect (the V1 plasma instrument does not work) show a monotonic decrease from mid-2005 to early 2010, when the radial speed reached zero. The radial speed has remained near zero through mid-2011. The flows in the T plane (parallel to the solar equator) oscillated about a mean of 40 km/s before late 2010, when they slowed to about 10 km/s (meridional flows are not measured by LECP). These very slow flows were a surprise - models show a flared HP with an outward radial flow component even at the HP. The observed flows are of the same magnitude as the LISM speed of 26.3 km/s Witte (2004). Krimigis et al. (2011) suggest that a boundary layer with flow roughly parallel to the HP boundary is present with a width of at least 4 AU. Such a boundary layer was proposed by Suess (1990), although the current paradigm had envisioned a relatively quick transition from heliosphere to LISM.

The V2 plasma experiment observes very different flow vectors from those at V1. The V2 flow magnitude remains roughly constant, but the flow direction has shifted as the flow turns toward the heliotail (Richardson & Wang 2011; Richardson 2011). Flow angles are larger in the RT than RN planes, with the average flow in the RT plane 55° from radial in 2011 and turning about 10°/year. The density in the heliosheath decreased by a factor of 2 from the termination shock (TS) crossing thru the end of 2010, then started to increase. The temperature has fallen by a factor of 3 across the heliosheath, much more than the expected adiabatic decrease. The magnetic field increases across the heliosheath if the change in the source field (as measured at 1 AU) is taken into account (Burlaga et al. 2009), consistent with model predictions (Burlaga & Ness 2010). The heliosheath field and plasma are highly variable. As solar minimum conditions with lower dipole tilts reached V2, fewer heliospheric current sheet (HCS) crossings were observed and V2 remained mostly in the same solar wind sector (Burlaga & Ness 2011). V1, although at higher heliolatitudes, continues to spend comparable amount of time in each sector although the HCS crossings are far apart. These observations are consistent with northward flow in the heliosheath carrying the HCS northward past the V1 heliolatitude (Borovikov et al. 2011). The large amount of time spent in each sector results from the very slow radial speeds observed at V1 convecting the sectors past V1 very slowly. Plasma parameters are fit well by Gaussian distributions (Richardson 2011) while the magnetic field distributions are sometimes Gaussians and sometimes power laws (Burlaga & Ness 2010, 2011).

The Voyager spacecraft will continue measuring heliosheath and hopefully LISM properties through 2025.

8.4. Energetic Particles

Energetic particles observed in the heliosheath can be divided into three main classes: termination shock particles at low energies, from a few keV/nuc to several MeV/nuc, anomalous cosmic rays, which are pickup ions accelerated to energies from a few to hundreds of MeV/nuc, and galactic cosmic rays (GCRs), which are accelerated elsewhere in the galaxy and which dominate the energy spectra above typically 50 to 100 MeV/nuc. The termination shock particle intensity peaks at the TS, indicating they are accelerated at the TS. A recent model suggests they are formed from the core population of pickup ions which are accelerated to the observed energies at the TS by a process similar to shock drift acceleration (Giacalone & Decker 2010). The source of the ACRs was also thought to be the TS before the Voyagers found no evidence of ACR acceleration at the TS crossings (Stone *et al.* 2005; Stone *et al.* 2008). The ACR source location and acceleration mechanism are now puzzles. The spectra of ACRs have continued to unfold as the two Voyager spacecraft penetrate further into the heliosheath (Cummings *et al.* 2011).

Suggestions for the ACR source include acceleration in the flanks or tail of the heliosphere (McComas & Schwadron 2006; Kota 2008), reconnection in the outer heliosphere (Lazarian & Opher 2009; Drake *et al.* 2010; Opher *et al.* 2011), or a stochastic pumping mechanism (Fisk & Gloeckler 2009). The galactic cosmic rays modulation boundary had been expected to occur at the HP, with relatively undisturbed GCR intensities in the LISM. However, Scherer *et al.* (2011) suggest that the GCR modulation boundary is beyond the heliopause , due both to modified diffusion in the outer heliosheath and confinement and cooling of these particles in the heliosphere. Caballero-Lopez *et al.* (2010) show that there is a large intensity difference in GCR electrons at V1 and V2, suggesting large hemispherical asymmetries.

References

Borovikov, S. N., Pogorelov, N. V., & Burlaga, L. F., Richardson J. D. 2011, *ApJ*, 728, L21

Burlaga, L. F., Ness, N. F., Acuña, M. H., Wang, Y.-M., & Sheeley, N. R. 2009, *J. Geophys. Res.*, 114, A06106

Burlaga, L. F. & Ness, N. F. 2010, *ApJ*, 725, 1306

Burlaga, L. F. & Ness, N. F. 2011, *ApJ*, 737, 35

Caballero-Lopez, R. A., Moraal, H., & McDonald, F. B. 2010 *ApJ*, 725,121

Cummings, A. C., Stone, E. C., McDonald, F. B., Heikkila, B. C., Lal, N., & Webber, W. R. 2011 *Proc. 32nd ICRC*, Beijing, China, SH3.1-0101

Drake, J. F., Opher, M., Swisdak, M., & Chamoun, J. N. 2010, *ApJ*, 709, 963

Fisk, L. A. & Gloeckler, G. 2009, *Adv. Space Res.*, 43, 1471

Florinski, V., Zank, G. P., & Heerikhuisen, J. 2010, *Proc. 9th Ann. Int. Astrophys. Conf., AIP Conf. Proc.*, 1302, 192

Funsten, H. O., *et al.* 2009, *Science*, 326, 964

Fuselier, S. A., *et al.* 2009, *Science*, 326, 962

Gamayunov, K., Zhang, M., & Rassoul, H. 2010, *ApJ*, 725, 2251

Giacalone, J. & Decker, R. B. 2010, *ApJ*, 710, 91

Grzedzielski, S., Bzowski, M., Czechowski, A., Funsten, H. O., McComas, D. J., & Schwadron, N. A 2010, *ApJ*, 715, L84

Heerikhuisen, J., *et al.* 2010, *ApJ*, 708, L126

Kota, J. 2008, *Proc. 30th Internat. Cosmic Ray Conf.*, Merida, Mexico, Vol. 1, 853

Krimigis, S. M., Mitchell, D. G., Roelof, E. C., Hsieh, K. C., & McComas, D. J. 2009, *Science*, 326, 971

Krimigis, S. M., Roelof, E. C., Decker, R. B., & Hill, M. E. 2011, *Nature*, 474, 359

Lazarian, A. & Opher, M. 2009, *ApJ*, 703, 8

Livadiotis, G., McComas, D. J., Dayeh, M. A., Funsten, H. O., & Schwadron, N. A. 2011, *ApJ*, 734, 1

McComas, D. J. & Schwadron, N. A. 2006, *Geophys. Res. Lett.*, 33, L04102

McComas, D. J., *et al.* 2009, *Science*, 326, 959

McComas, D. J., *et al.* 2010, *J. Geophys. Res.*, 115, A09113

McComas, D. J., Funsten, H. O., Fuselier, S. A., Lewis, W. S., Möbius, E., & Schwadron, N. 2011, *Geophys. Res. Lett.*, 38, L18101

Opher, M., Drake, J. F., Swisdak, M., Schoeffler, K. M., Richardson, J. D., Decker, R. B., & Toth, G. 2011, *ApJ*, 734, 71

Parker, E. N. 1961, *ApJ*, 134, 20

Pogorelov, N. V., *et al.* 2010, *Proc. Twelfth Int. Solar Wind Conf., AIP Conf. Proc.*, 1216, 559

Richardson, J. D. 2011, *ApJ*, 740, 113

Richardson, J. D. & Wang, C. 2011, *ApJ*, 734, L21

Scherer, K., Fichtner, H., Strauss, R. D., Ferreira, S. E. S., Potgieter, M. S., & Fahr, H.-J. 2011, *ApJ*, 735, 128

Schwadron, N. A., *et al.* 2009, *Science*, 326, 966

Stone, E. C., *et al.* 2005, *Science*, 309, 2017

Stone, E. C., *et al.* 2008, *Nature*, 454, 71

Suess, S. T. 1990, *Rev. Geophys.*, 28, 97

Washimi, H., Zank, G. P., Hu, Q., Tanaka, T., Munakata, K., & Shinagawa, H. 2011, *Mon. Not. R. Astron. Soc.*, 416, 1475

Witte, M. 2004, *A&A*, 426, 835

9. Closing Remarks

There have been enormous progress in the study of the Sun, the interplanetary plasma, and the heliosphere over the past three years largely because of the new instruments that provided unprecedented views from new vantage vantage points away from the Sun-Earth line. A large number of publications have been added to the scientific literature, but all of them could not be included in the report due to space limitations. The cited work is therefore a biased sample of what is available in the literature. More results are expected to be presented during the IAU general assembly in Beijing. These activities demonstrate that Commission 49 is a vital component of Division II and will continue to add new knowledge.

Natchimuthuk Gopalswamy
President of the Commission

Transactions IAU, Volume XXVIIIA
Reports on Astronomy 2009–2012   © International Astronomical Union 2012
Ian F. Corbett, ed.   doi:10.1017/S1743921312002700

DIVISION III   PLANETARY SYSTEMS SCIENCES

SCIENCES DES SYSTEMES PLANETAIRES

Division III provides a focus for astronomers studying a wide range of problems related to planetary systems, including the physical studies of planets, their satellites, small bodies (comets, asteroids and meteors), and including astrobiology and extrasolar planetary systems.)

PRESIDENT	Karen Meech
VICE-PRESIDENT	Giovanni Valsecchi
PAST PRESIDENT	Edward L. Bowell
BOARD	Dominique Bockelee-Morvan
	Alan Boss
	Alberto Cellino
	Guy Consolmagno
	Julio Fernandez
	William Irvine
	Daniela Lazzaro
	Patrick Michel
	Keith Noll
	Rita Schulz
	Jun-ichi Watanabe
	Makoto Yoshikawa
	Jin Zhu

DIVISION III COMMISSIONS

Commission 15	Physical Studies of Comets and Minor Planets
Commission 16	Physical Study of Planets and Satellites
Commission 20	Positions and Motions of Minor Planets, Comets and Satellites
Commission 22	Meteors, Meteorites and Interplaentary Dust
Commission 51	Biostronomy
Commission 53	Extrasolar Planets

DIVISION III WORKING GROUPS

Division III WG	Near Earth Objects
Division III WG	Small Bodies Nomenclature (CSBN)
Division III WG	Planetary System Nomenclature (WGPSN)

INTER-DIVISION WORKING GROUPS

Division I-III WG	Natural Satellites
Division I-III WG	Cartographic Coordinates and Rotational Elements of Planets and Satellites

TRIENNIAL REPORT 2009–2012

1. Introduction

Division III, with 1126 members, is the third largest of the 12 IAU Divisions, focusing on subject matter related to the physical study of interplanetary dust, comets, minor planets, satellites, planets, planetary systems and astrobiology. Within the Division are very active working groups that are responsible for planetary system and small body nomenclature, as well as a newly created working group on Near Earth Objects which was established order to investigate the requirements for international ground-and/or space-based NEO surveys to characterize 90% of all NEOs with diameters >40m in order to establish a permanent international NEO Early Warning System.

2. Developments within the past triennium

2.1. *Commission 15 – Physical Studies of Comets and Minor Planets*

Commission 15 has 2 working groups: Physical Studies of Asteroids, Physical Studies of Comets, and 4 task groups on Asteroid magnitudes, Cometary Magnitudes, Asteroid polarimetric Albedo Calibration, and on Geophysical and Geological Properties of Asteroids and Comet Nuclei. The material for Commission 15 has been prepared by Alberto Cellino.

During the last triennium there has been an effort to completely re-design the Commissin web page. This has been done by Mrs. S. Rasetti at the Observatory of Torino, in collaboration with the Commission President and Secretary. The new web page can be found at URL http://iaucomm15.oato.inaf.it/. This web page includes a link to a brand new Forum, which has been developed with the aim of possibly becoming a useful tool for Commission 15 members. The idea is that, by registering in the Forum, each member will have the possibility to write text containing a variety of useful information, and will be automatically informed whenever a new message is posted. In this way, it should be possible to create a kind of dedicated network to facilitate interactions between Commission 15 members, to encourage the exchange of ideas, the sharing of telescope time, and meeting organization. This Forum has been very recently improved in order to add some important facilities, and might hopefully become a tool appreciated by the community.

During the last triennium, Commission 15 devoted particular attention, through the activities of two dedicated Task Groups, to the subjects of obtaining a new, updated calibration of the relation between geometric albedo and some polarimetric properties, and the need of assessing and possibly improving the quality and accuracy of asteroid absolute magnitudes. The fundamental role played by the absolute magnitude in numerous areas of research is well known. Thus, the generally poor accuracy of the H values listed in major databases certainly constitutes a problem, with potential consequences on the results of a number of investigations concerning the size frequency distribution of the population, and inferences on the most likely values of asteroid collisional impact strengths. This topics are sufficiently important to be separately discussed below.

Activities of the Task Group on Asteroid Magnitudes — The Task Group on Asteroid Magnitudes (TGAM) was charged with investigating the quality of asteroid absolute magnitudes and the accuracy of the asteroid magnitude phase function and to suggest ways to improve both. Absolute magnitudes are important for deriving accurate sizes and albedos. Upcoming projects will produce high accuracy photometry on well over 100,000 asteroids. An accurate asteroid magnitude phase function is needed to reduce these observations.

Different possible options to improve the definitions of albedo and absolute magnitude were discussed. There has been some debate among the task group members concerning the problems related to the "zero phase angle problem" when dealing with physical parameters that are defined at zero phase angle (such as H and albedo). The prevailing opinion was that it is probably premature to propose drastic changes in definitions of H or albedo because this could introduce many problems. For instance, a definition of absolute magnitude as the brightness observed at unit distance from the Sun and the observer, and at a phase angle of, say, 10 degrees, would work for asteroids, but not for other, more distant solar system objects which are visible only at smaller values of phase angle. Similar considerations may apply to the definition of geomtric albedo. However, there was consensus that some limited improvement in the definition of the (H,G) system, for instance, are not only possible, but sorely needed. Typical accuracies of H value in available catalogues is about ±0.5 mag due to: (1) the often low signal to noise ratio of the measurements performed by the discovery surveys, (2) the crude photometric data reduction procedures employed by most current asteroid survey programs, which use unfiltered CCDs and do not use standard stars from photometric catalogs to calibrate the photometry, and (3) the ambiguity introduced by lightcurve effects.

In addition to these large random errors in H there are systematic errors as well. These have been noted previously (*e.g.*, Sec. 3.2.1 of Jurić *et al.* 2002 for a discussion of this particular issue, and Jedicke *et al.* 2002 for a broader treatment of issues regarding asteroid orbital element databases).

A possible way to improve the situation, in order to be ready to process properly the huge amount of new data which are going to be produced by present and imminent sky surveys, has been discussed in a paper by Muinonen *et al.* (2010). These authors have proposed the adoption of a new three-parameter photometric system (named H, G_1, G_2), which seems to produce better fits of the limited number of presently available high-quality phase-magnitudes curves, with respect to the currently adopted (H, G) system. The proposed system seems also potentially able to produce more accurate estimates of the absolute magnitude even in the vast majority of cases, when the sampling of the phase-magnitude curves is rather poor.

Activities of the Task Group on Asteroid Polarimetric Albedo Calibration — Polarimetry is an excellent technique to derive asteroid albedos, including small objects, and was used in the past also in asteroid taxonomy to distinguish between E, M, P classes, The empirical relations among the albedo and polarimetry parameters are therefore very important.

On the other hand, polarimetry has important advantages over thermal radiometry, the most used technique for size and albedo determination. Radiometric results are model-dependent, require observations in different IR bands, and generally suffer from poor knowledge of the absolute magnitude H. Radiometric albedos for small asteroids which are observed in only one IR band have uncertainties which may very high, up to 60%. In the past, different authors have used albedo-polarimetry relations obtained by several methods. Therefore IAU Commission 15 has recommended that astronomers converge to a unique choice of the albedo-polarimetric parameter empirical relations.

There is a general consensus that the well-known polarimetric slope-albedo relation must be urgently recalibrated using only high-quality V-band polarimetry of asteroids with accurate albedos. The best object list at present is the one by Shevchenko & Tedesco (2006), consisting of albedos derived from occultation and in-situ (four objects) size measurements, coupled with accurate estimates of absolute magnitudes obtained using one unique photometric system (H,G). Albedos determined using model-dependent fits to thermal spectra, and all results based on IRAS or single-wavelength radiometric observations, are not suitable for use as calibration objects, and should be no longer used for these purposes. Only albedos derived by using diameters from radiometric spectra fit using detailed thermophysical models can still be acceptable for calibration.

A reasonable road-map for the future is to obtain new data for albedo calibrations aimed at obtaining very accurate polarization measurements of the Shevchenko and

Tedesco target list. Some dedicated observing programs are presently ongoing and are starting to produce data which are expected to lead to a new, better caliobration of the polarization-albedo relation in the near future. On the other hand, some theoretical work is needed to explain the peculiar objects found during the last surveys (the so-called Barbarians, F-class, etc.).

Activites of the Task Group for Physical Properties of Near-Earth Objects — Since the IAU General Assembly in Rio de Janeiro, the Task Group for Physical Properties of Near-Earth Objects (TGNEO) has been involved in introducing a new three-parameter H, G_1, G_2 photometric function for asteroids (Muinonen *et al.* 2010), already mentioned in the previous Section 2.1. This was then followed by an application to photometric data available at the Minor Planet Center and carefully calibrated at Lowell Observatory (Oszkiewicz *et al.* 2011).

The three-parameter photometric phase function has been introduced as an improvement to the H, G photometric function approved at the IAU General Assembly in 1985. For the NEO cause, the H, G_1, G_2 photometric function allows for more accurate prediction of absolute magnitudes H based on sparse photometric data. Indirectly, this allows for more accurate size, geometric-albedo and Bond-albedo estimates to be derived.

The application of the new phase function has resulted in improved absolute magnitudes for over 500,000 asteroids, including more than 6,000 NEOs. Work is currently under way to correlate the individual G_1 and G_2 parameters with the individual geometric albedos, allowing for more accurate characterization of NEO sizes in particular.

2.2. *Commission 16 – Physical Study of Planets and Satellites*

Commission 16 had a face to face meeting of seven members during the American Astronomical Society's Division for Planetary Sciences conference in Pasadena, California in October 2010. At this meeting plans were made to update the Commission 16 web site for easier access to historical archives and ongoing Commission activities. Plans were also made to hold a Town Hall meeting to encourage new membership at the joint European Planetary Science and Division for Planetary Science Conference in October 2011 in Nantes, France. The Commission helped devel-op and sponsor several proposals for Symposia and Joint Discussions for the 2012 General Assembly in 2012.

2.3. *Commission 22 – Meteors, Meteorites and Interplanetary Dust*

After the GA in 2009, the president of Commission 22, J. Watanabe, started a Newsletter to inform members about commission business activities in a timely manner via e-mail. Since the first one was issued on Sept. 24, 2009, eleven newsletters have been released as of October 29, 2011.

New members were assigned to the Working Group on Meteor Shower Nomenclature, which is a new WG, changed from a task group at the last GA. New members have also been assigned to the WG on Professional-Amateur Cooperation in Meteors. All the members have been announced in the newsletter on the commission web site at http://www.iau-c22.org/.

The Meteoroids 2010 meeting was successfully held on May 24-28, 2010 in Colorado U.S.A. The business meeting of the organizing committee was held during this meeting. The next meeting will be held in Pozan, Poland in 2013. Detailed information will be released on the web: http://www.astro.amu.edu.pl/Meteoroids2013/. Commission 22 also proposed a joint discussion entitled "From Meteors and Meteorites to their Parent Bodies: Current status and future developments" at the next general assembly, 2012, in Beijing. This proposal has been approved as JD5.

2.4. *Commission 21 – Light of the Night Sky*

Since the last triennial report, and after extensive consultation with active members of Commission 21, it was decided to dissolve the Commission as it existed within Division III, and to reform it under the name of "Galactic and Extragalactic Background Radiation" under Commission IX.

2.5. *Working Group on Small Bodies Nomenclature*

The CSBN chaired by Jana Ticha continues its work and carries out its usual duties: collecting, judging, and approval of name proposals for minor planets as well as naming of comets. The CSBN still works in the close connection to the Minor Planet Center (both minor planets and comets) and the CBAT (comet names). Since the previous IAU triennium report (2008 July to 2011 October) 2140 minor planets, 6 satellites of minor planets and 546 comets were named. The total number of named minor planets is 16714 as of Oct. 12, 2011. Several batches of new names were the LINEAR discovery namings honor science students who are finalists in a series of science competitions and their teachers.

One of the CSBN goals is to consider names for various unusual objects such as NEOs, TNOs, minor planet satellites, binary bodies, targets of space missions, targets of detailed physical studies and other frequently cited objects. For the 2009-2012 triennium the most significant new names were for the TNO satellites (50000) Quaoar I = Weywot, (90482) Orcus I = Vanth, (120347) Salacia I = Actaea, main belt minor planet moons S/(216) 1 Alexhelios, S/(216) 2 Cleoselene and (702) Alauda I = Pichiüñëm.

In November 2010 the long-time Secretary of the WG SBN Brian G. Marsden unfortunately died. This was a very great loss to the small solar system body nomenclature work, of a wealth of his excellent knowledge, experience and insight. Gareth V. Williams began to serve as the new Secretary of the CSBN.

The CSBN also has been working to refine its naming guidelines for minor planets and satellites of minor planets and binary components of minor planets. Several members worked on the preparation of the web-based system for minor planet name proposals. The CSBN voting now all goes through the website, again in close cooperation with the MPC. The work of the WG will become much more efficient in the near future.

2.6. *Working Group for Planetary System Nomenclature*

From 1 September 2009 to 30 September 2011, 284 new names were assigned for planetary surface features (Mercury: 11, Venus: 14, the Moon: 3, Mars: 36, Phobos: 4, Dione: 16, Enceladus: 27, Rhea: 95, Titan: 26, Lutetia: 37, Vesta: 15). In addition the Jovian satellite Herse (Jupiter L) was named.

During this same time period the following additional actions have been taken:

- For Titan: Introduced the descriptor terms labyrinthus, lacuna and mons and their themes and amended the definition of the descriptor term lacus
- For the Moon: Updated the themes
- For Venus: Changed Metis Regio to Metis Mons, Ningal Undae to Ningal Lineae, Szlanya Dorsa to Szl-anya Lineae, Tezan Dorsa to Tezan Lineae, and dropped: Mnemosyne Regio, Lab Patera, and Lorelei. Added Alexandra to the origin information for Danilova crater.
- For Mars: Expanded the theme for large craters and clarified the theme for small craters. Changed Oenotria Scopulus to Oenotria Scopuli.
- For Mimas: Changed Tintagil Chasma to Tintagil Catena
- For Rhea: Changed Kun Lun Chasma to Kunlun Linea and Pu Chou Chasma to Puchou Catenae
- For Itokawa: Approved the theme
- For Lutetia: Approved naming theme for features

- For Vesta: Set the themes and pre-approved a list of names
- For Mercury: Approved names for 6 map quadrangles.

3. Division Science

3.1. *Commission 15 – Physical Studies of Comets and Minor Planets*

IAU Commission 15 is living in an epoch of frantic activities and exciting discoveries, for reasons which are in part related to the results of important space missions. Currently, two such missions, Rosetta and DAWN, are flying, while another one, Hayabusa, has just recently terminated its tasks, bringing back to Earth some samples of matter taken from the surface of asteroid (25143) Itokawa, and two additional misisons, EPOXI and StardustNExT have just completed their encounter science. Apart from space missions, however, physical studies of asteroids and comets are producing a large harvest of important results, based on purely theoretical studies and on data collected by means of remote observations.

As in the past, the triennial report of Commission 15 has included two main sections corresponding to the reports written by the Chairs of the two biggest Working Groups of this Commission: the Ricardo Gil-Hutton for the WG on Physical Studies of Asteroids (previously known as WG on Physical Studies of Minor Planets, a nomenclature which is now obsolete), and Dan Boice for the WG on Physical studies of Comets. In addition, important input came from a number of colleagues active in a number of Task Groups, including Dominique Bockelée-Morvan, Walter Huebner, Karri Muinonen, Gonzalo Tancredi, who also helped to produce the two main WG reports. A complete list of the members of the WGs of Commission 15 can be found in the web page http://www.iau.org/science/scientific_bodies/commissions/15/.

It must be noted that the number of publications issued in the last triennium in the fields of relevance for Commission 15 is simply too large (about 1,000 papers) to allow us to make an exhaustive review, and for this reason this document does not include a bibliography. This means that the few articles quoted in the rest of this document, which have been taken from the original WG reports, represent only a tiny sample, and are certainly not aimed at representing "the" most important papers published in the triennium. The full references noted in the report may be found from the ADS website (http://adsabs.harvard.edu/abstract_service.html). This will give readers a chance to appreciate the very intense activity in this field, and the outstanding quality of a very large fraction of these studies.

The very intense scientific activity carried out by Commission 15 members and more in general by the whole scientific community active in this field is summarized below. Following a long tradition, this summary is made separately for Comets and Asteroids, but we should take into account that the separation between these two categories of minor bodies has become increasingly less sharp as new discoveries are being done. The recent discovery of the so-called main-belt comets, for example, clearly suggests that our traditional categories do not reflect the complex interrelations between minor bodies orbiting at different heliocentric distances.

Comet Nuclei — Recent spacecraft flybys of comets continue to dramatically improve our understanding of the physical properties of cometary nuclei, particularly the results from Deep Impact (and its extended mission, EPOXI) and Stardust (and its extended mission Stardust-NExT). The extended missions, EPOXI encounter of comet 103P/Hartley 2 and Stardust-NExT return to comet 9P/Tempel 1, were undoubtedly the major events in the past three years concerning comet nuclei. EPOXI flew past comet 103P/Hartley 2, an unusually small but very active nucleus, on 4 November 2010, obtaining both images and spectra. Unlike large, relatively inactive nuclei, this nucleus is outgassing near perihelion primarily due to CO_2, which drags large pieces of ice off the nucleus. It also shows substantial differences in the relative abundance of volatiles from various parts of the nucleus. These results have strengthened our view that comet nuclei are of modest size (R <20 km), high porosity (density of only about 350 kg m^{-3}), low strength (tensile strength on order of 10^3 dyn cm^{-2}), and are chemically heterogeneous. The Stardust results show the need for mixing of microscopic solids from 1 AU to several tens of astronomical units at an early phase of the protoplanetary disk prior to the accretion of comets. The Stardust-NExT mission returned to comet Tempel 1 on 14 February 2011, and imaged the nucleus, identifying a plausible feature at the Deep Impact site. These missions continue to stimulate a large number of studies - observational, phenomenological, and theoretical - in the endeavor to interpret and understand the large body of accumulated data. The shape, topography, temperature distribution, spin state, composition, and activity pattern of the Hartley 2 nucleus were analyzed and discussed in papers published in *Astrophysical Journal Letters* (Volume 734, Issue 1, L1-8, June 2011) and in *Science* (Volume 332, pp. 1396-1400, June 2011), as well as a forthcoming special issue of *Icarus* in 2012. Due to space limitations here, only a few results can be included. For more details, see the excellent review of comet nuclei recently published by A'Hearn (2011).

Large flows with very smooth surfaces on the nucleus of Tempel 1 were seen in the Deep Impact images. There is some evidence that these layers are receding at the edges as predicted by the model of Britt *et al.* (2004) developed for the nucleus of comet Borrelly. The formation and morphologies of surface features appear to be driven by differential rates of sublimation erosion, which lead to active sites. Observations from the Stardust-NExT mission show that these layers receded in places by >10m between 2005 and 2011, further strengthening this view.

Another well-studied comet nucleus was 67P/Churyumov-Gerasimenko, the target of the Rosetta mission. Current mission status and science of Rosetta and many other spacecraft missions to small solar system bodies were presented at the B04 Event (Small Body Exploration: Past, Present, and Future Space Missions) at COSPAR 2010 in Bremen, Germany, during 21-23 July, 2010. A series of papers in published in a special issue of *Planetary and Space Science* describing its flyby of asteroid (2867) Steins (Vol. 58, Issue 9, July 2010, pp. 1057-1128). An up-to-date data set of the sizes and shapes of cometary nuclei is available from NASA's Planetary Data System (Paudel & Kolokolova 2010).

The latest analysis and results from SEPPCoN (Survey of Ensemble Physical Properties of Cometary Nuclei) of measuring thermal emissions of the sample nuclei have been published (Groussin *et al.* 2009; Licandro *et al.* 2009). This on-going survey involves studying 100 Jupiter-Family Comets (JFCs), about 25% of the known population, at both mid-infrared and visible wavelengths to constrain the distributions of sizes, shapes, spins, and albedos of this population. An important goal of SEPPCoN is to accumulate a large comprehensive set of high quality physical data on cometary nuclei in order to make accurate statistical comparisons with other minor-body populations such as Trojans, Centaurs, and Kuiper-belt objects. Information on the size, shape, spin-rate, albedo and color distributions is critical for understanding their origins and the evolutionary processes affecting them.

The size distribution of sun-grazing comets (members of the Kreutz family) has been investigated by Knight *et al.* (2010) using observations from the SOHO spacecraft coronagraphs. The estimated sizes are well fit by a power law of slope -2.2 over the range from several meters to several tens of meters, sharply steepening for larger sizes.

Modeling of comet nuclei focused on comet 67P/Churyumov-Gerasimenko, the Rosetta Mission target. A fully 3-dimensional model of comet nucleus evolution, including dust mantle formation was developed by Rosenberg and Prialnik (2009, 2010). Dust mantle thickness varies over the surface from 1 cm to about 10 cm. The water crystallization front advances inward in spurts, and its depth varies between 1 and several meters. Internal inhomogeneities affect both the surface temperature and the activity pattern of the comet. In particular, they may lead to outbursts at large heliocentric distances and also to activity on the night-side of the nucleus. Other groups have continued their quasi-three-dimensional thermal models for irregularly shaped cometary nuclei to interpret the current activity of comets in terms of initial characteristics, and to predict shape and internal stratification evolution of the nucleus (*e.g.*, Lasue *et al.* 2008; De Sanctis *et al.* 2009).

Gas coma, chemistry, plasma and tails — The study of compositional diversity among comets motivated several observational campaigns. Since A'Hearn *et al.* (1985), surveys of comets have provided taxonomies based on the abundance ratios of parent volatiles. To date, more than 20 molecules native to the nucleus ices are identified in comets. Measurements of nuclear spin ratios (in water, ammonia, and methane) and of isotopic ratios (D/H in water and HCN; $^{14}N/^{15}N$ in CN and HCN) have provided insights into the formation of the parent species. Identifications of abundant product species (*e.g.*, HNC) have provided evidence of gas-phase chemistry in the inner coma. Due to space limitations, only a few results are included here. For more details, see the excellent review of the chemistry of comets recently published by (Mumma & Charnley 2011).

The EPOXI results confirm that Comet Hartley 2 is enriched in CO_2 (nearly 20% relative to water) and that CO was unusually depleted (several tenths of a percent relative to water; Weaver *et al.* 2011). The production rate of water varied by a factor of two with nucleus rotation but CO_2 varied by a smaller factor, suggesting compositional heterogeneity of the nucleus. The production rates of other parent volatiles (H_2O, C_2H_6, HCN, CH_3OH) also vary with rotational phase (Dello Russo *et al.* 2011, Drahus *et al.* 2011, Mumma *et al.* 2011, Meech *et al.* 2011) but their abundance ratios remained constant during the EPOXI flyby. Mumma *et al.* (2011) suggest the presence of two separate ice phases (polar and apolar) in comet nuclei.

Mumma *et al.* (2011) obtained an ortho-para abundance ratio (OPR) in H_2O of 2.85 ± 0.20 in comet Hartley 2, consistent with statistical equilibrium (T_{spin} >32-55 K). Dello Russo *et al.* (2011) reported OPR=3.4±0.3. In H_2O^+ and NH_2, OPRs provided spin temperatures of >25 K and 33±3 K, respectively (Meech *et al.* 2011) and isotopic ratios for $^{12}C/^{13}C$ (95 ± 15) and $^{14}N/^{15}N$ (155 ± 25) in CN that were consistent with the mean values reported for other comets. Herschel achieved the first detection of HDO in a Jupiter-family comet (103P/Hartley 2; Hartogh *et al.* 2011). The D/H value of $(16.1\pm2.4)\times10^{-5}$ is the same as that in Earth's oceans and twice lower than values measured in Oort cloud comets. The reservoir of Earth-like water in the Solar System is significantly larger than previously thought.

Remote sensing can establish rough taxonomic classes of comets by measuring the volatile organic inventory of parent molecules at very low abundances (about 100 ppm relative to H_2O), including isotopic chemistry and nuclear spin ratios of molecules. More detailed compositional information is needed for the less-volatile organics. Stardust samples can provide this information for refractory organics but not for volatiles. The rendezvous of the Rosetta spacecraft with comet 67P/Churyumov-Gerasimenko in 2014 will measure the volatile fraction, while returned samples (such as the proposed Triple F and Comet Surface Sample Return missions) could provide this information as well.

Cometary atmosphere chemical modeling has been developed in several papers. Lederer *et al.* (2009a, 2009b) describe a three-dimensional, time-dependent Monte Carlo model developed to analyze the chemical and physical nature of a cometary gas coma including the necessary physics and chemistry to model comet Hale-Bopp at 1 AU from the Sun. Model calculations show that photoelectron impact excitation of CO and dissociative excitation of CO_2 can together contribute about 60-90% to Cameron-band emission in comet Hartley 2 (Bhardwaj and Raghuram 2011). Noteworthy are recent hydrodynamic simulations of the gas coma, which show that gas structures produced by nucleus composition inhomogeneities and nucleus shape and topography are indistinguishable. Boissier *et al.* (2010) find that an initial uniform surface flux of CO is assumed, the spiral structures created by the nucleus shape in the CO coma of comet Hale-Bopp are too faint to account for the observational data. Boice and Goldstein (2009) have applied their chemical model of cometary atmospheres to the plumes observed by the Cassini Mission at Enceladus. An extensive review of the expected plasma environment of comet 67P/Churyumov-Gerasimenko and current modeling capabilities has been given by Hansen *et al.* (2009).

Comets in the radio domain — Observations of comets in the radio and microwave domains are useful tools to probe the nucleus and dust thermal emission, and study the chemistry and kinematics of cometary atmospheres. The detection of the nucleus of 8P/Tuttle was achieved using the Plateau de Bure interferometer (Boissier *et al.* 2011). The abundances of a number of parent molecules (*e.g.*, HCN, CH_3OH, H_2S, CH_3CN) were measured in comets 73P, C/2002 T7, C/2002 X5, C/2002 V1 and C/2006 P1, the latter four while being at distances <0.5 AU from the Sun (Hogerheijde *et al.* 2009; Paganini *et al.* 2010; Biver *et al.* 2011). Analysis of interferometric CO data on Hale-Bopp does not confirm the presence of an extended source of CO gas, as suggested by IR data (Bockele-Morvan *et al.* 2010). Reviews of the compositional diversity of cometary atmospheres, based on radio observations, have been published (Crovisier *et al.* 2009). Periodic variations of emission line profiles, related to nucleus rotation, were detected in comets Hale-Bopp, C/2001 Q4, 73P and 103P and used to investigate the rotational state of their nuclei and the presence of active spots (Bockele-Morvan *et al.* 2009, Boissier *et al.* 2010, Biver *et al.* 2011, Drahus *et al.* 2010, 2011). A few papers have been published presenting results based on Herschel observations. A number of water rotational lines were detected in comets 81P, C/2008 Q3 and 103P (Hartogh *et al.* 2010, Val de Borro *et al.* 2010, Meech *et al.* 2011).

Comet dust and distributed sources — Our understanding of the complexity of cometary dust has significantly progressed during the past triennium: The Stardust mission brought for the first time dust samples from the coma of a comet (81P/Wild 2), and the Deep Impact mission released dust for remote observations from the subsurface of another comet (9P/Tempel 1). Analyses of these data have continued during these past 3 years. These missions have been complemented by ground observations, *i.e.*, spectroscopic observations providing information about the composition of the dust and polarimetric observations providing information about the physical properties of the dust.

Kelley & Wooden (2009) review the composition of JFC dust as inferred from infrared spectroscopy. They find that JFCs have 10μm silicate emission features roughly 20-25% above the dust continuum, similar to the weakest silicate features in Oort Cloud (OC) comets. Recent evidence suggests that grain porosity is different between JFCs and OC comets, but more observations and models of silicates in JFCs are needed. Models of observations from ground-based telescopes and the Spitzer Space Telescope have shown that JFCs have crystalline silicates with abundances similar to or less than those found in OC comets, although the crystalline silicate mineralogy of comets 9P/Tempel and C/1995 O1 (Hale-Bopp) differ from each other in Mg and Fe content. The heterogeneity of comet nuclei can be assessed with mid-infrared spectroscopy and evidence shows heterogeneous dust properties in the nucleus of comet 9P/Tempel. Models of dust formation, mixing in the solar nebula, and comet formation are constrained by the observed range of Mg and Fe content and the heterogeneity found in comet 9P/Tempel.

Recent successes of the Stardust Comet Sample Return and Hayabusa Asteroid Sample Return missions have demonstrated the powerful scientific insights that can be gained by bringing materials from extraterrestrial bodies to terrestrial laboratories for analysis. Such missions can establish the nature of extraterrestrial materials at a level of detail that can never be matched by in situ analyses, and analytical techniques can be used that do not suffer from the normal limitations of flight instrumentation. Analyses of the samples from the Stardust mission have shown the presence of a very wide range of olivine and low Ca-pyroxene compositions, possibly reflecting various formation regions in the protoplanetary nebula, and of refractory organic compounds. Carbonaceous matter in Stardust samples returned from comet 81P/Wild 2 is observed to contain a wide variety of organic functional chemistry (de Gregorio *et al.* 2011).

Remote observations of the light scattered by the cometary dust coma are of major importance for determining the physical properties of the particles. Light scattering and especially linear polarization observations allow comparison between different coma regions and different comets, to retrieve physical properties of the dust particles, and to characterize their evolution around perihelion passage. Hadamcik *et al.* (2010) find that polarization and intensity variations in the coma of 67P/Churyumov-Gerasimenko are reminiscent of those noticed for some comets such as comet 81P/Wild 2 and comet 9P/Tempel 1. The presence of rather large particles can thus be suggested before and just after perihelion and the ejection of post-perihelion smaller grains, eventually in fluffy aggregates. A strong seasonal effect related to the obliquity of the comet suggests that the different grains originate in different hemispheres of the nucleus. Another important result, as far as light scattering by cometary dust is concerned, was related to polarimetric observations of the comet C/2007 N3 (Lulin). Woodward *et al.* (2011) find that large, low-porosity, absorptive aggregate dust particles best explain both the polarimetric and the mid-infrared spectral energy distribution.

The Wide-Field Infrared Survey Explorer (WISE) was launched on 14 December 2009. WISE imaged more than 99% of the sky in the mid-infrared for a 9-month mission lifetime. In addition to its primary goals of detecting infrared galaxies and brown dwarfs, WISE detected over 155,500 Solar System bodies, 33,700 of which were previously unknown. Most of the new objects were Main Belt Asteroids, with emphasis on the discovery of Near Earth Asteroids. Comets observed by WISE included both long and short-period orbits, including Hartley 2. Over 120 comets were imaged by WISE, with discoveris of 20 new comets. Observations of comets in the WISE thermal-infrared bands are being used to provide comet nucleus size constraints, estimate coma dust temperature and particle size distributions, and derive comet trail grain sizes and β-parameters.

Cometary material origins and laboratory experiments — Measurements of the isotopic ratios and the nuclear spin temperatures in molecules are important to investigate origin of cometary materials (see section on Gas coma, chemistry, plasma, and tails). Kawakita and Kobayashi (2009) discussed constraints on the formation region of comets from their nuclear spin temperatures and D/H ratios. Relating to the origin of cometary materials, mechanisms of the outward transport of materials in the solar nebula are investigated theoretically (*e.g.*, Boss 2010; Ciesla 2009, 2011; and Hughes and Armitage 2010). Such outward transport is necessary to explain co-existence of icy materials with high temperature processed materials like crystalline silicates and CAIs in cometary grains (see Section Comet dust and distributed sources). Marboeuf *et al.* (2010) have shown that clathrate hydrates could exist in short-period comet nuclei, *i.e.*, that the thermodynamic conditions in their interiors allow the existence of clathrate hydrates in Halley-type comets.

Many laboratory studies were carried out. Pat-El *et al.* (2009) performed an experimental study of the formation of an ice crust and migration of water vapor in a comet's upper layers. Gundlach *et al.* (2011) showed that observed gas production rates of pure hexagonal water ice could be reproduced experimentally. The reduction of the gas production rate due to an additional dust layer on top of the ice surface was measured and compared to the results of another experimental setup in which the gas diffusion through dust layers at room temperature.

Many experimental studies were also performed on interstellar or cometary ice analogues (made from simple molecules such as H_2O, CO_2, and CO) irradiated by high-energy particles like UV-photons, electrons, and ions. Complex molecules could be formed in these ices at low temperatures (*e.g.*, Loeffler *et al.* 2011; Bennett et al. 2011; Hudson *et al.* 2009), while simple molecules or atoms were desorbed from the icy surfaces in some cases. Vigren *et al.* (2010) investigated dissociative recombination of protonated formic acid with implications for cometary chemistry.

Asteroids — More than 500 refereed papers dealing with the physical properties of the asteroids were published during the last triennium, between August, 2009 and October, 2011. The main areas covered in these publications were size distributions, masses and densities; photometry, shapes, and spin properties; radar, thermal IR, optical polarimetry, light-scattering phenomena; imaging, disk-resolved images, and binary systems; spectra, taxonomy, composition and space weathering; origins, impacts, families and evolutionary processes; space missions; asteroid-meteorite and asteroid-comet connections; near Earth asteroids (NEAs); trojan asteroids; and distant asteroids: Centaurs and TNOs.

Just as a few and certainly non-exhaustive examples of topics and papers of particular interest, we mention the discovery and continued characterization of the main belt comets (Hsieh & Jewitt (2006), Hsieh (2009-2011) Licandro *et al.* 2011), asteroid mass and density determinations (Zielenbach 2011), radar studies (Shepard *et al.* 2010), thermal IR (Lamy *et al.* 2010), polarimetry (Gil-Hutton & Cañada-Assandri 2011), water detection (Licandro *et al.* 2011), spectroscopy and composition (Reddy *et al.* 2010), space weathering (Willman *et al.* 2010), and light scattering (Muinonen *et al.* 2010). In addition, we also stress that the WISE mission, a satellite observing the sky at thermal IR wavelengths, has produced a major improvement in the field of thermal radiometry in the last triennium.

It is simply impossible to give a reasonable summary of all relevant activities in the space allocated to this report. The above-mentioned topics include all the traditional branches of investigations which make asteroid science so complex and interdisciplinary. In particular, the study of evolutionary processes, both physical and dynamical, are extremely important, since one of the ultimate goals of asteroid science is to be able to understand the properties of the primitive protoplanetary disk at the epoch of planetesimal growth, starting from the observable properties of the minor bodies as they are today. Processes including collisions and dynamical evolution triggered by physical properties, like the widely discussed Yarkovsky and YORP effects, have been the subjects of many investigations during the triennium. The traditional techniques of remote observation, including photometry and spectroscopy (both at visble and near-IR wavelengths), as well as polarimetry, high-resolution imaging and radar have been applied and the results have been described in many exciting papers. Significant advances have been made in the field of taxonomic classification, taking profit of the possibility to add near-IR reflectance spectra to the evidence coming from traditional spectroscopy and spectrophotometry at visible wavelengths. In this respect, there has been also an effort to produce practical tools to help observers in obtaining taxonomic classifications of object observed spectroscopically, by developing public facilities which can be found in the web. A good example is the Bus-De Meo spectrum classification facility available at http://smass.mit.edu/busdemeoclass.html. Other authors are also developing some similar tools, not limited to the field of spectroscopy and taxonomy, and this certainly is a very useful help for the activities in different branches of asteroid science.

3.2. *Meteors, Meteorites and Interplanetary Dust*

Significant advances were made as a result of meteorite falls. In particular, the Almahata Sitta meteorite was the first body to be observed and tracked in space, as asteroid 2008 TC3, prior to falling in the Nubian Desert of Northern Sudan on October 7, 2008. During a systematic search of the impact zone, some 600 fragments were recovered, most of which turned out to be ureilites, a rare meteorite type. A "Workshop on Asteroid 2008 TC3" was held at the University of Khartoum, Khartoum, Sudan, in December 2009, results of which were published in a special issue of *Meteoritics and Planetary Science* in 2010. Other recovered meteorite falls include the 9 April 2009 L6 ordinary chondrite Jesenice, Slovenia, the 26 September 2009 H5 ordinary chondrite Grimsby, Canada, the 28 February 2010 H5 ordinary chondrite Kosice, Slovakia, and the 13 April 2010 H5 ordinary chondrite Mason Gully, Australia. All recoveries were made as a consequence of the observed fireball, previously a rare event. This advance was a product of newly developed photographic and video camera fireball networks.

Significant advances in meteor astronomy were made from observations of meteor showers with radar and with security video cameras, networks of which provided precise orbit determinations. Some 83 meteor showers were added to the IAU C22 Working List of Meteor Showers. Particularly prolific was the Canadian Meteor Orbit Radar (CMOR), so far the best radar at separating showers from the sporadic background, and the video camera networks of SonotaCo in Japan, IMO in Europe, and CAMS in California. Results of CMOR included the detection of an outburst of Daytime Craterids, which were tentatively linked to hyperbolic comet C/2007 W1 (Boattini). CAMS detected the February eta Draconids, an outburst on February 4, 2011, caused by the dust trail of a yet undiscovered long period comet. Other unusual showers in this period included the 2009 and 2010 Orionids, now thought to be dust of 1P/Halley trapped in mean motion resonances. Newly established associations between meteor showers and parent bodies includes that of 169P/NEAT with the alpha-Capricornids. Thanks to the advanced application of theoretical dust trail calculations for meteor shower predictions, two observing campaigns were carried out during the 2009 Leonids and the 2011 Draconids. The Draconid shower was observed with two aircraft in a mission sponsored by CNRS and DLR. Both showers helped improve the models.

In part based on the meteor shower observations, the first dynamical model of the zodiacal cloud was constructed, showing the importance of (mostly dormant) Jupiter Family comets as a source of the interplanetary dust.

The re-entry of the spacecraft HAYABUSA over southern Australia on June 13, 2010, was observed as an artificial meteor experiment, both by ground-based observers and in a NASA sponsored airborne campaign. Scientific results have been published in the Special Issue "Re-entry of HAYABUSA spacecraft", Publication of Astronomical Society of Japan.

3.3. *Bioastronomy*

The primary activity of C51 is the triennial meeting. In 2011 for the first time this meeting was held jointly with a separate scientific society, the International Society for the Study of the Origin of Life – The International Astrobiology Society (ISSOL). The conference, called Origins 2011, http://www.origins2011.univ-montp2.fr/, took place July 3-8 in Montpellier, France. Thanks to an outstanding effort by the Scientific Organizing Committee (SOC), it was possible to retain the traditional C51 approach which supported only plenary sessions, so that all participants could fully participate in all sessions. The SOC had both nationality and gender balance, consisting of Alan Boss (USA), Andre Brack (France), Jose Cernicharo (Spain), Pascale Ehrenfreund (The Netherlands), Natalia Gontareva (Russia), Nils Holm (Sweden), Gerda Horneck (Germany), William Irvine (USA), Kensei Kobayashi (Japan), Ramanarayanan Krishna-murthy (USA), Antonio Lazcano (Mexico), Anny-Chantal Levasseur-Regourd (France), Claudio Maccone (Italy), Francois Raulin (France), Alan Schwartz (The Netherlands), and Janet Siefert (USA). The conference had 405 attendees, who gave 50 oral presentations and 320 poster presentations. Thanks to grants from NASA and ISSOL, 40 travel grants were made to young investigators, allowing them to attend the conference.

The success of Origins 2011 is evident from the decision of C51 and ISSOL to hold the next triennial conference also as a joint meeting. The C51 OC and the ISSOL Executive Committee agreed to accept an invitation from Japan, and the meeting will be in Nara, Japan, July 6-11, 2014. The Honorary Chair of the Local Organizing Committee is Norio Kaifu, President of the IAU, while the Chair is Kenji Ikehara, Director of the Nara Study Center and Past President of the Japanese Society for the Study of the Origin and Evolution of Life. The joint Chairs of the Scientific Organizing Committee will presumably be C51 OC member and presumed next Vice-President Sun Kwok from the University of Hong Kong (China) and Sandra Pizzarello (USA) from ISSOL. Current C51 President William Irvine plans to join ISSOL President David Deamer on a visit to the Nara venue in spring, 2012.

At the General Assembly in Beijing in 2012 C51 will be a primary sponsor of IAU Symposium 293, "Extrasolar Habitable Planets", with the Chair of the SOC being Nader Haghighipour from the University of Hawaii (USA); and also of Special Session 16, "Unexplained Spectral Phenomena in the Interstellar Medium", the SOC Chair being Sun Kwok from the University of Hong Kong (China). In addition, C51 is specifically supporting Special Session 14, "Communicating Astronomy with the Public for Scientists", and Special Session 15, "Data Intensive Astronomy".

4. Closing remarks

In spite of well known and widespread problems of financial support in these times of economic problems and consequent difficulties in getting dedicated manpower in many countries, the scientific community working in the field of the studies of comets and asteroids is extremely active and productive. This is a field of investigation which is very popular in many respects also in terms of public outreach, as a consequence of its intrinsic scientific importance and interest, as well as due to the fact of being strictly related to space missions which produce exciting results. These missions, which are based on the imagination and every-day activity carried out by a community of scientists from many countries, are also very useful to remind us that the mankind is able to obtain spectacular and breath-taking results when human genius is devoted to the study of Nature.

Karen Meech
president of the Division

Transactions IAU, Volume XXVIIIA
Reports on Astronomy 2009–2012   © International Astronomical Union 2012
Ian Corbett, ed.   doi:10.1017/S1743921312002712

COMMISSION 53   EXTRASOLAR PLANETS

EXTRASOLAR PLANETS

PRESIDENT   Alan Boss
VICE-PRESIDENT   Alain Lecavelier des Etangs
PAST PRESIDENT   Michel Mayor
ORGANIZING COMMITTEE   Peter Bodenheimer,
                                   Andrew Collier-Cameron,
                                   Eiichiro Kokubo,
                                   Rosemary Mardling,
                                   Dante Minniti,
                                   Didier Queloz

TRIENNIAL REPORT 2009-2012

1. Introduction

Commission 53 was created at the 2006 Prague General Assembly (GA) of the IAU, in recognition of the outburst of astronomical progress in the field of extrasolar planet discovery, characterization, and theoretical work that has occurred since the discovery of the first planet in orbit around a solar-type star in 1995. Commission 53 is the logical successor to the IAU Working Group on Extrasolar Planets (WGESP), which ended its six years of existence in August 2006. The founding President of Commission 53 was Michael Mayor, in honor of his seminal contributions to this new field of astronomy. The current President is Alan Boss, the former chair of the WGESP. The current members of the Commission 53 (C53) Organizing Committee (OC) began their service in August 2009 at the conclusion of the Rio de Janeiro IAU GA.

2. Exoplanet Definitions and Lists

The WGESP developed in 2001 a Working Definition of what is a 'planet', subject to change as we learn more about the population of extrasolar planets. The Working Definition was last modified in 2003 to address the question of objects found by imaging surveys in regions of active star formation. The current Working Definition can be found on the WGESP web pages located at:

http://www.dtm.ciw.edu/boss/iauindex.html

Note that this definition does not attempt to address the lower mass limit for the range of bodies that should be considered as planets, other than to say that the lower mass limit should be same as that used for our Solar System. In 2006, the IAU adopted a definition for Solar System planets, which can be found here:

http://www.iau.org/public/pluto/

where the definition of a planet is given as:

'A celestial body that (a) is in orbit around the Sun, (b) has sufficient mass for its self-gravity to overcome rigid body forces so that it assumes a hydrostatic equilibrium (nearly round) shape, and (c) has cleared the neighbourhood around its orbit.'

Unfortunately, for exoplanet systems, we can be sure of (a), but not of (b) [even for transiting systems], much less (c). Hence the above definition in practice cannot be applied to determine the lower mass limit for exoplanets. However, this is not problem for exoplanets, at least not to date. This is because, with the exception of certain pulsar planets, all the extrasolar planets discovered to date are more massive than the Earth. The C53 OC thus has not seen fit to modify this Working Definition, though the situation may well change in the future.

The WGESP maintained a list of planetary candidates that met its criteria for acceptance as planets up until its demise in August 2006. This list also established a criterion for discovery rights, namely the date of submission for publication in a refereed journal. This list can be found on the WGESP web pages. C53 has decided not to try to continue to maintain this list of planets, given the immense popularity and greater usefulness of the list maintained by Jean Schneider and his colleagues at the Extrasolar Planets Encyclopaedia web site:

http://exoplanet.eu/

As of June 2011, the Extrasolar Planets Encyclopaedia raised the upper mass limit for inclusion in the Encyclopaedia to a value of 25 Jupiter masses This inclusion formally violates the WGESP upper mass limit for an object to be called a 'planet' of 13 Jupiter masses (for solar metallicity). However, in the past the Encyclopedia has routinely listed objects more massive than the WGESP limit, e.g., the RV detection with m sin i = 14.4 Jupiter masses of HD 162020b. The Encyclopaedia's reasoning is that it is preferable to base the upper mass limit for exoplanets on the empirical evidence based on the census of low mass stars, brown dwarfs, and Jupiters, as argued in their paper (Schneider *et al.* 2011), which may be downloaded from this web page:

http://exoplanet.eu/README.html

Schneider *et al.* (2011) invoke the 'brown dwarf desert' as an empirical indicator of two different populations, the 'exoplanets' and the 'binaries' (see Figure 2a on page 3 of Schneider *et al.* 2011), and suggest that these two populations separate at a mass of about 25 Jupiter masses. This empirical approach to defining the upper mass limit for an exoplanet thus differs greatly from that of the WGESP, which was based on the ability of objects to undergo thermonuclear fusion of deuterium. The question of the definition of what constitutes an 'exoplanet' evidently is one that will continue to be discussed and debated.

3. Nomenclature

In the last several years the C53 OC discussed and debated at some length several issues regarding the nomenclature for newly discovered extrasolar planets. The first issue arose as a result of a detailed paper written by W. Lyra proposing a scheme for naming exoplanets by using a number of names from the classical (Greek, Latin) literature, rather than by the more mundane, but functional, current system of using the star's name (e.g. 51 Peg) followed by lower case letters, in order of discovery, e.g., 51 Peg b, c, d. [The use of proper names, such as those in use for asteroids and comets, was also considered.] The C53 OC decided against changing the current system of naming exoplanets, which is geared toward the clarity of astronomical databases of stars and exoplanets.

The second nomenclature issue dealt with the preferred means for naming exoplanets in systems of binary stars, e.g., Alpha Cen AB, where the planets could orbit either of the binary stars or could orbit both stars, i.e., a circumbinary planet. The C53 OC members discussed and voted upon several specific schemes for handling all possible combinations of binary and multiple star systems, but the C53 OC was unable to arrive at a consensus recommendation. Hence, no recommendation was made, and journal articles about exoplanets in binary systems are now published with nomenclatures agreed upon by the authors and journal editors involved.

4. Organizing Committee

The C53 Organizing Committee (OC) now has six members, one from the USA, two from Europe, one from Japan, one from Chile, and one from Australia, in accordance with IAU rules, which state that the OC should not have more than eight members and that OC members should be geographically diverse. Given that OC members are allowed to serve for two terms, the current OC is expected to serve again during 2012-2015.

5. Symposium and Special Session Sponsorship

C53 has been asked to support various proposals to hold IAU Special Sessions and Symposia, either at the 2012 IAU GA, or on their own. Many of these proposals were judged to be appropriate for C53 support. One of the successful proposals is of particular relevance for C53, namely IAU Symposium 293 on the Formation, Detection, and Characterization of Extrasolar Habitable Planets, to be held during the Beijing IAU GA.

6. New Members

As a recently formed IAU Commission, C53 continues to seek astronomers who wish to be recognized as C53 members. We encourage interested IAU members to ask to join C53, which can be accomplished simply by sending an e-mail to the C53 President or Vice President.

7. C53 Web Pages

The web pages for C53 are located at:

http://www.dtm.ciw.edu/boss/c53index.html

where a listing of the current members of C53 can be found, along with links to the WGESP web pages and to other items of interest.

8. Closing Remarks

C53 held its first Business Meeting during the IAU General Assembly in Rio de Janeiro on August 12, 2009. Several dozen current and prospective members of C53 attended and participated in the meeting. We look forward to holding our second C53 business meeting at the 2012 IAU GA and hope for an even larger attendance. At this meeting, we expect to welcome the selection of a new Vice President and to accept a number of new members of C53.

Alan P. Boss
President of the Commission

Transactions IAU, Volume XXVIIIA
Reports on Astronomy 2009–2012   © International Astronomical Union 2012
Ian Corbett, ed.   doi:10.1017/S1743921312002724

WORKING GROUP   NEAR-EARTH OBJECTS

NEAR-EARTH OBJECTS

CHAIR   **Alan W. Harris**
VICE-CHAIR   **Giovanni Valsecchi**
PAST CHAIR   **David Morrison**

TRIENNIAL REPORT 2009-2012

1. Introduction

Following its meeting in May, 2010, the IAU Executive Committee requested that a Working Group on NEOs within Div. III be re-activated and carry out the following activities:

a) investigate and formulate requirements for an international ground- and/or space-based NEO survey, to detect, track and characterize (optical/IR, radar) 90% of all NEOs with D >40 m and to establish as such a permanent International NEO Early Warning System; to submit to the President, Vice-President and OC of Division III by March 31, 2011, a progress report and by March 31, 2012, a final report on this matter, to be forwarded to the President and General Secretary of the IAU;

b) assemble a SOC in order to write and submit to the IAU Assistant General Secretary before December 1, 2010, a proposal for a GA IAU Symposium or a GA Special Session, to be held during the IAU XXVIII General Assembly, August 20-31, 2012 in Beijing, on theoretical and observational aspects of NEO research in general, and on requirements and other aspects of a permanent International NEO Early Warning System in particular;

c) prepare and submit to the IAU General Secretary by January 31, 2012, a Resolution for consideration by the IAU XXVIII General Assembly in Beijing, August 2012, asking for international action and support to establish an International NEO Early Warning System; such a Resolution, if accepted by the IAU XXVIII General Assembly, to be addressed to the IAU National Members, to the United Nations Committee on the Peaceful Uses of Outer Space (UN-COPUOS), and to the International Council for Science (ICSU).

2. New Working Group

Karen Meech, President of Divison III, with advice from a number of IAU members active in NEO studies, appointed the following members of the new WG:

PRESIDENT   Alan W. Harris
VICE-PRESIDENT   Giovanni Valsecchi
PAST PRESIDENT   David Morrison

Abell, Paul	USA	Consultant, human exploration
Beshore, Edward	USA	PI, Catalina Sky Survey (CSS)
*Chesley, Steve	USA	VP, C20
*Harris, Alan W.	USA	NEOWG Chair
*Harris, Alan W.	Germany	NEOMAP ESA
*Huebner, Walter	USA	
*Jedicke, Robert	USA	
Jones, Lynne	USA	Consultant, LSST
*Koschny, Detlef	Netherlands	ESA NEOs
*Larson, Steve	USA	CSS
*Mainzer, Amy	USA	Consultant, WISE NEO
*McMillan, Robert	USA	PI, Spacewatch
Meech, Karen	USA	Pres. Div III
*Milani, Andrea	Italy	
*Morrison, David	USA	NEOWG Past Chair
*Rickman, Hans	Sweden	
Sanchez, Salvador	Spain	Consultant
*Shustov, Boris	Russia	Chair NEO hazards, RAS
Spahr, Tim	USA	Director, Minor Planet Center
*Valsecchi, Giovanni	Italy	VP Div III, NEOWG Vice Chair
*Yoshikawa, Makoto	Japan	Pres. C20
Zhu, Jin	China	C20 OC

*Member of SpS SOC (see (b) below).

3. Progress

Some progress has been made on the above action items, which I describe as follows.

a) On the face of it, a "survey, to detect, track and characterize (optical/IR, radar) 90% of all NEOs with D >40 m" appears to be beyond reasonable or cost-effective capability from either ground or space. Expert studies done by the U.S. National Research Council and NASA instead recommend a combination of cataloging most (90%) of objects >140 m diameter, so that an impact from such an object could be predicted years or decades in advance, and mitigation measures considered. For smaller objects, down to 40 m or so in diameter, early warning surveys that could provide days or weeks warning in time for civil defense measures seem more appropriate. The WG would welcome further advice from the EC before undertaking further study or preparing a final report in 2012.

b) Upon consultation among the members of the WG, it was decided that a GA Special Session would be more appropriate than a full IAU Symposium. The SOC was composed as indicated (*) from the above list of WG members. Co-Chairs of the SOC are Valsecchi, Huebner, and Harris (USA). Editors of a proceedings were chosen: Milani, Harris (USA), Huebner, and Rickman. The proposal was accepted by the EC in its meeting in Prague in May, 2011. We record in this report the abstract of the proposal as accepted:

The Spaceguard Report, published in 1992, advocated discovery of 90% of Near-Earth Objects larger than 1 km in diameter. That goal has been essentially achieved, and a new goal has been proposed to reach 90% completion to a size of 140 m diameter. As we turn attention to smaller size NEOs, better and larger telescopic systems are required, data storage and processing becomes more complicated both by the shear numbers of objects but also the increased role of non-gravitational (radiation) forces on the objects, and appropriate mitigation responses are different for small (but potentially more frequent, and with less warning) impact events. Thus we propose a Special Session of 1-2 days duration to assess what has been achieved to date, and evaluate future plans for surveys, required physical observations, and mitigation scenarios.

We suggest that since the subjects to be addressed in the SpS bear directly on matters to be addressed in charge (a), perhaps a final report on those matters should be postponed until after the SpS.

c) The WG has not yet taken any action on the third item, due January, 2012.

Alan W. Harris
Chair of WG

Transactions IAU, Volume XXVIIIA
Reports on Astronomy 2009–2012
Ian Corbett, ed.

© International Astronomical Union 2012
doi:10.1017/S1743921312002736

DIVISION III / SERVICE
MINOR PLANET CENTER

DIRECTOR   Timothy B. Spahr
ASSOCIATE DIRECTOR   Gareth V. Williams
DIRECTOR EMERITUS   Brian G. Marsden

TRIENNIAL REPORT 2009-2012

1. Introduction

The activity of the Minor Planet Center continued generally to increase during the triennium. This report covers the period 2008 July 1 to 2011 October 12, and the phrase "during the triennium" in this report should be understood to mean this period.

During the triennium, the total number of observations in the MPC's files increased by 56%, from 55.4 million to 86.4 million. The next-generation Pan-STARRS (Panoramic Survey Telescope and Rapid Response System) program began to produce results, and the existing Catalina Sky Survey, Mount Lemmon Survey, Siding Spring Survey and Lincoln (Laboratory) Near-Earth Asteroid Research (LINEAR) programs continued to be very productive. In addition, the Wide-field Infrared Survey Explorer (WISE) spacecraft produced two million observations from low Earth orbit.

The number of numbered minor planets increased by 60%, from 189005 to 301841, during the triennium. The numbering of a minor planet should signify that the object has been sufficiently well observed to ensure that it is unlikely to be lost in the foreseeable future. The numbered minor planets represented 46% of the orbits in the MPC files in mid-2008 and 53% in 2011 October.

2. Publications and archiving

The permanent archiving of data continues to be done on a monthly basis, coinciding with the publication of the *Minor Planet Circulars* and the *Minor Planet Circulars Orbit Supplement*. As the traditional publication of the MPC, dating back to 1947, the former is a summary of MPC activity, the 13362 pages published during the triennium brings the total to 76678. The *Orbit Supplement*, first published in 2000, gives full details on the new numberings and the new identifications, the 68138 pages published during the triennium brings the total to 208366. The *Minor Planet Circulars Observation Supplement*, listing the bulk of the minor-planet observations, began publication in 1997 and are issued weekly, except around the time of *Minor Planet Circulars* production. The 142970 pages published during the triennium brings the total to 394976. All three permanent publications are now published only in machine-readable form, via PDFs downloadable from the MPC's website.

In addition, the *Minor Planet Electronic Circulars* (*MPEC*s), first issued in 1993, provide immediate information for newly-designated NEOs, TNOs and comets. A total of 6167 *MPEC*s were issued through the preparation of the 2011 October 12 batch of *MPEC*s, bringing the total issued since inception to 22222. The 'Daily Orbit Update' (DOU) *MPEC*s, prepared entirely automatically each night, tabulates all the orbits computed and identifications found at the MPC during the previous 24 hours. Each DOU issue also includes continuing observations of all NEOs, the automatic preparation precluding the crediting of the observers in a reliable fashion. The DOU issue is consistent with the intention that the *MPECs* are a temporary publication, for as long as further observations are made, orbit computations will always be improved.

The compromise of a staff member's work account by an external hacker on 2010 February 12 caused a major disruption. Internal access to the MPC's website, then hosted by the Center for Astrophysics Computation Facility, was cut off immediately as a security measure, meaning that no updates to the site were possible and that no circulars could be mailed out. A massive effort during that day meant that the external disruption was minimized: the DOU *MPEC* the next morning was issued on schedule (although there was a delay in its e-mail distribution) and issuance of discovery *MPECs* resumed the next day. Access to the NEOCP was possible without interruption thanks to mirror pages hosted on the script server machine, which continued to be updated throughout the crisis. It took three days to get the MPC's website restarted on an external host and some weeks to copy over the entire site. The MPC website continued to be hosted solely on the external site until 2011 January 19, when the new MPC-hosted website came on line. The two sites ran in parallel until 2011 June 7, when the external site was shut down. The new website has an interface to the MPC Database, a MySQL database that can be queried on-line, allowing user-definable access to most of the public data available from the MPC.

The MPC also hosts the Light Curve Database (LCDB), a repository of minor-planet light-curve observation files, on behalf of an outside group. The LCDB contains over a million individual observations in 12641 separate data files representing 1802 different minor planets. The LCDB can be queried via the MPC website and observers can upload new data sets through the DB interface.

The "NEOCP Blog" and "Daily Minor Planet Center" were brought on-line during the triennium to serve as conduits between the MPC and the observers. The "NEOCP Blog" deals exclusively with the NEOCP and allows observers to send reports on what they are planning to observe and what they actually do observe. Posts on the blog are of a transitory nature. The "Daily Minor Planet Center" is intended for longer articles of a permanent nature, written on a variety of topics by the MPC staff.

3. Near-Earth objects

Prior to the preparation of an *MPEC* documenting a discovery, alerts to possible NEOs and comets are issued on "The NEO Confirmation Page". This page has existed since 1996 and most new objects are posted to the page automatically, as are updates to the predictions as further observations are obtained. The NEOCP underwent a major upgrade in 2011 April, allowing observers more control over the selection of objects that are returned.

During the triennium, 2763 separate NEOs were discovered, and of these 299 are considered potentially hazardous asteroids (PHAs). These objects have minimum orbital intersection distances with Earth less than 0.05 AU and absolute magnitude H < 22. More than 97% of the discoveries are made by professional astronomers (over 99% of which were US-based or US-funded surveys), but there is still room for amateurs to contribute to the NEO effort by performing astrometric follow-up of NEOs and NEO candidates.

The discovery of 2008 TC_3 in 2008 October led to the first confirmed short-term prediction of an Earth impact. Small fragments of the object survived passage through the atmosphere to fall as meteorites in the Sudanese desert. As a direct result of this event, the MPC implemented an internal/external alert system, which sends alerts about potential very close Earth approaches to MPC staff members and, if an external alert is triggered, to a number of JPL scientists and NASA's NEO Observations Program Executive. Once an external alert is sent out, the object is not announced until a review of the situation is completed. The external alert system has been triggered about six times during the triennium.

Confusion of high-altitude artificial satellites as new NEOs continued to be a problem. The MPC tried to keep track of these objects using data from a variety of sources, but a number of NEO candidates that turned out to be untracked high-altitude objects were observed for only a short while before being lost again.

4. Comets & distant objects

Some of the NEO candidates appearing on the NEOCP turned out to be comets. The MPC worked in cooperation with the *Central Bureau for Astronomical Telegrams* (a service of Division XII/Commission 6) in the announcement of new comets. Routine astrometric follow-up of comets was handled by the MPC, with temporary publication of new observations and orbits generally weekly on special *MPEC*s and permanent archiving in the monthly *Minor Planet Circulars*. Almost 183000 comet observations were published during the triennium, bringing the total in the MPC archive to 595000.

Although the last triennium saw the discovery of only 112 new 'distant objects' (centaurs and transneptunian objects), bringing the total number known to 1549, some 52% of recent discoveries have already been observed at more than one opposition. Of the total known population (including the transneptunian dwarf planets) 234 have been numbered, while 39% have been seen at only one opposition.

5. Outer satellites of the giant planets

Because of their potential confusion with minor planets, the MPC continued to catalogue observations and to compute orbits for the outer satellites of the giant planets. Although many astrometric observations of known objects were received during the triennium (and published in the *Minor Planet Circulars*), there was little discovery activity. Two new satellites of Jupiter were reported from the 2010 opposition and both objects were recovered in 2011. One Jovian satellite observed at the 2003 opposition was rediscovered in 2009 and subsequently numbered and named as Jupiter L (Herse).

6. Personnel

Director Emeritus B. G. Marsden died on 2010 November 18, having continued working on MPC matters up until about a week before his death, with his last *MPEC* being issued late on November 10 and his last emails being sent on November 12. Former Associate Director C. M. Bardwell died on 2010 May 14. He began working at the MPC in 1957 and although he retired at the end of 1989, he continued to work for many years thereafter.

Director T. B. Spahr's workload continued to include normal (and abnormal) administrative duties. He worked principally with NEOs and made great inroads in extracting useful information from the file of One-Night Stand observations. He also had external teaching responsibilities. Associate Director, G. V. Williams, continued to be responsible for the bulk of the processing code, the processing of the comet and natural satellite observations, and the lion's share of the preparation of the permanent MPC publications. In addition, he served as secretary of (and MPC representative on) the *Committee on Small-Body Nomenclature* and as the MPC-CSBN representative on the *Working Group for Planetary System Nomenclature*. J. L. Galache was in charge of maintaining the MPC Blog, the Near-Earth Object Confirmation Page Blog, and bringing MPC communication front and center with modern social media technology. S. Keys wrote specialty software for the MPC, principally the MPC's NEO probability code, DIGEST2. This program, which is available for download to outside users, automatically selects from incoming astrometric data those objects that have the highest chance of being NEOs, allowing them to be posted on the NEO Confirmation Page. M. Rudenko, now moved to Harvard, was responsible for the setup and maintenance of the MPC's Linux systems, both the internal computational systems and the external web servers, and the databases.

Transactions IAU, Volume XXVIIIA
Reports on Astronomy 2009–2012
Ian F. Corbett, ed.

DIVISION IV   STARS

(Etoiles)

Division IV supports and coordinates astronomers studying the characteristics, interior and atmospheric structure, and evolution of stars of all masses, ages, chemical compositions, and multiplicity.

PRESIDENT	Christopher Corbally
VICE-PRESIDENT	Francesca D'Antona
PAST PRESIDENT	Monique Spite
BOARD	Martin Asplund, Corinne Charbonnel, Jose Angel Docobo, Richard O. Gray Nikolai E. Piskunov

DIVISION I COMMISSIONS

Commission 26	Double and Multiple Stars
Commission 29	Stellar Spectra
Commission 35	Stellar Constitution
Commission 36	Theory of Stellar Atmospheres
Commission 45	Stellar Classification

DIVISION IV WORKING GROUPS

Division IV WG	Abundances in Red Giants
Division IV WG	Massive Stars

INTER-DIVISION WORKING GROUPS

Division IV-V WG	Active B Stars
Division IV-V WG	Ap & Related Stars

TRIENNIAL REPORT 2009-2012

1. Introduction

This Division IV was started on a trial basis at the General Assembly in The Hague 1994 and was formally accepted at the Kyoto General Assembly in 1997. Its broad coverage of "Stars" is reflected in its relatively large number of Commissions and so of members (1266 in late 2011). Its kindred Division V, "Variable Stars", has the same history of its beginning. The thinking at the time was to achieve some kind of balance between the number of members in each of the 12 Divisions. Amid the current discussion of reorganizing the number of Divisions into a more compact form it seems advisable to make this numerical balance less of an issue than the rationalization of the scientific coverage of each Division, so providing more effective interaction within a particular field of astronomy. After all, every star is variable to a certain degree and such variability is becoming an ever more powerful tool to understand the characteristics of every kind of normal and peculiar star. So we may expect, after hearing the reactions of members, that in the restructuring a single Division will result from the current Divisions IV and V.

Division IV has strong connections with other Divisions such as Div.II (Sun & Heliosphere) and Div.VII (Galactic Systems). These connections will continue to be valued in the newly proposed Divisions, along with their reorganized Commissions and Working Groups.

2. Developments and conferences within the past triennium

Summaries of recent progress in Division IV's science will be found in the reports of its Commissions and Working Groups which follow. Here let us just note some highlights.

• Commission 26 (Double and Multiple Stars) has been active in observation, both speckle and traditional, and in the cataloging of fundamental data. It has had a broad outreach, especially during IYA2009. Its science included: a comparison of binary periods for Population I and II dwarfs, which was used to explain why exoplanets are rare around Pop.II dwarfs; the dynamics of multiple systems analyzed with respect to star formation and the dynamical evolution of stellar groups; preparations in the expectation that the Gaia mission will reveal even more binaries among Cepheids, with adverse effects on the current luminosity relations of these stars.

• The report of Commisson 29 (Stellar Spectra) remarks that over 8000 papers in its field were published just in refereed journals during the reporting period. This considerable activity is attributed both to continuous technology development and to closer integration of fields within astrophysics. Interest in exoplanets and their host stars has certainly been a driver, while the study of chemically peculiar stars has benefited from new spectropolarimeters and Doppler imaging techniques. In obtaining fundamental stellar parameters there has been considerable overlap of interest with other Commissions.

• Commission 35 (Stellar Constitution) has sponsored and participated in significant meetings; and we all know the value of such well-planned meetings in reflecting and promoting development within an area of astronomy. Hence there has been progress in massive star models and observations, in understanding the progenitors of type Ia supernovae and the explosion mechanism for core collapse, in investigating binarity in Planetary Nebulae, and in the transport processes in stars. Techniques and the application of helio- and asteroseismology are particularly flourishing, thanks both to ground-based and satellite data, while Surface Brightness Fluctuations are revealing ever more information on stellar populations in clusters and galaxies.

• Commission 45's report (Stellar Classification) is excited to announce the discovery of Y-type dwarfs. These fill the temperature gap between the coolest spectroscopically confirmed brown dwarfs and Jupiter. Further, the mass/age/spectral-type degeneracy in L-type brown dwarfs has been resolved by high-resolution spectroscopy. Spectral libraries of L-, T-, and now Y-type dwarfs are expanding, though curating all these is a problem under current discussion. Such curating of spectral and photometric surveys is an important activity of members of this Commission 45, and the future deep Galactic surveys will make this even more true.

• The Working Group on Abundances in Red Giants (WG-ARG) has a Newsletter with approximately 1000 subscribers. To these have gone announcements of specialist meetings, such as the second in the series "Why Galaxies Care About AGB Stars", 2010, in Vienna. The WG is working closely with members of Commission 35 in whose report many WG-ARG topics are included. The first comprehensive millimeter wavelengths studies are being carried out, and these are discovering dominant and unusual molecular species in the envelopes of giant stars. Multi-dimensional hydrodynamical studies are getting increasing attention as computing power increases.

• The Working Group on Massive Stars (WG-MS) has implemented an efficient automatic process to produce its Newsletter. Such newsletters are a valuable contribution to communication and so to progress in a WG's task area. Perhaps counter-intuitively, the WG-MS has been turning its attention to the infrared bands, since these have strong potential for circumstellar and atmospheric diagnostics. Also, new techniques and observations are allowing former assumptions on massive-star evolution to be undergoing revision.

• The Working Group on Active B Stars (WG-ABS) saw significant advances in our understanding of the activity and the evolution of B-type stars. New observations, many the result of large collaborations, and new stellar models that include rotation and magnetic fields in their computation were reported at IAUS-272, Paris, 2010. Noteworthy are interferometric observations with VLTI/MIDI, VLTI/AMBER, CHARA, and NPOI that have provided information on the geometry and structure of circumstellar disks and the degree of stellar flattening due to a star's rapid rotation. The MiMeS (Magnetism in Massive Stars) Project is yielding information on the origin, presence, and geometry of magnetic fields in B stars. The pulsational characteristics of B stars have been studied from photometry that continues to be accumulated from the Kepler and CoRoT spacecraft. The Spitzer SAGE survey of the Magellanic Clouds reveals a higher percentage of Be stars in the SMC and frequent transitions from a Be-phase to a non-emission state. The VLT-FLAMES surveys of OB stars in the Magellanic Clouds are providing fundamental parameters for thousands of objects.

• The Working Group on Ap & Related Stars (WG-Ap) announces new work on large-scale magnetic fields, especially the weak ones in stars. This WG shares with WG-ABS in the MiMeS Project's activity and results. MOST, CoRoT, and Kepler data continue to surprise researchers and promote the development of more consistent theoretical models. An important role is the WG-Ap's contact with the atomic physics community, mediating requests to the latter for new atomic data specific to the analysis and modeling of Ap and related stars.

3. Closing remarks

The coming triennium's prospects for this Division's members are more fully outlined in the respective reports that follow. Certainly members can look forward to a stimulating time at the XXVIII General Assembly in Beijing, China, where 7 Symposia, 4 Joint Discussions, and 12 Special Sessions have sponsorship from and/or direct relevance to this Division.

Since web links change, especially after a General Assembly, readers are advised to consult the IAU website http://www.iau.org for the most current links to this Division, its Commissions, and Working Groups, and so to its ongoing activities.

Christopher Corbally
president of the Division

Transactions IAU, Volume XXVIIIA
Reports on Astronomy 2009–2012
Ian Corbett, ed.

© International Astronomical Union 2012
doi:10.1017/S174392131200275X

COMMISSION 26

DOUBLE AND MULTIPLE STARS
ETOILES DOUBLE ET MULTIPLE

PRESIDENT	Jose A. Docobo
VICE-PRESIDENT	Brian D. Mason
PAST PRESIDENT	Christine Allen
ORGANIZING COMMITTEE	Frederic Arenou
	Yuri Balega
	Terry Oswalt
	Dimitri Pourbaix
	Marco Scardia
	Colin Scarfe
	Vakhtang S. Tamazian

TRIENNIAL REPORT 2009-2012

1. H. Abt, NOAO

Main Publications:

Abt, H. A. 2008, The Difference Between Metal-Poor and Metal-Rich Binaries, AJ, 135, 722

Abt, H. A. 2009, Stellar Rotation versus Duplicity in Open Cluster Early-Type Stars, PASP, 121, 1307

Abt, H. A. 2010, The Origin of the Exoplanets, PASP, 122, 1015

Abt also gave invited talks in the workshops held in Santiago de Compostela:
Double and Multiple Stars: Dynamics, Physics, and Instrumentation. December 10-11, 2009. "The Nature of Exoplanets".
Binaries Inside and Outside the Local Interstellar Bubble. February 10-11, 2011. "The age of the Local Interstellar Bubble".

2. David Dunham, President, International Occultation Timing Association

In 2008, the International Occultation Timing Association (IOTA) took over the role of collecting, analyzing, and archiving all observations of lunar occultations from the In- ternational Lunar Occultation Centre (ILOC) in Japan. In the course of our work, many close double stars were discovered, and other known close binaries measured. Recent Publications: Herald, D., Dunham, D., *et al.*, "New Double Stars from Aster- oidal Occultations, 1971-2008", Journal of Double Star Observations, Vol. 6, No. 1, pp. 88-96, 2010. Loader, B., *et al.*, "Lunar Occultation Observations of Known Double Stars - Report #1", Journal of Double Star Observations, Vol. 6, No. 3, pp. 176-179, 2010.

3. Z.Cvetković, Astronomical Observatory of Belgrade

Serbian astronomers have performed three series of CCD observations of visual double and multiple stars at the NAO Rozhen from 2009 to 2011. A total of 237 pairs were measured and the results were published in Cvetković *et al.* (*) and Cvetković *et al.* (+). Z.Cvetković announced 25 new/recalculated orbits and 3 linear solutions in Circulars IAU Commission 26 :
Circ. 167 - 7 orbits; Circ. 169 - 2 orbits; Circ. 170 - 6 orbits; Circ. 172 - 4 orbits; Circ. 174 - 6 orbits and 3 linear solutions.

We have started CCD observations of visual double and multiple stars at the Astronomical Station on the mountain of Vidojevica (South Serbia) with a 60 cm telescope.

Articles published in 2009-11: Publ. Astron. Obs. Belgrade: 2 articles in 2009 and 1 article in 2010. (*) Serbian Astronomical Journal: 3 articles in 2010. (+) Astronomical Journal: 1 article in 2010 and 2 in 2011. Astronomische Nachrichten: 1 article in 2010. New Astronomy: 1 article in 2010. Publ. of the Astronomical Society RudjerBokovic: 1 article in 2009.

4. P. Lampens, Royal Observatory of Belgium

Main Research lines and corresponding publications:

4.1. *A.1. Visual Binaries:*

Torres, K.B.V., *et al.* (2011) Spectra disentangling applied to the Hyades binary ThetaÂ² Tauri AB: new orbit, orbital parallax and component properties. A&A, 525, 50

4.2. *A.2. Eclipsing Binaries:*

Desmet, M., *et al.* (2010). Combined modelling of the interacting eclipsing binary AU Mon based on CoRoT and ground-based photometry and high-resolution spectroscopy MNRAS, 401, 418.

Lampens, P., *et al.* (2010) New Times of Minima of 36 Eclipsing Binary Systems Information Bulletin on Variable Stars, 5933, 1

Borkovits, T., *et al.* (2008) New and Archive Times of Minima of Eclipsing Binary Systems. Information Bulletin on Variable Stars, 5835, 1.

Wils, P., Lampens, P., Van Cauteren, P., Southworth, J. (2010). The Highly Active Low-Mass Eclipsing Binary BS Uma Information Bulletin on Variable Stars, 5940, 1

Ulas, B., Niarchos, P.G., Lampens, P., Liakos, A. The Algol-type eclipsing binaries RW CrB and VZ Leo: new RI photometric study and search for pulsations 2009, ApSS 319, 55

Rodriguez, E., *et al.* (2010) Delta Scuti-type pulsations in eclipsing binary systems: Y Cam MNRAS, 408, 2149

Hambsch, F.-J., *et al.* (2010) Detection of a Rapidly Pulsating Component in the Algol-Type Eclipsing Binary YY Boo Information Bulletin on Variable Stars, 5949, 1

Lampens, P., *et al.* (2011) Multi-site, multi-year monitoring of the oscillating Algol-type binary CT Her, A&A, 534, 111

5. V. Orlov, Astronomy Department, Saint Petersburg University, Russia

Research in progress and some relevant results:
1. Dynamics of multiple stars can reflect some features of star formation and following dynamical evolution of stellar groups. We consider multiple stars in the solar

neighbour- hood using by the catalog MSC (Multiple Star Catalogue). We have selected the systems which are near the stability boundary. These systems may be dynamically (sometimes physically) young. We discuss some hypotheses of origin of such systems and propose future observations for a number of systems in order to make more accurate some critical data.

2. Some physical and dynamical parameters of the multiple system HD 222326 are presented.

We suggest a new method for determination of individual radial velocities of wide binaries components, when difference of radial velocities is small.

3. The analysis of physical parameters, orbital elements, and dynamical stability for the multiple star HD 76644 is presented. The data of astrometric observations from the WDS Catalogue and original observations taken with the BTA 6-m Telescope of Special Astrophysical Observatory of the Russian Academy of Sciences and the RTT150 1.5-m Telescope at Tubitak National Observatory, Turkey, were used.

4. An analysis of the physical characteristics, orbital parameters, and dynamical stability of the multiple Be star HD 217675 (o And) is presented.

5. Criteria for stability of triple systems are studied and compared with the results of numerical simulations obtained for model triple systems and observed multiple stars. The results for the stability analyses using two new criteria - those of Aarseth and of Valtonen et al. - agree with the simulation results in 98% of cases. The last published version of the "Multiple-Star Catalog" of Tokovinin is analyzed to search for systems that may be unstable according to the two new criteria. More detailed studies are carried out for the HD 284419 (T Tau) system.

6. Oswalt and collaborators, Florida Institute of Technology.

Oswalt and collaborators Zhao and Holberg have been using fragile binaries to investigate several astrophysical problems.

Publications: 1) Johnston, K., Oswalt, T., Valls-Gabaud, D. 2011, "Orbital Separation Amplification in Fragile Binaries with Evolved Components," New Astronomy, accepted.
2) Zhao, J.K., Oswalt, T.D., Rudkin, M., Zhao, G., Chen, Y.Q., 2011, "The Chromospheric Activity, Age, Metallicity and Space Motions of 36 Wide Binaries", AJ 141, 107.
3)Zhao, J.K., Oswalt, T.D., Zhao, G. 2011, "Fragile Binary Candidates in the SDSS DR8 Spectroscopic Archive", AJ, accepted 11/8/11, 2011arXiv1111.2831Z.

Manuscripts submited:
1) Holberg, J.D. Oswalt, T.D., Barstow, M. 2011, "Observational Constraints on the Degenerate Mass-Radius Relation", submitted to AJ 11/2/11.
2) Zhao, J.K., Oswalt, T.D., Willson, L.A., Wang, Q., Zhao, G., 2011, "The Initial-Final Mass Relation among White Dwarfs in Wide Binaries", ApJ, submitted.

7. PISCO Group

M. Scardia and L. Pansecchi, INAF-Merate Observatory, Italy J.-L. Prieur, CNRS-Toulouse University, France R.W. Argyle, Cambridge Observatory, U.K.

During the period 2009-2011, we continued our speckle observation program of visual binaries with PISCO in Merate. The total number of measurements made since 2004 now exceeds 2300. Optical tests of the telescope were also done after this operation. We added a 32 mm eyepiece in the magnification wheel of PISCO. This allowed a new observing mode with a low magnification which increased the sensitivity. We are thus now able to observe 11th magnitude binaries when their separation is larger than 0.9 arcsec. In 2010 we started a collaboration with René Gili in Nice who observes visual binaries with the historical 74 cm refractor. A new speckle camera called PISCO2 was made in 2010 for this telescope. It is a simplified version of PISCO with the same optical concept.

Publications in refereed journals in 2009-2011:
Scardia, M., *et al.*, 2009, Astron. Nach., 330, 55-67
Prieur, J.-L., *et al.*, 2009, MNRAS, 395, 907-917
Scardia, M., *et al.*, 2010, Astron. Nach., 331, 286-299
Prieur, J.-L., *et al.*, 2010, MNRAS, 407, 1913-1925
Scardia, M., *et al.*, 2011, Astron. Nach., 332, 508-523

Other publications in 2009-2011
Argyle R.W., Scardia M., Prieur J.-L., Pansecchi L., 2009, Double and Multiple Stars: Dynamics, Physics, and Instrumentation, held 10-11 December 2009, in Santiago de Compostela (Spain) Eds: J.A. Docobo, V.S. Tamazian, Y.Y. Balega 2011, AIP Conf. Proc. 1346, 42-56
Argyle R.W., Scardia M., Prieur J.-L., Pansecchi L., 2010, Annual Meeting of the Royal Astronomical Society, held January 2010, in London (U.K.) Published in 2010, The Observatory, Vol 130, No 1217, pp 200-202
Sacrdia, M.; Prieur, J.-L.; Pansecchi, L.; Argyle, R. W. 15 orbits published in the Information Circulars n. 167, 168, 169, 170, 171 and 172.

8. The Main (Pulkovo) Astronomical Observatory of Russian Academy of Sciences

TOur team works in the Laboratory of Astrometry and Stellar Astronomy at The Main (Pulkovo) Astronomical Observatory of Russian Academy of Sciences. The observations of double stars have been continued in this period by means of the 26-inch Pulkovo observatory refractor. The results of observations are published on the Pulkovo site: http://puldb.ru.

Now we have 420 stars in the Pulkovo program. For more than 80 pairs there are the large series of observations since 1960 to 2005 years.

Publications:
N.A.Shakht, A.A.Kiselev, Planetary Space Sciences 2478, v.56, issue 14, p.p. 1903-1907, 2008.
N.A.Shakht, *et al.*, Astrofizika (Journal of Armenian Academy of Sciences), v.53, No 2, p. 257-267, 2010 (in Russian), also in Astrophysics, v.53, No2, 2010, (in English) STELLAR SPECTRA 5
V.A.Zakhozhaj, Yu.N.Gnedin, N.A.Shakht, Astrophysics v.53, N0 4, p.p.575-591, 2010.
N.A.Shakht, A.A.Kiselev, l.G.Romanenko, E.A.Grosheva. Isvestia GAO No 219. v. 4, p.p. 375-380, 2010 (in Russian).
Kiselev A.A., Romanenko L.G., Kalinichenko O. A., Astronomy Reports. v. 53. no. 2. p. 126-135, 2009.

Kiselev A.A., Romanenko L.G. and Gorynya N.A. Astronomy Reports. v. 53. no. 12. p. 1136-1145, 2009.
Kiselev A.A. and Romanenko L.G. Astronomy Reports. v. 55. n 6. p. 487- 496, 2011.

Conference presentations:
A. A. Kiselev, N. A. Shakht, E. A.Grosheva, L. G. Romanenko. Proc. of the Meeting "Earth Based Support to Gaia Solar system science, Beaulieu sur Mer, 27-28 Oct. 2008."
N. A. Shakht, E. A.Grosheva, D. L. Gorshanov, A. A. Kiselev, E. V. Polyakov, O. O. Vasil'kova. In Workshop "Astrometry now and in the future", Antalya, Turkey, September 12-13, 2011.

9. Ramon Maria Aller Astronomical Observatory (OARMA), Santiago de Compostela, Spain

9.1. *Research Projects.*

The main line of research at OARMA concerns binary and multiple stars. The Spanish Ministry of Research and Innovation supported our investigation during the past years through the following projects: Speckle interferometry, differential photometry, spectroscopy and fundamental astrophysical parameters of double and multiple stars (2008-2010); High precision astrometry and fundamental parameters of double and multiple stars (2011). Both were directed by Prof. J. A. Docobo. The research team was formed by J. A. Docobo, V. S. Tamazian, J. F. Ling, Y. Balega, N. Melikian, M. Andrade, J. F. Lahulla, I. Fernández, P. Campo. Information Circular IAU Comm. 26 (J. A. Docobo and J. F. Ling, eds.). From October 2008 to June 2011, 9 Information Circulars were published correspond- ing to Issues 166 to 174.

New orbits announced:
A total of 281 new orbits were announced in the Circulars, 140 of these belonging to stars of the Northern hemisphere and 141 to the Southern.
The following list shows the author(s) who announced them, together with the number of orbits contributed by each: Alzner-Argyle (5), Argyle-Alzner (1), Alzner-Argyle-Anton (1), Cvetković (28), Docobo (1), Docobo-Andrade (6), Docobo-Campo (6), Docobo-Ling (19), Docobo-Tamazian (14), Docobo-Tamazian-Kraus-Weigelt (1), Gili (1), Hartkopf-Mason (72),Mason (1),Mason-Hartkopf (40), Ling (14), Rica (9), Rica-Zirm (3), Roberts-Hartkopf (1), Roberts-Mason (1), Scardia-Prieur-Pansecchi-Argyle (19), Zirm (35), Zirm-Rica (3).
The circulars also include different information such as: new double stars, announcements, list of papers published on double and multiple stars, obituaries and notes. The Double Star Orbit Catalog (J. A. Docobo, J. F. Ling, P. Campo).

We have mantained our own Double Star Catalog which was updated on setember 23, 2011. We believe that this Catalog complements the USNO Catalog. Currently, 2151 orbits of 1701 binaries are included in it. The Catalog can be checked online at the following link: http://www.usc.es/astro/catalog.htm.

Publications in Journals (2009-11).
Docobo, J. A.; Ling, J. F. The Astronomical Journal, Volume 138, Issue 4, pp. 1159-1170 (2009). Kraus, S.; *et al.* Astronomy and Astrophysics, Volume 497, Issue 1, 2009,

pp. 195-207. Docobo, J. A.; Tamazian, V. S.; Balega, Y. Y.; Melikian, N. D. The Astronomical Jour- nal, Volume 140, Issue 4, pp. 1078-1083 (2010).
Melikian, N. D.; *et al.* Astrophysics, Volume 53, Issue 3, pp.373-386 (2010).
Melikian, N. D.; *et al.* Astrophysics, Volume 53, Issue 2, pp.202-211 (2010).
Melikian, N. D.; *et al.* Astrophysics, Volume 54, Issue 2, pp.203-213 (2011).
M. Andrade and J. A. Docobo. ICARUS: 2011, vol 215. Issue 2; 712-720 (2011).

Workshop Organization (J. A. Docobo, chairman; V. S. Tamazian, secretary).
In the last three years, the OARMA organized two international workshops on binaries in Santiago de Compostela (Spain).
Double and Multiple Stars: Dynamics, Physics, and Instrumentation. December 10-11, 2009. The proceedings were published by the AIP Series (AIP Conference Proceedings 1346. Ed. J. A. Docobo, V. S. Tamazian, Y. Y. Balega. ISBN 978-0-7354-0902-6. 2011).
Binaries Inside and Outside the Local Interstellar Bubble. February 10-11, 2011.The proceedings will be published in AIP also. Other Research Activities.
In this period, OARMA signed collaboration agreements with the Special Astrophysical Observatory (SAO) and the Byurakan Astrophysical Observatory (BAO), including onservation runs. In 2011, our EMCCD speckle camera was sent to the BAO in order to attach it to their 2.6 m telescope as stipulated in our agreement.

10. Laszlo Szabados, Konkoly Observatory, Budapest, Hungary

Main Research lines and corresponding publications:
Szabados, L., Kiss, Z. T., & Klagyivik, P. Binarity and Cluster Membership of Classical Cepheids. EAS Publications Series, Volume 45, 2011, pp.441-444.
Szabados, L. Type II Cepheid. Variable Stars, the Galactic halo and Galaxy Formation, Eds. C. Sterken, N. Samus & L. Szabados. Sternberg Astronomical Institute of Moscow University, Russia. 2010.
Csizmadia, Sz., Borkovits, T., Paragi, Zs. *et al.* Interferometric Observations of the Hierarchical Triple System Algol. 2009, ApJ, 705, 436 Klagyivik, P. & Szabados, L. Observational studies of Cepheid amplitudes. I. Period-amplitude relationships for Galactic Cepheids and interrelation of amplitudes. 2009, A&A, 504, 959.

11. US Naval Observatory

As of 1 October 2011 the repository of double star observations for Commission 26 (1964; Transactions of the IAU, Vol. 12B, 267; 1966), the Washington Double Star (WDS) Catalog, contained 817,749 means of 115,524 pairs. This represents an increase of 52,851 mean positions and 10,364 new pairs over the triennium. The most common means of WDS access other than the US Naval Observatory (USNO) website is the b/wds catalog on Vizier which can be utilized as an overlay in Aladin.
Those systems with more complete kinematic solutions can be found in the Sixth Catalog of Orbits of Visual Binary Stars (2,236 orbits of 2,127 systems) and the Catalog of Rectilinear Elements (1,267 linear solutions to optical pairs or long-period physical systems). Similar growth to that of the WDS was seen in these two catalogs, as well as the Third Photometric Magnitude Difference Catalog and the Fourth Catalog of Interferometric Measurements of Binary Stars, all of which are maintained at the USNO and represent

the suite of catalogs considered part of the WDS ensemble.

The primary double star observational technique used by the USNO continues to be speckle interferometry. Our primary speckle camera was used on the twin 4m NOAO telescopes at Cerro Tololo and Kitt Peak, as well as the 61in telescope of the USNO Flagstaff Station. In collaboration with Andrei Tokovinin we have also used HRCam on the SOAR 4.1m telescope in Chile. We also maintain a less complex secondary camera for use on the USNO 26in refractor in Washington. Over the triennium, 1,064 observations of astrophysically interesting systems have been obtained and published with large telescopes and 3,410 systems, mostly neglected pairs or service observing, have similarly been obtained and published with the secondary camera.

José A. Docobo
president of the Commission president of the Commission

Transactions IAU, Volume XXVIIIA
Reports on Astronomy 2009–2012   © International Astronomical Union 2012
Ian Corbett, ed.   doi:10.1017/S1743921312002761

COMMISSION 29

STELLAR SPECTRA

Spectres des etoiles

PRESIDENT	**Nikolai Piskunov**
VICE-PRESIDENT	**Katia Cunha**
PAST PRESIDENT	**Mudumba Parthasarathy**
ORGANIZING COMMITTEE	**Wako Aoki**
	Martin Asplund
	David Bohlender
	Kenneth Carpenter
	Jorge Melendez
	Silvia Rossi
	Verne Smith
	David Soderblom
	Glenn Wahlgren

TRIENNIAL REPORT 2009-2012

1. Introduction

Commission 29 consists of members of the International Astronomical Union carrying out theoretical and observational studies of stars using spectroscopy, developing instrumentation for spectroscopy and producing and collecting data for interpretation of spectra.

Specific interests range from classical abundance analysis to stellar evolution, dynamics and structure formation in and around stars. In the past three years the work of the Commission, carried out primarily by the organizing committee (OC), was focused on establishing the right balance and interaction format with other IAU commissions and working groups dealing with overlapping matters. The OC also dealt with routine matters that fall under the competence of the Commission. In particular, the OC has discussed and ranked proposals for several IAU sponsored meetings. The Commission participated in the organization of a number of Symposia, Joint Discussions, and Special Sessions held during the 26th IAU General Assembly at Rio in August 2009. Commission members themselves initiated and actively participated in organizing IAU sponsored meetings.

Commission 29 belongs to the IAU Division IV (Stars), dedicated to stellar analysis, stellar interior and atmospheric structure and evolution of stars of various masses, ages, and chemical compositions. Besides Commission 29, Division IV hosts Commissions 26 - Double and Multiple Stars, 35 - Stellar Constitution, 36 - Theory of Stellar Atmospheres, and 45 - Stellar Classification. The work conducted across commission competences is coordinated by working groups (WGs). Division IV has four working groups: WG on Active B Stars, WG on Massive Stars, WG on Red Giant Abundances, and WG on Chemically Peculiar and Related Stars.

2. Recent and approved IAU meetings endorsed by Commission 29

IAUS 272: Active OB stars - structure, evolution, mass loss, and critical limits, Paris, France, 19-23 July, 2010

IAUS 273: Physics of Sun and star spots, Los Angeles, USA, 23-26 August, 2010

IAUS 282: From Interacting Binaries to Exoplanets: Essential Modeling Tools, Tatranska Lomnica, Slovakia, 18-22 July, 2011

IAUS 294: Solar and astrophysical dynamos and magnetic activity, Beijing, China, 27-31 August 2012

SpS13: High-precision tests of stellar physics from high-precision photometry, Beijing China, 27-31 August, 2012

3. Scientific highlights

Stellar spectroscopy remains one of the most active fields of astrophysics with over 8000 papers published just in refereed journals during the reporting period. This can be attributed, in part, to continuous technology development (larger telescopes and better detectors help reaching more objects and with higher resolution) and, in part, to closer integration of fields within astrophysics (e.g. stars and planets through exoplanet research). Thus, in preparing this report we choose to present the advances in stellar spectroscopy in the form of highlights rather than producing an exhaustive report.

3.1. *Exoplanets and planet host stars*

The Kepler mission is producing hundreds of transiting exoplanet candidates each month, setting pressure on spectroscopic confirmation but also opening new opportunities for characterization. Multiple groups work on systematic spectroscopic follow-ups of the CoRoT and Kepler transit candidates. The task is not easy as the majority of targets are very faint; however, the combination of photometric light curve and spectroscopic characterization of the host star allows unambiguous determination of planetary mass and density (e.g. Leger 2009; Rouan 2011). Several papers described the methodology and results of chemical characterization of exoplanets. Today such analysis is only possible for either young (hot) planets in large orbits (e.g. Barman 2011) or for close-in massive planets using transit/eclipse spectroscopy (e.g. O'Donovan 2010; Moses 2011). Correlation between stellar chemical composition and the presence of planets remains a hot topic in stellar spectroscopy after the pioneering work by Fischer and Valenti (2005). Preliminary results extending their work to lower-mass stars have been presented at conferences and we expect new statistically sound surveys to be published soon.

3.2. *Fundamental stellar parameters*

The comparison of various analysis techniques indicates the main sources of uncertainty in determining fundamental stellar parameters (effective temperature, surface gravity and chemical composition) are the atomic and molecular data and the model atmospheres. This is a clear case where the interests of Commission 29 overlap with other commissions. Systematic studies of chemical composition were performed across various stellar associations or populations (e.g. Gonzalez 2011; Bensby 2010). The search for the oldest stars continues with the latest instrumentation (e.g. Caffau 2011; Hansen 2011). The most advanced simulations accounting for non-equilibrium effects and 3D dynamics in stellar atmospheres show progress, although the convergence between different groups and the helioseismology results are still somewhere in the future.

3.3. *Magnetic fields and structures in stellar atmospheres*

Studies of magnetic fields and application of Doppler imaging reached new qualitative levels. The new spectropolarimeters, such as ESPaDOnS, NARVAL and HARPSpol, when combined with advanced data analysis techniques reach unprecedented detection levels of 1-2 Gauss (e.g. Petit 2010; Kochukhov 2011a), making possible direct studies of activity on solar-type stars of various ages and stars hosting planets. The detection surveys are conducted by large international consortia. This work generated over one hundred refereed papers in the last three years.

Noticeable rotational modulation of spectral line profiles offers a possibility of investigating structures on stellar surfaces using Doppler imaging techniques. Such an approach was used for mapping chemical spots on chemically peculiar (CP) stars and temperature spots on cool stars. Series of Doppler images were used to detect and measure differential rotation (Korhonen 2011; Waite 2011). Temporal evolution of chemical spots on CP stars was unambiguously established. Often, when magnetic fields are detectable in these objects various flavours of magnetic Doppler imaging were used to establish the relation between the field and the spots (Kochukhov 2011b).

3.4. *New instrumentation*

The work on new spectroscopic instrumentation is progressing along three lines: stability, wavelength coverage and combination of techniques. The work on stability aims at reaching reproducibility of radial velocity measurements at the 10 cm/s level on the baseline of a few years. The focus is on stabilizing the light entering a spectrometer (e.g. by employing unconventional fibers to improve the scrambling), on better calibrations (gas cells, simultaneous wavelength references such as Fabry-Perot etalon or frequency comb), and on the analysis software (replacing the analysis of multiple short wavelength intervals by modelling of the entire focal plane). Several applications require a combination of high throughput and very large wavelength coverage, as answered by instruments such as the VLT X-shooter (Vernet 2010). Finally, we see a significant growth of results based on the combination of techniques. Studies of giant stars with spectrally-resolved interferometry (e.g. Paladini 2011) complemented by the theoretical 3D models (Chiavassa 2011) illustrate the productivity of such an approach.

4. Closing remarks

The current tendency for combining different techniques for the same objects and covering different objects with similar methods is not reflected by the interaction mechanisms between the commissions within IAU divisions or even between divisions. The question will be addressed during the GA in Beijing. Changes in the IAU division structure are urgently needed.

Nikolai Piskunov
president of the Commission

References

Barman, *et al.* 2011, *ApJ* 733, 65
Bensby, *et al.* 2010, *A&A* 521, 57
Caffau, *et al.* 2011, *A&A* 534, 4
Chiavassa, *et al.* 2011, *A&A* 535 22
Fischer & Valenti *ApJ* 622, 1102

Gonzalez, *et al.* 2011, *A&A* 530, 54
Hansen, *et al.* 2011, *A&A* 527, 65
Kochukhov, *et al.* 2011, *ApJ*, 732, 19
Kochukhov, *et al.* 2011, *ApJ*, 726, 24
Korhonen & Elstner 2011, *A&A*, 532, 106
Léger, *et al.* 2009, *A&A* 506, 287
Moses, *et al.* 2011, *ApJ* 737, 15
Rouan, *et al.* 2011, *ApJ* 741, 30
O'Donovan, *et al.* 2010, *ApJ* 710, 155
Palladinin, *et al.* 2011, *A&A* 533, 27
Petit, *et al.* 2010, *A&A* 523, 41
Vernet, *et al.* 2010, *SPIE* 7735, 50
Waite, *et al.* 2011, *MNRAS* 413, 1949

Transactions IAU, Volume XXVIIIA
Reports on Astronomy 2009–2012   © International Astronomical Union 2012
Ian Corbett, ed.   doi:10.1017/S1743921312002773

COMMISSION 35   **STELLAR CONSTITUTION**

CONSTITUTION DES ETOILES

PRESIDENT   Corinne Charbonnel
VICE-PRESIDENT   Marco Limongi
PAST PRESIDENT   Francesca D'Antona
ORGANIZING COMMITTEE   Gilles Fontaine, Jordi Isern,
                                John C. Lattanzio, Claus Leitherer,
                                Jacco Th. van Loon, Achim Weiss,
                                Lev R. Youngelson

COMMISSION 35 WORKING GROUPS

Div. IV / Commission 35 WG   Active B Stars
Div. IV / Commission 35 WG   Massive Stars
Div. IV / Commission 35 WG   Red Giant Abundances
Div. IV / Commission 35 WG   Chemically Peculiar and Related Stars

TRIENNIAL REPORT 2009-2011

1. The activity

Commission 35 consists of members of the International Astronomical Union whose research is concerned with the structure and evolution of stars in all parts of the H-R diagram. Their interests range from various aspects of stellar interior physics, such as convection, diffusion, rotation, magnetic fields, to asteroseismology and the prediction of the evolutionary and nucleosynthetic histories of stars that are of vital importance for our understanding of stellar populations and galactic chemical evolution.

After discussion with the OC and with many members of C35, the present President, Corinne Charbonnel, proposes to change the name of C35 to "Stellar structure and evolution". We shall ask reactions to C35 members in spring 2012, and we expect that this proposition will be accepted by IAU during the next GA in Beijing.

Commission 35 is part of Division IV of the International Union (Stars), which is concerned with the characterization, interior and atmospheric structure, and evolution of stars of all masses, ages, and chemical compositions. Division IV acts as the umbrella for Commissions 26 - Double and Multiple Stars, 29 - Stellar Spectra, 35 - Stellar Constitution, 36 - Theory of Stellar Atmospheres, and 45 - Stellar Classification. Division IV has established four Working Groups: WG on Active B Stars, WG on Massive Stars, WG on Red Giant Abundances, and WG on Chemically Peculiar and Related Stars.

The Commission home page (http://iau-c35.stsci.edu) is maintained by Claus Leitherer and contains general information on the Commission structure and activities, including links to stellar structure ressources that are made available by the owners. The The resources contain evolutionary tracks and isochrones from various groups, nuclear reactions, EOS, and opacity data, as well as links to main astronomical journals.

Members of IAU Commission 35 are encouraged to take advantage of a discussion board for communicating with their colleagues and friends. This service is moderated and maintained by the Organizing Committee of Commission 35 and hosted by http://iau-c35.stsci.edu/Blog/index.html.

As a routine activity, the Organizing Committee has commented on and ranked proposals for several IAU sponsored meetings. Our Commission participated in the organization of the following Symposia, Joint Discussions, and Special Sessions held at the XXVIth IAU General Assembly in Rio in August 2009: IAUS 262 "Stellar Populations - Planning for the Next Decade", JD3 "Neutron Stars: Timing in Extreme Environments", JD4 "Progress in Understanding the Physics of Ap and Related Stars", SpS7 "Young Stars, Brown Dwarfs, and Protoplanetary Disks", JD11 "New Advances in Helio- and Astero-Seismology", JD13 "Eta Carinae in the Context of the Most Massive Stars", SpS1 "IR and sub-mm spectroscopy: a new tool for studying stellar evolution".

Members of Commission 35 were involved in organization of other IAU sponsored meetings: IAUS 268 "Light Elements in the Universe", IAUS 271 "Astrophysical dynamics" from stars to galaxies", IAUS 272 "Active OB stars" structure, evolution, mass loss, and critical limits", IAUS 281 "Binary Paths to the Explosions of type Ia Supernovae", IAUS 283 "Planetary Nebulae: an Eye to the Future". Many other international meetings, in which members of the Commission were involved, were held in these years.

In the following we present some highlights that were published during the present triennial term and which concern the field of "Stellar Constitution".

2. Transport processes in stars (C. Charbonnel)

It is now widely recognised that the art of modelling stars strongly depends on our hability to model the various processes that transport angular momentum and chemicals in stellar interiors. This is a prerequisite to reproduce detailed data in various parts of the HR diagram, especially in the new area where asteroseismology revolutionises our research field (see §3.). In the recent years important advances were made to take into account into stellar models the physics of atomic diffusion, rotation, magnetic fields, internal gravity waves, mass loss, and of various hydrodynamical instabilities among which thermohaline convection. This is a challenging task since such mechanisms involve length and time-scales that differ by several orders of magnitude and impact on the stars both on dynamical and secular times. It implies that one needs to use and couple 1D, 2D, and 3D approaches to get a global picture of macroscopic MHD transport processes in stellar interiors. With the great endeavour made over the past two decades, work to go beyond the classical spherical picture of stars to get a global MHD understanding of their internal dynamics is now in a golden age. From dynamical to secular time scales, 1 to 3D approaches of the macroscopic transport processes indeed allow stellar modellers to enter into the details of the highly non-linear couplings between meridional circulation, differential rotation, turbulence, fossil and dynamo magnetic fields, and waves. Here we present some highlights concerning magneto-hydrodynamical processes in stellar radiative regions that appeared in the literature since 2009. Other related results are described in the other sections of this report.

Rotational mixing based on the prescription by Zahn, Maeder and collaborators was shown to explain the surface abundances of Li, Be, C, N, Na in low- and intermediate-mass stars (more massive that $\sim$1.5$M_\odot$) belonging to open clusters and to the field (e.g., Charbonnel & Lagarde 2010, Canto Martins et al. 2011). Diagnostic tools were designed to help disantangle the role of the various rotation-driven mechanisms in stars (Decressin et al. 2009), and large grids of rotating models and isochrones were computed (Brott et al. 2011; Ekström et al., in press; Lagarde et al., submitted). The effects of rotational mixing on the asterosismic properties of the Sun (Turck-Chièze et al. 2010) and of solar-type stars (Eggenberger *et al.* 2010) were examined in detail. The constraints on internal angular momentum transport in Solar-type stars that can be inferred from the spin-down of open cluster stars and from the solar rotation curve were critically examined; this lead to the conclusion that neither hydrodynamic mechanisms nor a revised and less efficient prescription for the Tayler-Spruit dynamo can reproduce both spin-down and the internal solar rotation profile by themselves (Denissenkov *et al.* 2010, Denissenkov 2010a). It is thus confirming the idea that in solar-type stars a successful model of angular momentum evolution must involve more than one mechanism. This has increased the urgency to describe the physics related to magnetic fields and internal gravity waves in stellar interiors.

Magnetic field (which are observed more and more extensively at stellar surfaces through spectropolarimetry) and their related dynamical effects are thought to be important in stellar interiors. Intense researches were devoted to understand their role in convective layers, in particular for the sun, as well as in radiative zones. Several papers were devoted to preparing the implementation in one-dimensional stellar evolution codes (i.e., the requested modifications to the structural equations of stellar evolution) of the different terms induced by the presence of a magnetic field (Duez *et al.* 2009, 2010), and to understand the fossil fields' origin, topology, and stability (the semi-analytical work of Duez & Mathis 2010). This is an important step to get a true MHD approach in stellar evolution.

Several works were devoted to study the transport of angular momentum by the internal gravity waves excited by turbulent motions at the border of stellar convective regions. 3D numerical simulations were aimed at determining the wave spectrum (Brun & Strugarek 2010). As far as mode excitation by turbulent convection is concerned, the influence of the Coriolis acceleration on the stochastic excitation of oscillation modes in rotating stars was studied through a perturbative analysis (Belkacem *et al.* 2009b). The transport by gravito-inertial waves has been studied in the case of a general differential rotation (Mathis 2009). Analytical models (MacGregor & Rogers 2011) and numerical simulations (Rogers & McGregor 2010, 2011) were carried out in order to investigate the interaction of internal gravity waves with magnetic fields in stellar radiative interiors. A global study of magneto-gravito-inertial waves in stellar radiation zones was achieved in the case of an axisymmetric toroidal magnetic field (Mathis & de Brye 2011). Also, nonlinear effects such as wave-braking at the center of the star were studied; signs appear of a possible acceleration of the central rotation of solar-type stars by this mechanism (Barker & Ogilvie 2010, 2011). In the near future, consequences for angular momentum transport and the case of general differential rotation and azimuthal magnetic field have to be studied.

Self-consistent stellar evolution models including atomic diffusion and radiative accelerations together with other macroscopic processes were published. The effects of atomic diffusion on internal and surface abundances of A- and F- pre-main-sequence stars with mass loss were studied in order to determine at what age the effects materialize, as well as to further understand the processes at play in HAeBe and young ApBp stars. Atomic diffusion in the presence of weak mass loss was shown to be able to explain the observed abundance anomalies of pre-main-sequence stars, as well as the presence of binary systems with metal-rich primaries and chemically normal secondaries such as V380 Ori and HD 72106. This is in contrast to turbulence models which do not allow for abundance anomalies to develop on the pre-main-sequence. The age at which anomalies can appear depends on stellar mass (Vick *et al.* 2011). A thorough study of the effects of mass loss on the internal and surface abundance of A- and F-type main sequence stars was undertaken in order to constrain mass loss rates and to elucidate some of the processes that compete with atomic diffusion in these stars (Vick *et al.* 2010, Michaud *et al.* 2011). However the current observational constraints (e.g. Gebran *et al.* 2010) do not allow to conclude that mass loss is to be preferred over turbulent mixing (induced e.g. by rotation) in order to explain the AmFm phenomenon, although internal concentration variations which could be detectable through asteroseismic tests should provide further information. Last but not least, the effect of atomic diffusion was studied during the relatively rapid red giant phase to determine the concentration variations it leads to and at what accuracy level it can be safely neglected (Michaud *et al.* 2010), as well as during the horizontal branch evolution (Michaud *et al.* 2011).

Several works were devoted to the study of the thermohaline double-diffusive instability that is expected to develop in stars under various circumstances. Its influence on (atomic) diffusion-induced iron accumulation was studied for A-F stars; it was shown that iron accumulation is still present when thermohaline convection is taken into account, but much reduced compared to when this physical process is neglected (Théado *et al.* 2009, Théado & Vauclair 2010). Metal-rich accretion and thermohaline instability were investigated in exoplanets-host stars, as well as their consequences on the light elements abundances (Théado & Vauclair 2011). The combined effects of thermohaline instability and rotation-induced mixing were studied for low- and intermediate-mass stars at various metallicities; these mechanisms were shown to account for the observational constraints very well over the whole investigated mass range for stars on the main sequence, on the red giant branch and clump, and on the early-AGB (Charbonnel & Lagarde 2010; see also Stancliffe *et al.* 2009, and Stancliffe 2010), and to account for new ^{3}He yields that reconcile Galactic requirements for the evolution of this light elements with the Big Bang nucleosynthesis (Lagarde *et al.* 2011). In super-AGB stars thermohaline mixing becomes important after carbon has been ignited off-center and it affects significantly the propagation of the flame; however during the subsequent thermally pulsing SAGB phase, the high temperature at the base of the convective envelope prevents the development of thermohaline instabilities associated with 3He burning as found in low-mass red giant stars (Siess 2009). Although these evolutionary studies sound promising, the question of the efficiency of the thermohaline instability in stellar interiors remains open. In particular the geometry of the instability cells that is deduced from the linear stability analysis and that is favoured by observational data for the red giants and by laboratory experiments is that of long thin fingers with an aspect ratio of the order of 6. This value turns out to be ~ 5 times higher than that obtained by current simulations of thermohaline convection (Denissenkov 2010b; Denissenkov & Merryfield 2011; Rosenblum *et al.* 2011; Traxler *et al.* 2011). However these simulations are still quite far from the stellar regime (i.e., in the best cases they are run at moderately low values of the Prandtl number), which calls for future numerical multi-dimensional simulations. Magneto-thermohaline mixing was studied as an alternative to thermohaline convection (Denissenkov *et al.* 2009). In view of the uncertainties affecting the current description of the mixing mechanisms at act in low-mass red giant stars (see also Cantiello & Langer 2010), the question of abundance anomalies on the RGB and AGB was revisited by means of parameterized non-convective mixing in post-processing calculations (Palmerini *et al.* 2011a,b).

Tidal dynamics of extended bodies in planetary systems and multiple stars was examined within the context of the recent discovery of a large number of extrasolar planets orbiting their parent stars at distances lower than 0.1 astronomical unit and the launch of dedicated space missions such as CoRoT and KEPLER; this is obviously also important when considering the position of inner natural satellites around giant planets in our Solar System and the existence of very close but separated binary stars. An important step beyond the traditional approach was made by Mathis & Le Poncin-Latiffe

(2009) who derived the dynamical equations for the gravitational and tidal interactions between extended bodies and associated dynamics and gave the conditions for applying this formalism.

3. Helio- and asteroseismology (J. Montalban & A. Noels)

Without underestimating the contribution from ground-based observations, which are essential, for instance, for mode identification of some stellar pulsators, the enormous progress recently made by asteroseismology is mainly due to the high quality light curves provided for a huge number of targets across the HR diagram by the space satellites CoRoT (Baglin *et al.* 06) and *Kepler* (Borucki *et al.* 2009). Both missions have the dual goal of detecting extra-solar planets by the transit technique (which implies high level of photometric sensitivity) and of observing stellar oscillations. CoRoT was launched on December 2007 and, although it was planned to operate during three years, its excellent results have led to the extension of the mission until March 2013. *Kepler* satellite was launched on March 2009 and it will operate during 3-4 years. The features of the telescopes, orbits, and observations are different and both missions provide complementary information. In particular, *Kepler* and CoRoT cover different regions of the Milky Way. *Kepler* observes a fixed (and more extended) field above the Galactic plane (b=$13°.5$) in the region of Cygnus-Lyra, while CoRoT alternates each 5 months observation fields in the direction of the Galactic center and of the anti-center, and not too far from the Galactic plane. For each observation field both space telescopes provide several thousands of light curves. These huge numbers together with the high photometric sensitivity and long time coverage of observations are at the origin of the exceptional oscillation data and results that we summarize below.

Oscillations stochastically excited by turbulence in convective envelopes (such as those observed in the Sun – solar-like oscillations) have been detected in several hundreds of dwarfs and sub-giants (more than 500) and in several thousands of red giants. The large number of targets involved in these studies has lead to what has been called "ensemble asteroseismology" (Chaplin 2010).

The detection of solar-like oscillations in G-K red giants has provided the most surprising results with a great impact on other astrophysical domains. Although stochastic oscillations in red giant stars were already known from ground-base observations (i.e., Frandsen *et al.* 2002), their non-radial character was definitely proven by CoRoT observations (de Ridder *et al.* 2009), and obliged to review the previous ideas concerning solar-like oscillation in this kind of objects. The oscillation spectra of about two thousand red giants showed a power excess with an almost gaussian shape centered on the frequency ν_{max} and a regular pattern of radial modes ($\Delta\nu$) (Hekker *et al.* 2009, Mosser *et al.* 2010, Bedding *et al.* 2010, Huber *et al.* 2010). Spectra of these stars seem to follow a "universal pattern" (Mosser *et al.* 2011, in agreement with the predictions of the asymptotic theory of acoustic modes) in which modes of the same angular degree are distributed along well-defined vertical ridges in the plane $\Delta\nu$ vs. $\nu/\Delta\nu$. The width of these ridges obtained from CoRoT and *Kepler* data show a dependence on the angular degree being the largest for dipole modes in general (Bedding *et al.* 2010, Huber *et al.* 2010, Mosser *et al.* 2011b) and in particular at a value of $\Delta\nu$ corresponding to the luminosity of the Red Clump (RC, see below). As for main sequence solar-like pulsators, it was possible to derive values of the small frequency separations ($\delta\nu_{01}$, $\delta\nu_{02}$) (Bedding *et al.* 2010, Huber *et al.* 2010) but contrarily to its usual interpretation, their mean values are not related to the properties of the central region (and thus to the stellar age), but to those of the envelope (Montalban *et al.* 2010a,b). The evolutionary state, however can be derived from other features of dipole modes which are sensitive to the contrast between central and mean densities. Theoretical predictions from stellar models and frequencies, as well as the good agreement between those and observational behaviours, led to suggest the scatter of dipole ridge in the echelle diagram as diagnostic of the evolutionary state (Montalban *et al.* 2010a,b). Moreover, as the observation time of red giant increased, the frequency resolution has been noticeably improved and it was possible to measure the properties of dipole mixed modes (Beck *et al.* 2011). Following adiabatic and non-adiabatic theoretical predictions (Montalban *et al.* 2010a, Dupret *et al.* 09), they should appear around the central peak that corresponds to the mode mostly trapped in the acoustic (outer) cavity. Bedding *et al.* (2011) and Mosser *et al.* (2011a) measured the properties of these modes in *Kepler* and CoRoT red-giants as well as their separation in period (what following the asymptotic theory is related to the central stratification), and found that at a given $\Delta\nu$ the value of the period spacing gathers the stars in two different groups. The comparison with theoretical models allowed their identification as H-shell burning stars ($\Delta P < 60$ s) and central-He burning ones ($\Delta P > 100$ s).

Important studies have also been done using the basic features of solar-like oscillation spectra : ν_{max} and $\Delta\nu$. In fact, these quantities are linked through scaling relations (Brown *et al.* 1991, Kjeldsen *et al.* 1995) to the values of global stellar parameters such as mass, radius, and effective temperature. It is worthwhile remarking that the values of stellar mass and radius resulting from scaling laws and simple features of oscillation spectra are model independent. Miglio *et al.* (2009) applied these scaling relationships to the stellar parameter distributions provided by stellar population models to characterize the population of red giants observed by CoRoT. They showed that the population was dominated by low-mass stars in the central-He burning phase (Red Clump stars), and they identified the values of $\nu_{max} \sim 35\mu$ and $\Delta\nu \sim 4\mu$ at which observational distributions peak (Hekker *et al.* 2009; Mosser *et al.* 2010), as those corresponding to low-mass stars at the red clump luminosity. Other studies of populations have been performed using the scaling laws, for red giants (Mosser *et al.* 2010, Kallinger *et al.* 2010, Hekker *et al.* 2011) and also for the 500 dwarfs and sub-giants pulsators detected by *Kepler* (Chaplin *et al.* 2011). The comparison of the mass and radius distributions to the predictions from stellar synthesis populations are very encouraging concerning the validity of these scaling relations and the perspectives of "ensemble" asteroseismology.

Solar-like oscillations in RGB (red giant branch) and RC stars in two stellar clusters have also been studied (Basu *et al.* 2011). In particular Stello *et al.* (2011) suggest to use seismic properties as membership diagnostic, and Miglio *et al.* (2011) used these data to test the scaling relations and estimate mass loss during the luminous RG phase.

Asteroseismology has finally entered a new era, and because of the unprecedented quality data, it is now possible to extend the techniques developed for the Sun to other stars and to perform individual studies of the internal structure. So, the deviations of frequency patterns with respect to the predictions of the asymptotic approximation might indicate the presence of sharps variations in the stellar structure since these features leave a periodic signature in the patter of oscillation modes (Gough 1990). As a striking example, the dependence of $\Delta\nu$ on frequency and the second frequency difference in the oscillation modes of HR 7349 (Carrier *et al.* 2010) make evident the presence of a sharp variation in the structure of this red giant. That was identified by Miglio *et al.* (2010) as the signal of the He second-ionization zone observed for the first time in a red giant. The prospect of obtaining precise enough frequencies to derive the helium abundance in the convective envelope of red giants like what was done in the Sun, has important obvious consequences that overtake the field of asteroseismology.

Seismic studies of individual dwarfs have also provided interesting results, mainly concerning the extension of the convective core and their evolutionary state (i.e., Deheuvels *et al.* 2010, Deheuvels & Michel 2010, Metcalfe *et al.* 2010). Moreover, Mazumdar & Michel (2010) detected the signal of HeII in addition to that of the bottom of the convective zone in the frequency spectrum of HD49933 (Appourchaux *et al.* 2009, Benomar *et al.* 2009), an active F-star observed by CoRoT.

The variability of the characteristics of the p-mode spectrum in the Sun shows a high level correlation with solar surface activity proxies. This kind of p-mode variations and its correlation with stellar activity has also been observed in HD49933 (Garcia *et al.* 2010). Although the origin of p-mode variability is not completely understood, the study of its relationship with solar activity and with activity cycles of stars with different rotation rates, depth of the convective envelope, and ages will significantly contribute to the understanding of the dynamo processes.

Finally, solar-like oscillations have also been detected in massive stars (Belkacem *et al.* 2009a, Degroote *et al.* 2010) such as expected by the theory (Belkacem *et al.* 2010). It is not still clear, however, whether the origin of these oscillations is the convective zone in the iron-peak region or the convective core.

Concerning the self-driven pulsators, significant changes have been observed in the pulsation properties of γ Doradus/δ Scuti stars. While the first ones are classified as pulsating with high order gravity modes and the second with low order acoustic modes and mixed modes, it has been shown that in fact a large fraction of these pulsators shows an hybrid character (Hareter *et al.* 2010, Uytterhoeven *et al.* 2011). The interest of these objects is evident since it allows to probe very different regions of the stars in a mass domain where physical processes such as convection, activity, rotation and chemical peculiarities present rapid changes. The analysis of these pulsators appears at the moment quite complicated given the high number of excited frequencies, Chapellier *et al.* (2011) identified 840 intrinsic frequencies in the CoRoT light curve of the γ Dor HD 49434.

In the same way that the bottom of the convective zone or the second ionization region of helium, the chemical gradient at the border of a receding convective core induces a periodic component in the period spacing of high-order gravity modes (i.e., Miglio *et al.* 2008). The amplitude and periodicity of this signature, which has been observed in a SPB (slowly pulsating B star) in the CoRoT field (Degroote *et al.* 2010), are directly related to the location and shape of the sharp variation opening the way to the characterization of extra-mixing processes in the central regions of these stars.

Evolved self-driven pulsators have also taken advantage from the long and continuous time coverage provided by space observations. First of all, hot sdB are predicted to pulsate with high order gravity modes (long period ones). These ones are very useful to probe the deepest regions of the star, but difficult to observe from ground-based observations. CoRot and *Kepler* have observed about 15 long period sdB's (Charpinet *et al.* 2010, Østensen *et al.* 2010, 2011) and the seismic analysis for three of these long period variables suggest that some kind of chemical mixing occurs outside the convective core (van Grootel *et al.* 2010a,b; Charpinet *et al.* 2011). The other evolved classical pulsators that have benefited from continuous time coverage are RR-Lyrae and in particular the study of the Blazhko effect (Poretti *et al.* 2010, Chadid *et al.* 2010, Kolenberg *et al.* 2011).

With the improvement of observations open questions and new problems arise. In particular, and as it was already the case more than 30 years ago, excited pulsations seem to require improvements in the opacity determinations in particular for the Fe-group opacity peak, where Ni appears to play a relevant role. The difficulties to explain observed frequencies have driven international collaboration involving theoretical opacity computation teams, laboratory measurements and stellar structure modelers (Turck-Chièze et al. 2011), as well as detailed studies of the role of these elements in the microscopic diffusion processes (Hu 2011, Hu et al. 2011).

At the moment, most of the data are being interpreted on the base of standard (or quasi) stellar modelling. However, at the same time that this first and fruitful analysis of CoRoT and Kepler data is underway, huge effort is being done to test the effect of non-standard processes such as rotation, mixing, magnetic fields, etc. on the oscillation spectra in order to be able to go deeper in the analysis and prospects (see §2.).

Finally it should be noticed that the connexion seismology-exoplanets is not reduced to the similarity in the observation requirements. In fact, the capability of seismology in deriving stellar global parameters is used by the Kepler team to derive planet radii, and to provide an estimation of the planetary system age (i.e., Batalha et al. 2011). An interesting example of the connexion planets-seismology, even if there is no transit, is the multiple system of sub-stellar objects around the γ Dor HR8799. The identification of the companions as exo-planets or brown dwarfs depends on the age of the system. As suggested by Moya et al. (2010) a seismic analysis of variable host star could provide an estimation of the age, and the space of possible solutions would be reduced if the inclination of the rotation stellar axis were known. Fortunately, seismology can also provide this information from spectroscopic identification of oscillation modes, such as Wright et al. (2011) did for HR8799.

4. Low- and intermediate-mass stars (C.Charbonnel)

We refer to §2. for discussions about the impact of non-standard processes in low- and intermediate-mass stars.

4.1. The Sun (C.Charbonnel)

During the last decade the use of 3D solar atmospheric models has lead to a revision downwards of the abundances of heavy elements (mainly oxygen; Asplund et al. 2005, Grevesse et al. 2010). This has lead to serious discrepancy between available seismic data and solar models constructed using these abundances, the main effect of the abundances on solar structure being through reduced opacities in the radiative interior. Considerable effort has been made in order to resolve this discrepancy both by modifying the solar models by changing some of the input physics like diffusion rates (see e.g., Basu & Antia 2008) as well as by independent determination of abundances. One of the solutions that had been proposed to alleviate the problem with the solar models constructed with AGS05 abundances was to increase the abundance of neon since its photospheric abundance is uncertain (Antia & Basu 2005; Bahcall et al. 2005). Modifications in solar models have not helped in resolving the discrepancy (but see e.g. Turck-Chièze et al. 2010 and references therein). Recently, independent 3D models have been used to calculate solar abundances of the key elements and calculated values are higher than the earlier estimates also obtained using 3D atmospheric models (Caffau et al. 2011a).

We would like to emphasize the review by Turck-Chièze & Couvidat (2011) that illustrates the importance of solar neutrinos in astrophysics, nuclear physics, and particle physics and discusses the many aspects related to the role of rotation and magnetism in solar physics.

4.2. Mass loss from red giant stars (J. van Loon)

At the start of 2009, mass loss from red giants posed the following challenges: how can oxygen-rich red giants drive a wind, given the low opacity of their dust? Is dust produced below the tip of the RGB? Does red-giant mass loss increase with metallicity? Some answers have been produced to these questions in the ensuing three years. While a carbon star in the metal-poor Local Group galaxy Sculptor was found to still be rather dusty (Sloan *et al.* 2009), little more has been revealed to settle the issue whether metal-poor carbon stars produce more or less dust than metal-rich counterparts. Much has been learnt about mass loss from RGB stars in globular clusters, though, with Origlia *et al.* (2010) finding dust production low on the RGB whereas Boyer *et al.* (2009, 2010) and McDonald *et al.* (2009, 2011a,b,c) find dust production to be confined to the tip of the RGB where RGB stars develop cool extended atmospheres and start to pulsate more vigorously. Momany *et al.* (2011), in an independent study, cast their verdict in favour of the latter. McDonald *et al.* (2010) further offer a new solution to the problem of the missing opacity in oxygen-rich red giant star winds: iron. There remains the question how do the warmer red giants and red supergiants drive a wind. It is very clear now that in metal-poor environments the red supergiants might not become cool enough to drive a dusty wind through combination of pulsation and radiation pressure, and that other mechanisms might need to be invoked such as those operating in chromospherically active stars (Cranmer & Saar 2011; Mauron & Josselin 2011; Bonanos *et al.* 2010) - which are by the way not yet established for certain.

Since the surprising discovery of a long tail of gas trailing the prototypical AGB star, Mira, bowshocks are being found around many red giant stars (Matthews *et al.* 2011; Mayer *et al.* 2011; Jorissen *et al.* 2011; Cox *et al.* 2011), which is not surprising giving their considerable ages and consequently large peculiar space motions with respect to the interstellar gas. These bowshocks offer a new tool to determine the stellar wind properties of the red giant and to quantify the erosion of the circumstellar envelope (with possible implications for mass-loss rate determinations).

4.3. *Magnetic fields in red giants (C. Charbonnel)*

The red giant branch is a crucial phase for understanding the interplay of magnetic fields and stellar evolution. During the last 3 years, a pilot study of more than 50 red giant stars (mainly selected as presenting evidence of magnetic activity) was carried out with the spectropolarimeters NARVAL@OHP and ESPaDOnS@CFHT. The first direct detection of the magnetic field of Betelgeuse was made; this magnetic field may be associated to the giant convection cells that could enable a "local dynamo" (Aurière *et al.* 2010). Almost all the Zeeman detected sample stars are located at the base of the RGB or at the clump. Importantly the strenght of the magnetic fields was found to be related to the rotational period. The fast rotators host magnetic fields of several to tens Gauss (Konstantinova-Antova *et al.* 2008, 2009). Magnetic fields as weak as one-half G were also detected, e.g. in Pollux (Aurière *et al.* 2009); they correspond to the activity level of solar twins, or even weaker (Petit *et al.* 2008). For EK Eri, it was concluded that the outstanding magnetic field of about 200 Gauss originates from the remnant of a strongly magnetic Ap star (Aurière *et al.* 2008, 2011). Konstantinova-Antova *et al.* (2010) measured the magnetic fields at a level of around a Gauss on several M-type giants among which Ek Boo which is either at the onset of the thermal pulse phase on the asymptotic giant branch, or at the tip of the first red giant branch. Some of these M stars are known to be rotating fast for their class, and they are presumably also intermediate-mass AGB stars. In this way, a new class of magnetically active stars was unveiled. More observations like these will answer the question whether these stars are a special case, or magnetic activity is, rather, more common among M giants than expected. An unusually high lithium content was reported in HD 232 862, a field giant classified as a G8II star that hosts a magnetic field (Lèbre *et al.* 2009).

4.4. *Asteroseismology for red giants (J. van Loon)*

Another field rapidly encroaching upon the study of red giant stars is asteroseismology. First expected to reveal mainly the constitution of main sequence stars like the Sun, perhaps shining light on the age-old problem of convection, it has actually become feasible to use the oscillations of RGB stars to determine their masses (Hekker *et al.* 2011a,b,c; Miglio *et al.* 2011). This already opens many ways in which to calibrate stellar models, and one may envisage that in future the detailed oscillation spectrum will become available to study the internal profile of the sound speed much as this was done for the Sun. Much more about Asteroseismology can be found in §3.

4.5. *AGB and super-AGB stars (J. Lattanzio, J. van Loon, P. Ventura, M. Lugaro, C.Charbonnel)*

Research on AGB stars has continued, becoming more increasingly sophisticated and quantitative but still facing some substantial hurdles, usually associated with convection, convective borders, and other mixing processes. Recent work has significantly advanced our predictions for very detailed neutron capture nucleosynthesis; new models have been published of the s-process in AGB stars both improved and more self-consistent (Church *et al.*2009, Cristallo *et al.* 2009, 2011) and for a large range of metallicities (Bisterzo *et al.* 2010, 2011). These efforts will help us to understand the composition of the elements heavier than iron in Carbon-Enhanced Metal-Poor stars and the production of these elements in low-metallicity environments.

In connection to the s-process, the first observations of Zr and Rb in putative massive AGB stars (M > 4 Msun) in the Galaxy (Garcia-Hernandez *et al.* 2006; Garcia-Hernandez *et al.* 2007) and in the MC (Garcia-Hernandez *et al.* 2009) provided the first observational evidence that the Ne22(α,n)Mg25 reaction is the main neutron source in these stars, as opposed to the C13(α,n)O16, which is the main neutron source in AGB stars of lower masses.

Another area that is starting to give up its secrets is the evolution of Super-AGB stars, those which ignite carbon in the core but continue on to experience thermal pulses also (Ventura & D'Antona 2009, 2010; Siess 2007, 2009, 2010; Doherty *et al.*2010). The yields from these sources were used to discuss the possibility that these stars played a role in the self-enrichment of globular clusters, and thus in the formation of multiple populations, as indicated by recent photometric and spectroscopic evidences (see e.g., D'Ercole *et al.* 2010 and references therein).

There has always been a lot of activity in the field of Planetary Nebula research, including abundance studies and mineralogy which reflect on the nuclear processing and dust production during the thermal-pulsing part of the AGB evolution (see e.g. Stasinska *et al.* 2010). The white dwarf mass function has also been a powerful constraint on the AGB evolution. Recently, a new issue has emerged: white dwarf debris discs (Jura *et al.* 2009a,b; Farihi *et al.* 2009, 2010a,b; Kilic *et al.* 2011). Planetary companions are being detected to stars low on the RGB. Where in previous years the focus was on stellar companions and the formation of equatorial density enhancements in the outflows from the AGB progenitor of the PN, possibly due also to shaping by magnetic fields, it has now become necessary to include the effect of planetary systems and debris discs (Farihi *et al.* 2011).

As the AGB phenomenon spans timescales from a dozen Gyr to as short as dozens of Myr, they are powerful tracers of a galaxy's evolution. Tonini *et al.* (2009, 2010) and Henriques *et al.* (2011) have demonstrated that incorporating thermal-pulsing AGB stars in population synthesis codes that are used to model the integrated properties of galaxies makes a notable difference to the interpretation of the evolution of high-redshift galaxies.

Javadi *et al.* (2011a,b) used the pulsating AGB star population to reconstruct the star formation history of the central part of the Local Group spiral galaxy Triangulum (M33). Applying our knowledge of AGB stars to studies of galaxies is a powerful way to expose and make progress in areas of uncertainty: Javadi *et al.* found that super-AGB stars must evolve to high luminosities and low effective temperatures in order to yield plausible star formation histories in M33. With extremely large telescopes and the IR-optimised JWST on the horizon, the application of AGB populations to the cosmic evolution awaits a golden age.

Detailed models of the evolution of isotopic ratios in the Galaxy cannot be complete/correct without including AGB yields (and this point is related to the Vienna conference series). Kobayashi *et al.* (2011) have produced these predictions with the latest AGB yields and GCE models.

GCE of isotopic ratios will also help improving our understanding of the evolution of the Galaxy and dust formation when coupled to the composition of stardust grains. Nittler (2009) used the composition of stardust oxide grains to derive a very shallow age-metallicity relationship and to infer that the 18O/17O of the Sun is typical for its age and location (To do this we just need basic predictions after the first dredge-up) The new models will help making these calculations even more accurate and then, at least in terms of age-metallicity relationships, we can compare the results with stellar surveys (GAIA!).

4.6. *Binary central stars of Planetary Nebulae (B. Miszalski, O. De Marco, A. Acker, H. Boffin, R. Corradi, T. Hillwig, D. Jones, T. Moffat, R. Napiwotzki, J. Nordhaus, P. Rodriguez-Gil, M. Santander-Garcia)*

One of the longest standing astrophysical problems concerns finding the mechanism responsible for shaping the extraordinary variety of nebula morphologies (Balick & Frank 2002). At several recent international meetings researchers in the field have conceded that there is no current theory that can quantitatively predict AGB mass-loss and shaping and hence explain the non-spherical PNe that dominate the PN population. The community therefore rallied around binary central stars as the most promising solution to the problem (for a review, De Marco 2009). During the last three years great strides have been made on both theoretical and observational fronts to determine the extent to which binarity is responsible for the formation and evolution of PNe.

From an observational perspective, after decades of slow progress we have now more than tripled the number of known close binary central stars to more than 40 post-common envelope systems (Miszalski *et al.* 2011a); we have independently refined the close binary fraction of Bond (2000) to be at least 17 ± 5 % (Miszalski *et al.* 2009a); identified strong tendencies for close binaries to be associated with equatorial rings and fast collimated polar outflows (Miszalski *et al.* 2009b, 2011b; Corradi *et al.* 2011) and we have identified the binary as the shaping agent in at least 6 systems where the nebula inclination matches the orbital inclination (Jones *et al.* 2011). In addition several efforts are underway to detect binaries of wider separation to finally determine the total binary fraction of PNe, predicted to be higher than expected from the main-sequence binary fraction (Moe & De Marco 2006, 2011).

On the theoretical front, Soker (2006) and Nordhaus & Blackman (2006) demonstrated that no current theory can explain mass-loss shaping which leads to non-spherical PNe. At the same time, efforts have intensified to demonstrate how common-envelope interactions can shape planetary nebulae (see in particular, Passy *et al.* 2011). These efforts not only try to determine how binary interactions can shape PNe, but also show that PNe are fundamental tools to constrain common-envelope simulations. This has repercussions for the validity of these simulations in more general contexts, such as the search for a mechanism to explain Type-Ia supernovae.

Our endeavours have blossomed into an exciting and highly active field of research as demonstrated by having, from 2009 onwards, (a) the two highest cited papers in the field (De Marco 2009; Miszalski *et al.* 2009a), and (b) more than 40 publications. Although we are not yet able to resolve the debate statistically, we would emphasise that binaries can no longer be ignored in the field of PNe and that with the current momentum in the field we are enthusiastically optimistic that future discoveries will bring us much closer to this goal in the medium term.

Conferences focused on the topic of binarity in planetary nebulae
1. Asymmetric Planetary Nebulae 1 (APNI), Oranim, Israel, Aug 8-11, 1994
2. APNII, MIT, Cambridge, USA, Aug 3-6, 1999
3. APNIII, Mount Rainier, Seattle, USA, Jul 28-Aug 3, 2003
4. APNIV, La Palma, Spain, 18-22 Jun, 2007
5. Rochester workshop on Asymmetric Planetary Nebulae, Jun 17-19, 2009
6. APNV, Bowness-on-Windermere, UK, 20-25 Jun, 2010
7. APNVI, Mexico City, Mexico, Summer 2013 (planned)
and their conference summaries:
1. Kahn, 1995, AnIPS, 11, 282, Closing summary
2. Schwarz, 2000, ASPC, 199, 457, Conference Impression
3. Habing, 2004, ASPC, 313, 575, Summary
4. Zijlstra, APNIV conference, 2007, Asymmetric Planetary Nebulae: what are we learning?
5. De Marco *et al.* 2011, APNV conference, 19, The Rochester White Paper: A Roadmap for Understanding Aspherical Planetary Nebulae
6. Kastner, APNV conference, 2011, APN V: A Highly Skewed and Biased conference Summary

5. Massive Stars (M.Limongi & C. Leitherer)

5.1. *Massive Star Models (M. Limongi & G. Meynet)*

A good understanding of the evolution of massive stars is required in order to shed light on many topical subjects like the nucleosynthesis and the UV outputs of the first stellar generations, the origin of the Carbon Enhanced Metal Poor stars, the anticorrelations observed in globular clusters, the properties of the Galactic and the Magellanic Clouds Wolf-Rayet stars, the final fate of massive stars and how they explode as core collapse supernovae of different types, the rotation rate of young pulsars and black holes that are produced after the explosion, the nature of the progenitors of the long Gamma Ray Bursts, the origin of the Be-type stars, and so on. In the few lines below, we shall focus on only a few themes on which substantial works have been made during the 2009-2012 period.

5.1.1. *Observational highlights*

First, many new observations have been collected to test the theoretical models using technics like photometry, spectroscopy, interferometry, asteroseismology and spectropolarimetry.

The use of multiobjects spectrographs has allowed spectrocospic analysis of large populations of massive stars. Hunter *et al.* (2009) provide chemical compositions for about 50 Galactic and about 100 SMC early B-type stars in the frame of the VLT-FLAMES survey. Penny & Gies (2009) provide velocity values for 97 Galactic, 55 SMC, and 106 LMC O-B type stars from archival FUSE observations. Huang & Gies (2010) have obtained from moderate dispersion spectra the rotation for more than 230 cluster and 370 field B stars. They find that lower mass B stars are born with a larger proportion of rapid rotators than higher mass B stars. The spectroscopy of more than 800 massive stars in 30 Doradus (LMC) has been obtained by Evans et al. (2011).

The Sloan Extension for Galactic Understanding and Exploration (SEGUE) Survey obtained $\sim 240,000$ moderate-resolution spectra from 3900 Å to 9000 Å of fainter Milky Way stars ($14.0 < g < 20.3$) of a wide variety of spectral types, both main-sequence and evolved objects, with the goal of studying the kinematics and populations of our Galaxy and its halo (Yanny et al. 2009). For stars with signal-to-noise ratio > 10 per resolution element, stellar atmospheric parameters are estimated, including metallicity, surface gravity, and effective temperature. The hundreds of stars in this and in the previous Milky Way halo surveys (HK survey, Beers et al. 1992; Hamburg/ESO (HES) survey, Christlieb et al. 2008) identified as extremely metal-poor stars (EMP) have been followed up with high resolution spectroscopic observations. They provide a means of probing the earliest phases of the evolution of the Milky Way and supernovae (SNe) in the early universe. Extremely important was the discovery of the Galactic halo star SDSS J102915+172927 (Caffau et al. 2011b) with a very low metallicity ($Z \leqslant 6.9 \cdot 10^{-7}$, which is $4.5 \cdot 10^{-5}$ times that of the Sun), [Fe/H] $= -4.99$, and a chemical pattern typical of the classical extremely metal-poor stars - that is, without enrichment of carbon, nitrogen and oxygen. Such a discovery confirms the previous suggestion, came with the discovery of the EMP HE 0557-4840 with [Fe/H] $= -4.75$ (Norris et al. 2007), that the metallicity distribution function of the Galactic halo does not have a gap between [Fe/H] $= -4.0$, where several stars are known, and the two most metal-poor stars, at [Fe/H] $= -5.3$. Moreover, since this star does not show enhancements of C, N and O, this suggests that the C-richness is not ubiquitous at metallicities below [Fe/H] $= -5.0$, as it was previously though. In addition to that, the very low global metallicity points toward the crucial role of dust for the formation of such an extremely metal-poor stars. Caffau et al. (2011b) also claim that stars similar to SDSS J102915+172927 are probably not very rare and they expect 5−50 stars of similar or even lower metallicity than SDSS J102915+172927 to be found among the candidates accessible from the VLT, and many more in the whole SDSS sample.

Asteroseismology represents an extraordinary window on the interior of stars. Solar-like oscillations in V1449 Aql, which is a large-amplitude (β Cephei) pulsator have been detected by Belkacem et al. (2009a). Miglio et al. (2009) have used asteroseismic observations to distinguish convective core extension due to overshooting and the extension due to rotation. Seismic diagnostics of rotation for massive stars have also been studied by Goupil & Talon (2009). The stability of g modes in rotating B-type stars has been addressed by Lee & Saio (2011).

Interferometry allows to determine the shape of fast rotating stars and to detect anisotropic distributions of circumstellar matter due for instance to polar winds or to equatorial disk around Be stars. Spectropolarimetry provides data on the surface magnetic field of early-type stars. An update survey of these areas of research can be found in the IAU Symposium 279, "Active OB Stars: Structure, Evolution, Mass-Loss, and Critical Limits" (Neiner et al. 2011) and in the recent review by Walder et al. (2011), "Magnetic Fields in Massive Stars, Their Winds, and Their Nebulae".

5.1.2. *Theoretical developments*

To interpret all these observational data, also theoretical efforts have been made in the last three years in order to refine as much as possible the theoretical models as well as to include new physical processes like, e.g., stellar rotation and magnetic fields.

Massive stars evolve through all the nuclear burning stages until an iron core is formed. This core is surrounded by the typical onion structure where all the zones keep track of the various nuclear burning occurred either in the core or in shells. The key aspects of the evolution leading to such a presupernova structure are the following. 1) The neutrino losses due to the pair production become efficient starting from the core C ignition hence the evolutionary lifetimes of the advanced burning stages reduce dramatically. 2) All the advanced nuclear burning are activated by few key reactions that release light particles which can be captured by almost all the nuclei present in the plasma, hence an increasingly large number of processes become efficient. 3) The nuclear energy generation, which in the core H and core He burning phases depends essentially on both the temperature and density, becomes very sensitive to the chemical composition and not only to the temperature and density. 4) Because of the efficient neutrino pair production, the mixing turnover times become comparable to the evolutionary lifetimes hence a time dependent treatment of convection becomes necessary. In addition to that, the convective turnover times become of the same order of magnitude of the typical nuclear burning lifetimes hence the coupling between convection and nuclear burning become important. For all these reasons the computation of the presupernova evolution of massive stars requires special attention to the following points: 1) the equations describing the physical structure of the stars should be coupled to those describing the chemical composition of the matter due to the nuclear burning and to those for the mixing due to the various convective processes; 2) a very extended nuclear network, including as much isotopes as possible, should be adopted and coupled to the above mentioned equations; 3) a time dependent convective algorithm should be adopted to describe the efficiency of mixing as a function of the time. Different approaches have been followed by the main groups working on this field in the course of the years. Generally, the equations describing the physical structure of the star are solved separately to those describing the chemical evolution and the nuclear energy generation is computed by means of a reduced network (Weaver *et al.* 1978 and subsequent work) or in a tabular form (Nomoto & Hashimoto 1988 and subsequent works). In both cases a quasi- or full- nuclear statistical equilibrium approximation is adopted above a critical temperature (of the order of $\sim 3 \cdot 10^9$ K). Convection is treated as a diffusive process and coupled only to the nuclear burning. Chieffi & Limongi were the first to couple both the structure equations to the chemical evolution adopting a large network and removing any kind of nuclear statistical approximation (Chieffi, Limongi & Straniero 1989). All these computations produced extended set of presupernova models and corresponding explosive chemical yields that were mainly used for galactic chemical evolution computations as well as supernova light curve and spectra simulations. The results obtained reproduced the main observational constraints although many open questions remained among which the reproduction of (1) the evolution of the abundances of the various elements as a function of the metallicity in the Milky Way (Francois *et al.* 2004), (2) the abundance pattern observed in extremely metal-poor stars (Chieffi & Limongi 2002, Umeda & Nomoto 2002, Cayrel *et al.* 2004), (3) the abundance pattern of the peculiar C-rich ultra metal-poor stars (Tominaga *et al.* 2007). The only way to find a reasonable fit to these observations was to adopt specific fine tuned assumptions on both the models and yields. In the last three years, because of the increasing power of the computer machines, there have been an improvement on the computational techniques. The american group refined significantly the treatment of the nucleosynthesis, still separated by the structure equations, by means of an adaptive reaction network including up to ∼ 900 nuclear species. Isotopes were added and removed as necessary to follow the nuclear reaction flow, with decisions based upon a conservative set of assumptions regarding abundances and flows in the neighborhood (Heger & Woosley 2010). Big effort has been done by Chieffi & Limongi, in the last three years, in order to couple all the equations together and solve them simultaneously in order to avoid any kind of approximation either for the nuclear energy generation and for the coupling between convection and nuclear burning (Chieffi & Limongi 2011). They adopted a diffusive treatment for the mixing, increased the size of the nuclear network to ∼ 300 isotopes and used sophisticated state of the art numerical techniques for solving very large sparse systems. This is the best numerical solution that can be achieved, at present, in 1D stellar evolution codes. Computation of extended grid of presupernova models for different metallicities are under way (Chieffi & Limongi 2011). In spite of these strong efforts made by these groups the "old" discrepancies between the standard theoretical predictions (based on classical models taken at face values) and the observed abundances in extremely metal-poor (normal and C-rich) stars still remain. At present no galactic chemical evolution model using these very recent grid of stellar yields have been carried on hence we have to wait for these results to judge the impact of the improvement in the presupernova models.

The investigation of the potential role of rotation on massive star evolution, started more than 10 years ago (Maeder & Meynet 2000 and subsequent works, Heger, Langer & Woosley 2000 and subsequent works), continued in the last three years and is still under way (see also the recent review by Maeder & Meynet 2011 and references therein).

Let us recall that the effects of rotation on massive star evolution can be classified in four categories: 1) axial rotation deforms the star and therefore has an impact on the hydrostatic configuration; 2) axial rotation triggers many instabilities in the stellar interior driving the transport of the angular momentum and of the chemical species; 3) axial rotation has an impact on the way stars are losing mass through radiative winds and through mechanical mass losses; 4) rotation may in some circumstances activate dynamo mechanisms and thus have an impact on the magnetic field which in its turn has an impact on the rotation of stars.

Through these different effects rotation modifies in a significant way many outputs of the stellar models (evolutionary tracks, lifetimes, evolutionary scenarios, nucleosynthesis, and so on). One of the most important effects for the evolution comes from rotational instabilities inducing efficient mixing of the chemical elements. Comparisons between the changes of the surface abundances predicted by the rotating stellar models and the observations are delicate because the observed sample of stars often mix stars of different initial masses and ages. From such comparisons, some authors argue that a significant fraction of stars do not follow the trend predicted by the rotating single star models (see e.g. Brott et al. 2011). Other underline that in order for such a relation to be tested great care should be taken to reduce the range of initial masses and ages in the sample of stars used for the comparison (Maeder et al. 2009). Probably to make progresses in that field, careful analysis of stars in stellar clusters will be needed. Stars that do not follow the expected trends can be due either to close binary evolution and/or magnetic field (see e.g. de Mink et al. 2009; Meynet et al. 2011).

Present stellar models show that all other parameters being kept fixed (initial mass, initial rotation), rotational mixing is more efficient at low metallicity. This more efficient mixing allows diffusion of elements between the He-burning core and the H-burning shell at very low metallicity. This boosts the production of some isotopes like ^{13}C, ^{14}N, ^{22}Ne

and the s-process elements and may have an impact on the early phases of the chemical evolution of galaxies (Cescutti & Chiappini 2010; Chiappini et al. 2011). It has also been proposed that the observed "primary-like" evolution of Be can be related to the fact that Galactic Cosmic Rays are accelerated from the wind material of rotating massive stars (enriched by rotational mixing), when the supernova explosion occurs (Prantzos 2010). Also fast rotating massive stars have longer main-sequence lifetimes, they keep a bluer position in the HR diagram and thus may have an impact on the UV outputs.

The first presupernova rotating models were computed by Heger, Langer & Woosley (2000). In this paper they included the centrifugal force following Kippenhahn & Thomas (1970) in the approximation of Endal & Sofia (1976) and the rotational mixing processes discussed in Endal & Sofia (1978), following the work of Pinsonneault et al. (1989). This was an approach different from the one developed by the Geneva group (see above) in the same years. One of the main results they obtained was that applying the presupernova specific angular momentum of the iron core to a neutron star with a typical radius of ~ 10 km the newly forming young pulsar would rotate with a period of the order of ~ 1 ms, which is one or even two order of magnitudes larger than the rotational periods of known young pulsars. Other presupernova models including rotation were computed by Hirschi, Meynet & Maeder (2004) although their nuclear network adopted for the advanced burning phases was quite small and included only the alpha isotopes. Attempts to include additional phenomena in order to slow down the spin up of the iron core were performed by Heger, Woosley & Spruit (2005) which included, in an approximate way, the effect of magnetic breaking. They found find that magnetic torques decrease the final rotation rate of the collapsing iron core by about a factor of 30–50 when compared with the nonmagnetic counterparts. However they clearly pointed out that their results were affected by a number of uncertain parameters. Hence their results could not be considered as well settled.

The study of the interactions between magnetic field and rotation is currently under way and, probably, will receive even higher attention in the next years. Ud-Doula and Owocki (2009) have examined the angular momentum loss of magnetic hot stars with a line-driven stellar wind and a rotation-aligned dipole magnetic field. These authors give typical spin-down times of the order of 1 Myr for several known magnetic massive stars. The discovery of rotational braking in the magnetic Helium-strong star Sigma Orionis E by Townsend et al. (2010) has provided a nice support to the theoretical developments by Ud-Doula and Owocki (2009). Lau et al. (2011) estimate the spin-down time-scale of rapidly non-convective stars hosting an α-Ω dynamo. They find that the spin-down time scale could be only a small fraction of the main-sequence lifetime.

Very interesting constraints on the question of the magnetic field and rotation will come probably from complementary observations in asteroseimology, spectropolarimetry and spectroscopy. Questions as the following ones could be more precisely addressed: do stars with strong surface magnetic fields rotate as solid body in their interior? What role does play a surface magnetic field to slow down the central regions? What is the role, if any, of magnetic field in the evolution leading to the long gamma ray bursts? So a very interesting period is ahead of us.

A big effort was done by Limongi & Chieffi in order to study the effect of rotation in their more recent set of models (Chieffi & Limongi 2011). They included the effect of rotation in their latest most updated version of the stellar evolution code (FRANEC) following the two proposed schemes, i.e., the one adopted by Heger & Woosley (2000) and the one proposed by the Geneva group (see above). A first set of models with and without rotation have been already computed and will be published very soon. Preliminary results confirm the previous finding of Heger & Woosley (2000) that these models have even much more angular momentum in the collapsing iron core than a neutron star can possibly carry, i.e., too fast rotating pulsars will be produced by these models. This means that still a lot of work must be done in order to revisit the mechanisms of the angular momentum transport (including the interaction between rotation and magnetic fields) during the presupernova evolution of massive stars.

5.2. Massive Binary Evolution (S. de Mink)

In the last three years there has been an increasing interest in the importance of binarity for the evolution of massive stars. This is partially driven by new determinations of the high close binary fraction in massive stars, such as performed by Sana & Evans (2011). These authors compiled the most complete dataset of the binary properties of young massive stars in nearby clusters. A striking result is the strong preference of for binaries with orbital periods of several days and less. The authors convincingly argue that this cannot be attributed to observational biases alone, as was previously thought. Questions about the role of binaries where also raised in the context of the puzzling trends of observed rotation rates and surface abundances found in the VLT-FLAMES survey of massive stars (Hunter et al. 2008; Evans et al. 2011), even for stars that appear to be single (de Mink et al. 2011).

Studies of individual massive binary systems are now providing us with ever more accurate parameters of the most massive stars, enabling critical tests for evolutionary models in the upper parts of the HR diagram (Pavlovski & Southworth 2009; Pavlovski et al. 2009; Ritchie et al. 2009; Ritchie et al. 2010; Clark 2011). Record holders include R136, host of the most massive O supergiants (Taylor et al. 2011), and R145 (Schnurr et al. 2009) hosting a hydrogen-rich Wolf-Rayet star for which minimum masses near 100 $M_\odot$ have been derived.

Significant progress has also been made with respect to modeling the evolution of stars in binaries. Cantiello et al. (2007) and de Mink et al. (2009) explored the consequences of rotationally induced mixing on stars in binaries. Eldridge et al. (2008), Yoon et al. (2010) and Claeys et al. (2011) focussed on the consequences of binary interaction on the progenitors of supernova. Eldridge et al. (2011) undertook an ambitious population synthesis study of runaway stars as the progenitors of supernova and gamma-ray bursts.

5.3. Existence of Extremely Massive Stars in R136 (P. Crowther)

Until recently, a general consensus had been reached that the upper stellar mass limit was close to 150 $M_\odot$ (Figer 2005). Crowther et al. (2010) reopened this debate, by attributing initial masses of up to ~ 300 $M_\odot$ for the brightest stars within R136, the central ionizing cluster of the 30 Doradus star forming region in the LMC. Crowther et al. attributed higher stellar masses than previous estimates for these stars using new infrared VLT spectroscopic and photometric observations, together with high stellar temperatures from spectral analyses and contemporary evolutionary models for very massive stars. Supporting evidence for their results was provided by consistent spectroscopic and dynamical mass estimates for NGC 3603-A1, a massive eclipsing binary system located in a similar star cluster within the Milky Way (Schnurr et al. 2008).

If the stellar mass limit were to exceed $\sim 150 M_\odot$, this would open up the possibility of pair-instability supernova – hitherto thought to be restricted to Population III stars – within the local universe. Indeed, Gal-Yam et al. (2009) have attributed SN 2007bi, an extremely bright supernova,to the pair-instability supernova of a metal-poor $\sim 200 M_\odot$ star. The high luminosities and powerful stellar winds of very massive stars would be anticipated to dominate the early appearance and feedback from high-mass star clusters. Indeed, Hoversten & Glazebrook (2011) argue that stars more massive than 120 $M_\odot$ are required to reproduce the observed Hα to continuum ratio of Sloan Digital Sky Survey galaxies.

5.4. *Luminous Blue Variable-type Eruptions (N. Smith)*

Luminous Blue Variable (LBV)-like eruptions have been recognized as playing a more dominant mass-loss role than previously thought for two reasons. One reason is that the inclusion of far-IR observations (sampling cool dust) has shown that LBV shells contain far more mass that previously known (e.g., Smith & Owocki 2006). In some extreme cases, an LBV can shed as much as 15-20 $M_\odot$ in a single eruption lasting only a few years, and these can happen multiple times in a star's life. A large number of new dust shells around LBVs and related stars have been discovered in recent surveys with Spitzer, showing that the phenomenon is widespread (Wachter *et al.* 2010; Gvaramadze *et al.* 2011). The second reason is that a large amount of work on the steady line-driven winds of hot O-type and Wolf-Rayet stars has revealed that their winds are very clumpy, and that standard mass-loss rates that have been used for decades are therefore too high by factors of perhaps 3 – 10 (Bouret *et al.* 2005; Fullerton *et al.* 2006; Puls *et al.* 2006). This indicates that while steady, metallicity-dependent winds are less important in removing a massive star's H envelope, LBV eruptions (which may be independent of metallicity) are probably the dominant mode of mass loss (Smith & Owocki 2006).

Studies of a particular type of supernova called Type IIn supernovae (the "n" is for narrow H lines) reveal that eruptive LBVs might be the progenitors of some of the most luminous supernovae known (Smith *et al.* 2007). This is a surprise, since LBVs are generally expected to shed their H envelope and evolve to the Wolf-Rayet phase before exploding (Heger *et al.* 2003). These Type IIn supernovae represent about 8 – 9% of all core-collapse supernovae (Smith *et al.* 2011a), but they may be the most massive 8 – 9% of stars that explode. The key observation is that the circumstellar material into which the supernova blast wave expands is extremely dense and massive, requiring eruptions of large amounts of mass just a few years or decades prior to core-collapse (Smith *et al.* 2007, 2008, 2010; Ofek *et al.* 2007; Woosley *et al.* 2007; Chevalier & Irwin 2011). The only known analog to produce the required amount of mass loss in such a short time is the eruptions of LBVs. There have been three cases where a IIn progenitor star candidate was identified in pre-explosion archival images, and those sources are consistent with very massive blue LBV-like stars (Gal-Yam & Leonard 2009; Smith *et al.* 2011b; Smith *et al.* 2011c). Therefore, it would seem that LBV-like eruptions can be an immediate precursor to a core-collapse supernova for very massive stars.

Recent years have seen the discovery of a substantial number of extragalactic eruptions that resemble LBV eruptions (i.e. fainter than supernovae and similar spectra), partly due to the increased emphasis on transient studies and the different methods employed in supernova surveys. These have yielded quite surprising results. One of the most interesting is that the family of eruptive outbursts that has been referred to as LBVs is surprisingly diverse, and some of them appear to come from rare classes of progenitor stars that are not necessarily the most massive stars (Thompson *et al.* 2009). This diverse class of transients is referred to variously as LBV eruptions, η Car analogs, supernova impostors, intermediate-luminosity optical transients, and luminous red novae (see Smith *et al.* 2011b and references therein for a recent summary). The underlying physical mechanism remains unknown, and various possibilities have been proposed, including super-Eddington LBV eruptions, binary merger outbursts or stellar collisions, electron capture supernovae, nuclear flashes, and "failed" supernovae (underluminous supernova from fallback to a black hole). Some of the progenitor stars of these outbursts certainly are what we would call classical LBVs, but some appear to be heavily dust enshrouded red stars with luminosities that suggest initial masses as low as 8 $M_\odot$ (Thompson et al. 2009; Prieto et al. 2009). Yet, the observed outburst properties are all quite similar and resemble known LBVs.

5.5. CNO Anomalies in Massive Stars (D. Lennon)

Our understanding of the origin of CNO anomalies in main-sequence massive stars has undergone what might be described as a paradigm shift in recent years. This was prompted in part by the VLT-FLAMES Survey of Massive Stars which found little evidence for a correlation between nitrogen enrichment and stellar rotation (Hunter et al. 2009). Their work presented surface abundances of fast rotators for the first time. However the most extreme cases of nitrogen enrichment were found among the slowest rotators, challenging our ideas concerning rotational mixing and leading to new estimates of the efficiency of rotational mixing and convective overshooting (Brott et al. 2011). The picture is further complicated by the realization that some nitrogen enriched slowly rotating B-type main-sequence stars in the Galaxy are also magnetic (Morel 2009) although the link between magnetism and nitrogen enrichment is as yet unclear. What is now evident however is that for those OB stars with CNO anomalies we are indeed seeing the products of CNO-cycled matter mixed to their surfaces, as shown by Przybilla et al. (2010) in a very careful analysis of Galactic B-type stars. A quantitative understanding of CNO anomalies in more massive early O-type stars has been hindered by limitations in non-LTE models of the line formation of higher ionization stages of nitrogen in particular. It is therefore gratifying to see crucial progress being made on this front, Rivero et al. (2011) presenting a comprehensive analysis of the N III line formation problem.

Hunter et al. (2009) also found that among their fast rotators there exists a wide range of nitrogen abundances, a result which was reflected in related work by Dunstall et al. (2011) for Be stars. The lack of a strong correlation between rotational velocity and nitrogen enhancements in massive stars, together with almost bimodal distribution of rotational velocities, has led to speculation that binaries may play a more important role than was previously thought in the evolution of massive star populations (de Mink et al. 2009a, even to the extent of perhaps explaining abundance anomalies in globular clusters (de Mink et al. 2009b). As discussed above, Sana & Evans (2011) emphasize the importance of the binary channel in massive star evolution, showing that approximately half OB stars in open clusters are in fact spectroscopic binaries. Additionally, it is suggested that binary interaction may produce a system which is difficult to distinguish from a single star (de Mink et al. 2011) which implies that understanding the origin of CNO peculiarities in an apparently single star may hinge on distinguishing between single and binary star evolutionary histories. Complicating the picture even further, Pavlovski et al. (2011) analysed disentangled spectra of components of high-mass binaries and have so far found no trace of CNO cycled material on their surfaces.

In summary, the processes leading to the origin of CNO anomalies in OB stars are much more complex than thought only a few years ago. Ongoing large scale observing programs such as the VLT-FLAMES Tarantula Survey (Evans et al. 2011), the VLT-FLAMES survey of massive binaries in Westerlund I (Ritchie et al. 2009), and MiMeS Project (Wade et al. 2011) surveying for magnetic fields in massive stars (see below) offer the potential to shed light on this problem.

5.6. Magnetism in Massive Stars (S. Owocki & G. Wade)

Hot, massive OB stars lack the vigorous subsurface convection thought to drive the dynamo central to the magnetic activity cycles of the sun and other cool stars. But building on pioneering detections of strong (kG) fields in the chemically peculiar Ap and Bp stars

(Babcock 1947; Borra & Landstreet 1980), new generations of spectropolarimeters have
revealed organized (often significantly dipolar) magnetic fields ranging in strength from
0.1 to 10 kG in several dozen OB stars (e.g. Donati *et al.* 2002, 2006; Hubrig *et al.* 2006;
Grunhut *et al.* 2009; Martins *et al.* 2010; Petit *et al.* 2011). A consortium known as
MiMeS (for Magnetism in Massive Stars) is carrying out both a survey for new detections, and monitoring known magnetic OB stars with high resolution spectroscopy and
polarimetry (Wade & the MiMeS Collaboration 2010). The observed field characteristics
favor a fossil origin over active dynamo generation (e.g., Donati & Landstreet 2009).

Much current research focuses on the effects of the magnetic field on stellar mass loss,
rotation, and evolution. Theoretical models and MHD simulations (e.g. Townsend &
Owocki 2005; Ud-Doula *et al.* 2008, 2009) indicate magnetic trapping of stellar wind outflow can feed a circumstellar magnetosphere, resulting in rotationally modulated Balmer
line emission that matches closely that which is observed (e.g., Townsend *et al.* 2005;
Oksala *et al.* 2011). Moreover, eclipses by co-rotating magnetically bound clouds induce
a photometric variation that, with extended monitoring, allows precise measurement of
the stellar rotation, and even its secular slowing due to angular momentum loss from
the magnetized wind. For the B2V magnetic star σ Ori E, the inferred spindown time of
1.3 Myr is in very good agreement with theoretical predictions based on MHD models
(Townsend *et al.* 2010). And the discovery of very long rotation periods (years to decades)
in two O-type magnetic stars (Howarth *et al.* 2007; Nazé *et al.* 2001) provides further
evidence that magnetic braking can strongly affect the rotational evolution of magnetic
massive stars. Larger population samples are needed to understand the origin of these
strong stellar fields, to determine their long-term evolution, and to clarify their potential
role for massive-star evolutionary end-states as SNe, pulsars, and magnetars.

6. Supernovae (M. Limongi, A. Tornambè, A. Mezzacappa)

A substantial fraction of the huge number of papers published on supernovae during
the past three years, i.e. more than 5000, are related to the two main still open questions:
(1) the nature of the progenitor of the thermonuclear supernovae (SN Ia) and (2) the
comprehension of the explosion mechanism for core collapse supernovae.

The nature of the progenitors of type Ia supernovae still remains elusive in spite of
the huge efforts to identify the(ir) evolutionary path(s) produced in the last three years
by several authors. Needless to recall the pivotal role of type Ia SNe in cosmology and
therefore the need to clarify in detail all their evolutionary and explosive properties.
The historical progenitor models are essentially two: the single degenerate (SD) model,
consisting of a normal star accreting H on a CO white dwarf (WD) (see Kobayashi &
Nomoto 2009 for an updated version of this model) and the double degenerate model
(DD), consisting of two WDs which merge due to gravitational radiation emission. Both
SD and DD models suffer drawbacks. One problem for the DD scenario largely addressed
in the last years arises when the two dwarfs strongly interact in a very short time during the merging. A number of authors have followed the dynamical evolution of such
a system in order to identify the outcome and to asses how much the classical view of
the secondary dwarf dissolved in a disk, which provides mass to an almost untouched
primary, has to be modified (Loren-Aguilar, Isern and Garcia-Berro 2009, Pakmor *et al.*
2011). This aspect deserves however further attention. Stellar rotation has been claimed
to be the pivotal mechanism leading to a self-regulated accretion rate during the merging of the two components which, in turn, tightly points toward an explosive outcome.
All along this evolutionary path an additional source of gravitational waves emission has
been also found to exist. These studies were carried on in the assumption of the solid body
rotation (Tornambè *et al.*2004). Extension to differentially rotating accreting WDs

suggests to revisit the evolutionary path toward the explosion. Such a differentially rotating structure may still remain stable even if the mass largely overcomes the Chandrasekhar limit (which is ~ 1.5 $M_\odot$ for a solid body rotation). Therefore, supra-stable structures may be produced during the accretion phase. These structures will become suddenly unstable when the accretion phase ends and the angular momentum is redistributed all along the accreted structure. In this scenario there is a change of perspective in the description of the evolutionary path toward the degenerate explosion: we are no more facing a degenerate structure which is driven to the critical limit by mass accretion but, on the contrary, a smooth process in which the critical mass is safely and largely bypassed until the structure become suddenly unstable when the accretion phase ends (Piersanti *et al.* 2009). Pfannes *et al.* (2010a,b) mimicked the explosion of a fast rotating massive WD obtaining results which, in the case of C-detonation, provide a good agreement with the observations. Maeda *et al.* (2010) explain the diversity of the spectral evolution of type Ia SN in the framework of an asymmetric explosion which is the natural outcome of the explosion of a fast rotating structure. In the case of a SD progenitor, rotation has also been invoked to suppress the strong nuclear pulses occurring when the accreted hydrogen is converted into He and then He into C that, in turn, would avoid the growth to the Chandrasekhar limiting mass because of strong mass loss occurring during the pulses. All in all, while rotation helps to trace the good route to the explosion, no robust effect to discriminate theoretically DD from SD models as progenitors of SNe Ia has been found yet. Interesting observational results seem however to point toward progenitors mostly composed by DD systems (Gifanoy & Bodgan 2010). The reason is that the observed X-ray flux from nearby elliptical galaxies is a factor of $\sim 30 - 50$ less than the one predicted in the case of the H-accretion SD scenario. Several observational approaches have been suggested to discriminate the nature of the progenitor model (among them, Fryer *et al.* 2010, Bianco *et al.* 2011, Livio & Pringle 2011, Marsh 2011) which should be exploited in the near future. At present, the diversity of the type Ia SNe has been emphasized in some details by means of HST observations in a GO international project whose results will soon appear on ApJ (Xiaofeng Wang *et al.* 2011). Once again non-asymmetric explosion are claimed to explain observations. A very interesting observational result comes from the observations of the near-by SN Ia in M101 (2011fe). According to Li *et al.* (2011) the hypothesis that a red giant star could have been present in the progenitor system has to be ruled out from the analysis of the archive frames obtained prior to the explosion occurred in August 2011. The progenitor system was therefore made of two WDs, i.e. consistent with the DD scenario, or, at most, by a WD plus a Main Sequence (or very little evolved) star companion. None of the described results (neither observational nor theoretical) seem to be able, however, to challenge the cosmological use of the SN Ia, as it was done so far.

The past several years have seen notable progress toward ascertaining the core collapse supernova explosion mechanism. Three independent groups worldwide (Bruenn *et al.* 2009, Marek & Janka 2009, and Suwa *et al.* 2010) have now reported neutrino-driven explosions, beginning with a range of stellar progenitor masses from 11 to 25 $M_\odot$. In all cases, the standing accretion shock instability (SASI; Blondin, Mezzacappa, & De-Marino 2003) couples with neutrino heating to power the explosions and thus plays a central role. These results were obtained in two-dimensional simulations. Key now will be an extension of these models to three dimensions. All of the above are multi-physics models. In particular, they all deploy multi-frequency neutrino transport. In addition, approximate general relativity is included in some of the models, as well as state of the art neutrino weak interactions and industry-standard nuclear equations of state. While Wilson's delayed-shock mechanism (Bethe & Wilson 1985) remains operative, significant variations on this theme have been reported in the above models. In particular, the explosions develop over a significantly longer period of time than originally reported by Wilson, with weak explosion energies at times significantly less than one second after stellar core bounce and still growing as late as one second after bounce. The SASI leads to continued long-term accretion to deep layers above the proto-neutron star surface, which in turn leads to efficient neutrino heating of the accreted material, some of which becomes unbound and joins the ejecta. The neutrino transport approximations used in the above models in particular, the ray-by-ray approximation (Buras *et al.* 2003) must be replaced by two- and three-dimensional multi-frequency, and ultimately multi-angle and multi-frequency, neutrino transport. In three spatial dimensions, multi-angle and multi-frequency neutrino transport will require sustained exascale supercomputing resources to complete. The approximation to general relativity (Marek *et al.* 2006) used in the above models must also be replaced, especially if peculiar core collapse supernovae are to be studied. The ability to evolve black-hole-forming collapse will be essential in order to understand systems with significant rotation and magnetic field generation and their connection with hypernovae and gamma-ray bursts. With non-parameterized explosions now in hand, the opportunity presents itself to compute the gravitational wave signatures from core collapse supernovae, through explosion (Yakunin *et al.* 2010). While these computations are based on two- and not three-dimensional models, they mark the first time all phases of core collapse supernova gravitational wave emission have been delineated in non-parameterized models, providing more accurate prediction of the development of the gravitational wave amplitudes and more accurate prediction of the time scales over which the amplitudes develop. These computations must now be repeated once comparable three-dimensional models are in hand.

7. Stellar populations (E. Brocato)

Stars are the basic elements of the light emitted by stellar clusters and galaxies, so that the challenge of de-convolving the light emitted by such distant objects have been caught by stellar population synthesis models which, typically, foresee the expected integrated photometric colors and spectra. An innovative method to use the light emitted by stars to unveil information on the age and metallicity of stellar populations in unresolved galaxies is the multi-band measurements of their Surface Brightness Fluctuations (SBF). As suggested by stellar population synthesis models (Worthey 1993; Brocato *et al.* 1998), the last years have been made clear that SBF are an independent and effective tool to derive information on the features of stellar populations, in particular when optical and near-IR high-quality images are observed (e.g. Blakeslee 2009; Cantiello *et al.* 2011). An exiting new branch of this technique discloses the possibility of deriving the age of stellar clusters both on theoretical and observational bases (Raimondo 2009; Whitmore *et al.* 2011).

8. Closing remarks

We wish to thank the researchers who actively contributed to this report, and apologize to those whose work is not quoted here. Progress has been made in many other fields within our remit that unfortunately we cannot all do justice here. As a matter of fact, C35 researchers are extremely active and their work has tremendous impact on all the areas of astrophysics and cosmology. It is thus an impossible task to present an exhaustive review of all the results.

Corinne Charbonnel
president of the Commission 35

References

Antia, H. M. & Basu, S., 2005, *ApJ*, 620, L129

Aprilia, Lee U., & Saio H., 2011, *MNRAS*, 412, 2265

Appourchaux, T., Samadi, R., & Dupret, M.-A., 2009, *A&A*, 506, 1

Asplund, M., Grevesse, N., & Sauval, A. J., 2005, in *Cosmic Abundances as Records of Stellar Evolution and Nucleosynthesis*, ASP Conf. Series, 336, 25

Aurière, M., Konstantinova-Antova R., Petit P. *et al.*, 2008, *A&A*, 491, 499

Aurière, M., Wade, G., Konstantinova-Antova R., *et al.*, 2009, *A&A*, 504, 231

Aurière, M., Donati, J. F., Konstantinova-Antova R., *et al.*, 2010, *A&A*, 516, L2

Aurière, M., Konstantinova-Antova R., Petit, P., *et al.*, 2011, *A&A*, in press

Babcock, H. W. 1947, *ApJ*, 105, 105

Bahcall, J., *et al.*, 2005, *ApJ*, 631, 1281

Baglin, A., Michel, E., Auvergne, M., & The COROT Team, 2006, Proceedings of SOHO 18/GONG 2006/HELAS I, Beyond the spherical Sun (ESA SP-624)

Balick & Frank, 2002, *ARA&A*, 40, 439

Barker, A. J. & Ogilvie, G. I., 2010, *MNRAS*, 404, 1849

Barker, A. J. & Ogilvie, G. I., 2011, *MNRAS*, 417, 745

Basu, S. & Antia, M. H., 2008, *Phys. Rep.*, 457, 217

Basu, S., Grundahl, F., Stello, D., *et al.*, 2011, *ApJ*, 729, L10

Batalha, N. M., Borucki, W. J., Bryson, S. T., *et al.*, 2011, *ApJ*, 729, 27

Beck, P. G., Bedding, T. R., Mosser, B., *et al.*, 2011, *Science*, 332, 205

Bedding, T. R., Mosser, B., Huber, D., *et al.*, 2011, *Nature*, 471, 608

Bedding, T. R., Huber, D., Stello, D., *et al.*, 2010, *ApJ*, 713, 176

Belkacem K., *et al.*, 2009, *Sci*, 324, 1540

Belkacem K., *et al.*, 2009b, *A&A*, 494, 191

Belkacem K., Dupret, M. A., & Noels, A., 2010, *A&A*, 510, 6

Beers, T. C., Preston, G. W., & Shectman, S. A. 1992, *AJ*, 103, 1987

Benomar, O., Baudin, F., Campante, T. L., *et al.*, 2009, *A&A*, 507, L13

Bethe, H. A. & Wilson, J. R. 1985, *ApJ*, 295, 14

Bianco, F. B. *et al.* 2011, *ApJ*, 741, 20

Bisterzo, S., Gallino, R., Straniero, O., *et al.*, 2010, *MNRAS*, 404, 1529

Bisterzo, S., Gallino, R., Straniero, O., *et al.*, 2011, *MNRAS* in press (arXiv:1108.0500)

Blakeslee, J. 2009, *AIPC*, 111, 27

Blondin, J. M., Mezzacappa, A., & De Marino, C. 2003, *ApJ*, 584, 971

Bonanos, A. Z., Lennon, D. J., Köhlinger, F., *et al.*, 2010, *AJ*, 140, 416

Bond, 2000, *ASPC*, 199, 115

Borra, E. F. & Landstreet, J. D. 1980, *ApJS*, 42, 421

Borucki, W. J., *et al.*, 2009, IAU Symposium 253, 289

Bouret, J. C., Lanz, T., & Hillier, D. J. 2005, *A&A*, 438, 301

Boyer, M. L., van Loon, J. Th., McDonald, I., *et al.*, 2010, *ApJ*, 711, L99

Boyer, M. L., McDonald, I., van Loon, J. Th., *et al.*, 2009, *ApJ*, 705, 746

Brocato, E. *et al.* 1998, *Mem. SAIt*, 69, 155

Brott, I., Evans, C. J., Hunter, I., *et al.* 2011, *A&A*, 530, A116

Brott, I., de Mink, S. E., Cantiello, M., Langer, N., de Koter, A., Evans, C. J., Hunter, I., Trundel, C., & Vink, J. S., 2011, *A&A*, 530, A115

Brown, T. M., Gilliland, R. L., Noyes, R. W., & Ramsey, L. W., 1991, *ApJ*, 368, 599

Bruenn, S. W., Mezzacappa, A., Hix, W. R., Blondin, J. M., Marronetti, P., Messer, O. E. B., Dirk, C. J., & Yoshida, S. 2009, *Journ. Phys. Conf. Ser.*, 180, 012018

Buras, R., Janka, H. Thomas., Keil, M. T., Raffelt, G. G., & Rampp, M. 2003, *ApJ*, 587, 320

Caffau, E., *et al.* 2011a, *Solar Physics*, 268, 255

Caffau, E., *et al.* 2011b, *Nature*, 477, 69

Cantiello, M. *et al.* 2011, *A&A*, 532, 154

Cantiello, M. & Langer, N., 2010, *A&A*, 521, A9

Cantiello, M., Yoon, S. C., Langer, N., & Livio, M. 2007, *A&A*, 465, L29

Canto Martins, B. L. Lèbre, A. Palacios, A., de Laverny, P., Richard, O., Melo, C. H. F., Do Nascimento, Jr., J. D., & de Medeiros, J. R., 2011, *A&A*, 527, A94

Carrier, F., De Ridder, J., Baudin, F., *et al.*, 2010, *A&A*, 509, 73

Cayrel, R., *et al.* 2004, *A&A*, 416, 1117

Cescutti G. & Chiappini C., 2010, *A&A*, 515, A102

Chadid, M., Benkő, J. M., Szabó, R., *et al.*, 2010, *A&A*, 510, 39

Chapellier, E., Rodríguez, E., Auvergne, M., *et al.*, 2011, *A&A*, 525, 23

Chaplin, W. J., 2010, Astronomische Nachrichten, 331, 1090

Chaplin, W. J., *et al.*, 2010, *ApJ*, 713, L169

Chaplin, W. J., Kjeldsen, H., Christensen-Dalsgaard, J., *et al.*, 2011, *Science*, 332, 213

Charbonnel, C. & Lagarde, N., 2010, *A&A*, 522, A522

Charpinet, S., Green, E. M., Baglin, A., *et al.*, 2010, *A&A*, 516, L6

Charpinet, S., van Grootel, V., Fontaine, G., *et al.*, 2011, *A&A*, 530, A3

Chevalier, R. A. & Irwin, C. M. 2011, *ApJL*, 729, L6

Chiappini, C., Frischknecht, U., Meynet, G., Hirschi, R., Barbuy, B., Pignatari, M., Decressin, T., & Maeder, A. 2011, *Nature*, 472, 454

Chieffi, A. & Limongi, M. 2002, *ApJ*, 577, 281

Chieffi, A. & Limongi, M. 2011, *ApJ*, submitted

Chieffi, A., Limongi, M., & Straniero, O. 1998, *ApJ*, 502, 737

Christlieb, N., Schoerck, T., Frebel, A., Beers, T. C., Wisotzki, L., & Reimers, D. 2008, *A&A*, 484, 721

Church, R. P., Cristallo, S., Lattanzio, J. C., Stancliffe, R. J., Straniero, O., & Cannon, R. C., 2009, *PASA*, 26, 217

Claeys, J. S. W., de Mink, S. E., Pols, O. R., Eldridge, J. J., & Baes, M. 2011, *A&A*, 528, A131

Clark, J. S., Ritchie, B. W., Negueruela, I., *et al.* 2011, *A&A*, 531, A28

Corradi *et al.* 2011, *MNRAS*,410,1349

Cox, N. L. J., Kerschbaum, F., van Marle, A.-J., *et al.*, 2011, A&A in press (arXiv:1110.5486)

Cranmer, S. R. & Saar, S. H., 2011, *ApJ*, 741, 54

Cristallo, S., Piersanti, L., Straniero, O., Gallino, R., Dominguez, I., Abia, C., DiRico, Gi., Quintini, M., & Bisterzo, S., 2011, in press (arXiv:1109.1176)

Cristallo, S., Straniero, O., Gallino, R., Piersanti, L., Domguez, I., & Lederer, M. T., 2009, *ApJ*, 696, 797

Crowther, P. A., Schnurr, O., Hirschi, R., Yusof, N., Parker, R. J., Goodwin, S. P., & Kassim, H. A. 2010, *MNRAS*, 408, 731

Decressin, T., Mathis, S., Palacios, A., *et al.* 2009, *A&A*, 495, 271

Degroote, P., Aerts, C., Baglin, A., *et al.*, 2010, Nature, 464, 259

Deheuvels, S., *et al.*, 2010, *A&A*, 514, 31

Deheuvels, S. & Michel, E., 2010, *A&A*, Astronomische Nachrichten, 331, 929

Deheuvels, S. & Michel, E., 2011, *A&A*, in press (arXiv:1109.1191)

De Marco, O., 2009, PASP, 121, 316

De Marco, O., Farihi, J., & Nordhaus, J., 2009, JPhCS, 172, 2031

De Marco *et al.*, 2011, *MNRAS*, 411, 2277

de Mink, S. E., Cantiello, M., Langer, N., *et al.* 2009a, *A&A*, 497, 243

de Mink, S. E., Langer, N., & Izzard, R. G. 2011, Bulletin de la Societe Royale des Sciences de Liege, 80, 543

de Mink, S. E., Pols, O. R., Langer, N., & Izzard, R. G. 2009, *A&A*, 507, L1

Denissenkov, P. A., 2010a, *ApJ*, 719, 28

Denissenkov, P. A., 2010b, *ApJ*, 723, 563

Denissenkov, P. A. & Merryfield, W. J., 2011, *ApJ*, 728, L8

Denissenkov, P. A., Pinsonneault, M., & MacGregor, K. B., 2009, *ApJ*,696, 1823

Denissenkov, P. A., Pinsonneault, M., Terndrup, D. M., & Newsham, G., 2010, *ApJ*, 716, 1269

D'Ercole, A., D'Antona, F., Ventura, P., *et al.*, 2010, *MNRAS*, 407, 854

De Ridder, J., *et al.*, 2009, *Nature*, 459, 398

Doherty, C., Siess, L., Lattanzio, J. C., & Gil Pons, P., 2010, *MNRAS*, 401, 1453

Donati, J. F., Babel, J., Harries, T. J., *et al.* 2002, *MNRAS*, 333, 55

Donati, J. F., Howarth, I. D., Jardine, M. M., *et al.* 2006, *MNRAS*, 370, 629

Donati, J. F. & Landstreet, J. D. 2009, *ARA&A*, 47, 333

Duez, V., Mathis, S., & Turck-Chièze, S., 2009, *MNRAS*, 402, 271

Duez, V., Braithwaite, J., & Mathis, S., 2010, *ApJLetters*, 724, L34

Duez, V. & Mathis, S., 2010, *A&A*, 517, A58

Dunstall, P. R., Brott, I., Dufton, P. L., *et al.* 2011, arXiv:1109.6661

Dupret, M.-A., Belkacem, K., Samadi, R., *et al.*, 2009, *A&A*, 507, 57

Eggenberger, P., *et al.*, 2010, *A&A*, 519, A116

Ekström, S., *et al.*, 2011, *A&A*, in press (arXiv:1110.5049)

Eldridge, J. J., Izzard, R. G., & Tout, C. A. 2008, *MNRAS*, 384, 1109

Eldridge, J. J., Langer, N., & Tout, C. A. 2011, *MNRAS*, 414, 3501

Endal, A. S. & Sofia, S. 1976, *ApJ*, 210, 184

Endal, A. S. & Sofia, S. 1978, *ApJ*, 220, 279

Evans, C. J., Taylor, W. D., Hénault-Brunet, V., *et al.* 2011, *A&A*, 530, A108

Evans C. J., *et al.*, 2011, *A&A*, 530, A108

Farihi, J., Burleigh, M. R., Holberg, J. B., Casewell, S. L., & Barstow, M. A., 2011, *MNRAS*, 417, 1735

Farihi, J., Barstow, M. A., Redfield, S., Dufour, P., & Hambly, N. C., 2010b, *MNRAS*, 404, 2123

Farihi, J., Jura, M., Lee, J.-E., & Zuckerman, B., 2010a, *ApJ*, 714, 1386

Farihi, J., Jura, M., & Zuckerman, B., 2009, *ApJ*, 694, 805

Figer, D. F. 2005, *Nat*, 434, 192

Francois, P., Matteucci, F., Cayrel, R., Spite, M., Spite, F., & Chiappini, C. 2004, *A&A*, 421, 613

Frandsen, S., Carrier, F., Aerts, C., *et al.*, 2002, *A&A*, 394, L5

Fryer, C. L. *et al.* 2010, *ApJ*, 725, 296

Fullerton, A. W., Massa, D. L., & Prinja, R. K. 2006, *ApJ*, 637, 1025

Gal-Yam, A. & Leonard, D. C. 2009, *Nature*, 458, 865

Gal-Yam, A., Mazzali, P., Ofek, E. O., *et al.* 2009, *Nature*, 462, 624

García, R. A., Mathur, S., Salabert, D., *et al.*, 2010, *Science*, 329, 1032

Garcia-Hernandez, D. A., *et al.*, 2006, *Science*, 314, 5806

Garcia-Hernandez, D. A., *et al.*, 2007, *A&A*, 462, 711

Garcia-Hernandez, D. A., *et al.*, 2000, *ApJ*, 705, L31

Gebran, M., Vick, M., Monier, R., & Fossati, L., 2010, *A&A*, 523, A71

Gifanoy, M. & Bodgan, A 2010, *Nature*, 463, 924

Gough, D. O., 1990, Progress of Seismology of the Sun and Stars

Goupil M. J. & Talon S., 2009, *CoAst*, 158, 220

Grevesse, N., Asplund, M., Sauval, A. J., & Scott, P., 2010, *Astrophysics and Space Science*, 328, 179

Grunhut, J. H., Wade, G. A., Marcolino, W. L. F., *et al.* 2009, *MNRAS*, 400, L94

Gvaramadze, V. V., Kniazev, A. Y., Fabrika, S., *et al.* 2010, *MNRAS*, 405, 520

Hareter, M., Reegen, P., Miglio, A., *et al.*, 2010, Proceedings of the 4th HELAS International Conference (arXiv:1007.3176)

Hekker, S., Gilliland, R. L., Elsworth, Y., Chaplin, W. J., De Ridder, J., Stello, D., Kallinger, T., Ibrahim, K. A., Klaus, T. C., & Li, J., 2011c, *MNRAS*, 414, 2594

Hekker, S., Basu, S., Stello, D., *et al.*, 2011b, *A&A*, 530A, 100

Hekker, S., Elsworth, Y., De Ridder, J., *et al.*, 2011a, *A&A*, 525A, 131

Hekker, S., Kallinger, T., Baudin, F., *et al.*, 2009, *A&A*, 506, 465

Heger, A. & Woosley, S. E. 2010, *ApJ*, 724, 341

Heger, A., Langer, N., & Woosley, S. E. 2000, *ApJ*, 528, 368

Heger, A., Fryer, C. L., Woosley, S. E., Langer, N., & Hartmann, D. H. 2003, *ApJ*, 591, 288

Heger, A., Woosley, S. E., & Spruit, H. C. 2005, *ApJ*, 626, 350

Henriques, B., Maraston, C., Monaco, P., Fontanot, F., Menci, N., De Lucia, G., & Tonini, C., 2011, *MNRAS*, 415, 3571

Hillwig *et al.*, 2010, *AJ*, 140, 319

Hirschi, R., Meynet, G., & Maeder, A. 2004, A&A, 425, 649

Hoversten, E. A. & Glazebrook, K. 2011, in UP2010: Have Observations Revealed a Variable Upper End of the Initial Mass Function?, eds. M. Treyer, T. Wyder, J. Neill, M. Seibert, & J. Lee, (San Francisco: ASP), ASP Conf Ser. 440, 251

Howarth, I. D., Walborn, N. R., Lennon, D. J., et al. 2007, MNRAS, 381, 433

Hrivnak et al., 2010, ApJ, 709, 1042

Hrivnak et al., 2011, ApJ, 734, 25

Hu, H., 2011, ASP Conference Series, the 61st Fujihara Seminar: Progress in solar/stellar physics with helio- and asteroseismology (arXiv:1109.0121)

Hu, H., Tout, C. A., Glebbeek, E., & Dupret, M.-A., 2011, MNRAS in press, (arXiv:1108.1318)

Huang W., Gies D. R., & McSwain M. V., 2010, ApJ, 722, 605

Huber, D., Bedding, T. R., Stello, D., et al., 2010, ApJ, 723, 1607

Hubrig, S., Briquet, M., Schöller, M., et al. 2006, MNRAS, 369, L61

Hunter, I., Brott, I., Langer, N., et al. 2009, A&A, 496, 841

Hunter, I., Brott, I., Lennon, D. J., et al. 2008, ApJL, 676, L29

Javadi, A., van Loon, J. Th., & Mirtorabi, M. T., 2011a, MNRAS, 411, 263

Javadi, A., van Loon, J. Th., & Mirtorabi, M. T., 2011b, MNRAS, 414, 3394

Jones et al., 2010a, MNRAS, 408, 2312

Jones et al., 2010b, MNRAS, 401, 405

Jones et al., 2011, IAUS 283, arXiv:1109.2801

Jorissen, A., Mayer, A., van Eck, S., et al., 2011, A&A, 532A, 135

Jura, M., Muno, M. P., Farihi, J., & Zuckerman, B., 2009b, ApJ, 699, 1473

Jura, M., Farihi, J., & Zuckerman, B., 2009a, AJ, 137, 3191

Kallinger, T., Mosser, B., Hekker, S., et al., 2010, A&A, 422, A1

Kilic, M., Patterson, A. J., Barber, S., Leggett, S. K., & Dufour, P., 2011, MNRAS Letters in press (arXiv:1110.3799)

Kippenhahn, R. & Thomas, H. C. 1970, in IAU Colloq. 4, Stellar Rotation, ed. A. Slettebak (Dordrecht : Reidel), 20

Kjeldsen, H., Bedding, T. R., Viksum, M., & Frandsen, S., 1995, textitAJ, 109, 1313

Kobayashi, C. & Nomoto, K. 2009, ApJ, 707, 1466

Kobayashi, C., Karakas, A., & Umeda, H., 2011, MNRAS, 414, 3231

Kolenberg, K., Bryson, S., Szabo, R., et al., 2011, MNRAS, 411, 878

Konstantinova-Antova, R., Aurière M., Iliev, I., et al., 2008, A&A, 480, 475

Konstantinova-Antova, R., Aurière M., Charbonnel, C. et al., 2010, A&A, 524, A57

Lagarde, N., Charbonnel, C., Decressin, T., & Hagelberg, J., 2011, A&A, in press (arXiv:1109.5704)

Lagarde, N., Decressin, T., Charbonnel, C., Eggenberger, P., Ekström, S., & Palacios, A., 2011, submitted to A&A

Lau H. H. B., Potter A. T., & Tout C. A., 2011, MNRAS, 415, 959

Lèbre, A., Palacios, A., Do Nascimento, J. D., et al., 2009, A&A, 504, 1011

Li, W., et al. 2011, in press

Livio, M. & Pringle, J. E. 2011, ApJ, 740, 18

Loren Aguilar, P., Isern, J., & Garcia-Berro, E. 2009, A& A, 500, 1193

MacGregor, K. & Rogers, T., 2011, Solar Physics, 270, 417

Maeda, K., et al. 2010, Nature, 466, 82

Maeder, A. & Meynet, G. 2000, ARA&A, 38, 143

Maeder A., Meynet G., Ekström S., Georgy C., 2009, CoAst, 158, 72

Marek, A., Dimmelmeier, H., Janka, H. Th., Mueller, E., & Buras, R. 2006, A& A, 445, 273

Marek, A. & Janka, H. T. 2009, ApJ, 694, 664

Marsh, T. R. 2011, Classical and Quantum Gravity, 28, 094019

Martins, F., Donati, J. F., Marcolino, W. L. F., et al. 2010, MNRAS, 407, 1423

Matthews, L. D., Libert, Y., Gérard, E., Le Bertre, T., Johnson, M. C., & Dame, T. M., 2011, AJ, 141, 60

Mathis S., 2009, A&A, 506, 811

Mathis S. & de Brye, N., 2011, A&A, 526, A65

Mathis S. & Le Poncin-Latiffe C., 2009, *A&A*, 497, 889

Mauron, N. & Josselin, E., 2011, *A&A*, 526A, 156

Mayer, A., Jorissen, A., Kerschbaum, F., *et al.*, 2011, *A&A*, 531, L4

McDonald, I., van Loon, J. Th., Sloan, G. C., Dupree, A. K., Zijlstra, A. A., Boyer, M. L., Gehrz, R. D., Evans, A., Woodward, C. E., & Johnson, C. I., 2011c, *MNRAS*, 417, 20

Mazumdar, A., & Michel, E., 2010, Proceedings of the HELAS-IV International Conference (arXiv:1004.2739)

McDonald, I., Boyer, M. L., van Loon, J. Th., *et al.*, 2011b, ApJS, 193, 23

McDonald, I., Boyer, M. L., van Loon, J. Th., & Zijlstra, A. A., 2011a, *ApJ*, 730, 71

McDonald, I., Sloan, G. C., Zijlstra, A. A., Matsunaga, N., Matsuura, M., Kraemer, K. E., Bernard-Salas, J., & Markwick, A. J., 2010, *ApJ*, 717, L92

McDonald, I., van Loon, J. Th., Decin, L., Boyer, M. L., Dupree, A. K., Evans, A., Gehrz, R. D., & Woodward, C. E., 2009, *MNRAS*, 394, 831

Metcalfe, T. S., *et al.*, 2010, *ApJ*, 723, L213

Meynet G., Eggenberger P., Maeder A., 2011, *A&A*, 525, L11

Michaud, G., Richer, J., & Richard, O., 2010, *A&A*, 510, A104

Michaud, G., Richer, J., & Richard, O., 2011, *A&A*, 529, A60

Miglio A., Montalbán J., Eggenberger P., Noels A., 2009, CoAst, 158, 233

Miglio, A., *et al.*, 2011, MNRAS in press (arXiv:1109.4376)

Miglio, A., Montalban, J., Baudin, F., *et al.*, 2009, *A&A*, 503, L21

Miglio, A., Montalban, J., Carrier, *et al.*, 2010, *A&A*, 520, L6

Miglio, A., Montalban, J., Noels, A., & Eggenberger, P., 2008, *MNRAS*, 386, 1487

Miszalski *et al.*, 2009a, *A&A*, 496, 813

Miszalski *et al.*, 2009b, *A&A*, 505, 249

Miszalski *et al.*, 2011a, APNV conference, 328

Miszalski *et al.*, 2011b, *MNRAS*, 413, 1264

Moe & De Marco, 2006, *ApJ*, 650, 916

Moe & De Marco, 2011, IAUS 283, in press

Momany, Y., Saviane, I., Smette, A., Bayo, A., Girardi, L., Marconi, G., Milone, A. P., & Bressan, A., 2011, A&A in press (arXiv:1110.2458)

Montalbán, J., Miglio, A., Noels, A., *et al.*, 2010a, *ApJ*, 721, L182

Montalbán, J., Miglio, A., Noels, A., *et al.*, 2010a, *Astronomische Nachrichten*, 331, 1010

Mosser, B., Barban, C., Montalban, J., *et al.*, 2011a, *A&A*, 532, A86

Mosser, B., Belkacem, K., Goupil, M. J., *et al.* 2010, *A&A*, 517, A22

Mosser, B., Belkacem, K., Goupil, M. J., *et al.* 2011b, *A&A*, 525, L9

Moya, A., Amado, P. J., Barrado, D., *et al.*, 2010, *MNRAS*, 406, 566

Nazé, Y., Vreux, J.-M., & Rauw, G. 2001, *A&A*, 372, 195

Neiner, C., Wade, G., Meynet, G., & Peters, G., 2011, IAUS, 272

Nittler, L. R., 2009, *PASA*, 26, 271

Nomoto, K. & Hashimoto, M. 1988, *Phys. Rep.*, 163, 13

Nordhaus & Blackman, 2006, *MNRAS*, 370, 2004

Nordhaus *et al.*, 2010, *MNRAS*, 408, 613

Norris, J. E., Christlieb, N., Korn, A. J., Eriksson, K., Bessell, M. S., Beers, T. C., Wisotzki, L., & Reimers, D. 2007, *ApJ*, 670, 774

Ofek, E. O., Cameron, P. B., Kasliwal, M. M., *et al.* 2007, *ApJL*, 659, L13

Oksala, M. E., Wade, G. A., Marcolino, W. L. F., *et al.* 2010, *MNRAS*, 405, L51

Origlia, L., Rood, R. T., Fabbri, S., Ferraro, F. R., Fusi Pecci, F., Rich, R. M., & Dalessandro, E., 2010, *ApJ*, 718, 522

Østensen, R. H., Silvotti, R., & Charpinet, S., *et al.*, 2010, *MNRAS*, 409, 1470

Østensen, R. H., Silvotti, R., & Charpinet, S., *et al.*, 2011, *MNRAS*, 414, 2860

Pakmor, R., ,Hachinger, S., Roepke, F. K., & Hillebrandt, W. 2011, *A& A*, 528, 117

Palmerini, S., La Cognata, M., Cristallo, S., & Busso, M., 2011a, *ApJ*, 729, 3

Palmerini, S., Cristallo, S., Busso, M., Abia, C., Uttenthaler, S., Gialanella, L., & Maiorca, E. , 2011b, *ApJ*, 741, 26

Pavlovski, K. & Southworth, J. 2009, *MNRAS*, 394, 1519

Pavlovski, K., Southworth, J., Tamajo, E., & Kolbas, V. 2011, Bulletin de la Societe Royale des Sciences de Liège, 80, 714

Pavlovski, K., Tamajo, E., Koubský, P., et al. 2009, MNRAS, 400, 791

Passy et al., 2011, MNRAS, in press, arXiv:1107.5072

Penny L. R., Gies D. R., 2009, ApJ, 700, 844

Petit, V., Massa, D. L., Marcolino, W. L. F., Wade, G. A., & Ignace, R. 2011, MNRAS, 412, L45

Pfannes, J. M. M., Niemeyer, J. C., & Schmidt, W. 2010a, A& A, 509, 75

Pfannes, J. M. M., Niemeyer, J. C., Schmidt, W., & Klingenberg, C. 2010b, A&A, 509, 74

Piersanti, L., Tornambè, A., Straniero, O., & Dominguez, I. 2009, in: L.A. Antonelli, E. Brocato, M. Limongi, N. Menci, G. Raimondo, and A. Tornambè (eds.), Probing stellar populations out to the distant universe: cefalu 2008, AIP Conference Proceedings, Volume 1111, pp. 259–266

Pinsonneault, M. H., Kawaler, S. D., Sofia, S., & Demarque, P. 1989, ApJ, 338, 424

Poretti, E., Paparo, M., & Deleuil, M., et al., 2010, A&A, 520, 108

Prantzos N., 2010, IAUS, 268, 473

Prieto, J. L., Sellgren, K., Thompson, T. A., & Kochanek, C. S. 2009, ApJ, 705, 1425

Przybilla, N., Firnstein, M., Nieva, M. F., Meynet, G., & Maeder, A. 2010, A&A, 517, A38

Puls, J., Markova, N., Scuderi, S., et al. 2006, A&A, 454, 625

Raimondo, G., 2009, ApJ, 700, 1247

Ritchie, B. W., Clark, J. S., Negueruela, I., & Crowther, P. A. 2009, A&A, 507, 1585

Ritchie, B. W., Clark, J. S., Negueruela, I., & Langer, N. 2010, A&A, 520, A48

Rivero Gonzalez, J. G., Puls, J., & Najarro, F. 2011, arXiv:1109.3595

Rodriguez-Gil et al., 2010, MNRAS, 407, 21

Rogers, T. & MacGregor, K.,, 2010, MNRAS, 401, 191

Rogers, T. & MacGregor, K., 2011, MNRAS, 410, 946

Rosenblum, E., Garaud, P., Traxler, A., & Stellmach, S., 2011, ApJ, 731, 66

Sana, H. & Evans, C. J. 2011, IAU Symposium, 272, 474

Santander-Garcia et al., 2010, A&A, 519, 54

Santander-Garcia et al., 2011, APNV conference, 259

Schnurr, O., Casoli, J., Chené, A.-N., Moffat, A. F. J., & St-Louis, N. 2008, MNRAS, 389, L38

Schnurr, O., Moffat, A. F. J., Villar-Sbaffi, A., St-Louis, N., & Morrell, N. I. 2009, MNRAS, 395, 823

Siess, L., 2007, A&A, 476, 893

Siess, L., 2009, A&A, 497, 463

Siess, L., 2010, A&A, 512, 10

Sloan, G. C., Matsuura, M., Zijlstra, A. A., Lagadec, E., Groenewegen, M. A. T., Wood, P. R., Szyszka, C., Bernard-Salas, J., & van Loon, J. T.h., 2009, Sci, 323, 353

Smith, N., Chornock, R., Li, W., et al. 2008, ApJ, 686, 467

Smith, N., Chornock, R., Silverman, J. M., Filippenko, A. V., & Foley, R. J. 2010, ApJ, 709, 856

Smith, N., Li, W., Filippenko, A. V., & Chornock, R. 2011a, MNRAS, 412, 1522

Smith, N., Li, W., Foley, R. J., et al. 2007, ApJ, 666, 1116

Smith, N., Li, W., Miller, A. A., et al. 2011, ApJ, 732, 63

Smith, N., Li, W., Silverman, J. M., Ganeshalingam, M., & Filippenko, A. V. 2011b, MNRAS, 415, 773

Smith, N. & Owocki, S. P. 2006, ApJl, 645, L45

Soker, 2006, PASP, 118, 260

Stancliffe, R. J.,, 2010, MNRAS, 403, 505

Stancliffe, R. J., Church, R. P., Angelou, G. C., & Lattanzio, J. C., 2009, MNRAS, 396, 2313

Stasinska, G. et al., 2010, A&A, 511, 44

Stello, D., Meibom, S., & Gilliland, R. L., et al., 2011, ApJ, 739, 13

Suwa, Y., Kotake, K., Takiwaki, T., Whitehouse, S. C., Liebendrfer, M., & Sato, K. 2010, PASJ, 62, L49

Taylor, W. D., Evans, C. J., Sana, H., et al. 2011, A&A, 530, L10

Théado, S., & Vauclair, S., 2010, *Astrophysics and Space Science*, 328, 209

Théado, S., & Vauclair, S., 2011, *ApJ*, in press (arXiv:1109.4238)

Théado, S., Vauclair, S., Alécian, G., & Le Blanc, F., 2009, *ApJ*, 704, 1262

Thompson, T. A., Prieto, J. L., Stanek, K. Z., *et al.* 2009, *ApJ*, 705, 1364

Tominaga, N., Umeda, H., & Nomoto, K. 2007, *ApJ*, 660, 516

Tonini, C., Maraston, C., Thomas, D., Devriendt, J., & Silk, J., 2010, *MNRAS*, 403, 1749

Tonini, C., Maraston, C., Devriendt, Thomas, D., J., & Silk, J., 2009, *MNRAS*, 396, L36

Tornambè, A., Piersanti, L., Iben, I. Jr., & Gagliardi, S. 2004, *Mem. SAIt*, 75, 178

Tornambè, A., Piersanti, L., Iben, I. Jr., & Gagliardi, S. 2004, *Mem. SAIt*, 75, 178

Townsend R. H. D., Oksala M. E., Cohen D. H., Owocki S. P., & Ud-Doula A., 2010, *ApJ*, 714, L318

Townsend, R. H. D. & Owocki, S. P. 2005, *MNRAS*, 357, 251

Townsend, R. H. D., Owocki, S. P., & Groote, D. 2005, *ApJL*, 630, L81

Traxler, A., Garaud, P., & Stellmach, S., 2011, *ApJ*, 728, L29

Turck-Chièze, S., & Couvidat, S., 2011, *Reports on Progress in Physics*, 74, pp.086901

Turck-Chièze, S., Loisel, G., & Gilles, D., *et al.*, 2011, Journal of Physics: Conference Series, 271, pp.012035

Turck-Chièze, S., Palacios, A., Marques, J. P., & Nghiem, P. A. P., 2010, *ApJ*, 715, 1539

Ud-Doula A., Owocki S. P., & Townsend R. H. D., 2009, *MNRAS*, 392, 1022

Ud-Doula, A., Owocki, S. P., & Townsend, R. H. D. 2008, *MNRAS*, 385, 97

Umeda, H. & Nomoto, K. 2002, *ApJ*, 565, 385

Uytterhoeven, K., Moya, A., & Grigahcene, A., *et al.*, 2011, *A&A*, 534, 125

van Grootel, V., Charpinet, S., & Fontaine, G., *et al.*, 2010a, *A&A*, 524, 63

van Grootel, V., Charpinet, S., & Fontaine, G., *et al.*, 2010b, *ApJ*, 718, L97

Ventura, P., & D'Antona, F., 2009, *A&A*, 499, 835

Ventura, P., & D'Antona, F., 2010, *MNRAS*, 402, 72

Walder R., Folini D., & Meynet G., 2011, SSRv, 125

Vick, M., Michaud, G., Richer, J., & Richard, O., 2010, *A&A*, 521, A62

Vick, M., Michaud, G., Richer, J., & Richard, O., 2011, *A&A*, 526, A37

Vick, M., Richer, J., & Michaud, G., 2011, *A&A*, 534, 18

Wachter, S., Mauerhan, J. C., Van Dyk, S. D., *et al.* 2010, *AJ*, 139, 2330

Wade, G. A., Alecian, E., Bohlender, D. A., *et al.* 2011, IAU Symposium, 272, 118

Wade, G. A., the MiMeS Collaboration 2010, arXiv:1012.2925

Walder, R., Folini, D., & Meynet, G. 2011, *SSRv*, 125

Weaver, T. A., Zimmerman, G. B., & Woosley, S. E. 1978, *ApJ*, 225, 1021

Witt *et al.*, 2009, 693, 1946

Whitmore, B. C. *et al.* 2011, *ApJ*, 729, 78

Woosley, S. E., Blinnikov, S., & Heger, A. 2007, *Nature*, 450, 390

Worthey, G. 1993, *ApJ*, 409, 530

Wright, D. J., Chené, A.-N., De Cat, P., *et al.*, 2011, *ApJ*, 728, L20

Xiaofeng Wang *et al.* 2011, *ApJ*, in press

Yakunin, K. M. *et al.* 2010, *Classical and Quantum Gravity*, 28, 194005

Yanny, B., *et al.* 2011, *A&A*, 530, 108

Yoon, S. C., Woosley, S. E., & Langer, N. 2010, *ApJ*, 725, 940

Transactions IAU, Volume XXVIIIA
Reports on Astronomy 2009–2012   © International Astronomical Union 2012
Ian Corbett, ed.   doi:10.1017/S1743921312002785

COMMISSION 45   **SPECTRAL CLASSIFICATION**

CLASSIFICATION STELLAIRE

PRESIDENT   Richard O. Gray
VICE-PRESIDENT   Birgitta Nordström
PAST PRESIDENT   Sunetra Giridhar
ORGANIZING COMMITTEE   Adam J. Burgasser, Laurent Eyer,
       Ranjan Gupta, Margaret M. Hanson,
       Michael J. Irwin, Caroline Soubiran

TRIENNIAL REPORT 2009-2012

1. Introduction

This report gives an update on developments since the last General Assembly in Rio de Janeiro. Classification – both photometric and spectral – continues to play a vital role in stellar astrophysics and stellar surveys. During the past three years, rapid progress has been made in the classification of brown dwarfs, with the discovery of the first Y dwarfs and the introduction of a near-IR classification system for M- and L-dwarfs. The number of known L-dwarfs now exceeds 1000, and so peculiar types are beginning to show up. For instance, there is now enough material to define a low-gravity spectral sequence for the L0 – L5 dwarfs. In addition, a number of unusually blue L-dwarfs are now known. Large-area surveys, always of interest to Commission 45, have proliferated during this period, including RAVE, SEGUE, and WISE with many more in the planning stages.

2. Developments in the Classification of Brown Dwarfs

2.1. *The Discovery of Y Dwarfs*

As of early 2011, the latest brown dwarfs (T-type) known had $T_{\rm eff} \sim 500 - 700$K; there thus existed a gap of nearly 400 K between the coolest spectroscopically confirmed brown dwarfs and Jupiter ($T_{\rm eff} \sim 124$ K), suggesting the existence of a cooler class of brown dwarfs. Targeted searches for companions to nearby stars (Luhman *et al.* 2011, ApJ, 730, L9; Liu *et al.* 2011, ApJ, 740, 108) turned up two very cool dwarfs with $T_{\rm eff} \sim 300-400$ K, but the faintness of those objects coupled with proximity to their primary stars frustrated spectroscopic observations. One of the science goals of the Wide-field Infrared Survey Explorer (WISE), which has recently completed a survey of the entire sky in four infrared bands (W1, W2, W3, W4) was the detection of ultra-cool brown dwarfs. The W1 (3.4μm) and W2 (4.6μm) bands were specifically designed to sample the deep CH_4 absorption band at 3.3 μm and the "opacity-free" 4.7μm region in the spectra of cool brown dwarfs. Kirkpatrick *et al.* (2011, arXiv) report on the discovery of 100 new brown dwarfs from WISE data, and Cushing *et al.* (2011, arXiv) focus on the 7 coolest dwarfs of that sample. Six of those seven are identified as the first spectroscopically verified members of a new spectral class, the Y-type dwarfs. Those Y dwarfs are distinguished from the T dwarfs spectroscopically by two features. First, a T dwarf is easily identified by the presence of prominent flux peaks in the J and H bands. When plotted in units of f_λ, the J-band peak is always higher than the H-band peak in T dwarfs. However, in these six ultracool dwarfs, the J- and H-band peaks are nearly equal in height. In addition, the Y dwarfs show evidence of NH_3 absorption in the blue wing of the H-band peak, at 1.53μm. The presence of NH_3 absorption has been suggested as a trigger for a new spectral class (cf. Kirkpatrick 2008, ASPCS, 384).

2.2. Gravity Classification in L Dwarfs

Brown dwarfs are sub-stellar objects with insufficient mass to sustain hydrogen burning in their cores. As a consequence, they cool with time and evolve through the MLT spectral sequence. This implies that, unlike main-sequence stars, the spectral type of a brown dwarf is not uniquely related to its mass or age, as, for instance, an early-L dwarf could be either an old low-mass star, or a young low-mass brown dwarf. This mass/age/spectral-type degeneracy has plagued the study of brown dwarfs, and it has only recently been that high-resolution spectra of L dwarfs have become available that enable the detection of gravity-sensitive features. Cruz et al. (2009, AJ, 137, 3345) have published the first preliminary low-gravity optical spectral sequence for L0 to L5 dwarfs, which should be considered an expansion of the spectral classification scheme for L dwarfs first published by Kirkpatrick et al. (1999, ApJ, 519, 802). The features that can be used to gravity type L dwarfs are the pressure-broadened wings of prominent lines of the alkali metals, and metal-oxide and metal-hydride bands (VO, TiO, CrH, and FeH).

2.3. A Near-IR Classification Sequence for M and L Dwarfs

It is remarkable that up until now, a near-IR classification sequence along with supporting spectral standards has not been established for the M and L dwarfs, although there is such a system for the T dwarfs. Generally, we expect that optical and near-IR spectral types should be in good agreement. However, since the two spectral regions sample different layers of the atmosphere, some differences are to be expected, and those differences can give us insight into brown-dwarf physics. Kirkpatrick et al. (2010, ApJS, 190, 100), utilizing new spectroscopic data, constructed a standard sequence for M0 – L9, consistent with the earlier T-dwarf sequence. This project also detected a number of peculiar objects including unusually blue and red L dwarfs, and low-gravity and low-metallicity objects.

2.4. Blue L Dwarfs – Adam J. Burgasser

The number of identified members of the L-dwarf spectral class, first defined in 1999 (see review by Kirkpatrick 2005, ARA&A, 43, 195) has now surpassed 1000, and many rare and unusual members of this class are being uncovered. These include "unusually blue" L dwarfs, or UBLs, of which there are currently about two dozen examples (Cruz et al. 2003, AJ, 126, 2421; Knapp et al. 2004, AJ, 127, 3553; Chiu et al. 2006, AJ, 131, 2722; Cruz et al. 2007, AJ, 133, 439; Folkes et al. 2007, MNRAS, 378, 901; Burgasser et al. 2008, ApJ, 674, 451; Kirkpatrick et al. 2010, ApJS, 190, 100; Bowler et al. 2010, ApJ, 710, 45; Schmidt et al. 2010, AJ, 139, 1045). UBLs have near-infrared colors and/or spectral energy distributions that are significantly bluer for a given optical spectral type, with offsets of ΔJ-K $\lesssim$ 0.5–1.0 mag compared to the mean color of that subtype. The distinguishing spectral features of UBLs appear largely at near-infrared wavelengths, whereas their optical spectra are minimally affected. This segregates UBLs from L-type subdwarfs, which have distinct optical and near-infrared spectra compared to local field dwarfs, largely due to metallicity effects (e.g., Burgasser et al. 2007, ApJ, 657, 494). In addition to bluer colors and spectral slopes, UBLs exhibit stronger water absorption at 1.4 μm and 1.8 μm and weaker CO absorption at 2.3 μm. Near-infrared classifications of their spectra, using the schemes of Geballe et al. (2002, ApJ, 564, 466) or Kirkpatrick et al. (2010, ApJS, 190, 100), are typically 1–3 subtypes later than optical classifications.

Kirkpatrick *et al.* (2010, ApJS, 190, 100) have described a method for identifying UBLs through direct comparison to near-infrared spectral standards. There is as yet no formal system for formally designating a source as a UBL, but various authors have adopted prefixes of "b" before the optical spectral class as a signifier (e.g., "bL5").

Considerable progress in understanding the UBL population was made in 2010 with the discovery of the nearby UBL SDSS J141624.08+134826.7 (Schmidt *et al.* 2010, AJ, 139, 1045; Bowler *et al.* 2010, ApJ, 710, 45; Burningham *et al.* 2010, MNRAS, 404, 1952; Kirkpatrick *et al.* 2010, ApJS, 190, 100) and its well-separated, unusually blue T dwarf common proper motion companion ULAS J141623.94+134836.3 (Burningham *et al.* 2010, MNRAS, 404, 1952; Scholz *et al.* 2010, A&A, 510, L8; Burgasser *et al.* 2010, AJ, 139, 2448). Spectral model fits indicate stronger H_2 absorption for both sources and reduced cloud opacity for the UBL primary (late-type T dwarfs do not typically exhibit significant cloud opacity; e.g., Tsuji *et al.* 1996, A&A, 308, L29). The common spectral and photometric peculiarities of these sources suggest older age and higher surface gravities as underlying causes, with thinner clouds in the primary being a secondary effect. Nevertheless, the distinct near-infrared spectra of UBLs as compared to other L dwarfs, and analogous spectral discrepancies among unusually red and possibly "thick cloudy" L dwarfs (e.g., Looper *et al.* 2008, ApJ, 686, 528) indicate the need for an additional classification dimension for L dwarfs that may encompass cloud and/or age characteristics. *Additional input for this report was provided by Brendon Bowler (U. Hawaii), Michael Cushing (U. Toledo), Mark Marley (NASA Ames) and Sarah Schmidt (U. Washington).*

2.5. *Ultracool Dwarf Near-Infrared Spectral Libraries – Adam J. Burgasser*

Ever-larger samples of near-infrared spectra of low-temperature dwarfs have been acquired by researchers over the past decade, and several of these researchers have created on-line spectral libraries to distribute these data to the community. As a reference, this report provides an (incomplete) list of online spectral libraries containing sizeable (>50) datasets of spectra for the coolest spectral classes, the M, L, and T dwarfs. A more comprehensive list of spectral libraries across all stellar classes is maintained by David Montes (Universidad Complutense de Madrid):
http://www.ucm.es/info/Astrof/invest/actividad/spectra.html

2.5.1. *The IRTF Spectral Library*

The IRTF Spectral Library is a homogenous compilation of stellar spectra observed with the NASA IRTF SpeX spectrograph (Rayner *et al.* 2003, PASP, 115, 362), based primarily on data published in Cushing *et al.* (2005, ApJ, 623, 115) and Rayner *et al.* (2009, ApJS, 185, 289). The data were acquired at a common resolving power of $\lambda/\Delta\lambda$ ≈ 2000 and span 0.8-2.5 μm, with a number of spectra extending to 5.2 μm. All data were reduced using the SpeXtool package (Cushing *et al.* 2003, PASP, 116, 362; Vacca *et al.* 2003, PASP, 115, 389), and have signal-to-noise ratios S/N > 100 for $\lambda < 4$ μm. The spectra are photometrically calibrated using 2MASS photometry. The library primarily consists of solar-metallicity, late-type stars with spectral types F through M and luminosity classes I through V; there are also AGB stars, carbon and S stars, and L and T dwarfs (Wolf-Rayet, O, B, and A star spectra are forthcoming). Data for Jupiter, Saturn, Uranus, and Neptune are also available. This library is maintained by Michael Cushing (U. Toledo) and John Rayner (U. Hawaii) and was last updated in December 2009. http://irtfweb.ifa.hawaii.edu/ spex/IRTF_Spectral_Library

2.5.2. *The SpeX Prism Spectral Libraries*

The SpeX Prism Spectral Libraries are a compilation of roughly 1000 low-resolution, near-infrared spectra of primarily M, L and T dwarfs, also obtained with the IRTF SpeX spectrograph. These data were acquired using SpeX's prism-dispersed mode, yielding 0.65-2.55 μm spectra with $\lambda/\Delta\lambda \approx 75$-$200$, depending on the slit width used. The SpeX Prism Libraries are a heterogenous compilation, incorporating data from several dozen publications; hence, the resolution, photometric calibration and quality of the data vary (these are indicated on the webpage). Spectra are organized into separate libraries according to spectral class, luminosity class, spectral standards, and other groupings. This library is maintained by Adam Burgasser (UC San Diego) and was last updated in October 2010. http://www.browndwarfs.org/spexprism

2.5.3. *The NIRSPEC Brown Dwarf Spectroscopic Survey*

The NIRSPEC Brown Dwarf Spectroscopic Survey (BDSS) compiles both moderate- and high-resolution spectra for late-type M, L, and T dwarfs obtained with the Keck-II Near-Infrared Spectrometer (NIRSPEC; McLean *et al.* 1998, SPIE, 3354, 566; McLean *et al.* 2000, SPIE, 4008, 1048), with most of the data published in McLean *et al.* (2003, ApJ, 596, 561); McLean *et al.* (2007, ApJ, 658, 1217); and Rice *et al.* (2010, ApJS, 186, 63). Spectra were acquired with the same slit, dispersion and filter combinations, so this is a homogenous dataset. All sources in the moderate-resolution sample were observed with the NIRPSEC-3 filter, spanning 1.14-1.38 μm at a resolution $\lambda/\Delta\lambda \approx 2000$. Several sources were also observed in various filters spanning 0.95–2.43 μm. The spectra are flux-calibrated and (for broader-band spectra) stitched together using 2MASS photometry. High-resolution data were also obtained with the NIRSPEC-3 filter in 8 orders spanning 1.16-1.32 μm at $\lambda/\Delta\lambda \approx 20{,}000$. All data in the NIRSPEC BDSS were reduced using the REDSPEC package. This library is maintained by Gregory Mace (UC Los Angeles), Ian McLean (UC Los Angeles) and Emily Rice (CUNY College of Staten Island), and was last updated in August 2010. http://bdssarchive.org

2.5.4. *DwarfArchives M Dwarf Spectroscopic Library*

DwarfArchives, an online database of photometry and astrometry for known late M, L and T dwarfs, provides a separate catalog of red optical spectra for over 500 M dwarfs. Data are drawn from a variety of instruments and sources (including unpublished data), and are primarily low- to moderate-resolution spanning 6300–9000 Å. DwarfArchives is maintained by Chris Gelino (Caltech/Spitzer Science Center), J. Davy Kirkpatrick (Caltech/IPAC) and Adam Burgasser (UC San Diego), and was last updated in February 2011. http://DwarfArchives.org

2.5.5. *Ultracool Dwarf Catalog*

The Ultracool Dwarf Catalog, like DwarfArchives, compiles astrometric, photometric and spectroscopic data for over 850 M, L and T dwarfs. The heterogenous spectral data are drawn from various sources, and include red optical spectra from individual follow-up programs (mainly related to DENIS search programs) and the SDSS survey, and near-infrared spectra from the literature. The available spectral range for each source is indicated on the main table. The catalog was produced by Juan Cabrera, Elena Cenizo and Eduardo Martín (Instituto de Astrofsica de Canarias; Centro de Astrobiologa), and published in Martín *et al.* (2005, AN, 326, 1026). The site was last updated in 2007. http://www.iac.es/galeria/ege/catalogo_espectral

2.5.6. L and T Dwarf Data Archive

Sandy Leggett (Gemini Observatory) hosts a catalog of red optical and near-infrared spectra of L and T dwarfs based primarily on data reported in Chiu *et al.* (2006, AJ, 131, 2722), Golimowski *et al.* (2004, AJ, 127, 3516) and Knapp *et al.* (2004, AJ, 127, 3553). The spectra were obtained with a variety of instrumentation (including outside sources), and include both low- and moderate-resolution data over various ranges of spectral coverage and quality. The site was last updated in January 2006.
http://staff.gemini.edu/ sleggett/LTdata.html

2.5.7. Low-resolution Near-infrared Spectral Library of M-, L-, and T-dwarfs

Testi (2009, A&A, 503, 639) published a sample of 54 low-resolution ($\lambda/\Delta\lambda \approx 100$), near-infrared (0.85–2.45 μm) spectra of M, L, and T dwarfs obtained with the TNG Near Infrared Camera and Spectrograph (NICS; Baffa *et al.* 2001, A&A, 378, 722), using that instrument's Amici prism mode. The data are homogenous in spectral coverage and resolution, but vary in S/N. Spectral data are served through Vizier at the catalog link:
http://vizier.u-strasbg.fr/viz-bin/VizieR?-source=J/A%2BA/503/639

2.5.8. Keck LRIS Spectra of Late-M, L, and T Dwarfs

I. Neill Reid (STScI) has compiled a set of low-resolution ($\lambda/\Delta\lambda \approx 1800$), red-optical spectra (0.6-1.0 μm) of late-type M, L, and T dwarfs obtained with the Keck Low Resolution Imaging Spectrometer (LRIS; Oke *et al.* 1995, PASP, 107, 375) between 1997 and 1999. The site also contains red optical spectra of a handful of M subdwarfs. The site was last updated in December 2000. http://www.stsci.edu/ inr/lris.html

2.5.9. Improving Access to Spectral Data

These online spectral libraries span a wide range of quality, completeness and accessibility, suggesting work is needed to devise and implement more efficient and uniform methods of curating and disseminating spectral datasets (this issue was discussed specifically at the 2009 joint ESA-Constellation workshop on brown dwarf formation in Noordwijk, the Netherlands). There are several issues relevant here, including the time and funding needed for data producers to consolidate, format and curate their data; standardization of data products and formats; and optimal dissemination platforms for the data. As these issues are relevant to both stellar classification and Virtual Observatory initiatives, a joint workshop between Commissions 5 and 45 may be of benefit. *Additional input for this report was provided by Kelle Cruz (Hunter College), Michael Cushing (U. Toledo), Eduardo Martín (Centro de Astrobiolog'a), Ian McLean (UC Los Angeles), I. Neill Reid (STScI) and Emily Rice (CUNY College of Staten Island).*

3. Spectroscopic and Photometric Surveys

Caroline Soubiran

Spectral and photometric surveys provide the raw materials for stellar classification, and cross-correlation is an effective technique for isolating peculiar and astrophysically interesting objects. The following is a partial list of recent surveys, most of which provide not only spectra and/or photometry, but also derived stellar parameters:

RAVE (RAdial Velocity Experiment, http://www.rave-survey.aip.de/rave/) is a spectroscopic survey at $R = 7500$, wavelength range 8400-8800Å, magnitude range $9 < I < 12$. The Third Data Release (Siebert *et al.* 2011, AJ, 141, 187S) provides stellar parameters for 39,833 stars.

SEGUE (Sloan Extension for Galactic Understanding and Exploration, http://www.sdss3.org/surveys/segue2.php) is a spectroscopic survey at $R \sim 2000$, wavelength range 3820-9200Å, magnitude range $14 < g < 20.3$, providing atmospheric parameters for ~250,000 stars (Yanny *et al.* 2009, AJ, 137, 4377, Lee *et al.* 2011, AJ, 141, 90).

The Geneva-Copenhagen survey of the Solar neighbourhood (Nordstrom *et al.* 2004, A&A, 418, 989, Holmberg *et al.* 2009, A&A, 501, 941) provides effective temperatures and metallicities from uvbyβ photometry for 16,682 nearby F and G dwarfs with $V < 8.6$. Query at http://cdsarc.u-strasbg.fr/viz-bin/Cat?V/130

PASTEL is a regularly updated bibliographical compilation of atmospheric parameters (Soubiran *et al.* 2010 A&A, 515. 111). Latest version has 31724 records including 6527 stars with (T_{eff}, $\log g$, [Fe/H]) relying on high-resolution spectroscopy. Query at http://cdsarc.u-strasbg.fr/viz-bin/Cat?B/pastel

4. Catalogues, Atlases, & Search Engines

B.A. Skiff (Lowell Obs) continues to build a comprehensive catalogue of published spectral classifications. That catalog can be found at the URL: http://cdsarc.u-strasbg.fr/viz-bin/Cat?B/mk. The latest version contains 452890 entries, but still lags the current literature by a number of years. The Skiff catalog has served as the source of spectral types for an innovative search engine/spectral class encoding system that has been developed to make archives such as the VO and NASA's MAST searchable in terms of detailed spectral types (Smith *et al.* 2011, IVOA Design note 2011 Oct 24). The system encodes spectral classes into a digital format of the form TT.tt.LL.PPPP, where TT and tt refer to spectral type and subtype, LL to luminosity class, and PPPP to possible spectral peculiarities. Archive centers can utilize this system to quantify classes of formerly arbitrary spectral classification strings found in classification catalogs. The encoding system will also allow users to request archived data based on spectral class ranges. The encoding system may also have applications to automatic classification of stellar spectra. See http://www.ivoa.net/Documents/latest/SpectClasses.html

The previous triennial report (2006 – 2009) listed a number of large-scale spectral classification catalogs and stellar spectral atlases, most of which are still active. The reader is referred to that document for that listing.

5. Closing remarks

Stellar classification, the mandate of IAU Commission 45, is an area of long-standing interest to the IAU. While stellar classification is a traditional discipline, it has been at the forefront of astronomical research for over a century. Most recently, classification has played a crucial role in the discovery and characterization of brown dwarfs, leading to the establishment, now, of three new stellar classes, the L, T, and Y dwarfs. We are convinced that stellar classification will be of increasing importance in the future, especially with the advent of deep Galactic surveys, such as Pan Starrs, LSST, *Gaia* and others. Because of this, we expect the relevance of our Commission to increase with the years.

It is a pleasure to thank the contributors to this report, and for the cooperation of the organizing committee over the past 3 years.

Richard Gray
President of the Commission

Transactions IAU, Volume XXVIIIA
Reports on Astronomy 2009-2012
Ian Corbett, ed.

© International Astronomical Union 2012
doi:10.1017/S1743921312002797

WORKING GROUP on ABUNDANCES IN RED-GIANTS

CHAIR	John C. Lattanzio
VICE-CHAIR	Jacco van Loon
BOARD	Pavel Denissenkov
	Paul Hauschildt
	Josef Hron
	Thomas Lebzelter
	Maria Lugaro
	Chris Tout
	Robert F. Wing
	Ernst Zinner
	Lucy Ziurys

TRIENNIAL REPORT 2009-2011

1. Introduction

The WGARG was created in 2001 to oversee the rapid growth of the quantitative determination and understanding of the abundance patterns seen in red-giant stars. As the field progresses we are regularly reminded of how broad and multi-disciplinary is this area of research.

2. Activities in the past triennium

2.1. *Ongoing activities*

The WG webpage recently moved to a new address:

http://users.monash.edu.au/~johnl/wgarg/

Our monthly newsletter is edited by the Vice-Chair Jacco van Loon and Albert Zijlstra. It is called "The AGB Newsletter" and currently has approximately 1000 subscribers.

2.2. *Specialist meetings*

The WG assisted the Vienna group in the organization of the second meeting in the "Why Galaxies Care About AGB Stars" series. This took place in August 2010 at the University of Vienna. As for the first meeting of this small series in 2006, the IAU working group on Abundances in Red Giants was actively involved in the discussion on the topical focus and supported the conference by promoting it actively via its members and its mailing list. Several members of the working group played a key role in the scientific and local organization. The conference was very successful with more than 170 participants discussing current research in the area of AGB stars and the relation to stellar populations. A focus of the meeting was set on the most recent and forthcoming instrumental developments. A considerable number of members of our IAU working group participated in this conference which opened the possibility to discuss and exchange most recent observational and modelling results both within the scientific program and in informal talks. The proceedings of the conference, edited by Franz Kerschbaum, Thomas

Lebzelter, and Robert Wing, have been published as volume 445 of the Astronomical Society of the Pacific Conference Series.

3. Developments within the past triennium

3.1. *Deep-Mixing in red-giants*

The search for understanding of deep-mixing in red-giants continues. While thermohaline mixing has many promising attributes, eg Stancliffe (2010); Cantiello & Langer (2010); Charbonnel & Lagarde (2010); Angelou *et al.* (2011), the first multi-dimensional numerical simulations are not as supportive (Denissenkov (2010), Denissenkov & Merryfield (2010), Traxler *et al.* (2011)) and remind us that although progress has been made, a full understanding still eludes us (see also Wachlin *et al.* (2011)).

3.2. *Opacity for AGB envelopes*

Another significant development was the calculation of opacity tables for the low temperature envelopes of red giants, but with the effects of dredge-up included in the envelope composition (Lederer & Aringer (2009), Ventura & Marigo (2009), Marigo & Aringer (2009)). The extra opacity sources can significantly alter the envelope structure of these stars, with feedback on the mass-loss and hence the dredge-up and final evolution. Models which include these new opacities are just making their way into the literature.

3.3. *Mass-loss*

To avoid duplication, the WG report on this topic is included in the report from Commission 35. Please see that section in Commission 35 report.

3.4. *Multiple populaton in globular clusters*

To avoid duplication, the WG report on this topic is included in the report from Commission 35. Please see that section in Commission 35 report.

3.5. *Observations*

For the first time, comprehensive observational studies are being carried out at millimeter wavelengths to examine the molecular composition of the envelopes of oxygen-rich supergiant/hypergiant stars. A 1 mm spectral line survey of VY Canis Majoris across the frequency range 210-285 GHz has been conducted using the Submillimeter Telescope (SMT) of the Arizona Radio Observatory (ARO; Tenenbaum et al. 2010a,b). This work has demonstrated that the chemistry of this envelope is dominated by SO_2, SiO, and SiS, as well as H_2O, although certain carbon-bearing molecules are also abundant, such as HCN and CS (Ziurys *et al.* 2009). Unusual species such as AlO, AlOH, and PO (Tenenbaum & Ziurys 2009, 2010; Tenenbaum *et al.* 2007) are also present in VY CMa. A similar observational study is currently being conducted of NML Cygnus.

3.6. *Super-AGB Stars*

To avoid duplication, the WG report on this topic is included in the report from Commission 35. Please see that section in Commission 35 report.

3.7. *Hydrodynamics*

One area that is having increasing impact is multi-dimensional hydrodynamical studies. These have been referred to above in relation to the thermohaline mixing mechanism, but continued progress in understanding "normal" convection is being made regularly (Arnett *et al.* (2009), Arnett *et al.* (2010), Arnett & Meakin (2011)). Such codes also have application in unusual cases, such as when the mixing and nuclear burning timescales are similar (Stancliffe *et al.* (2011)). We can expect more improvements in this area in the future as increasing computing power enables us to tackle previously impractical problems.

4. Looking Forward

The WG is closely involved in organizing the 2012 edition of the "Nuclei in the Cosmos" series, which will take place in August in Cairns, Australia. A number of satellite meetings on related topics are also expected to be held about the same time. One that the WG is planning will be dedicated to the subject of multi-dimensional hydrodynamics and the effect of such studies on our discipline.

5. Closing remarks

The study of red giants is fundamental to many areas of modern astrophysics. These stars are not only intrinsically interesting, exhibiting many different and complex phenomena, but they are also crucial to many other branches of astrophysics because of their ubiquity, high intrinsic luminosity, and important role in nucleosynthesis. We expect the WG to be active for many years to come.

John C. Lattanzio
Chair of Working Group

References

Arnett, W. D., Meakin, C., & Young, P., 2009, *ApJ* 690 1715
Arnett, W. D., Meakin, C., & Young, P., 2010, *ApJ* 710 1619
Arnett, W. D. & Meakin, C. 2011, *ApJ* 733 78
Angelou, G., Church, R. P., Stancliffe, R. J., Lattanzio, J. C., & Smith, G. H., 2011, *ApJ* 728, 79
&antiello, M. & Langer, N., 2010, *A&A* 521, 9
Charbonnel, C. & Lagarde, N., *A&A* 522, 10
Denissenkov, P. A., 2010, *ApJ* 723, 563
Denissenkov, P. A. & Merryfield, W. J., 2011, *ApJ* 727, 8
Lederer, M. T. & Aringer, B., 2009, *A&A* 494, 403
Marigo, P. & Aringer, B., 2009, *A&A* 508, 1539
Stancliffe, R. J., 2010, *MNRAS* 403, 505
Stancliffe, R. J., *et al.*, 2011, *ApJ* in press.
Tenenbaum, E. D., Dodd, J. L., Milam, S. N, Woolf, N. J., & Ziurys, L. M., 2010a, *Ap.J. Suppl.*, 190, 348
Tenenbaum, E. D., Dodd, J. L., Milam, S. N, Woolf, N. J., & Ziurys, L. M., 2010b, *Ap.J. (Letters)*, 720, L102
Tenenbaum, E. D., Woolf, N. J., & Ziurys, L. M., 2007, *Ap.J. (Letters)*, 666, L29
Tenenbaum, E. D. & Ziurys, L. M., 2010, *Ap.J.(Letters)*, 712, L93
Tenenbaum, E. D. & Ziurys, L. M., 2009, *Ap.J.(Letters)*, 694, L59
Traxler, A., Garaud, P., & Stellmach, S., 2011, *ApJ* 728, 29
Ventura, P. & Marigo, P., 2009, *MNRAS* 399, 54
Wachlin, F. C., Miller Bertolami, M. M., & Althaus, L. G., 2011, *A&A* 533, 139
Ziurys, L. M., Tenenbaum, E. D., Pulliam, R. L., Woolf, N. J., & Milam, S. N., 2009, *Ap.J.*, 695, 1604

Transactions IAU, Volume XXVIIIA
Reports on Astronomy 2009–2012   © International Astronomical Union 2012
Ian Corbett, ed.   doi:10.1017/S1743921312002803

DIVISION IV-V / WORKING GROUP
ACTIVE B STARS

CHAIR	Geraldine J. Peters
VICE-CHAIRS	Carol E. Jones & Richard D. Townsend
PAST CHAIR	Juan Fabregat
BOARD	Karen S. Bjorkman, M. Virginia McSwain,
	Ronald Mennickent, Coralie Neiner,
	Philippe Stee

TRIENNIAL REPORT 2009-2012

1. Introduction

The Working Group on Active B Stars (WGABS) was re-established under IAU Commission No. 29 at the IAU General Assembly in Montreal, Quebec, Canada in 1979. Its main goal is to promote and stimulate research and international collaboration in the field of active B stars. Originally known as the Working Group on Be Stars, its name was changed at the 22nd IAU General Assembly in The Hague, Netherlands in 1994 when the research interests of the group were broadened to include activity in all B stars, especially pulsating OB stars, interacting binaries, stellar winds, and magnetic fields.

2. Developments within the past triennium

The *Be Star Newsletter* † has continued to be the main source of information on new discoveries, ideas, manuscripts, and meetings on active B stars. G. Peters, D. Gies, and D. McDavid continue, respectively, as Editor-in-Chief, Technical Editor, and Webmaster. News items and abstracts are published online soon after they are accepted. Since 2000 all contributions have been refereed. When sufficient material is accumulated an issue is formed and a paper version is distributed. Issue No. 39 was published in June 2009, and contains 20 research notes and advertisements of interest to the active B star community and 32 abstracts of papers. The current working issue No. 40 contains 24 research notes and 34 abstracts received by 20 October 2011.

IAUS-272 (Active OB Stars: Structure, Evolution, Mass-Loss, and Critical Limits) was held in Paris, France from 19-23 July 2010. The key topics included the internal structure of active OB stars (pulsations, rotation, magnetism, transport processes), evolution of OB stars (formation, binaries, late evolutionary stages including magnetars and GRBs), the circumstellar environment (disks, magnetospheres, Be phenomenon, and winds), OB stars as extreme condition test beds (critical rotation, mass loss, radiation fields), normal OB stars for calibration purposes (fundamental parameters, astronomical quantities), and populations of OB stars (studies, tracers of galactic structure, cosmic history). The symposium was supported by IAU Divisions IV (Stars) & V (Variable Stars), Commissions 27 (Variable Stars), 36 (Theory of Stellar Atmospheres), and 42 (Close Binary Stars),

† http://http://astsun.astro.virginia.edu/~dam3ma/benews/

and Working Groups on Active B Stars, Ap and Related Stars, and Massive Stars. There were over 150 participants. The Proceedings, edited by Coralie Neiner, Gregg Wade, Georges Meynet, and Geraldine Peters, were published in July 2011.

High resolution interferometry in the optical and the infrared is continuing to be a valuable probe for circumstellar material surrounding massive stars especially for the disk-like structures that surround Be stars. Improvements in this observational technology now allows angular scales below 1 milli-arcsecond to be reached and in turn has resulted in advances in our understanding of these disk structures (Tycner 2011). For example, interferometry has been used to model individual stars (Štefl *et al.* 2011, Schaefer *et al.* 2011, Delaa *et al.* 2011), compute disk sizes in various wavelength regimes (Millan-Gabet *et al.* 2010, Meilland *et al.* 2008), infer the disk density structure (Tycner *et al.* 2008), find binary systems (Koubský *et al.* 2010, Meilland *et al.* 2008), and estimate disk mass (Kraus *et al.* 2011). New data combined with other observables has led to successful dynamic models that follow disk structure over time to be developed and tested (cf. Carciofi *et al.* 2009). Key to continuing progress will be further development of theoretical models tightly constrained by observations. Interferometric imaging still suffers from sparse (u,v) plane coverage, and theoretical images play a critical role in interpreting the observations. Several groups have developed this modeling capacity (cf. Carciofi *et al.* 2009, Kervella *et al.* 2009, Jones *et al.* 2008, Schaefer *et al.* 2011) and further refinements of these models are expected. Theoreticians and observers from this active area of research will meet in Foz do Iguaçu, Brazil in February 2012 at a European Southern Observatory workshop entitled, *Circumstellar Dynamics at High Resolution*.

The origin of magnetic fields in B stars and their role in generating short-term activity and controlling the evolution is currently a topic of widespread interest. The MiMeS (Magnetism in Massive Stars) Project (Wade *et al.* 2011) is an international collaborative effort that has been awarded 1230 hours of observing time on the Canada-France-Hawaii Telescope and the Telescope Bernard Lyot from 2008-12 to obtain high resolution spectropolarimetry with the ESPaDOnS and Narval instruments. Data on 25 targeted stars and ∼200 survey stars are being modeled by contemporary codes (e.g. Townsend *et al.* 2007, ud-Doula *et al.* 2008).

The *Kepler*, *CoRoT*, and *MOST* spacecraft continue to produce high-quality photometry of active B stars (especially the β Cep, Be, and SPB stars) over long baselines (Balona *et al.* 2011, De Cat *et al.* 2011, Gutiérrez-Soto *et al.* 2011). Along with spectroscopic information these observations reveal a rich set of pulsational frequencies and modes from which constraints can be placed on the internal rotation, metallicity, and the nature of sub-photospheric convection.

The *Fermi Gamma-ray Space Telescope* is providing a wealth of data on a rare group of Be/X-ray binaries that exhibit very high energy emission in the MeV-TeV range. Ongoing studies are investigating their stellar/compact companion interactions, identify the mechanism for particle acceleration, and determine the nature of the compact companions (e.g. Aragona *et al.* 2009, McSwain *et al.* 2010).

Large surveys are yielding information on the nature of entire classes of active B stars, and the extent of the variability in individual objects. The *Spitzer* SAGE survey of the Magellanic Clouds (Bonanos *et al.* 2011) that includes ∼5000 OB stars from which Be stars can be clearly identified reveals a higher percentage of Be stars in the SMC and frequent transitions from a Be-phase to a non-emission state . The VLT-FLAMES surveys of OB stars in the Galaxy and Magellanic Clouds are providing fundamental parameters for thousands of OB stars (Evans *et al.* 2005, 2006, Lennon *et al.* 2011).

Stellar models and evolutionary tracks for B stars continue to become more representative of actual stars with the inclusion of rotation, magnetic fields, and lower metallicities (cf. Zahn 2011, Ekström *et al.* 2011a,b, 2008) in their computation. Evolutionary models for rotating stars predict an enhancement in the nitrogen abundance due to meridional circulation and turbulent diffusion of CNO-processed material from the stellar core to the photosphere. The effect becomes more prominent at very high rotational velocities. Several investigations have failed to confirm an elevated N abundance in Be stars in general (Brott *et al.* 2011, Dunstall *et al.* 2011, Peters 2011) and thus do not offer support for the presence of critical rotation in these objects. Substantial progress was made on the computation of evolutionary models for Algol binaries with B-type primaries (Van Rensbergen *et al.* 2011) taking into account mass loss from the system from spin-up of the mass gainer and the creation of an accretion hot spot due to the impacting gas stream. Computations of 561 models (Van Rensbergen *et al.* 2008) are available through the CDS.

3. Closing remarks

The WGABS has seen significant advances in our understanding of the activity and the evolution of B-type stars during the past triennium. Large collaborations have produced substantial databases from ground-based telescopes and spacecraft that reveal information on disk geometry and structure, the nature of magnetic fields, pulsation characteristics, and fundamental stellar parameters. Evolutionary models for single stars and Algol binaries are increasingly representing actual stars, though more attention must be given to predicting the N abundances and distribution of mass ratios for the binaries. Mass-loss, rotation, and accretion in hot stars are important processes that still lack sufficient understanding. They affect the evolution of massive stars and, in turn, these stars govern the evolution of their parent galaxies. Circumstellar disks and winds offer unique laboratories for studying these key phenomena.

Geraldine J. Peters & Carol E. Jones
Chair and Co-Chair of the Working Group on Active B Stars

References

Aragona, C., McSwain, M. V., & De Becker, M. 2010, *ApJ*, 724, 306

Balona, L. A. & 35 co-authors 2011 *MNRAS*, 413, 2403

Bonanos, A. Z. & 10 co-authors 2011 in: C. Neiner, G. Wade, G. Meynet, G. Peters (eds.), *Active OB Stars: Structure, Evolution, Mass-Loss, and Critical Limits*, Proc. IAU Symposium 272, Paris, France, 19-23 July 2010 (Cambridge: University Press), p. 254

Brott, I., Evans, C. J., Hunter, & 9 co-authors 2011, *A&A*, 530, A16

Carciofi, A. C., Okazaki, A. T., Le Bouquin, J.-B., & 5 co-authors 2009, *A&A*, 504, 915

De Cat, P., Uytterhoeven, K., Gutiérrez-Soto, J., Degroote, P., & Simón-Díaz, S. 2011, in: C. Neiner, G. Wade, G. Meynet, & G. Peters (eds.), *Active OB Stars: Structure, Evolution, Mass-Loss, and Critical Limits*, Proc. IAU Symposium 272, Paris, France, 19-23 July 2010 (Cambridge: University Press), p. 433

Delaa, O., Stee, Ph., Meilland, A., & 18 co-authors 2011, *A&A*, 529, 87

Dunstall, P. R., Brott, I., Dufton, P. L., Lennon, D. J., Evans, C. J., Smartt, S. J., & Hunter, I. 2011, *astro-ph, arXiv:1109.6661*

Ekström, S., Meynet, G., Maeder, A., & Barblan, F. 2008, *A&A*, 478, 467

Ekström, S., Georgy, C., Meynet, G., Maeder, A., & Granada, A. 2011a, in: C. Neiner, G. Wade, G. Meynet, & G. Peters (eds.), *Active OB Stars: Structure, Evolution, Mass-Loss, and Critical Limits*, Proc. IAU Symposium 272, Paris, France, 19-23 July 2010 (Cambridge: University Press), p. 62

Ekström, S. & 10 co-authors 2011b, *astro-ph, arXiv:1110.5049*, http://obswww.unige.ch/Recherche/evol/-Database-

Evans, C. J. & 25 co-authors 2005, *A&A*, 437, 467

Evans, C. J., Lennon, D. J., Smartt, S. J., & Trundle, C. 2006, *A&A*, 456, 623

Gies, D. R., Touhami, Y. N., & Schaefer, G. H. 2011, in: C. Neiner, G. Wade, G. Meynet, G. Peters (eds.), *Active OB Stars: Structure, Evolution, Mass-Loss, and Critical Limits*, Proc. IAU Symposium 272, Paris, France, 19-23 July 2010 (Cambridge: University Press), p. 390

Gutiérrez-Soto, J., & 6 co-authors 2011, in: C. Neiner, G. Wade, G. Meynet, & G. Peters (eds.), *Active OB Stars: Structure, Evolution, Mass-Loss, and Critical Limits*, Proc. IAU Symposium 272, Paris, France, 19-23 July 2010 (Cambridge: University Press), p. 451

Jones, C. E., Tycner, C., Sigut, T. A. A., Benson, J. A., & Hutter, D. J. 2008, *ApJ*, 687, 598

Kervella, P., Domiciano de Souza, A., Kanaan, S., Meilland, A., Spang, A., & Stee, Ph. 2009, *A&A*, 493, 53

Koubský, P., Hummel, C. A., Harmanec, P., Tycner, C., & 7 co-authors 2010, *A&A*, 517, 24

Kraus, S., Monnier, J. D., Che, X., Schaefer, G. H., & 10 co-authors 2011, *astro-ph*, arXiv:1109.3447

Lennon, D. J., & 30 co-authors 2011 in: C. Neiner, G. Wade, G. Meynet, G. Peters (eds.), *Active OB Stars: Structure, Evolution, Mass-Loss, and Critical Limits*, Proc. IAU Symposium 272, Paris, France, 19-23 July 2010 (Cambridge: University Press), p. 296

McSwain, M. V., Grundstrom, E. D., Gies, D. R., & Ray, P. S. 2010, *ApJ*, 724, 379

Meilland, A., Millour, F., Stee, Ph., & 5 co-authors 2008, *A&A*, 488, 67

Millan-Gabet, R., Monnier, J. D., Touhami, Y., 6 co-authors, the CHARA group 2010, *ApJ*, 723, 544

Peters, G. J. 2011, in: C. Neiner, G. Wade, G. Meynet, G. Peters (eds.), *Active OB Stars: Structure, Evolution, Mass-Loss, and Critical Limits*, Proc. IAU Symposium 272, Paris, France, 19-23 July 2010 (Cambridge: University Press), p. 101

Schaefer, G. H., & 5 co-authors 2011, in: C. Neiner, G. Wade, G. Meynet, G. Peters (eds.), *Active OB Stars: Structure, Evolution, Mass-Loss, and Critical Limits*, Proc. IAU Symposium 272, Paris, France, 19-23 July 2010 (Cambridge: University Press), p. 424

Štefl, S., Carciofi, A. C., & 7 co-authors 2011, in: C. Neiner, G. Wade, G. Meynet, G. Peters (eds.), *Active OB Stars: Structure, Evolution, Mass-Loss, and Critical Limits*, Proc. IAU Symposium 272, Paris, France, 19-23 July 2010 (Cambridge: University Press), p. 430

Townsend, R. H. D., Owocki, S. P., & Ud-Doula, A. 2007, *MNRAS*, 382, 139

Tycner, C., Jones, C. E., Sigut, T. A. A., & 4 co-authors 2011, *ApJ*, 689, 461

Tycner, C. 2011, in: C. Neiner, G. Wade, G. Meynet, G. Peters (eds.), *Active OB Stars: Structure, Evolution, Mass-Loss, and Critical Limits*, Proc. IAU Symposium 272, Paris, France, 19-23 July 2010 (Cambridge: University Press), p. 337

ud-Doula, A., Owocki, S. P., & Townsend, R. H. D. 2008, *MNRAS*, 385, 97

Van Rensbergen, W., De Greve, J. P., De Loore, C., & Mennekens, N. 2008, *A&A*, 487, 1129 (see also CDS, catalog: cats/J/A+A/487/1129)

Van Rensbergen, W., De Greve, J. P., Mennekens, N., Jansen, K., & De Loore, C. 2011, *A&A*, 487, 1129

Wade, G. A., & 19 co-authors 2011, in: C. Neiner, G. Wade, G. Meynet, G. Peters (eds.), *Active OB Stars: Structure, Evolution, Mass-Loss, and Critical Limits*, Proc. IAU Symposium 272, Paris, France, 19-23 July 2010 (Cambridge: University Press), p. 118

Zahn, J.-P. 2011, in: C. Neiner, G. Wade, G. Meynet, G. Peters (eds.), *Active OB Stars: Structure, Evolution, Mass-Loss, and Critical Limits*, Proc. IAU Symposium 272, Paris, France, 19-23 July 2010 (Cambridge: University Press), p. 14

Transactions IAU, Volume XXVIIIA
Reports on Astronomy 2009–2012   © International Astronomical Union 2012
Ian Corbett, ed.   doi:10.1017/S1743921312002815

DIVISIONS IV-V / WORKING GROUP
Ap & RELATED STARS

CHAIR	Gautier Mathys
PAST CHAIR	Margarida Cunha
BOARD	Michael Dworetsky
	Oleg Kochukhov
	Friedrich Kupka
	Francis LeBlanc
	Richard Monier
	Ernst Paunzen
	Olga Pintado
	Nikolai Piskunov
	Jozef Ziznovsky

TRIENNIAL REPORT 2009-2012

1. Scientific highlights

The purpose of the Working Group on Ap and Related Stars (ApWG) is to promote and facilitate research about stars in the spectral type range from B to early F that exhibit surface chemical peculiarities and related phenomena. This is a very active field of research, in which a wide variety of new developments have taken place since 2009, as illustrated by the following selected highlights.

The evolutionary context of the large-scale organised magnetic fields of Ap stars, which have been known for more than 60 years to be one of their most salient features, is starting to be outlined with the recent detection and study of rotationally modulated magnetic fields in their progenitors, the Herbig Ae/Be stars (Alecian et al. 2009; Hubrig *et al.* 2011a) and the identification and characterisation of magnetic late-type supergiants that are their potential descendants (Grunhut et al. 2010; Aurière *et al.* 2011). On the other hand, the sequence of hotter early-type magnetic stars is becoming increasingly populated through the works of the MiMeS (Magnetism in Massive Stars) collaboration (e.g., Wade *et al.* 2011) and of other teams (e.g., Hubrig *et al.* 2011b), putting the magnetism of Ap stars in a new perspective. The discovery of sub-Gauss magnetic fields with large-scale structure in the A0V star Vega (Lignières *et al.* 2009; Alina et al. 2011) and in the hot Am star Sirius (Petit et al. 2011) raises the possibility that all tepid main-sequence stars may be magnetic to a certain level. However, unprecedented stringent limits have been set on the mean longitudinal magnetic fields of Am and HgMn stars (e.g., Aurière *et al.* 2010; Makaganiuk *et al.* 2011), indicating that they must be at least one order of magnitude weaker than those of magnetic Ap stars.

Arlt & Rüdiger (2010) have proposed Tayler instability as a possible mechanism of generation of magnetic fields in A stars while Ferrario *et al.* (2009) have suggested that magnetic Ap stars could have formed from the merging of two protostars.

Kepler observations have allowed the non-radial pulsation modes of hundreds of A-F stars to be classified, revealing the existence of a class of "hybrid" stars showing both

γ Dor and δ Sct pulsations (Uytterhoeven *et al.* 2011). They have also led to the first detections of γ Dor and δ Sct pulsations in Ap stars; in particular, the A5p star KIC 8677585 is the first star to show both the rapid oscillations typical of Ap stars and γ Dor pulsations (Balona *et al.* 2011). δ Sct and γ Dor pulsations have been found in about 200 Am stars from the analysis of SuperWASP observations (Smalley *et al.* 2011); such pulsations were also studied with Kepler (Balona *et al.* 2011). A new rapidly oscillating Ap (roAp) star discovered with Kepler, KIC 10195926, has been found to pulsate in two modes with different axes, unlike any roAp star known so far (Kurtz et al. 2011). These results can be expected to represent only the first steps of a major breakthrough in our knowledge of the pulsational properties of tepid stars, made possible in particular by the unprecedented amount of data that are becoming available from the space missions CoRoT and Kepler. In parallel, the pioneering satellite MOST continues to deliver datasets allowing the pulsations of rapidly oscillating Ap (roAp) stars to be studied in great detail (e.g., Gruberbauer et al. 2011). Bigot & Kurtz (2011) have computed theoretical light curves of such stars in the framework of the revisited oblique pulsator model, matching well the observations. Khomenko & Kochukhov (2009) have performed simulations of magnetoacoustic pulsations in their atmospheres. Sousa & Cunha (2011) have shown that π rad jumps that have been seen in the pulsation phase of some spectral lines of roAp stars, may be due to the fact that that their surface is not spatially resolved in the observations rather than reflect the presence of a pulsation node in their outermost layers.

Additional observational results worth noting include the determination of the angular diameter of the Ap star γ Equ (one of the prototypes of the class) and derivation of its fundamental parameters from interferometric observations (Perraut et al. 2011), the observation of debris disks around main-sequence A stars (Smith & Wyatt 2011), the first spectroscopic verification of an extragalactic Ap star (Paunzen et al. 2011), the investigation of the correlation of the surface brightness distribution of an Ap star (as inferred from CoRoT observations) with its surface chemical abundance inhomogeneities and its magnetic field (Lüftinger *et al.* 2010), the inference from the analysis of spectropolarimetric observations of surprisingly complex magnetic structures in the Ap star α^2 CVn (Kochukhov & Wade 2010) and in the Bp star HD 37776 (Kochukhov *et al.* 2011), and the characterisation of the long-term modulation of the rotation periods of Ap and Bp stars (Mikulášek *et al.* 2011).

On the theoretical front, Vick *et al.* (2010) have developed self-consistent evolution models of Am stars including atomic diffusion and mass loss. Théado *et al.* (2009) have demonstrated the importance of thermohaline convection in the computation of the abundance variations induced by atomic diffusion. Model atmospheres including abundance stratification at equilibrium have been constructed by Alecian & Stift (2010) and LeBlanc *et al.* (2009). Alecian *et al.* (2011) have computed numerical simulations of time-dependent diffusion in stellar atmospheres. Stift *et al.* (2011) have illustrated the importance of taking into account variations of the structure of the atmospheres across the stellar surface for mapping of abundance inhomogeneities on the surface of Ap stars.

2. The role of the Working Group on Ap and Related Stars

The above summary is by no means exhaustive. But it shows how the theme of the Working Group transcends the borders between various fields of astrophysics. It ranges across the subjects of a large number of IAU Commissions: in particular, stellar photometry and polarimetry (C25), variable stars (C27), stellar spectra (C29), stellar constitution (C35), and theory of stellar atmospheres (C36). Ap and related stars display the most extreme manifestations of stellar physical processes such as magnetism and selective diffusion, and their understanding requires the consideration of a broad spectrum of mechanisms, including stellar pulsation, radiative acceleration and winds, accretion, and binarity. The role of the ApWG in providing a meeting point and a channel of communication for the members of a community involving such a wide range of scientific expertise, is essential. One of the ways in which the WG fulfils this role is via its ApN newsletter (http://apn.arm.ac.uk/newsletter/). This newsletter is used both to announce news and as a public forum, where registered members can submit requests or announcements, and initiate discussions or participate in them.

The ApWG also serves as a contact point with the atomic physics community, to compile and communicate to the latter requests for new atomic data that are needed for the analysis and modelling of Ap and related stars.

3. Scientific meetings

To follow up on the two science meetings that it organised during the previous triennium (*CP#Ap Workshop* in Vienna, Austria, in 2007 and Joint Discussion on *Progress in understanding the physics of Ap and related stars* at the IAU XXVIIth General Assembly in Rio de Janeiro, Brazil, in 2009), the ApWG has now started plans for a symposium entitled *Putting A Stars into Context*, to take place in 2013. A number of other conferences and workshops including sessions relevant to the physics and study of Ap and related stars took place or will take place between August 2009 and August 2012. They include: *Seismological Challenges for Stellar Structure* (Lanzarote, Spain, February 2010); *Magnetic Fields From Core Collapse to Young Stellar Objects* (London, Ontario, Canada, May 2010); *IAU Symposium 272: Active OB Stars: Structure, Evolution, Mass Loss and Critical Limits* (Paris, France, July 2010); *10th International Colloquium on Atomic Spectra and Oscillator Strengths for Astrophysical and Laboratory Plasmas* (Berkeley, California, USA, August 2010); *IAU Symposium 273: Physics of Sun and Star Spots* (Ventura, California, USA, August 2010); *Magnetic Stars 2010* (Nizhny Arkhyz, Russia, August 2010); *Progress in Solar/Stellar Physics with Helio- and Asteroseismology* (Hakone, Japan, March 2011); *Stellar Polarimetry: From Birth to Death* (Madison, Wisconsin, USA, June 2011); *4th Kepler Asteroseismic Science Consortium workshop: From unprecedented data to revolutionary science* (Boulder, Colorado, USA, July 2011); *7th Potsdam Thinkshop: Magnetic Fields in Stars and Exoplanets* (Potsdam, Germany, August 2011); *The First Kepler Science Conference* (Moffett Field, California, USA, December 2011). Members of the ApWG feature both as organisers and as invited speakers of these meetings.

4. Conclusion

The community interested in Ap and related stars has been very active scientifically over this triennium. Considerable progress has been made, opening new perspectives in several areas, such as magnetic fields (with the novel possibility to follow their evolution from pre- to post-main-sequence stages, the developing knowledge of their extension to hotter stars, and the realisation that they may be ubiquitously present in all A stars, with different strengths) and pulsation (with an upheaval of the overall picture of pulsation in A and B stars resulting from the unprecedented wealth of exquisitely detailed observations obtained e.g. by the space missions CoRoT and Kepler). Developments in these, and related fields, will assuredly continue in the next triennium, and the Working Group on Ap & Related Stars will play a critical role in ensuring that effective interactions take place between the various constituents of this research community.

Gautier Mathys
chair of Working Group

References

Alecian, E., Wade, G. A., Catala, C., *et al.* 2009, *MNRAS*, 400, 354
Alecian, G. & Stift, M. J. 2010, *A&A*, 516, A53
Alecian, G., Stift, M. J., & Dorfi, E. A. 2011, *MNRAS* (in press)
Alina, D., Petit, P., Lignières, F., *et al.* 2011, arXiv:1107.5639
Arlt, R. & Rüdiger, G. 2011, *AN*, 332, 70
Aurière, M., Wade, G. A., Lignières, F., *et al.* 2010, *A&A*, 523, A40
Aurière, M., Konstantinova-Antova, R., Petit, P., *et al.* 2011, arXiv:1109.5570
Balona, L. A., Cunha, M. S., Kurtz, D. W., *et al.* 2011, *MNRAS*, 410, 517
Balona, L. A., Ripepi, V., Catanzaro, G., *et al.* 2011, *MNRAS*, 414, 792
Bigot, L. & Kurtz, D. W. 2011, arXiv:1110.0988
Ferrario, L., Pringle, J. E., Tout, C. A., & Wickramasinghe, D. T. 2009, *MNRAS*, 400, L71
Gruberbauer, M., Huber, D., Kuschnig, R., *et al.* 2011, *A&A*, 530, A135
Grunhut, J. H., Wade, G. A., Hanes, D. A., & Alecian, E. 2010, *MNRAS*, 408, 2290
Hubrig, S., Mikulášek, Z., González, J. F., *et al.* 2011, *A&A*, 525, L4
Hubrig, S., Schöller, M., Kharchenko, N. V., *et al.* 2011, *A&A*, 528, A151
Khomenko, E. & Kochukhov, O. 2009, *ApJ*, 704, 1218
Kochukhov, O. & Wade, G. A. 2010, *A&A*, 513, A13
Kochukhov, O., Lundin, A., Romanyuk, I., & Kudryavtsev, D. 2011, *ApJ*, 726, 24
Kurtz, D. W., Cunha, M. S., Saio, H., *et al.* 2011, *MNRAS*, 414, 2550
LeBlanc, F., Monin, D., Hui-Bon-Hoa, A., & Hauschildt, P. H. 2009, *A&A*, 495, 937
Lignières, F., Petit, P., Böhm, T., & Aurière, M. 2009, *A&A*, 500, L41
Lüftinger, T., Fröhlich, H.-E., Weiss, W. W., *et al.* 2010, *A&A*, 509, A43
Makaganiuk, V., Kochukhov, O., Piskunov, N., *et al.* 2011, *A&A*, 525, A97
Mikulášek, Z., Krtička, J., Henry, G. W., *et al.* 2011, *A&A*, 534, L5
Paunzen, E., Netopil, M., & Bord, D. J. 2011, *MNRAS*, 411, 260
Perraut, K., Brandão, I., Mourard, D., *et al.* 2011, *A&A*, 526, A89
Petit, P., Lignières, F., Aurière, M., *et al.* 2011, *A&A*, 532, L13
Smalley, B., Kurtz, D. W., Smith, A. M. S., *et al.* 2011, arXiv:1107.0246
Smith, R. & Wyatt, M. C. 2011, *A&A*, 515, A95
Sousa, J. C. & Cunha, M. S. 2011, *MNRAS*, 414, 2576
Stift, M. J., Leone, F., & Cowley, C. R. 2011, arXiv:1110.1291
Théado, S., Vauclair, S., Alecian, G., & LeBlanc, F. 2009, *ApJ*, 704, 1262
Uytterhoeven, K., Moya, A., Grigahcene, A., *et al.* 2011, arXiv:1107.0335
Vick, M., Michaud, G., Richer, J., & Richard, O. 2010, *A&A*, 521, A62
Wade, G. A., Howarth, I. D., Townsend, R. H. D., *et al.* 2011, *MNRAS*, 416, 3160

Transactions IAU, Volume XXVIIIA
Reports on Astronomy 2009–2012   © International Astronomical Union 2012
Ian Corbett, ed.   doi:10.1017/S1743921312002827

DIVISION IV / WORKING GROUP
on MASSIVE STARS

CHAIR	Joachim Puls
VICE-CHAIR	Claus Leitherer
PAST CHAIR	Stan Owocki
BOARD	Paul Crowther
	Margaret Hanson
	Artemio Herrero
	Norbert Langer
	Stan Owocki
	Gregor Rauw
	Nicole St-Louis
	Richard Townsend

TRIENNIAL REPORT 2010-2012

1. Introduction and background

Our Working Group (WG) studies massive, luminous stars, both individually and in resolved and unresolved populations, with historical focus on early-type (OB) stars, A-supergiants, and Wolf-Rayet stars. Our group also studies lower mass stars (e.g., central stars of planetary nebulae and their winds) which display features similar or related to those present in massive stars, and thus may improve our understanding of the physical processes occurring in massive stars. In recent years, massive red supergiants that evolve from hot stars have been included into our activities as well. We emphasize the role of massive stars in other branches of astrophysics, particularly regarding the First Stars, long duration Gamma-Ray bursts, formation of massive stars and their feedback on star formation in general, pulsations of massive stars, and starburst galaxies.

The goal of this Working Group is to focus research in the massive star community and to help communicating the most recent results. Minor and major conferences shall be proposed, (co-)sponsored, and/or help in the application process shall be provided. Most prominent are the 'massive star conferences' covering major topics in massive star research, held about every five years, with many of them under the auspices of the International Astronomical Union.

By-laws generated and approved by the Organization Committee (OC) and other activities of the Working Group are posted on an offical website at:

http://www.astroscu.unam.mx/massive_stars

The webmaster for this site is Raphael Hirschi (University of Keele, UK).

2. Developments within the past triennium

Over the past two years, and projected to 2012, WG activities focussed on issues related to the "Massive Star Newletter", and on co-sponsorship and preparation of various conferences/workshops. In particular, the OC called for proposals regarding Joint Discussions (JD) and Special Sessions (SpS) to be held during the IAU GA 2012 in Beijing, and helped with the preparation of the corresponding events. The follow gives further details.

2.1. *The Massive Star Newletter*

Among other features, our web portal offers an automatic abstract submission interface. This allows the members of the Working Group to submit recent abstracts and circulates newly received abstracts to registered members, both individually as well as collected within *The Massive Star Newsletter* (together with other topics of interest such as finished thesis work, upcoming conferences, and job offers).

The Newsletter, edited by Philippe Eenens (Universidad de Guanajuato, Mexico) and Raphael Hirschi, continues to be the main means of communication and science propagation in our Working Group. As of October 2011, 124 issues of the Newsletter have been published. Back issues are posted on the Working Group website:

http://www.astroscu.unam.mx/massive_stars/news.php

During the past two years, the WG has helped Raphael Hirschi and Philippe Eeenens in switching from a 'manual' production of the Newsletter to an automatic process.

2.2. *Workshop – Hot and Cool: Bridging Gaps in Massive Star Evolution†*

Held November 10-12, 2008, on the campus of the California Institute of Technology, and hosted by the NASA Herschel Science Center.

The topic of this workshop was a result of intense (and still ongoing!) discussions within the WG to bridge the gap between and bring together researchers studying stars in the upper blue and red parts of the Hertzsprung-Russell diagram (HRD). While morphologically separated, stars occupying these extremes of the HRD are intimately related via evolution, as well as both having atmospheric properties affected by extension and mass outflow. At cosmological scales, like in distant starburst galaxies, the historical distinction between blue and red stellar populations becomes obsolete, and understanding the complex relation between the red and blue parts of the HRD is mandatory.

The meeting, chaired by Claus Leitherer, member of our OC, was intended to investigate similarities and differences between hot and cool stars, identify the most challenging questions in evolutionary connections, and discuss latest theoretical and observational progress, to provide a more unified understanding of massive stars and luminous extragalactic stellar populations. Major topics included:

- Evolutionary connections
- Late and end phases (post-RSG and blue-loop stars, supernovae from red vs. blue progenitors, Gamma-ray bursts)
- Atmospheres
- Mass loss (rates, physics, episodic mass loss, pulsations)
- Interacting winds and circumstellar environments

Further information can be found in the workshop proceedings edited by Leitherer *et al.* (2010).

2.3. *JD2 at IAU GA 2012 – 'Very Massive Stars in the Local Universe'*

This 1.5 day JD, to be held during the IAU GA 2012 in Beijing, is co-sponsored by our WG, and will be chaired by Jorick Vink (Armagh, Ireland).

† Technically, this workshop falls outside the reporting period, but was not covered in the last report, because being held at the end of 2008.

Recent findings indicating the existence of very massive stars (VMS) up to ~ 300 M$_\odot$ in the local Universe (Gal-Yam *et al.* 2009, Crowther *et al.* 2010) might lead to a paradigm shift for the stellar upper-mass limit. The JD intends to discuss the status of the data, as well as the far-reaching implications of such objects.

The determination of both current and final masses of the most massive stars shall be discussed, to reach consensus between observers and theorists on how to identify and quantify the importance of the dominant physical processes. The objects may evolve almost chemically homogeneously, implying that the detailed mixing processes (e.g., rotation, magnetic fields) could be less relevant compared to 10-50 M$_\odot$ stars. Instead, the evolution and death of VMS is likely dominated by mass loss. Major topics of this JD include

- Weighing the most massive stars from their binary motions
- Mass determinations from stellar spectroscopy and model atmosphere analysis
- Formation of the most massive stars
- Mass loss mechanisms, incl. eruptions of Luminous Blue Variables
- Stellar structure and evolution modeling
- The fate of the most massive stars (over cosmological time)
- Mass and energy return to the interstellar medium (ISM)

Further information at the website `http://www.arm.ac.uk/IAU/index.html`

2.4. *SpS5 at IAU GA 2012 – 'The IR view of massive stars: the main sequence and beyond'*

This 1.5 day SpS, to be held during the IAU GA 2012 in Beijing, is co-sponsored by our WG, and will be chaired by Yaël Nazé (Liège, Belgium).

Even though a major fraction of astronomers still relies on the optical domain, this is certainly going to change, as most current and future instruments are dedicated to the infrared, from the near- to the far-IR bands. While this domain is a known 'must' for (very) low-mass stars, the infrared emission of high-mass stars has been often neglected. Many advantages of the infrared need to be stressed, however, like its strong potential for circumstellar material and atmosphere diagnostics, and its insensitivity to obscuration. Its interest with respect to the first generation of stars is also well known.

In this SpS, it is intended to discuss the results obtained for massive stars from existing IR facilities (VLTs/VLTI, Spitzer, Herschel, CRIRES, GAIA, ...) as well as tools for interpreting IR data (e.g. atmosphere modeling) and observing capabilities of future facilities (ELTs, JWST, ...). Major topics to be discussed are

- Obscured and distant clusters in the IR
- Stellar and wind parameters from the IR
- Matter ejection and feedback

Further information at the website
`http://www.gaphe.ulg.ac.be/IAU_XXVIII/index.html`

3. Future outlook

Our WG is presently involved in the preparation of the upcoming 'massive star conference', planned to be held in 2013 in Rhodes (Greece), with the title "Massive Stars: from alpha to Omega".

Other upcoming meetings on massive stars include:

– "Circumstellar Dynamics at High Resolution", to be held February 27 - March 2, 2012, in Foz do Iguaçu, Brazil.
Weblink: `http://www.eso.org/sci/meetings/2012/csdyn.html`

– "The Evolution of Massive Stars and Progenitors of GRBs", to be held June 17 - July 1, 2012, at the Aspen Center for Physics, Aspen, CO. Further details at
`http://casa.colorado.edu/~emle6425/aspen/`

In the recent few years, but even more in the near future, exciting new results on massive stars have been and will be obtained by spectropolarimetry, interferometry and asteroseismology. These techniques probe both the physics of the surfaces and interiors of massive stars, allowing unprecedented constraints of key physical parameters such as rotation, magnetic fields, and chemical composition.

Large-scale, high-quality surveys of resolved stellar populations (e.g the VLT-FLAMES survey on massive stars and the Tarantula survey), as well as highly detailed investigations of individual systems (e.g. η Car), have exposed key weaknesses in our assumptions on massive-star evolution. It has become clear now that understanding stellar populations is a considerable challenge that will require substantial efforts to resolve. We envisage interesting times for the massive star community.

Acknowledgements

The OC gratefully acknowledges logistic support for several teleconferences by STScI. Many thanks to the Institute of Astronomy of UNAM (Mexico) for support with their web server.

Joachim Puls
chair of Working Group

References

Crowther, P. A., Schnurr, O., Hirschi, R., *et al.* 2010, *MNRAS*, 408, 731
Gal-Yam, A., Mazzali, P., Ofek, E. O., *et al.* 2009, *Nature*, 462, 624
Leitherer, C., Bennett, P., Morris, P., & van Loon, J. (editors) 2010, *Astronomical Society of the Pacific Conference Series*, 425

Transactions IAU, Volume XXVIIIA
Reports on Astronomy 2009–2012   © International Astronomical Union 2012
Ian F. Corbett, ed.   doi:10.1017/S1743921312002839

DIVISION V   VARIABLE STARS

ETOILES VARIABLES

Division V deals with all aspects of stellar variability including intrinsic variability and variability caused by a companion in a binary system. In the case of intrinsic variability the analysis of pulsating stars, surface inhomogeneities, stellar activity, and oscillations are considered. For close binaries, detached eclipsing binaries are studied as well as interacting systems. Contact and semi-detached binaries, or those with compact components like cataclysmic variables and X-ray binaries, are examined within the context of the physics of accretion processes.

PRESIDENT   Steven D. Kawaler
VICE-PRESIDENT   Ignasi Ribas
PAST PRESIDENT   Alvaro Gimenez
BOARD   Michel Breger,
                       Edward F. Guinan,
                       Gerald Handler,
                       Slavek M. Rucinski

DIVISION V COMMISSIONS

Commission 27   Variable Stars
Commission 42   Close Binary Stars

DIVISION V WORKING GROUPS

INTER-DIVISION WORKING GROUPS

Division IV-V WG   Active B-type Stars
Division IV-V WG   Ap and Related Stars

TRIENNIAL REPORT 2009-2012

1. Introduction

Division V on *Variable Stars* consists of Commission 27, also called *Variable Stars*, and Commission 42, *Close Binary Stars*. The former deals with stars whose variations are intrinsic, whereas in the latter the variations are caused by the interactions between the components in the binary or multiple star system. There may be cases where the assignment of an object to one of the two Commissions may be in doubt. For example, the observation of pulsating stars in eclipsing binaries within nearby galaxies, or the relation between some types of oscillation modes and membership to binary systems, continue to be widely discussed.

The report of the Division for the triennium is more extensively documented in the reports of each of the two Commissions.

2. Variable stars

The progress of studies on variable stars has been reviewed in a series of international meetings in almost all the domains of interest for Division V, and listed in the C27 Triennial report. The dominant milestone during this period was the successful launch and deployment of NASA's *Kepler* spacecraft. Though designed to search for Earth-sized planets around Sun-like stars, the photometric and operating characteristics of *Kepler* make it an ideal asteroseismic observatory. A large number of Division members are involved in analysis of *Kepler* data on a variety of pulsating stars – those known about prior to the launch, and many kinds of stellar variability not anticipated prior to being seen in *Kepler* data.

The study of solar-like oscillations in main sequence stars, sub-giants, and giants has blossomed with the availability of *Kepler* data on thousands of targets. On the main sequence, these data have refined the parameters of planetary host stars (in particular the radius, but also the age and mass), paying back handsomely on the investment made by the *Kepler* Mission in asteroseismology. The detailed analysis of solar-like oscillations in red giants has provided a new tool to determine the core structure of these stars, and identify giant stars that have hydrogen shell burning from those stars with helium-burning cores. In the classical variable star area, RR Lyrae stars show evidence of period doubling in Blazhko stars in the exquisite *Kepler* light curves.

Additional results from space-based astronomy, along with many significant advances using ground–based data, are discussed in the report of Commission 27.

3. Close binary stars

As with the intrinsic variable star field, close binary stars have been the subject of, or components of, several international meetings over the past triennium, principally IAU Symposium 282 in 2011, "From Interacting Binaries to Exoplanets: Essential Modeling Tools." Please see the Commission 42 report for further details.

Concerning close binaries, there have also been important new advances. A probable instance of a stellar merger caught "in the act," and a significant nova outburst of V407 Cyg are discussed in the Commission 42 triennial report. As a measure of the overlapping interest of the two Commissions within Division V, significant progress in observing and analyzing pulsating stars within binary systems has been made over the past three years, using ground–based and space–based facilities.

As might be expected, the *Kepler* Mission has been a tremendous asset to the study of eclipsing, reflection–effect, and ellipsoidal variables with periods ranging from hours to months. The central mission of *Kepler*, in fact, is to discover eclipsing binaries, albeit with a substellar companion doing the eclipsing. Discoveries here include fascinating systems such as eccentric binaries experiencing tidally-forced brightenings and oscillations, binary systems with low–mass white dwarfs that produce brightness variations in their (brighter) companions through Doppler beaming, and circumbinary planetary systems.

Steven Kawaler
president of the Division

Transactions IAU, Volume XXVIIIA
Reports on Astronomy 2009–2012
Ian Corbett, ed.

© International Astronomical Union 2012
doi:10.1017/S1743921312002840

COMMISSION 27

VARIABLE STARS
ÉTOILES VARIABLES

PRESIDENT	Gerald Handler
VICE-PRESIDENT	Karen R. Pollard
PAST PRESIDENT	Steven Kawaler
ORGANIZING COMMITTEE	Margarida S. Cunha, Katalin Olah,
	Katrien Kolenberg, C. Simon Jeffery,
	Márcio Catelan, Laurent Eyer,
	Timothy R. Bedding, S. O. Kepler,
	David Mkrtichian

TRIENNIAL REPORT 2009-2012

1. Introduction

As research on variable stars continues at an ever growing pace, this report can only give a selection of research highlights from the past three years, with a rigorously abbreviated bibliography. The past triennium has been dominated by results of the CoRoT (Astronomy & Astrophysics 2009) and *Kepler* (Gilliland *et al.* 2010) space missions, stemming from their unprecedented photometric accuracies and large time bases.

Numerous conferences related to the field have taken place; the following list certainly is incomplete. Two of the biennial meetings on stellar pulsation were held in Santa Fe, USA (2009) and Granada, Spain (2011). Kepler Asteroseismic Science Consortium Workshops took place in Aarhus, Denmark (2010) and Boulder, USA (2011); CoRoT symposia in Paris, France (2009) and Marseille, France (2011). The Kukarkin Centenary Conference was held in Zvenigorod, Russia (2009) and IAU Symposium 272 on active OB stars in Paris (France, 2010). IAU Symposium 273 dealt with the Physics of Sun and Star Spots (Ventura, USA, 2010), IAU Symposium 286 with Comparative Magnetic Minima (Mendoza, Argentina, 2011). IAU Symposium 285 (New Horizons in Time Domain Astronomy, Oxford, UK, 2011) had substantial interest for C27. The American Association of Variable Star Observers held its 100th annual meeting 2011 in Boston, USA.

2. Science highlights

2.1. *Stellar activity*

A comprehensive review of the origin and properties of starspots was published by Strassmeier (2009), with special emphasis on the the possibility of detecting exoplanets around spotted host stars. The existence of spots for discovering planets and in studying possible star-planet interactions has recently been recognised.

Several papers have been devoted to the active star CoRoT-7 and its planet(s). The review by Pont *et al.* (2011) highlights the importance of a realistic treatment of both the activity and the uncertainties which are applicable to most small-planet candidates observed by CoRoT and *Kepler*. The combined presence of activity and additional errors precludes a meaningful search for additional low-mass companions.

Deming *et al.* (2011) developed a method to correct planetary radii for the presence of both crossed and uncrossed star spots. The exo-Neptune HAT-P-11b transits nearly perpendicular to the stellar equator, and the authors related the dominant phases of star-spot crossings to active latitudes on the star. Precise transit measurements over long durations may allow one to construct a butterfly diagram to probe the cyclic evolution of magnetic activity on the active K-dwarf planet-host star.

Osten *et al.* (2010) observed a large stellar flare and fluorescence from the dMe star EV Lac. The size of the flare, in terms of its peak X-ray luminosity, exceeded the non-flaring stellar bolometric luminosity, providing important constraints on the time-scales for energy storage and release in a stellar context.

Kolláth & Oláh (2009) and Oláh *et al.* (2009) demonstrated that time-frequency distributions provide useful tools for analysing the observations of active stars whose magnetic activity varies with time. Their technique applied to sunspot data revealed a complicated, multi-scale evolution in solar activity. Time variations in the cycles of 20 active stars based on decade-long photometric or spectroscopic observations show that stellar activity cycles are generally multiple and variable (see also Sect. 2.2).

Korhonen *et al.* (2009) presented simultaneous low-resolution longitudinal magnetic field measurements and high-resolution spectroscopic observations of the cool single giant FK Com. The maxima and minima in the mean longitudinal magnetic field are both detected close to the phases where cool spots appear on the stellar surface.

2.2. *Solar-like oscillations*

Oscillations in the Sun are excited stochastically by convection, as also they are in other stars with convective envelopes. CoRoT has contributed substantially to the list of main-sequence and subgiant stars with solar-like oscillations (e.g. Mathur *et al.* 2010 and references therein). A problem particularly present in F-stars (e.g. Benomar et al. 2009) was illuminated by CoRoT data: the short-mode lifetimes cause blending of some oscillation modes, thereby hampering mode identification.

A major achievement by CoRoT came from observations of hundreds of G- and K-type red giants showing clear oscillation spectra that are remarkably solar-like, with both radial and non-radial modes (e.g. De Ridder *et al.* 2009, Mosser *et al.* 2011). CoRoT continues to produce excellent results on red giants, main-sequence and subgiant stars.

The *Kepler* mission carried out a survey targeting more than 2000 main-sequence and subgiant stars for one month each. Solar-like oscillations were detected, and clear measurements of the large frequency separation for about 500 stars were made (Chaplin et al. 2011) – an increase by a factor of ~20 over previous results. Some of these stars have been studied individually. The first results on two main-sequence stars and one subgiant provided evidence for mixed-mode oscillations in the latter (Chaplin *et al.* 2010).

Kepler detected solar-like oscillations in thousands of red giants (e.g. Hekker *et al.* 2011 and references therein), including some in open clusters (e.g. Basu *et al.* 2011), enabling asteroseismic studies that would not have been thought possible only a few years ago. For example, the gravity-mode period spacings in red giants provide a means to disentangle the evolutionary phases of hydrogen- and helium-burning in red-giant stars (Bedding et al. 2011), an otherwise difficult task given the similarities in the mass, luminosity and radius of these two groups. Miglio *et al.* (2010) described the first detection of the seismic signature of the helium second-ionization region in red-giant stars, opening up the interesting possibility of determining seismically the helium content of their envelopes.

García *et al.* (2010) presented the first strong evidence for cyclic frequency variations associated with the presence of a stellar magnetic-activity cycle in a star other than the Sun. Seismic signatures of stellar activity cycles, in combination with additional information such as differential rotation, extent of convective envelope, etc., have the potential to increase substantially our understanding of mechanisms for magnetic-field generation and evolution. In the solar case, Fletcher et al. (2010) hypothesized a second dynamo based on quasi-biennial solar oscillation frequency variations.

Solar-like and heat-engine oscillations are not mutually exclusive. Both may in fact operate in a star provided that the surface convection layer is thin (Samadi et al. 2002). There is evidence from Kepler data for solar-like oscillations in at least one δ Scuti star (Antoci et al. 2011). There has even been a suggestion that sub-surface convection in B-type stars could excite solar-like oscillations (Cantiello et al. 2009, Belkacem et al. 2010). Meanwhile, Degroote et al. (2010) suggested that stochastically-excited oscillations are revealed in CoRoT photometry of the O-type star HD 46149.

2.3. Classical and heat-driven main sequence pulsators

The enigmatic Blazhko effect (amplitude/phase modulation) turns out to be very common in RR Lyrae stars, as ground-based (Jurcsik et al. 2009), CoRoT (Szabó et al. 2010a), and Kepler results confirm (e.g. Kolenberg et al. 2010). The pulsations in at least some RR Lyrae stars are remarkably stable (Nemec et al. 2011).

Space photometry data of RR Lyrae stars reveal previously unseen features in Blazhko stars, such as period doubling (Kolenberg et al. 2010) and additional modulations, which are not fully understood yet (Benkő et al. 2010, Guggenberger et al. 2011). This is also the case for RR Lyr itself, successfully observed by Kepler (Kolenberg et al. 2011). Period doubling has been traced to a 9:2 resonance between the fundamental mode and the ninth radial overtone (Szabó et al. 2010b), confirmed by models of Buchler & Kolláth (2011). Their results may even explain the irregular amplitude modulation which recent observations reveal. At last we may come closer to an explanation for the Blazhko effect.

RR Lyrae stars continue to fulfill their role as indicators, not only of distance but also of tracers of galaxy formation histories. They are increasingly being used by large-scale surveys such as SDSS (e.g. Sesar et al. 2010). In particular, studies of the so-called Oosterhoff dichotomy, which until recently were confined to the Milky Way and its nearest neighbours (e.g. Catelan 2009), can now probe greater distances (e.g., Fiorentino et al. 2010), and even include globular clusters as far away as M31 (Clementini et al. 2009).

Engle et al. (2009) reported X-ray emission of three bright Cepheids observed with XMM-Newton and Chandra. Despite differences in spectral type and pulsation properties, the Cepheids have similar X-ray luminosities and soft-energy distributions. Such high energy could arise from warm winds, shocks or pulsationally-induced magnetic activity.

Herschel images of Mira (Mayer et al. 2011) reveal broken arcs and faint filaments in the ejected material of the primary star. Mira's IR environment appears to be shaped by the complex interaction of its wind with its companion, the bipolar jet and the ISM. High-angular resolution Chandra imaging by Karovska et al. (2011) indicated focused-wind mass accretion, a "bridge" between Mira A and Mira B, indicating gravitational focusing of the Mira A wind whereby components exchange matter directly as well as by wind accretion. That greatly helps explain accretion processes in symbiotic systems and other detached and semi-detached interacting systems.

The "Cepheid Mass Problem" (the mismatch between masses computed from evolutionary tracks and hydrodynamic pulsation calculations for classical Cepheids) is a fundamental test of stellar-evolution models. The problem may be related to convective overshoot and possible mass loss (e.g. Neilson et al. 2011). Marengo et al. (2010) discovered an infrared nebula and a bow shock around δ Cephei and its hot companion, supporting the hypothesis that δ Cephei may be currently losing mass.

The *Kepler* characterization of the variability in A- and F-type stars (Grigahcène *et al.* 2010, Uytterhoeven *et al.* 2011) revealed a large number of hybrid δ Sct/γ Dor pulsators, thereby opening up an exciting new channel for asteroseismic studies.

Intriguingly, Kurtz *et al.* (2011) presented the first example of a star that oscillates around multiple pulsation axes. Evidence for the presence of torsional modes was also given (also a first). Sousa and Cunha (2011) gave the first theoretically-based explanation for the diversity found observationally in the atmospheric behaviour of the oscillations of roAp stars; that may have important consequences for the study, based on seismic data, of atmospheres of roAp stars.

2.4. *Pulsation in hot subdwarf and white dwarf stars*

Pulsating hot subdwarf stars include the V1093 Her variables, subdwarf B (sdB) stars exhibiting gravity mode-pulsations, V361 Hya variables, sdB stars exhibiting pressure-mode pulsations, DW Lyn variables, hybrids showing both V1093 Her and V361 Hya type variability, and a unique subdwarf O pulsator. Pulsating white dwarf stars include the GW Vir variables, the hottest white or pre-white dwarfs. Among cooler stars they also include the helium-rich V777 Her (DBV) stars, the carbon-rich hot DQV pulsators and the classical hydrogen-rich ZZ Ceti (DAV) stars.

Among hot subdwarfs the first *Kepler* survey found only one V361 Hya variable in the field, but several V1093 Her variables were identified (Østensen *et al.* 2010). Those discoveries spawned an industry of more detailed analyses and follow-up surveys (e.g. Reed et al. 2010, Pablo *et al.* 2011, and references therein).

Detailed asteroseismic modelling for a number of sdB stars has mostly been based on the more mature CoRoT data, although *Kepler* data are also now having an impact (e.g. Charpinet *et al.* 2011, and references therein). *Kepler* has had less influence in the pulsating white dwarf arena; however, representatives of the ZZ Cet and V777 Her classes in the *Kepler* field (Hermes *et al.* 2011, Østensen *et al.* 2011) have now been found.

Larger-scale surveys continue to find more variables in *all* of the classes described above. The recent literature contains too many examples of surveys to be listed individually; it is the sum of contributions, rather than any individual publication, which makes this a progressive field of research. The possibility that any of these discoveries might represent something really new is investigated in follow-up observations of individual stars. Two pulsating sdB stars have attracted particular attention: CS 1246 is a radial pulsator in a close binary; fortnightly variations in the ephemeris are caused by light-travel time delays as the star orbits a $0.12 \, M_{\odot}/\sin i$ companion (Barlow *et al.* 2011). Long-period variability has been confirmed in the sdB star LS IV$-14°116$ (Green *et al.* 2011), an object that is chemically extremely peculiar (Naslim *et al.* 2011). It has been suggested that the pulsations may be excited by the ϵ mechanism in He-burning shells (Miller Bertolami *et al.* 2011), possibly making it the first pulsator known to be excited in this way.

The asteroseismic properties of white dwarfs of all four types (DAV, DBV, DQV and DOV) have been explored, with the DBV prototype GD 358 coming under close scrutiny (e.g. Montgomery *et al.* 2010), partly in the wider quest to characterise convection physics in DAV and DBV white-dwarf atmospheres using non-linearities in the light curves. Evidence that the DOV prototype GW Vir rotates as a solid body (Charpinet *et al.* 2009) addresses the long-standing angular-momentum question for white dwarfs: angular momentum must be removed at an earlier phase of evolution.

3. Catalogues and data archives

The flow of new variable-star discoveries makes compilation of new Name-Lists in the GCVS system a very complicated task. Thus, the 80th Name-List will contain more than

6000 new variables and is being published in three parts (part 1: published, part 2: end 2011, part 3: 2012). Samus *et al.* (2009) have published a catalogue of accurate equatorial coordinates for variable stars in globular clusters.

The Div. V WG for Spectroscopic Data Archives was wound up in 2010 March. In operation since 1992, it had accomplished – or witnessed – some valuable changes in attitudes towards archiving observations of spectra. Having fulfilled its prime objective in those achievements, a new bottleneck in the form of creating archives of *reduced* spectra was recognised, in particular for echelles. Efforts to reorient the WG towards "Pipeline Reductions" did not have the right audience or the right platform for adequate revitalization. The mission to design a dependable future for astronomy's heritage of spectra on photographic plates has been actively taken over by the Task Force for the Preservation and Digitization of Photographic Plates (PDPP, Comm. 5).

4. Projected future of the Commission and its science

Currently, Commission 27 is dominated by people working on stellar pulsation. Given the coming changes in the structure of the IAU, we need to consider whether to keep the current comprehensive Commission, albeit with a better balance between its different fields, or to propose a subdivision following the nomenclature of variable stars.

Concerning science, we are awaiting the launch (foreseen in 2013) of the Gaia mission. It will provide astrometry, photometry, spectrophotometry and spectroscopy of $\sim 10^9$ objects with $6 < V < 20$ including variable stars, over a projected mission length of 5 years, with an average number of measurements of ~ 70 per object. BRITE-Constellation comprises six nanosatellites to be launched sequentially in 2012 - 2013, aiming at variable stars with $V < 4$. It will be the first using at least two photometric filters. Regrettably, the PLATO mission has not been selected for implementation in the near future.

The VISTA ESO Public Survey (Minniti *et al.* 2010) will provide information on about 10^6 variable stars in several infrared passbands, and going several magnitudes deeper than 2MASS. The survey totals $\sim$1929 hours of observations, spread over $\sim$5 years, and includes $\sim$35 known globular clusters and hundreds of open clusters towards the Galactic bulge and an adjacent part of the disk. The (predominantly) spectroscopic SONG global telescope network is still progressing, and aims to make precise radial-velocity measurements with a strong focus on investigating solar-like oscillations.

Acknowledgments

The OC thanks Elizabeth Griffin and Nikolai N. Samus for their contributions.

Gerald Handler
President of the Commission

References

Antoci, V., Handler, G., Campante, T. L., *et al.*, 2011, *Nature*, 477, 570
Astronomy & Astrophysics, 2009, Special Issue, Vol. 506
Barlow, B. N., Dunlap, B. H., Clemens, J. C., *et al.*, 2011, *MNRAS*, 414, 3434
Basu, S., Grundahl, F., Stello, D., *et al.*, 2011, *ApJ*, 729, L10
Bedding, T. R., Mosser, D., Huber, D., *et al.*, 2011, *Nature*, 471, 608
Belkacem, K., Dupret, M. A., & Noels, A. 2010, *A&A*, 510, A6
Benkő, J., *et al.* 2010, *MNRAS*, 409, 1585
Benomar, O., Baudin, F., Campante, T. L., *et al.* 2009, *A&A*, 507, L13

Buchler, J. R. & Kolláth, Z., 2011, *ApJ*, 731, 24
Cantiello, M., Langer, N., Brott, I., *et al.* 2009, *A&A*, 499, 279
Catelan, M., 2009, *Ap&SS*, 320, 261
Chaplin, W. J., Appourchaux, T., Elsworth, Y., *et al.*, 2010, *ApJ* 713, L169
Chaplin, W. J., Kjeldsen, H., Christensen-Dalsgaard, J., *et al.*, 2011, *Science*, 332, 213
Charpinet, S., Fontaine, G., & Brassard, P., 2009, *Nature*, 461, 501
Charpinet, S., van Grootel, V., Fontaine, G., *et al.*, 2011, *A&A*, 530, A3
Clementini, G., Contreras, R., Federici, L., *et al.* 2009, *ApJ*, 704, L103
De Ridder, J., Barban, C., Baudin, F., *et al.*, 2009, *Nature* 459, 398
Degroote, P., Briquet, M., Auvergne, M., *et al.*, 2010, *A&A*, 519, A38
Deming, D. *et al.* 2011, *ApJ* 740 33
Engle, S. G., Guinan, E., Evans, N., & DePasquale, J., 2009, *BAAS* 41, 303 (# 433.12)
Fiorentino, G., Monachesi, A., Trager, S. C., *et al.*, 2010, *ApJ*, 708, 817
Fletcher, S. T., Broomhall, A.-M., Salabert, D., *et al.*, 2010, *ApJ*, 718, L19
García, R. A., Mathur, S., Salabert, D., *et al.*, 2010, *Science*, 329, 1032
Gilliland, R. L., Brown, T. M., & Christensen-Dalsgaard, J., *et al.*, 2010, *PASP* 122, 131
Green, E. M., Guvenen, B. O'Melly, C. J., *et al.*, 2011, *ApJ*, 734, 59
Grigahcène, A., Antoci, V., Balona, L. A., *et al.*, 2010, *ApJ* 713, L192
Guggenberger, E., *et al.*, 2011, *MNRAS* 415, 1577
Hekker, S., Gilliland, R. L., Elsworth, Y., *et al.*, 2011, *MNRAS*, 414, 2594
Hermes, J. J. & Mullally, Fergal, Østensen, R. H., *et al.*, 2011, *ApJ*, 741, L16
Jurcsik, J., *et al.*, 2009, *MNRAS* 393, 1553
Karovska, M., de Val-Borro, M., Hack, M. *et al.*, 2011, *BAAS* 43 (#228.03)
Kolenberg, K., *et al.*, 2010, *ApJ* 713, L198
Kolenberg, K., *et al.*, 2011, *MNRAS* 411, 878
Kolláth, Z. & Oláh, K. 2009, *A&A* 501 695
Korhonen, H. *et al.* 2009, *MNRAS* 395, 282
Kurtz, D. W., Cunha, M. S., Saio, H., *et al.*, 2011, *MNRAS* 414, 2550
Marengo, M., Evans, N. R., Barmby, P., *et al.* 2010, *ApJ*, 725, 2392
Mathur, S., García, R. A., Catala, C., *et al.*, 2010, *A&A*, 518, A53
Mayer, A., Jorissen, A., Kerschbaum, F., *et al.*, 2011, *A&A* 531, L4
Miglio, A., Montalbán, J., Carrier, F., *et al.*, 2010, *A&A*, 520, L6
Miller Bertolami, M. M., Córsico, A. H., & Althaus, L. G., 2011, *ApJ*, 741, L3
Minniti, D., Lucas, P. W., Emerson, J. P., *et al.*, 2010, *NewA*, 15, 433
Montgomery, M. H., Provencal, J. L., Kanaan, A., *et al.*, 2010, *ApJ*, 716, 84
Mosser, B., Belkacem, K., Goupil, M. J., *et al.*, 2011, *A&A*, 525, L9
Naslim, N., Jeffery, C. S., Behara, N. T., & Hibbert, A., 2011, *MNRAS*, 412, 363
Nemec, J. M., Smolec, R., Benkő, J. M., *et al.* 2011, *MNRAS*, 417, 1022
Neilson, H. R., Cantiello, M., & Langer, N. 2011, *A&A*, 529, L9
Oláh, K. *et al.* 2009, *A&A*, 501 703
Osten, R., *et al.* 2010, *ApJ*, 721 785
Østensen, R. H., Silvotti, R., Charpinet, S. *et al.*, 2010, *MNRAS*, 409, 1470
Østensen, R. H., Bloemen, S., Vučković, M., *et al.*, 2011, *ApJ*, 736, L39
Pablo, H., Kawaler, S. D., & Green, E. M., 2011, *ApJ*, 740, L47
Pont, F. *et al.* 2011, 411, 1953
Reed, M. D., Kawaler, S. D., Østensen, R. H., *et al.*, 2010, *MNRAS*, 409, 1496
Samadi, R., Goupil, M.-J., & Houdek, G. 2002, *A&A*, 395, 563
Samus, N. N., Kazarovets, E. V., Pastukhova, E. N., *et al.*, 2009, *PASP* 121, 1378
Sesar, B., Ivezić, Ž., & Grammer, S. H., 2010, *ApJ*, 708, 717
Sousa, J. C. & Cunha, M. S., 2011, *MNRAS*, 414, 2576
Strassmeier, K. G., 2009, *Astron. Astrophys. Rev.* 17, 250
Szabó, R., Paparo, M., Benkő, J., *et al.* 2010a, *AIPC* 1170, 291
Szabó, R., Kolláth, Z., Molnaár, L., *et al.*, 2010b, *MNRAS* 409, 1244
Uytterhoeven, K., Moya, A., Grigahcène, A., *et al.*, 2011, *A&A*, 534, A125

Transactions IAU, Volume XXVIIIA
Reports on Astronomy 2009–2012   © International Astronomical Union 2012
Ian Corbett, ed.   doi:10.1017/S1743921312002852

COMMISSION 42

CLOSE BINARY STARS
ETOILES BINARIES SERREES

PRESIDENT	Ignasi Ribas
VICE-PRESIDENT	Mercedes T. Richards
PAST PRESIDENT	Slavek Rucinski
ORGANIZING COMMITTEE	David H. Bradstreet, Petr Harmanec,
	Janusz Kaluzny, Joanna Mikolajewska,
	Ulisse Munari, Panagiotis Niarchos,
	Katalin Olah, Theodor Pribulla,
	Colin D. Scarfe, Guillermo Torres

TRIENNIAL REPORT 2009-2012

1. Introduction

The present report covers the main developments in the field of close binaries during the triennium 2009-2012. In addition to scientific publications, there have been several opportunities for direct interaction of researchers working on close binaries. A number of meetings focused on more or less specific topics have taken place during this past years but the highlight for Commission 42 is arguably IAU Symposium 282 held in 2011 in Slovakia. The meeting exploited a strong connection in the methodology and tools used by close binary studies and the rapidly advancing field of exoplanet research. After all, exoplanetary systems are mostly discovered and studied using techniques employed by analyses of close binaries for decades. Modelling of exoplanet radial velocity curves and transiting planet light curves are just particular cases of single-lined and eclipsing binary systems, respectively, with very unequal component properties. As shown by IAU Symposium 282, the synergies between the two fields are strong and potentially very useful. Found below is a summary of the main scientific topics and conclusions from this very successful Symposium.

Meetings of interest to close binary research will continue to happen during 2012. We were very happy to see strong interest on the scientific aspects of our Commission through numerous sponsorship and support requests received for the IAU General Assembly in Beijing. Our Commission members will be glad to discover a broad range of Symposia, Joint Discussions, and Special Sessions in which close binaries play an important role in the science programme.

The present report includes an account of the activities of the Bibliography of Close Binaries publication. BCB gathers references to all publications containing results that are of interest to our Commission. Bringing BCB to light every six months entails significant work from the Editor-in-Chief (C. D. Scarfe) and a team of 10 editors. We have investigated the diffusion of this work and discovered with pleasure that the impact of BCB on close binary research is very significant, with a large number of downloads of every issue. This has given us all reasons to push forward with producing BCB twice yearly for the benefit of our community. In representation of C42, I would like to thank the team of people who invest time and effort in making BCB a reality. Further details on BCB for the past triennium are provided below.

And, finally, this report includes contributions from C42 Organizing Committee members that have kindly accepted to provide brief reviews on topics they think can be of interest to the rest of the C42 community. This includes an analysis of the observations of a binary merger, a report on the impact of the Kepler space mission to studies on close binaries, an account of the current status of pulsating binary observations, and a description of the 2010 eruption of V407 Cyg. I am sure that many other events and developments took place during 2009–2012 that are of importance to our field, but the short contributions included here do provide an interesting view on a variety of aspects and demonstrate that close binary research is active and lively field.

2. Report on IAU Symposium 282 (M. T. Richards & I. Hubeny)

IAU Symposium 282 entitled "From Interacting Binaries to Exoplanets: Essential Modeling Tools" was held in Tatranská Lomnica, Slovakia, from July 18–22, 2011. This symposium was designed to build a bridge between the exoplanet and binary star communities to discuss the many techniques in common use. This was the first joint meeting between these groups and it is significant that 31 countries were represented. IAU S282 was sponsored by four divisions: V – Variable Stars, III – Planetary Systems, IV – Stars, and IX – Optical & IR Techniques, as well as 7 Commissions: C25, C27, C29, C36, C42, C53, and C54. Commission 42 played a central role in the planning of this meeting. There were 8 sessions with equal weight to interacting binaries and exoplanets, with less emphasis on brown dwarfs. The main topics were (1) Multiwavelength Photometry and Spectroscopy of Interacting Binaries, (2) Observations and Analysis of Exoplanets and Brown Dwarfs, (3), Imaging Techniques, (4) Model Atmosphere Codes, (5) Synthetic Light Curves, Velocity Curves, Spectra of Binary Stars, and Spectra of Binaries with Accretion Structures, (6) Techniques for Analysis and Disentangling of Spectra and Light Curves, (7) Formation and Evolution Models, and (8) Hydrodynamic Simulations (see http://www.astro.sk/IB2E/).

The opening lectures reviewed the modeling tools as well as several open problems in the field. As a result, an IAU Resolution will be drafted to adopt updated astrophysical parameters and constants to improve the accuracy of fundamental parameters. These include the use of GM(Sun) since this product is more accurate than the product of the separate quantities. Several review talks were presented about existing and future ground-based and space-based observational instruments devoted to a study of close binaries, with an overview of observational techniques and results. While the CoRoT and Kepler instruments were designed primarily for exoplanet studies, Kepler is already having an impact on close binary research. We also heard about the planned Gaia and LSST projects. Didier Queloz, a co-discoverer of the first exoplanet orbiting a solar-type star, summarized the present status of exoplanet search with an emphasis on the radial velocity techniques, and outlined the expected development of the field in the near future. The remaining talks concentrated both on exoplanets and brown dwarfs, emphasizing mostly the transiting planets observed by CoRoT and Kepler, including an introduction to brown dwarfs in binary systems. There were several reviews of imaging techniques: interferometry, Doppler tomography, polarimetry, vortex coronagraphy, nulling interferometry, adaptive optics, and direct imaging. Several talks summarized the predominant model atmospheres codes for computing LTE as well as non-LTE model atmospheres, model atmospheres of exoplanets, extended atmospheres with stellar winds, and 3-dimensional hydrodynamic simulations of stellar atmospheres. Some interesting new results were presented, including a possible use of accurately determined limb darkening coefficients for constraining basic stellar parameters. The new variants of the classical methods and programs for solving light curves were discussed, as well as codes for computing the spectrum of a complex close binary system by modeling in detail both stars and the circumstellar accretion structures around them. We were delighted to see a statistical survey and analysis of the many close binaries already discovered by the Kepler mission.

There were talks on two independent methods of spectral disentangling of the components of the binary system with proposed applications to exoplanetary spectra. Other highlights included a talk on the history of the Rossiter-McLaughlin effect followed by applications of the technique to transiting planets and low mass eclipsing binaries. This method represents the closest methodological connection between close binary and exoplanetary research. The effect was used earlier to identify critical properties of close binary systems, and it is now providing an analogously rich source of information in the case of exoplanets. There were reviews of the stellar evolution of the components of a close binary system, as well as dynamical models of formation and evolution of exoplanets, a summary of non-conservative effects in the evolution of binaries, and a synthesis of simultaneous modeling of atmospheres and the global evolution of exoplanets. The final session dealt with close binary and exoplanet dynamics, including 3D hydrodynamic simulations of their atmospheres of these objects, and the mass transfer between them. We saw several simulations that showed the ever-increasing power of current numerical simulations to provide a detailed picture of mass transfer in the case of young binaries with low mass companions, common envelope binaries, and how magnetic fields influence cataclysmic variables and polars. One simulation also considered the meteorology of exoplanet atmospheres via models of global atmospheric circulations and transport of energy from the day to the night side for close-in giant exoplanets.

Panel discussions were held at the end of each day to summarize the lectures and to make proposals for enhancements to current techniques; this was a special feature of the symposium led by influential astronomers who have contributed to the development of important modeling tools or who have provided insightful reviews of these developments. There was a public outreach event on the last day of the conference.

3. Close Binary Research from the Perspective of BCB (C. D. Scarfe)

This report brings up to date the one that was included in C42's last triennial report, in 2008. It covers BCB issues 86 (June 2008) through 92 (June 2011), with reference to the 14 issues covered in 2008.

BCB has maintained its average size, but the lack of an increase can probably be attributed in part to the editors' careful attention to avoiding the inclusion of papers which are primarily within the purview of other Commissions, most frequently Commissions 26 and 53. We also continue to place papers which discuss similar work on four or more systems in the 'Collections of Data' section, to avoid as much as possible having multiple entries for the same paper in the 'Individual Stars' section, as we had begun to do three years ago.

We continue to try as much as possible to use variable star nomenclature, or names which involve equatorial co-ordinates, with cross-references to the names actually used by authors, when those differ from the above. And since there are now many extragalactic close binaries, several of them having co-ordinate-based names, we have abandoned the use of a separate sub-section of the 'Individual Stars' section for those distant objects.

We continue to keep up-to-date the coding system that appears on the inside cover of each issue, and welcome suggestions for its further improvement.

The average number of authors per paper continues to increase slowly, but much of the increase is due to an increase in the number of papers with very large numbers of listed 'authors'. In the most recent seven issues, there are nine entries with over 100, two of which have over 200. Apart from those papers the distribution of the number of authors per paper is roughly Poissonian, with every possible number represented up to 24, and few gaps up to 43. If the entries with over 50 'authors' are omitted, the mean number of authors per paper is just under 5, with the most frequently occurring number being 3. The 'Individual Stars' section continues to have a higher average number of authors and the 'General' section a lower number, while the 'Collections of Data' section averages the same as the combined total. The total number of entries in the last seven issues is 2557, an average of 365 per issue.

The list of objects discussed most frequently continues to evolve. V1357 Cyg and V1487 Aql continue their record of having at least one paper in every issue, and although the total number for V1487 Aql has decreased to 18 for the seven issues, V1357 Cyg again has more papers (26) than any other object. In the most recent 21 issues there are a total of 103 papers on V1357 Cyg; V1487 Aql is still in second place with 80, but others are gaining on it. In the past seven issues, there are 24 entries on η Car, 22 on V821 Ara, 21 on V1521 Cyg, 20 on V1343 Aql, and 19 on V615 Cas. RS Oph has 18 like V1487 Aql, and V4580 Sgr has 17. No other object has more than 13. Interest in KV UMa has declined from 37 papers in 14 issues to 5 in the most recent seven issues, and that in PSRJ0737-3039 has decreased from its peak of 12 papers in a single issue to none in any the most recent five issues.

Preparing each issue is time-consuming for all the editors, but we are advised by Andras Holl, who maintains the Commission web site, that each issue is downloaded about 500 times in the six months after it appears, so clearly our efforts are worthwhile. I would like to thank all the contributing editors for their work, in particular Yasuhisa Nakamura, whose home is in Fukushima, Japan, and who contributed as usual to the most recent issue despite the great earthquake of March 2011 and the events that followed it.

4. Observations of a binary merger: V1309 Sco (S. Rucinski)

Tylenda *et al.* (2011) collected very convincing evidence of a first ever observed contact-binary coalescence. In 2008, an inconspicuous ($I \simeq 17$) star in the direction of the Galactic Centre ($l = 359.8$, $b = -3.1$) exploded reaching $I \simeq 6.8$. Nova Sco 2008, also called V1309 Sco, showed rather unusual properties similar to "red-nova" or V838 Mon-type eruptions (Mason *et al.* 2010) which had been linked to possible mergers of binary components (Tylenda & Soker 2006).

Photometry collected by the OGLE project during 10 years starting in 2001 (Tylenda *et al.*) very clearly established that V1309 Sco looked originally like a contact binary with a period 1.44 days. Its orbital period shortened before the outburst by an incredibly large amount of 1.2% in five years 2002 – 2007. Not only the relatively long period, but also the red colour very clearly indicated that it was not a W UMa binary: After accounting for the heavy reddening in this direction, its colour corresponded to that of a K1-2 red giant. During a few months in 2007, the light curve underwent a dramatic evolution: It lost its double-eclipse, contact-binary character progressively looking more like that of a single, spotted star. Within about half a year before the outburst by some 10 magnitudes, the time-scale of the luminosity increase strongly shortened from the initially almost constant $\simeq 27$ days to single days when the OGLE project lost it due to detector saturation.

After the outburst, no periodicity was detected, but the star is now very red and probably shrouded in a dust envelope (Tylenda 2011) so that its photosphere is invisible. No observations exist to check if any periodicity is present in the infra-red.

The magnitude and temporal development of the V1309 Sco outburst can be very well explained by energy released in a merger of a small mass-ratio binary, $q \leqslant 0.1$, with a helium core of mass $0.12M_\odot$ and radius $0.05R_\odot$ spiralling in the envelope of a $1.5M_\odot$ K-type giant (Tylenda 2011). Such a scenario is well described by evolutionary models of Stępień (2011). He proposes that in this particular case, a close binary (period $2.2 - 3.3$ days, original masses $1.1 - 1.3$ and $0.5 - 0.9M_\odot$) progressively lost mass and angular momentum, to enter after a few Gyr into a dynamical mass exchange and a strong mass-ratio reversal. In the last stages before coalescence the binary may have looked like a contact one, but its primary was a red giant and the contact stage lasted very short.

The long period of the progenitor binary corresponds very well to an abrupt discontinuity in the period distribution of contact binaries which was observed in the same direction before (Rucinski 1998). Some of the "long-period contact" binaries beyond the 1.5 day edge may be in fact not contact systems, but semi-detached binaries with red giants and small companions.

The excellent coverage of the whole event is to a large extent due to the persistence and dedication of the OGLE team. It very clearly shows how useful are surveys continuing in time and retaining high systematic consistence.

Most sincere thanks are due to Drs. R. Tylenda and K. Stępień for their detailed comments and suggestions.

5. Impact of the Kepler Mission on close binary studies (G. Torres & K. Oláh)

In March of 2009 NASA launched the Kepler Mission (Borucki *et al.* 2010) in an effort to detect, characterize, and determine the frequency of Earth-size planets around other stars, using the transit method. In its 3.5-year mission the spacecraft has been monitoring about 160,000 stars almost continuously in a 115 square degree field in the direction of the constellation Cygnus, with a photometric precision that is measured in *parts per million* rather than milli-magnitudes. In addition to the many planetary transit signals detected, a large number of eclipsing binaries have turned up, of which only a fraction were previously known. At the time of this writing the latest catalogue released by the Mission contains 2177 systems detected during the first four months of observation (Slawson *et al.* 2011, online version), and includes a morphological classification of the light curves (detached, semi-detached, overcontact, ellipsoidal) as well as estimates of the orbital parameters for each binary. A total of 1273 systems are detached, 152 are semi-detached, 469 are overcontact, 139 are ellipsoidal variables, and 144 are uncertain or unclassified systems. The database is available online at `http://keplerebs.villanova.edu/`. This wealth of information is likely to keep interested astronomers busy for years to come.

Because of the long duration of the observations, the nearly continuous coverage, and the high photometric precision afforded by the spacecraft, a number of unusual binaries have been detected that have merited special study. One rather spectacular example is KOI-54 (HD 187091), a pair of A stars in very eccentric ($e = 0.8335$) 41-day orbit oriented nearly face-on ($i = 5.5\,\mathrm{deg}$) that shows periodic *brightenings* caused by the tidal distortions of the stars during each periastron passage (Welsh *et al.* 2010). Additionally, the light curve shows clear evidence of pulsations in one or both objects excited by the dynamical tides at periastron (Fuller & Lai 2011). Several other unusual systems have been uncovered by Kepler in which the relative eclipse depths show that the more compact, less luminous object is hotter than its stellar host (KOI-71, KOI-84, KIC 10657664, KOI-1224; Rowe *et al.* 2010, van Kerkwijk *et al.* 2010, Carter *et al.* 2011, Breton *et al.* 2011). These have turned out to be low-mass hot white dwarfs that are thermally bloated (up to ~10 times their degenerate radii). In all cases the extremely high precision of the observations makes it possible to detect and measure the effects of Doppler beaming in the light curves, which provides valuable additional information on the physical parameters.

Although circumbinary planets were suggested in a few cases from stellar eclipse timings, the first direct evidence of such a system was discovered by Doyle *et al.* (2011). The star KIC 12644769 was first found to be an eclipsing binary, later tertiary eclipses were detected by Doyle *et al.* (2011) indicating a third body in the system. From the depth of the tertiary eclipses and on the magnitude of the eclipsing timing variation, the third body proved to have 0.333±0.016 Jupiter massses and 0.754±0.003 Jupiter radii, i.e., a planet. The host binary is on an eccentric orbit of 41 deays, with components of 0.690 and 0.203 solar masses. The circumbinary planet is thus named as Kepler 16(AB)-b. Winn *et al.* (2011) determined the rotation period of Kepler 16A as 35.1±1 days, and found that the stellar orbit, the planetary orbit and the rotation of the primary are closely aligned. To reach this configuration two alternative scenarios are suggested. The system is about 2–4 Gyr old, and the stars agree with the evolutionary models of low-mass stars.

Aside from individual studies of such systems, other interesting areas of research made possible by this remarkable data set include the search for eclipse timing variations, which may reveal the presence of third bodies or other astrophysical effects.

6. Pulsating binaries (or oEA stars) (P. G. Niarchos)

Mkrtichian *et al.* (2004) introduced the oEA (oscillating EA) stars as the (B)A-F spectral type mass-accreting main-sequence pulsating stars in semi-detached Algol-type eclipsing binary systems. The oEA stars are the former secondaries of evolved, semi-detached eclipsing binaries which are (still) undergoing mass transfer and form a newly detected class of pulsators close to the main-sequence. The study of a pulsating star (e.g., δ Scuti star) in a binary system provides the possibility to accurately measure its fundamental stellar parameters, greatly helping mode identifications and thus the application of asteroseismology. Several papers have been published during the last 2–3 years on oEA stars. Some of them are mentioned below.

Mkrtichian (2010), gave a review of the present status of the pulsation studies of close binary systems, focusing on pulsating gainers of semi-detached Algol type eclipsing binaries. Zhou (2010) published a catalogue containing 89 systems and distinguished them according to their pulsational properties. This catalog is intended to be a collection of pulsating binary stars across the Hertzsprung-Russell diagram. Soydugan *et al.* (2011) published also a similar list including 43 cases of systems including a δ Scuti component. Although few pulsating EBs are known (Mkrtichian *et al.* 2007), Kepler will lead to a major increase in this number.

The following oEA systems were studied: RZ Cas (Tkachenko *et al.* 2009), Y Cam (a three-continent multi-site photometric campaign by Rodr-guez et al. 2010), CoRoT 102931335 (Damiani *et al.* 2010), CT Her (multi-site, multi-year photometric monitoring, by Lampens *et al.* 2011), KIC10661783 (Kepler photometry, by Southworth *et al.* 2011), sdB KPD 1930+2752 (Reed *et al.* 2011). New discoveries of oEA systems were reported by Soydugan *et al.* (2009), Soydugan *et al.* (2010), Liakos & Niarchos (2009), Liakos & Niarchos (2011).

7. The 2010 eruption of V407 Cyg (U. Munari)

The symbiotic binary Mira V407 Cyg erupted into a spectacular nova outburst in March 2010, which was happily detected and eagerly studied at any wavelength, from Fermi-LAT gamma-rays at energies >100 MeV, through X-rays (Swift) and UV/optical/IT, down to radio range (single dish, E-VLA, VLBI, maser). In normal novae, the initial high energy flash goes undetected and the subsequent expansion of the ejecta follows a simple ballistic path. In V407 Cyg, the dense wind of the Mira was instantaneously ionized by the initial flash and begun glowing following recombining with a time-scale of 4 days, while the underlying fast expanding nova ejecta were progressively decelerated by the collision with the Mira's wind, from an initial >1500 km s^{-1} velocity to ~100 km s^{-1} in less than 200 days. Something similar, but at lower energies given the thinner circumstellar medium, had been observed only once before, the 2006 outburst of the symbiotic recurrent nova RS Oph. This outburst is triggering a far-reaching modeling effort that, considering the current strong momentum, will takes year to complete.

A selection among the >30 already published papers includes Orlando et al. (2011), Munari *et al.* (2011), Razzaque *et al.* (2011), Deguchi *et al.* (2011), Lü *et al.* (2011) and Abdo *et al.* (2010).

Ignasi Ribas
president of the Commission

References

Abdo, A. A., *et al.* 2010, *Science*, 329, 817

Borucki, W. J. *et al.* 2010, *Science*, 327, 977

Breton, R. P., Rappaport, S. A., van Kerkwijk, M. H., & Carter, J. A. 2011, *ApJ*, submitted (arXiv:1109.6847)

Carter, J. A., Rappaport, S., & Fabrycky, D. 2011, *ApJ*, 728, 139

Damiani, C., Cardini, D., Maceroni, C., & Debosscher, J. 2010, Binaries - Key to Comprehension of the Universe, 435, 41

Deguchi, S., *et al.* 2011, *PASJ*, 63, 309

Doyle, L. R., *et al.*, 2011, *Science*, 333, 1602

Fuller, J. & Lai, D. 2011, MNRAS, submitted (arXiv:1107.4594)

Lampens, P., *et al.* 2011, *A&A*, 534, A111

Liakos, A. & Niarchos, P. 2009, Communications in Asteroseismology, 160, 2

Liakos, A. & Niarchos, P. 2011, Communications in Asteroseismology, 162, 51

Lü, G., Zhu, C., Wang, Z., Huo, W., & Yang, Y. 2011, *MNRAS*, 413, L11

Mason, E., Diaz, M., Williams, R. E., Preston, G., & Bensby, T., 2010, *A&A*, 516, A108

Mkrtichian, D. E. 2010, Binaries - Key to Comprehension of the Universe, 435, 277

Mkrtichian, D. E., Kusakin, A. V., Rodriguez, E., *et al.* 2004, *A&A*, 419, 1015

Munari, U., *et al.* 2011, *MNRAS*, 410, L52

Orlando, S. & Drake, J. J. 2011, *MNRAS*, in press (arXiv:1109.5024)

Razzaque, S., *et al.* 2011, *PhysRevD*, 82, 3012

Reed, M. D., *et al.* 2011, *MNRAS*, 412, 371

Rodríguez, E., *et al.* 2010, *MNRAS*, 408, 2149

Rowe, J. F. *et al.* 2010, *ApJ*, 713, L150

Rucinski, S. M., 1998, *AJ*, 115, 1135
Slawson, R. W. *et al.* 2011, *AJ*, 142, 160
Southworth, J., *et al.* 2011, *MNRAS*, 414, 2413
Soydugan, E., Soydugan, F., Şenyüz, T., Püsküllü, Ç., & Demircan, O. 2011, *NewA*, 16, 72
Soydugan, E., *et al.* 2010, Binaries - Key to Comprehension of the Universe, 435, 331
Soydugan, E., *et al.* 2009, Information Bulletin on Variable Stars, 5902, 1
Stępień, K., 2011, *A&A*, 531, A18
Tkachenko, A., Lehmann, H., & Mkrtichian, D. E. 2009, *A&A*, 504, 991
Tylenda, R., 2011, private communication
Tylenda, R. & Soker, N., 2006, *A&A*, 451, 223
Tylenda, R., *et al.*, 2011, *A&A*, 528, A114
van Kerkwijk, M. H., Rappaport, S. A., Breton, R. P., Justham, S., Podsiadlowski, P., & Han, Z., 2010, *ApJ*, 715, 51
Welsh, W. F. *et al.*, 2010, *ApJS*, 197, 4
Winn, J. N., *et al.*, 2011, *ApJ*, 741, L1
Zhou, A.-Y., 2010, arXiv:1002.2729

Transactions IAU, Volume XXVIIIA
Reports on Astronomy 2009–2012
Ian F. Corbett, ed.

© International Astronomical Union 2012
doi:10.1017/S1743921312002864

DIVISION VI INTERSTELLAR MATTER

MATIÈRE INTERSTELLAIRE

Division VI provides a focus for astronomers studying a wide range of problems related to the physical and chemical properties of interstellar matter in the Milky Way and other galaxies.

PRESIDENT	You-Hua Chu
VICE-PRESIDENT	Sun Kwok
PAST PRESIDENT	Thomas J. Millar
BOARD	Dieter Breitschwerdt, Michael G. Burton,
	Sylvie Cabrit, Paola Caselli,
	Elisabete M. de Gouveia Dal Pino,
	Neal J. Evans, Thomas Henning,
	Mika J. Juvela, Bon-Chul Koo,
	Michal Rozyczka, Laszlo Viktor Toth,
	Masato Tsuboi, Ji Yang

DIVISION VI COMMISSIONS

Commission 34	Interstellar Matter

DIVISION VI WORKING GROUPS

Division VI WG	Astrochemistry
Division VI WG	Planetary Nebulae

TRIENNIAL REPORT 2009-2012

1. Introduction

Division VI, consisting of one Commission (Commission 34) and two Working Groups (Astrochemistry WG and Planetary Nebulae WG), has 972 members whose theoretical, observational, and experimental research interests cover a wide spectrum of activities associated with the study of the interstellar medium (ISM) in the Universe. As such, the Division has close links with Division VIII, IX, and X. The ISM and stars, the two major visible components of a galaxy, are coupled to each other through star formation, stellar feedback, and gravitational potential; thus, the Division is also closely linked to Division VII.

Our report on activity since 2009 is divided into four sections, covering our view of past and future developments of the Division, important meetings and conferences, relevant proceedings and monographs, and a list of important review articles published in the reporting period. Reports of the Working Groups are provided separately.

2. Developments

2.1. *Scientific advances*

Recent advances in observing facilities have made it possible to study the physical structure and processes of the interstellar medium in the Galaxy and nearby galaxies with higher angular resolution in a wide range of wavelengths. Most notably, the dust component of the interstellar medium can be studied with unprecedented sensitivity and angular resolution in mid- to far-infrared by the Spitzer Space Telescope, the Herschel Space Observatory, and the Planck Satellite. Numerous surveys were made for the Galactic plane, e.g. Spitzer's GLIMPSE and MIPSGAL, and Herschel's Hi-GAL, and for the Magellanic Clouds, e.g. SAGE and S^3C. These surveys have been used to analyze the abundance and distribution of polycyclic aromatic hydrocarbons, very small grains, and big grains. Planck, in addition to its mission goal in cosmology, mapped the coldest dust in the Galaxy. Besides these large-scale mappings, Spitzer and Herschel observations have been used to discover molecules, e.g. fullerines, and study astrochemistry.

From the ground, the International Galactic Plane Survey (IGPS) combines many radio telescope surveys from around the world to map the interstellar gas and dust in the galaxy, while the HI Nearby Galaxy Survey (THINGS) maps out the interstellar gas in nearby galaxies. The Bolocam Galactic Plane Survey (BGPS) and the APEX Telescope Large Area Survey of the Galaxy (ATLASGAL) are mapping out the colder dust in the Galaxy. The Atacama Large Millimeter Array (ALMA) will provide a clear view of dust and star formation in the Galaxy and nearby galaxies. The polarimetric capabilities of Institut de Radioastronomie Millimetrique (IRAM) in lines, Planck in continuum, and the upcoming ALMA in mm-submm wavelengths are opening a new window to probe magnetic fields. The Wisconsin Hα Mapper (WHAM) has mapped the distribution and velocities of warm ionized gas in the Galaxy from the north and is extending to the south.

At the high-energy end, Chandra and XMM-Newton Observatories have been used to investigate the distribution and physical properties of the 10^6 K hot ionized gas in star forming regions, such as the Carina Nebula, as well as diffuse fields in the Galaxy, Magellanic Clouds, and nearby galaxies. The Far UV Spectroscopic Explorer (FUSE) and the Hubble Space Telescope (HST) STIS and COS have been used to probe the 10^5 K hot gas at interfaces and in the Galactic halo. The Fermi Gamma-Ray Observatory has resolved the gamma-ray emission and revealed the sites of cosmic-ray production in the Galaxy as well as the Magellanic Clouds, and revealed the existence of huge bubbles in the Galactic Center of mysterious origin.

Progress has also been made in the numerical modeling of local and global conditions of the ISM, its morphology and its time-dependent evolution, owing to a rapid development of suitable hard- and software. It is now possible to follow the full non-linear evolution of a plasma by solving the hydro- or MHD equations in high resolution simulations with adaptive mesh refinement. One of the key results of the past years was to recognize, and quantitatively describe, the role of compressible turbulence in the ISM and its impact on the distribution of gas into phases, the mixing of chemically enriched material, the volume and mass filling factors of the ISM plasma, and its heating and cooling history, amongst others.

Advances have also been made in the treatment of magnetic field in star formation. To address the magnetic flux problem in the collapse of a molecular cloud to form stars, the ambipolar diffuse mechanism was suggested. Recently, an alternative mechanism to transport magnetic flux in the early stage of star formation is proposed – magnetic reconnection in the presence of turbulence. 3-D MHD simulations of ISM clouds with consideration of a gravitational field provided by embedded stars and introducing turbulence indicate that magnetic reconnections can indeed reduce the magnetic field built up at the center. It has also been shown that, with the considering of effects of the self-gravity in the clouds, the turbulent reconnection transport may be efficient enough to make an initially subcritical cloud that becomes supercritical and collapse. 3-D MHD simulations show that turbulent magnetic reconnection can also transport magnetic flux to the outskirts and form a protostellar disk. (Contributed by E. M. de Gouveia Dal Pino.)

To promote scientific communications and collaborations, the Division has supported or organized several Symposia and Special Sessions, covering topics of ISM in general, star formation, and magnetic fields in the IAU General Assembly at Beijing in August 2012. These Symposia and Special Sessions are listed in Section 3.

2.2. *Restructuring of Divisions and Commissions*

The Division members are mostly in favor of restructuring the IAU Divisions. It is logical to merge Division VI Interstellar Matter and Division VII Galactic System into Division H. The general concern is the title of Division H. Among the responses, the majority think that the title "Interstellar Matter & Local Universe" can be fine-tuned to better represent the constituents of Division H.

It has been noticed that Laboratory Astrophysics is conspicuously absent in the IAU structure. Laboratory work has made essential contributions to, for example, astrochemistry and plasma astrophysics that are particularly relevant to Commission 34. While there is no specific commission on Laboratory Astrophysics, it should at least be mentioned in "Note b" that states what Division B will cover.

There is also concern that some Commissions in other Divisions are very narrowly focused and may continue to exist for historical purposes. It is questioned whether some small Commissions with similar focus could be merged into a reasonably sized Commission. The draft document for the IAU Restructuring did not provide sizes of membership for the Commissions, making it difficult to judge whether some Commissions might benefit from mergers.

Scientists working on space infrared missions identify themselves more strongly with Division B, and question why Division D is not included in Division B.

Finally, the two Division Working Groups, Astrochemistry and Planetary Nebulae, have both proposed to change their status from Working Group to Commissions. Both Working Groups have organized successful IAU Symposia every ~5 years, and both have large numbers of active researchers in the subjects. The Division Organizing Committee fully support these initiatives to convert Astrochemistry and Planetary Nebulae Working Groups to Commissions.

3. Meetings and conferences

Many conferences devoted in whole or in part to the scientific interests of Division members were held in the reporting period. Below, we list some of the most significant meetings:

- *Assembly, Gas Content and Star Formation History of Galaxies*, 21 – 24 September 2009, Charlottesville, Virginia, USA
- *Ten Years of Science with Chandra*, 22 – 25 September 2009, Boston, Massachusetts, USA
- *Planetesimal Formation*, 28 – 30 September 2009, Cambridge, UK
- *Herschel Space Observatory: Discovering the Cold Universe*, 2 – 4 October 2009, Thessaloniki, Greece

- *Interstellar Matter and Star Formation - A Multi-Wavelength Perspective*, 8 – 10 October 2009, Hyderabad, India
- *Reionization to Exoplanets: Spitzer's Growing Legacy*, 26 – 28 October 2009, Pasadena, California, USA
- *2009 Fermi Symposium*, 2 – 5 November 2009, Washington, DC, USA
- *From Circumstellar Disks to Planetary Systems*, 3 – 6 November 2009, Garching, Germany
- *5th Korean Astrophysics Workshop on Shock Waves, Turbulence, and Particle Acceleration*, 18 – 21 November 2009, Pohang, Korea
- *Plasmas in the Laboratory and in the Universe: Interactions, Patterns, and Turbulence*, 1 – 4 December 2009, Como, Italy
- *Infrared Emission, Interstellar Medium and Star Formation*, 22 – 24 February 2010, Heidelberg, Germany
- *Multi-Phase Interstellar Medium and Dynamics of Star Formation*, 28 February – 2 March 2010, Nagoya, Japan
- *Starbursts Near and Far*, 12 March 2010, London, UK
- *From First Light to Newborn Stars: A Science Symposium Celebrating 50 years of our National Observatory*, 14 – 17 March 2010, Tucson, Arizona, USA
- *Galaxies and their Masks - KC Freeman 70th birthday fest*, 11 – 17 April 2010, Namibia
- *Herschel First Results Symposium (44th ESLAB Symposium 2010)*, 4 – 7 May 2010, Noordwijk, The Netherlands
- *Rotation Measure Analysis of Magnetic Fields in and around Radio Galaxies*, 10 – 14 May 2010, Riccione, Italy
- *Magnetic Fields on Scales from Kiloparsecs to Kilometres: Properties and Origin*, 17 – 21 May 2010, Krakow, Poland
- *Magnetic Fields: From Core Collapse to Young Stellar Objects*, 17 – 19 May 2010, London, Ontario, Canada
- *Science with ALMA Band 5 (163 - 211 GHz)*, 24 – 25 May 2010, Rome, Italy
- *IAU Symposium 270: Computational Star Formation*, 31 May – 4 June 2010, Barcelona, Spain
- *PAHs and the Universe: A Symposium to Celebrate the 25th Anniversary of the PAH Hypothesis*, 31 May – 4 June 2010, Toulouse, France
- *Ultraviolet Universe - 2010*, 31 May – 4 June 2010, St. Petersburg, Russia
- *The Dynamic ISM: a celebration of the Canadian Galactic Plane Survey*, 6 – 10 June 2010, Naramata, BC, Canada
- *Cosmic Magnetism - From Stellar to Intergalactic Scales*, 7 – 11 June 2010, Kiama, Australia
- *A Universe of Dwarf Galaxies*, 14 – 18 June 2010, Lyon, France
- *EPoS 2010 The Early Phase of Star Formation*, 14 – 18 June 2010, Ringberg Castle, Germany
- *Asymmetrical Planetary Nebulae V*, 20 – 25 June 2010, Lake District, UK
- *The Multi-Wavelength View of Hot, Massive Stars (39th Liége International Astrophysical Colloquium)*, 12 – 16 July 2010, Liége, Belgium
- *The Infrared/X-ray Connection in Galaxy Evolution*, 12 – 15 July 2010, London, UK
- *New Insights into the Physics of Supernova Remnants and Pulsar Wind Nebulae. 38th COSPAR Scientific Assembly 2010 Event E19*, 18 – 25 July 2010, Bremen, Germany

- *IAU Symposium 272 - Active OB Stars: Structure, Evolution, Mass Loss and Critical Limits*, 19 – 23 July 2010, Paris, France
- *Molecules in Galaxies*, 26 – 30 July 2010, Oxford, UK
- *Astronomy & Astrophysics in Antarctica*, 16–20 August 2010, Xi'an, China
- *Immersion Grating Infrared Spectrometer Science Workshop*, 26 – 27 August 2010, Seoul, Korea
- *25th Summer School and International Symposium on the Physics of Ionized Gases*, 30 August – 3 September 2010, Donji Milanovac, Serbia
- *IAU Symposium 274: Advances in Plasma Astrophysics*, 6 – 10 September 2010, Giardini-Naxos, Italy
- *Herschel and the Formation of Stars and Planetary Systems*, 6 – 9 September 2010, Gothenborg, Sweden
- *Great Barriers in High Mass Star Formation*, 13 – 17 September 2010, Townsville, North Queensland, Australia
- *Conditions and Impact of Star Formation: New Results with Herschel and Beyond* , 19 – 24 September 2010, Zermatt, Switzerland
- *Charge Exchange Meeting*, 29 September 29 – 1 October 2010, Madrid, Spain
- *WittFest: Origins & Evolution of Dust*, 10 – 12 October 2010, Toledo, Ohio, USA
- *Science with the Hubble Space Telescope - III*, 11 – 14 October 2010, Venice, Italy
- *Galaxy Evolution: Infrared to Millimeter Wavelength Perspective*, 25 – 29 October 2010, Guilin, China
- *Stormy Cosmos: The Evolving ISM from Spitzer to Herschel and Beyond*, 1 – 4 November 2010, Pasadena, California, USA
- *Spiral Structure in the Milky Way: Confronting Observations and Theory*, 7 – 10 November 2010, Copiapó, Chile
- *Kinetic Processes in Plasma: Instabilities, Turbulence and Transport*, 8 – 11 November 2010, Bochum, Germany
- *Massive Galaxies Over Cosmic Time 3: The Role of Gas and Dust*, 8 – 10 November 2010, Tucson, Arizona, USA
- *The Submillimeter Universe: The CCAT View*, 12 – 13 November 2010, Ithaca, New York, USA
- *Observing with ALMA: Early Science*, 29 November – 1 December 2010, Grenoble, France
- *Star Formation under Extreme Conditions: the Galactic Center*, 6 – 9 December 2010, Besancon, France
- *The Millimeter and Submillimeter Sky in the Planck Mission Era*, 10 – 14 January 2010, Paris, France
- *ALMA: Extending the Limits of Astrophysical Spectroscopy*, 15 – 17 January 2011, Victoria, BC, Canada
- *Herschel and the Characteristics of Dust in Galaxies*, 28 February – 3 March 2011, Leiden, Netherlands
- *2nd CARMA Symposium* , 28 February – 3 March 2011, Berkeley, California, USA
- *Star Formation across Space and Time: Frontier Science with the LBT and Other Large Facilities*, 31 March – 3 April 2011, Tucson, Arizona, USA
- *Assembling the Puzzle of the Milky Way*, 17 – 22 April 2011, Le Grand-Bornand, France
- *The 5th Korea-Mexico Joint Workshop*, 20 – 22 April 2011, Goheung, Korea
- *A Decade of Exploration with the Magellan Telescopes*, 25 – 28 April 2011, Pasadena, California, USA

- *IAU S280:The Molecular Universe*, 30 May – 3 June 2011, Toledo, Spain
- *Frontier Science Opportunities with the James Webb Space Telescope*, 6 – 8 June 2011, Baltimore, Maryland, USA
- *Star Formation Summer School*, 15 – 18 June 2011, Taipei, Taiwan
- *Cosmic Ray and their Interstellar Environment*, 26 June – 1 July 2011, Montpellier, France
- *Multiwavelength Views of the ISM in High-Redshift Galaxies*, 27 – 30 June 2011, Santiago, Chile
- *From Dust to Galaxies*, 27 – 30 June 2011, Paris, France
- *The X-ray Universe 2011*, 27 – 30 June 2011, Berlin, Germany
- *Recent Advances in Star Formation : Observations and Theory (Part of the Silver Jubilee celebration of the Vainu Bappu Telescope)*, 28 June – 1 July 2011, Bangalore, India
- *EWASS2011 - Special Session SPS2: Massive Stars Formation*, 4 July 2011, Saint-Petersburg, Russia
- *Sixth NAIC/NRAO School on Single Dish Radio Astronomy*, 10 – 16 July 2011, Green Bank, West Virginia, USA
- *Four Decades of Research on Massive Stars: A Scientific Meeting in the Honour of Anthony F.J. Moffat*, 12 – 14 July 2011, Lanaudière, Québec, Canada
- *Galaxy Formation*, 18 – 22 July 2011, Durham, UK
- *IAU Symposium No.283: Planetary Nebulae: An Eye to the Future* 25 – 29 July 2011, Tenerife, Spain
- *The 11th Asian Pacific Regional IAU Meeting*, 26 – 29 July 2011, Chiang Mai, Thailand
- *32nd. International Cosmic Ray Conference*, 11 – 18 August 2011, Beijing, China
- *Magnetic Fields in the Universe: from Laboratory and Stars to Primordial Structures III*, 21 – 27 August 2011, Tatra Mountains, Poland
- *Jan65: Magnetic Fields and the Cosmos*, 24 – 26 August 2011, Nijmegen, The Netherlands
- *IAU Symposium 284:The Spectral Energy Distribution of Galaxies (SED2011)*, 5 – 9 September 2011, Preston, UK
- *The Starburst-AGN Connection under the Multiwavelength Limelight* 14 – 16 September 2011, Villafranca del Castillo, Spain
- *FIR2011: Star Formation and Feedback in Galaxies as Revealed by Far Infrared and Submillimeter Wavelengths*, 14 – 16 September 2011, London, UK
- *6th IRAM 30m Summer School, Star formation - Near and Far*, 23 – 30 September 2011, Sierra Nevada, Spain
- *Through the Infrared Looking Glass: A Dusty View of Galaxy and AGN Evolution*, 2 – 5 October 2011, Pasadena, California, USA
- *Formation and Early Evolution of Very Low Mass Stars and Brown Dwarfs*, 11 – 14 October 2011, Garching, Germany
- *A Workshop on the Emerging, Multi-wavelength View of the Galactic Centre Environment*, 17 – 20 October 2011, Heidelberg, Germany
- *Midwest Astrochemistry Meeting 2011*, 21 – 22 October 2011, Urbana, Illinois, USA
- *Galaxy Mergers in an Evolving Universe*, 23 – 28 October 2011, Hualien, Taiwan
- *Science with Parkes at 50 years Young*, 31 October – 4 November 2011, Parkes Telescope, NSW, Australia
- *High Energy Astroparticle Physics 2011 - Gamma-Ray Universe: Fermi to CTA*, 13 – 15 November 2011, Tsukuba, Japan

- *6th Korean Astrophysics Workshop:Fundamental Processes of Astrophysical Turbulence*, 16 – 19 November 2011, Pohang, Korea
- *IAU Symposium 287: Cosmic Masers: From OH to H_0*, 29 January – 3 February 2012, Stellenbosch, South Africa
- *The Second AKARI Conference, Legacy of AKARI: A Panoramic View of the Dusty Universe*, 27 – 29 February 2012, Jeju island, Korea
- *Circumstellar Dynamics at High Resolution*, 27 February – 2 March 2012, Foz do Iguaçu, Brazil
- *IAU Symposium 279 - Death of Massive Stars: Supernovae & Gamma-Ray Bursts*, 12 – 16 March 2012, Nikko, Japan
- *From Atoms to Pebbles: Herschel's View of Star and Planet Formation*, 20 – 23 March 2012, Grenoble, France
- *Cosmic-ray Induced Phenomenology in Star-Formation Environments (2nd Workshop of the Sant Cugat Forum on Astrophysics)*, 16 – 19 April 2012, Sant Cugat, Barcelona, Spain
- *RTS 2012- Resolving The Sky - Radio Interferometry: Past, Present and Future*, 18 – 20 April 2012, Manchester, UK
- *First Stars IV - From Hayashi to the Future*, 21 – 25 May 2012, Kyoto, Japan
- *The Physics of Feedback Processes and their Role in Galaxy Evolution*, 10 June – 1 July 2012, Aspen, Colorado, USA
- *The Labyrinth of Star Formation*, 18 – 22 June 2012, Crete, Greece
- *Ultraviolet Astronomy: HST and Beyond*, 18 – 21 June 2012, Kauai, Hawaii, USA
- *7th International Conference on Numerical Modeling of Space Plasma Flows - ASTRONUM-2012*, 24 – 29 June 2012, Big Island, Hawaii, USA
- *Centenary Symposium 2012: Discovery of Cosmic Rays*, 26 – 28 June 2012, Denver, Colorado, USA
- *PoS 2012 The Early Phase of Star Formation - Assembling Pieces of the Missing Paradigm*, 1 – 6 July 2012, Tegernsee, Germany

The Division is supporting a number of Symposia and Special Sessions and is the coordinating Division for Special Sessions 12 and 16 at the IAU XXVIII General Assembly in Beijing, August 2012.

- *IAU Special Session 3: Galaxy Evolution Through Secular Processes*, 12 – 16 August 2012, Beijing, China
- *IAU Symposium 288: Astrophysics from Antarctica*, 20 – 24 August 2012, Beijing, China
- *IAU Symposium 292: Molecular Gas, Dust, and Star Formation in Galaxies*, 20 – 24 August 2012, Beijing, China
- *IAU Special Session 4: New era for studying interstellar and intergalactic magnetic fields*, 20 – 23 August 2012, Beijing, China
- *IAU Special Session 8: Calibration of star-formation rate measurements across the electromagnetic spectrum*, 27 – 30 August 2012, Beijing, China
- *IAU Special Session 12: Modern views of the interstellar medium* 27 – 31 August 2012, Beijing, China
- *IAU Special Session 16: Unexplained spectral phenomena in the interstellar medium*, 27 – 28 2012, Beijing, China

4. Proceedings and monographs

4.1. Conference proceedings

- *The Magellanic System: Stars, Gas, and Galaxies.* Proc. IAU Symposium No. 256, Eds. J. Th. van Loon & J. M. Oliveira (Cambridge: CUP), 2009
- *Cosmic Magnetic Fields: From Planets, to Stars and Galaxies.* Proc. IAU Symposium No. 259, Eds. K. G. Strassmeier, A. G. Kosovichev & J. E. Beckman (Cambridge: CUP), 2009
- *AKARI, a Light to Illuminate the Misty Universe.* Eds. T. Onaka, G. J. White, T. Nakagawa, & I. Yamamura (San Francisco: Astronomical Society of the Pacific Conference Series, vol 418), 2009
- *The Dynamic Interstellar Medium: A Celebration of the Canadian Galactic Plane Survey.* Eds. R. Kothes, T. L. Landecker, & A. G. Willis (San Francisco: Astronomical Society of the Pacific Conference Series, vol 438), 2010
- *Asymmetric Planetary Nebulae 5.* Eds. A. A. Zijlstra, F. Lykou, I. McDonald, & E. Lagadec (Jodrell Bank Centre for Astrophysics), 2011
- *Computational Star Formation.* Proc. IAU Symposium No. 270, Eds. J. Alves, B. Elmegreen, & V. Trimble (Cambridge: CUP), 2011
- *Active OB stars: structure, evolution, mass loss, and critical limits.* Proc. IAU Symposium No. 272, Eds. C. Neiner, G. Wade, G. Meynet, & G. Peters (Cambridge: CUP), 2011
- *The multi-wavelength view of hot, massive stars.* Eds. G. Rauw, M. De Becker, Y. Nazé, J.-M. Vreux, P. Williams (Liège: Société Royale des Sciences de Liège, Bulletin, vol. 80), 2011

4.2. Research monographs

- *Physics of the Interstellar and Intergalactic Medium.* Bruce T. Draine (Princeton: Princeton University Press), 2011

5. Review articles

Recent invited reviews on interstellar matter published in the Annual Reviews of Astronomy and Astrophysics or the Publications of the Astronomical Society of the Pacific have included:

- Kalberla, P. M. W., & Kerp, J. 2009, *ARA&A*, 47, 27, *The HI Distribution of the Milky Way*
- Zweibel, E. G., & Yamada, M. 2009, *ARA&A*, 47, 291, *Magnetic Reconnection in Astrophysical and Laboratory Plasmas*
- Herbst, E., & van Dishoeck, E. F. 2009, *ARA&A*, 47, 427, *Complex Organic Interstellar Molecules*
- Churchwell, E., et al. 2009, *PASP*, 121, 213, *The Spitzer/GLIMPSE Surveys: A New View of the Milky Way*
- De Marco, O. 2009, *PASP*, 121, 316, *The Origin and Shaping of Planetary Nebulae: Putting the Binary Hypothesis to the Test*
- Calzetti, D., & Kennicutt, R. C. 2009, *PASP*, 121, 937, *The New Frontier: Galactic-Scale Star Formation*
- Henning, T. 2010, *ARA&A*, 48, 21, *Cosmic Silicates*
- Adams, F. C. 2010, *ARA&A*, 48, 47, *The Birth Environment of the Solar System*
- Mann, I. 2010, *ARA&A*, 48, 173, *Interstellar Dust in the Solar System*

- Dullemond, C. P., & Monnier, J. D. 2010, *ARA&A*, 48, 205, *The Inner Regions of Protoplanetary Disks*
- Bastian, N., Covey, K. R., Meyer, M. R. 2010, *ARA&A*, 48, 339, *A Universal Stellar Initial Mass Function? A Critical Look at Variations*
- Fukui, Y., & Kawamura, A. 2010, *ARA&A*, 48, 547, *Molecular Clouds in Nearby Galaxies*
- Smith, I. W. M. 2011, *ARA&A*, 49, 29, *Laboratory Astrochemistry: Gas-Phase Processes*
- Williams, J. P. & Cieza, L. A. 2011, *ARA&A*, 49, 67, *Protoplanetary Disks and Their Evolution*
- Kotera, K., & Olinto, A. V. 2011, *ARA&A*, 49, 119, *The Astrophysics of Ultrahigh-Energy Cosmic Rays*
- Armitage, P. J. 2011, *ARA&A*, 49, 195, *Dynamics of Protoplanetary Disks*
- Frisch, P. C., Redfield, S., & Slavin, J. D. 2011, *ARA&A*, 49, 237, *The Interstellar Medium Surrounding the Sun*

You-Hua Chu
president of the Division

Transactions IAU, Volume XXVIIIA
Reports on Astronomy 2009–2012
Ian Corbett, ed.

© International Astronomical Union 2012
doi:10.1017/S1743921312002876

DIVISION VI / COMMISSION 34 / WORKING GROUP
ASTROCHEMISTRY

TRIENNIAL REPORT 2009-2012

1. Introduction

The study of molecules in space, known as astrochemistry or molecular astrophysics, is a rapidly growing field. Molecules exist in a wide range of environments in both gaseous and solid form, from our own solar system to the distant early universe. To astronomers, molecules are indispensable and unique probes of the physical conditions and dynamics of regions in which they are detected, especially the interstellar medium. In particular, the many stages of both low-mass and high-mass star formation are better understood today thanks to the analysis of molecular observations. Molecules can also yield a global picture of the past and present of sources. Moreover, molecules affect their environment by contributing to the heating and cooling processes that occur.

Molecular observations are currently being used to study the interstellar gas and dust in diffuse interstellar clouds, dense molecular clouds, protostellar objects, maser and star-forming regions, envelopes of evolved stars, protoplanetary disks, (exo-)planetary atmospheres and comets. In addition to our own Milky Way Galaxy, molecules and solid-state features are routinely detected in interstellar regions of external galaxies ranging from the nearby Magellanic Clouds to distant starbust galaxies at redshifts of more than 6. Indeed, the role of molecules in the study of the early universe, including the formation of the first stars, is a critical one. Molecules are also key probes of galactic assembly processes, e.g., mergers vs. smooth accretion, and molecular gas provides the fuel for star formation and black hole activity. The role of molecules in astronomy has grown to such an extent that it is no longer an exaggeration to refer to a sizable portion of the universe as 'The Molecular Universe'.

To chemists, the synthesis of molecules from simple atoms to complex species is of increasing interest. The distant goal of learning about the onset of pre-biotic processes is coming closer to home through studies of cometary samples and meteorites, as well as of circumstellar disks, which are the precursors of planets. The processes that produce these species are unusual and exciting since interstellar space provides a laboratory with conditions not readily simulated on Earth. Thus, astrochemistry attracts scientists from a number of different disciplines. Astrochemistry directly connects with even newer fields such as astrobiology and the study of exoplanets and their atmospheres. As a result, astrochemistry has many adherents throughout the world, and the potential for huge growth as new and more powerful telescopes search the sky for molecules. Large-scale international cooperation occurs in the construction and exploitation of expensive telescopes, such as space-based Herschel and JWST, air-borne SOFIA, and ground-based ALMA, for which the initial scientific cases were largely driven by astrochemists.

The IAU working group (WG) on Astrochemistry in Division VI (new Division H), Commission 34 (Interstellar Matter) was established about 30 years ago by Alexander Dalgarno and led by him for the first two IAU Symposia. From 1991–1999, the WG was chaired by David Williams, with Ewine van Dishoeck as vice-chair. From 1999-now, the WG is chaired by Ewine van Dishoeck with Eric Herbst as vice-chair.

The current WG members come from 13 countries and represent different disciplines (see http://www.strw.leidenuniv.nl/iau34/ for overview). The WG is responsible for fostering interest in the subject and ensuring the continuation of meetings on the subject under the auspices of the IAU. So far, six IAU-sponsored meetings have been held, roughly every 5-6 years: IAU Symposium 120, in Goa, India 1985 on 'Astrochemistry'; IAU Symposium 150 on 'Astrochemistry of Cosmic Phenomena' in Brazil 1991, attended by 120 scientists from 19 countries; IAU Symposium 178, 'Molecules in Astrophysics: Probes and Processes' in Leiden, The Netherlands 1996, with 231 participants from 27 countries; IAU Symposium 197 'Astrochemistry: From Molecular Clouds to Planetary Systems' in Sogwipo, Korea 1999, with 262 participants from 25 countries; IAU Symposium 231 'Astrochemistry: Recent Successes and Current Challenges' in Asilomar, USA in 2005 with 300 participants from 26 countries, and IAU Symposium 280 'The Molecular Universe' in Toledo in 2011 with 440 participants from 30 countries. Worldwide, the number of astrochemists is expected to be at least double this number. The activity and growth of the field is further illustrated by a number of other international meetings and workshops around the world, including numerous meetings having to do with planning for the Herschel and ALMA telescopes, as well as dedicated networks on astrochemistry on a regional and continental basis. The purpose of the IAU symposia has been to ensure that the many aspects of astrochemistry are reviewed, and that the entire community (both astronomers and chemists/physicists) is brought together.

2. Developments within the past triennium

During the 2009-2012 triennium, the WG concentrated on the proposal, planning and organization of IAU Symposium 280 on 'The Molecular Universe' in beautiful Toledo, Spain, which took place from 30 May - 3 June 2011 at the newly renovated Technological Campus of the University of Castilla-La Mancha. The IAU WG acted as the Scientific Organizing Committee of the symposium. The Local Organizing Committee, chaired by J. Cernicharo, organized both the scientific and structural aspects of the meeting very well, including a delightful banquet and preceding concert. The cultural mecca that is

Toledo added a sense of awe and excitement to the symposium. The large size of the meeting did not interfere with the proceedings in any way. A large number of questions were asked of speakers, who, given their relative youth and diversity, brought many different viewpoints to the proceedings. The three dedicated 2.5-hr. poster sessions were very well attended and enriched the experience of the participants. The sessions were enlivened by tapas and by a variety of beverages. Informal conversations at intermissions and during the poster sessions were many and spirited. The large number of younger scientists at the meeting was quite impressive, and confirmed that the field of Astrochemistry is entering a period of rapid growth led by new and exceedingly powerful telescopes.

The scientific program of the symposium consisted of 41 invited and review talks, 32 contributed talks, and 323 poster presentations. The SOC democratically proposed and elected the speakers. In the oral program were three sessions on new results from the Herschel Space Observatory labeled 'Herschel hot results', as well as a panel discussion entitled 'On to ALMA'. Three awards were given to the best posters in each of the three sessions from personal funds by Ewine van Dishoeck, as well as a prize for the best proposal in the ALMA session. During the third poster session, there were also computer demonstrations of databases. The abstracts for all contributions to the symposium can be found on the NASA Astrophysics Data System and on the conference website: http://www.cab.inta-csic.es/molecular_universe/show-abstracts.php, where actual poster presentations have been uploaded. A number of video interviews and highlights of the poster sessions can be found on the IAU Symposium 280 YouTube Channel (http://www.youtube.com/user/IAUsymposium280). Invited and review talks will appear in the symposium volume, edited by J. Cernicharo and R. Bachiller.

The symposium started with a session on star formation. This field has become broader since the last astrochemistry symposium, and observational talks concerning stages of both low-mass and high-mass star formation were given, as was a theoretical talk on a new class of models that combines hydrodynamics with chemical simulations. The first session of hot results from Herschel emphasized observations of water vapor, molecules in protostellar shocks, and a wide spectral survey toward Orion KL.

Astrochemistry certainly extends to planetary studies. This session started with a review talk on the chemistry of the solar system, including the origin of water on Earth, which was followed by talks on comets, meteorites, and the atmospheres of Titan and Saturn. The power of sample return missions to solar system bodies was emphasized.

The second day started with a session on evolved stars, in which supernova chemistry was also discussed. Talks on the molecular evolution from AGB stars to planetary nebulae, the role of time-dependent anionic chemistry (involving negatively-charge molecules) in IRC+10216, and the detection of fullerenes in assorted environments rounded out the session. Complex molecules were subsequently discussed in a variety of objects, along with current gas-grain simulations as well as possible future simulations involving the use of stochastic methods to improve the surface chemistry occurring in icy mantles.

Astrochemistry is based on the laboratory and theoretical study of basic atomic and molecular processes, and two sessions were held on this subject. The first concerned gas-phase processes, where a review talk was given on gas-phase reactions as a function of temperature, followed by a talk concerning the theory of low-temperature reactions, and one on experimental studies on the rates of reactions involving anions and how they relate to the observations of such species in various sources. Surface processes in the laboratory and in space were discussed in the second session (on day 3). Much progress has been made during the last decade, but there is still a great need for further laboratory studies before robust interstellar chemical simulations including surface processes can be constructed. The second day of the meeting ended with another Herschel hot topic session, highlighted by the report of an unambiguous detection of interstellar O_2.

The topic of protoplanetary disks on day three demonstrated vividly the phenomenal developments in observations at a variety of wavelengths ranging from the millimeter to the far-UV and an emphasis on interferometry. Modeling was also discussed, as was the chemical history of molecules from the hot core to the disk stage.

Although most of astrochemistry still revolves around galactic sources, the field of extragalactic astrochemistry will receive a big boost with the onset of ALMA observations. So, it was quite appropriate to have a session on extragalactic astrochemistry, which was held on day 4, including talks on the early universe chemistry and on extragalactic line surveys. It is impressive to now see spectra of extragalactic sources with similar complexity to those found in galactic star-forming regions three decades ago!

Next in line was the explosive topic of exoplanets and their atmospheres. Talks on observations, atmospheric models and their chemistry, as well as biomarkers of habitable worlds were included. The final session on day 4 concerned the tools of analysis and databases, including how to reduce the problem of unidentified lines in hot cores, on various analysis tools for spectral surveys, and on database uses.

The last day started with a session on diffuse clouds and photon-dominated regions (PDRs). The role of turbulence in diffuse clouds was discussed. Overview talks on both PDRs and XDRs as well as diffuse interstellar bands were given. A number of aspects of the PAH hypothesis were touched upon. Finally, the complex nature of the central molecular zone of our galaxy, as seen through H_3^+, was explored. Next came the third of the Herschel sessions, which included talks on observations of diffuse clouds in the spiral arms of the Milky Way, and carbon chemistry in translucent clouds. The detection of the reactive ions OH^+ and H_2O^+ in a variety of sources was an exceptionally interesting topic. The oral program was concluded with a thoughtful summary of the field, past, present, and future, by John Black.

3. Closing remarks

Based on the growth of the field, the sustained activity of the working group for over 25 years, and exciting future prospects, the WG proposes to elevate Astrochemistry to a proper commission under Division VI (new Division H). Among the activities that can be enhanced as a full-fledged commission are broader advertising of the role of astrochemistry and its multiple uses in astronomy; emphasizing the large amount of new spectra coming from new telescopes and the need for improved molecular data; expanding the training of astrochemistry among astronomers; and stimulating further interdisciplinary activities. By the time of the next astrochemical symposium, planned for 2017, much progress will have been made thanks to Herschel and ALMA, and the field will have grown both in size and, we trust, in understanding.

Ewine F. van Dishoeck
chair of Working Group
Eric Herbst
vice-chair of Working Group

Reference

Cernicharo, J. & Bachiller, R. 2012, *The Molecular Universe*, IAu Symposium 280, Toledo, Spain, 30 May - 3 June 2011, *CUP*, in press

Transactions IAU, Volume XXVIIIA
Reports on Astronomy 2009–2012   © International Astronomical Union 2012
Ian Corbett, ed.   doi:10.1017/S1743921312002888

DIVISION VI / COMMISSION 24 / WORKING GROUP
PLANETARY NEBULAE

CHAIR	Arturo Manchado
PAST CHAIR	Mike Barlow
BOARD	You-Hua Chu
	Romano Corradi
	Shuji Deguchi
	Orsola de Marco
	Adam Frank
	Amanda Karakas
	Karen Kwitter
	Xiawei Liu
	Alberto Lopez
	Roberto Mendez
	Quentin Parker
	Miriam Peña
	Letizia Stanghellini
	Albert Zijlstra

TRIENNIAL REPORT 2009-2011

1. Introduction

The aims of this Working Group are:

• To ensure that scientific symposia on planetary nebulae take place regularly, ideally every 5 years. These symposia would preferably be sponsored by the IAU;

• To organize and coordinate the Joint Discussions on the subject at the IAU General Assemblies. These discussions should address topics of interest not only to our Division VI but to other Divisions as well; and

• To maintain a Web page with general information about the WG, the activities related to planetary nebulae, and the future meetings and symposia.

2. Meetings and Conferences

• IAU Symposium 283, "Planetary Nebulae: an Eye to the Future" Puerto de La Cruz, Tenerife, Spain, 25-29 July 2011, http://www.iac.es/congreso/iaus283/

• "Asymmetric Planetary Nebulae V", Lake District, United Kingdom 20-25 June 2010, http://www.jodrellbank.manchester.ac.uk/meetings/APN5/

• "Towards understanding Asymmetric Planetary Nebulae: strategic research collaboration", Rochester NY 17-19 June 2009, http://www.pas.rochester.edu/ afrank/Pn_Meeting_P

• "Legacies of the Macquarie/AAO/Strasbourg Hα Planetary Nebula project", February 16-18th 2009, Sydney, Australia http://www.physics.mq.edu.au/astronomy/mash-workshop/

3. Proceedings and Monographs

3.1. *Conference Proceedings*

• IAU Symposium 283, "Planetary Nebulae: an Eye to the Future", 2012, (ed. A. Manchado, L. Stanghellini, D. & D. Schönberner, eds.) (in press)

• Asymmetric Planetary Nebulae V, ed. A. Zijlstra, F. Lykon, E. Lgadec and I. McDonald http://www.jodrellbank.manchester.ac.uk/meetings/APN5/proceedings.html

4. Developments within the past triennium

The Working Group on Planetary Nebulae (PNWG) planned and organized the IAU Symposium 283, "Planetary Nebulae: an Eye to the Future", http://www.iac.es/congreso/iaus283/ that took place on July 25-29 2011 at Puerto de la Cruz of Tenerife in the Canary Islands, Spain. One hundred and fifty seven participants from 26 countries from the five continents interacted and discussed the many different aspects and facets of the planetary nebulae field. This meeting followed the time-honored tradition of having an IAU symposium on planetary nebulae every 5 years. The symposium proceedings are edited by A. Manchado, L. Stanghellini and D. Schoenberner. The main topics of the Symposium were: new results from observations; the stellar evolutionary connection; aspects of the planetary nebula phase; the central stars; the population of Galactic, extragalactic, and intra-cluster planetary nebulae; and future endeavours in the field.

Some of the PNWG members have been organizing interim meetings in the field of planetary nebulae. Albert Zijlstra, Romano Corradi and Alberto López organized the "Asymmetrical Planetary Nebulae V" conference, in June 2010, held in Bowness-on-Windermere, UK. (http://www.astrophysics.manchester.ac.uk/apn5.html)

Adam Frank organized the workshop "Towards understanding Asymmetric Planetary Nebulae: strategic research collaboration", in June 2009, in Rocherter, NY, USA http://www.pas.rochester.edu/ afrank/Pn_Meeting_Page.htm

Quentin Parker, organized the workshop "Legacies of the Macquarie/AAO/Strasbourg Hα Planetary Nebula Project", in February 2009, in Sydney, Australia http://www.physics.mq.edu.au/astronomy/mash-workshop/

The SFWG webpage (http://www.iac.es/proyecto/PNgroup/wg/index.html) is maintained and updated with links to past and forthcoming meetings on PNe, and useful databases related with the PN field can also be found there. It has recently been updated with the "Kinematic Catalogue of Galactic Planetary Nebulae" consisting of high-resolution ($6 - 11.5$ km s^{-1}) spectra for about 600 planetary nebulae.

5. IAU reorganization

During the last IAU Symposium 283, the PNWG held a business meeting, where the members voted in great majority to request an upgrade to an IAU Commission. The reasons are as follows: Our WG has been very active over the last 44 years, having organized eight IAU Symposia. Research on PNe has undergone vigorous growth in recent years, the number of published papers has more than doubled in the last 5 years. There has been an impressive diversity in scientific topics and observational techniques in PN research. In addition, PNe consist of material ejected by intermediate- and low-mass stars. The formation of PNe is in many ways similar to the formation of ring nebulae around massive stars, such as Wolf-Rayet (WR) stars and luminous blue variables (LBVs). Both PNe and ring nebulae around WR stars contain hot, X-ray-emitting plasma originating from shocked fast stellar winds. In fact, researchers who study WR ring nebulae and LBV nebulae frequently study PNe as well, as the study of these objects leads to a better understanding of their formation and physical conditions.

These objects represent the stellar material returned into the ISM, yet they are still closely associated with their central stars. Researchers of circumstellar nebulae have very little overlap with Commission 34 members who study the ISM in general.

Therefore, the PNWG proposed to upgrade the PNWG to an IAU Commission of "Circumstellar Nebulae" to include both PNe and ejecta nebulae around massive stars within the current Division VI or future Division H of the IAU.

Arturo Manchado
Chair of the Planetary Nebulae Working Group

Transactions IAU, Volume XXVIIIA
Reports on Astronomy 2009–2012   © International Astronomical Union 2012
Ian F. Corbett, ed.   doi:10.1017/S174392131200289X

DIVISION VII   THE GALACTIC SYSTEM

SYSTÈME GALACTIQUE

Division VII provides a forum for astronomers studying the Milky Way as a galactic system, and its constituents.

PRESIDENT   Despina Hatzidimitriou
VICE-PRESIDENT   Rosemary Wyse
PAST PRESIDENT   Ortwin Gerhard
BOARD   Giovanni Carraro
                   Bruce Gordon Elmegreen
                   Birgitta Nordström

DIVISION VII COMMISSIONS

Commission 33   Structure & Dynamics of the Galactic System
Commission 37   Star Clusters & Associations

INTER-DIVISION WORKING GROUPS

TRIENNIAL REPORT 2009-2012

1. Introduction

Division VII provides a forum for astronomers studying the Milky Way as a galactic system, as well as its constituents. It acts as an umbrella for two commissions, Commission 33 and Commission 37.

Commission 33 focuses on the structure and dynamics of the Galactic System, providing a paradigm for the processes involved in the formation and evolution of the stellar and gaseous components of spiral galaxies in general.

Commission 37 studies star clusters and associations, as test particles for deciphering the formation and evolution of the galactic system as a whole, but also as important astrophysical objects in their own right.

2. Developments within the past triennium

2.1. *Observations - Surveys*

There has been a great revival of Milky Way research, in recent years, partly in preparation for the forthcoming *GAIA* mission. Several large surveys and experiments, either photometric or spectroscopic, in several wavelength regimes, have been recently completed, are in progress, or are planned for the near future: the Sloan Extension for Galactic Understanding and Exploration 2 (SEGUE-2), the RAdial Velocity Experiment (RAVE), the Apache Point Observatory Galactic Evolution Experiment (APOGEE), the HERMES Galactic archaeology survey, the ESO-GAIA spectroscopic survey, the LAMOST Experiment for Galactic Understanding and Evolution (LEGUE), the SkyMapper survey, the Young stellar object variability (YSOVAR) survey with Spitzer, the VISTA Variables in The Via Lactea (VVV) public survey, the UKIDSS near infrared galactic plane and galactic clusters survey, the JCMT Legacy Survey (at 450 ìm and 850 ìm), the APEX telescope ATLASGAL 870 micron survey, the Herschel Hi-GAL survey (between 70 and 500 microns), the Caltech sub-millimeter Observatory BGPS Bolocam 1.1 mm survey, the Spitzer GLIMPSE 360 mid-IR survey, the UKIRT Widefield Infrared Survey for molecular hydrogen, the International Galactic Plane Survey which combines many radio telescope surveys from around the world and the Spitzer and Herschel guaranteed time projects aimed at mapping nearly all of the molecular clouds within 500 pc of the Sun. The results of these surveys combined with the GAIA mission results and the also awaited Large Synoptic Survey Telescope (LSST) contribution are expected to revolutionize our perception of the MW galaxy and our understanding of how galaxies form and evolve, in general.

2.2. *Theory - Simulations*

Simulations, both pure N-body and hydrodynamic, are increasing in resolution and dynamic range and are now capable of following the formation of a Milky Way analogue galaxy within a proper cosmological framework and with fairly realistic modeling of the complex physics of gas dynamics, star formation and chemical evolution. They can now make robust, detailed predictions to be compared with the comprehensive large observational data-sets from the surveys mentioned above. At the same time, simulations have shown that, contrary to assumptions often made, the Milky Way's morphology is strongly affected by even low-mass minor mergers predicted to be common throughout the Universe.

Simulations are also growing in importance in studies of cluster formation and evolution. Complete N-body simulations of globular clusters are now possible, and simulations of star formation in the cluster environment reproduce the stellar initial mass function, mass segregation, hierarchical structure, the formation efficiency and overall time scale.

2.3. *Meetings and publications*

The triennial reports by the organizing committees of Commissions 33 and 37 that are presented in this volume, describe the main developments and activities related to Division VII. Briefly, during the reporting period, there have been over 30 international conferences and workshops either entirely dedicated to or partly related to the galactic system and/or to star clusters and associations. More than 2000 scientific papers have been published in refereed journals, reporting on new observational or theoretical insights into the Galactic System and its constituents (including star clusters and associations).

Inspection of the subjects of the 10 most cited articles (with "Milky Way" in the title) per reporting year (i.e. 30 in all), reveals the focus of the current efforts of the scientific community: Among these 30 papers, almost half are related to the MW dwarf satellites and their role in its evolution. Results from large surveys account for the next largest most cited group of papers, while the rest are related to different MW components (e.g. the nuclear cluster, gaseous clouds, sub-halos, thick disk, spiral arms). A similar analysis of the publications related to star clusters, show that more than a third of the 30 most cited papers (10 per reporting year) are related to the presence of multiple stellar generations in star clusters, while another third study different aspects of metal enrichment in clusters.

2.4. *Business Matters*

The Working Group of Division VII, the "Galactic Center", which was created during the IAU XXVI General Assembly in Prague (2006) has been discontinued during the

Business Meeting of Division VII at the IAU XXVII General Assembly in Rio Janeiro (2009).

2.5. *Closing Remarks*

The remarkable international effort on large multiwavelength observational surveys both ground based and from space, along with the great advances in theory and simulations, promise that the next decade will see a revolution in our undestanding of the formation and evolution of our Galaxy.

Information on Division VII can be found on the Division web page:
$http : //www.iau.org/science/scientific\_bodies/divisions/VII/$.

Transactions IAU, Volume XXVIIIA
Reports on Astronomy 2009–2012   © International Astronomical Union 2012
Ian Corbett, ed.   doi:10.1017/S1743921312002906

COMMISSION 33

STRUCTURE AND DYNAMICS
OF THE GALACTIC SYSTEM
STRUCTURE ET DYNAMIQUE DU
SYSTÈME GALACTIQUE

PRESIDENT	Rosemary Wyse
VICE-PRESIDENT	Birgitta Nordström
PAST PRESIDENT	Ortwin Gerhard
ORGANIZING COMMITTEE	Joss Bland-Hawthorn, Sofia Feltzing, Burkhard Fuchs, Dante Minniti

TRIENNIAL REPORT 2009-2012

1. Introduction

Research on the structure and dynamics of the Galactic System covers a large dynamic range of spatial scales and timescales and investigates the evolution of gas, stars and dark matter. Much recent activity, not just in Europe, has focused on preparing for the data from the upcoming ESA astrometric mission Gaia, scheduled for launch in 2013. Several ongoing (plus planned) wide-area ground-based surveys, both spectroscopic and photometric, are providing the necessary very large datasets for robust determination of the joint position-chemical-kinematic distribution functions of Galactic stellar populations. The time domain adds another dimension, the goal of ongoing and future massive photometric surveys (e.g. LSST). The dynamical evolution of disks has been the focus of much recent theoretical research, stimulated by the possibility of radial migration of stars and gas under transient dynamical perturbations.

Recent discoveries of several satellite galaxies within the distance of the LMC perhaps argues for a broadening of the definition of the Galactic System. The increased capabilities of ever-more powerful computers has made feasible the detailed simulation of individual galaxies within a general cosmological framework, providing predictions for typical galaxies like the Milky Way. Very wide-ranging research is of relevance for the Galactic System. The overlap with the research undertaken by members of Division VIII is increasing.

In what follows, we list only the meetings of most direct relevance to the Galactic System (see also http://www2.cadc-ccda.hia-iha.nrc-cnrc.gc.ca/meetings/) and give a brief summary of publication statistics.

2. Past Meetings

- The Galactic Bulge, August 22 - September 11, 2011, Aspen, USA
- Assembling the Puzzle of the Milky Way, April 17 - 22, 2011, Le Grand-Bornand, France

- RR Lyrae Stars, Metal-Poor Stars and the Galaxy, January 23 - 25, 2011, Carnegie Observatories, Pasadena, USA
- Star Formation under Extreme Conditions: the Galactic Center Dec 6 - 9, 2010, Observatoire de Besançon, France
- Chemistry, Dynamics and Structure of the Milky Way : Summer School and Workshop on Galactic Studies with the LAMOST Surveys, July 5 - 23, 2010, Beijing, China
- IAU Symposium 271: Astrophysical Dynamics: From Stars to Galaxies, June 21 - 25, 2010, Nice, France
- Nordic-Baltic Astronomy Summer School on Star Formation in the Milky Way and Nearby Galaxies, June 8 - 18, 2009, Tuorla Observatory, Turku, Finland
- Gaia: At the Frontiers of Astrometry, June 7 - 11, 2010, Paris, France
- The Dynamic ISM: a celebration of the Canadian Galactic Plane Survey, June 6 - 10, 2010, Naramata, Canada
- The Sixth Harvard-Smithsonian Conference on Theoretical Astrophysics, Sponsored by Raymond and Beverly Sackler: Dynamics from the Galactic Center to the Milky Way Halo, May 10 - 13, 2010, Cambridge, USA
- Stellar Populations in the Cosmological Context, May 3 - 6, 2010, Baltimore, USA
- Galactic Center Workshop 2009 : The Galactic Center: A Window to the Nuclear Environment of Disk Galaxies Oct 19 - 23, 2009, Shangai, China
- B.V. Kukarkin Centenary Conference: Variable Stars, the Galactic Halo and Galaxy Formation, Oct 12 - 16, 2009, Zvenigorod (Moscow Region), Russia
- The Milky Way and the Local Group - Now and in the Gaia Era, Aug 31 - Sep 4, 2009, Heidelberg, Germany
- The Galactic Plane, in Depth and Across the Spectrum (IAU Special Session 8), Aug 11 - 14, 2009, Rio de Janiero, Brazil

3. Upcoming Meetings

- A Workshop on the Emerging, Multi-wavelength View of the Galactic Centre Environment, Oct 17 - 20, 2011, Heidelberg Convention Center, Heidelberg, Germany
- The 3rd Subaru International Conference: The First NAOJ Symposium, Galactic Archaeology, Nov 1 - 4, 2011, Shuzenji, Japan
- A Window to the Formation of the Milky Way, May 20 - June 9, 2012, Aspen, USA
- International Astronomical Union General Assembly, August 20 - 31, 2012, Beijing, China

4. Publications

As noted above, the research into the Milky Way overlaps that of other commissions. Use of the key words 'Milky Way', in the abstract field of the SAO/NASA Astrophysical Data Service interface, results in 1,087 papers published in refereed journals in the period from January 2009 to September 2011.

Reviews related to the Galactic System that appeared in the SAO/NASA Astrophysical Data Service from 2009 to September 2011 include:

- Soderblom, D.R., 'The Ages of Stars', 2010, ARA&A, 48, 581
- Bastian, N., Covey, K.R., & Meyer, M.R., 'A Universal Stellar Initial Mass Function? A Critical Look at Variations', 2010, ARA&A, 48, 339
- van der Kruit, P. & Freeman, K.C., 'Galaxy Disks', 2011, ARA&A, 49, 301

• Several reviews in the proceedings of 'The Galaxy Disk in Cosmological Context', IAU Symposium, Volume 254. Edited by J. Andersen, J. Bland-Hawthorn and B. Nordström

Rosemary Wyse & Birgitta Nordström
President and Vice President of the Commission

Transactions IAU, Volume XXVIIIA
Reports on Astronomy 2009–2012
Ian F. Corbett, ed.

© International Astronomical Union 2012
doi:10.1017/S1743921312002918

COMMISSION 37

STAR CLUSTERS AND ASSOCIATIONS
AMAS STELLAIRES ET ASSOCIATIONS

PRESIDENT
VICE-PRESIDENT
PAST PRESIDENT
ORGANIZING COMMITTEE

Bruce Elmegreen
Giovanni Carraro
Despina Hatzidimitriou
Gary Da Costa, Richard de Grijs
LiCai Deng, Charles Lada
Young-Wook Lee, Dante Minniti,
Ata Sarajedini, Monica Tosi

1. Introduction

Research on star clusters and associations includes the observation and theory of stellar groupings as they form and evolve, cluster disruption, stellar interactions inside clusters, and star formation in dense environments. In what follows, we list past, present and future meetings (http://www2.cadc-ccda.hia-iha.nrc-cnrc.gc.ca/meetings/), publications statistics and important surveys, reviews, and databases about clusters.

2. Past Meetings

• 6th IRAM 30m Summer School, Star formation - Near and Far, September 23 - 30, 2011, Pradollano, Sierra Nevada, Spain
• From Star Clusters to Galaxy Formation - The Virtual Universe, September 20 - 23, 2011, University of Heidelberg, Germany
• Recent Advances in Star Formation : Observations and Theory (Part of the Silver Jubilee celebration of the Vainu Bappu Telescope), June 28 - July 1, 2011, Indian Institute of Astrophysics, Bangalore, India
• Star Formation Summer School, June 15 - 18, 2011, Academia Sinica Institute of Astronomy and Astrophysics, Taiwan
• Stellar Clusters and Associations, May 23 - 27, 2011 Granada, Spain
• Dynamics of Low-Mass Stellar Systems : From Star Clusters to Dwarf Galaxies, April 4 - 8, 2011, Santiago, Chile
• Star Formation under Extreme Conditions: the Galactic Center, December 6 - 9, 2010, Observatoire de Besançon, France
• 4th East Asian Numerical Astrophysics Meeting, November 2 - 5, 2010, Academia Sinica Institute of Astronomy and Astrophysics, Taiwan
• MODEST-10: Encounters and interactions in dense stellar systems - modeling, computing, and observations, August 30 - September 3, 2010, National Astronomical Observatories - Chinese Academy of Sciences (NAOC) and The Kavli Institute for Astronomy and Astrophysics at Peking University (KIAA), China
• IAU Symposium 271: Astrophysical Dynamics: From Stars to Galaxies, June 21 - 25, 2010, Nice, France
• UP: Have Observations Revealed a Variable Upper End of the Initial Mass Function?, June 21 - 25 2010, Sedona, Arizona, USA

- IAU Symposium 270: Computational Star Formation, May 31 - June 4, 2010, Barcelona, Spain
- Infrared Emission, Interstellar Medium and Star Formation, February 22 - 24, 2010, MPI for Astronomy, Heidelberg, Germany
- International Workshop Double and Multiple Stars: Dynamics, Physics, and Instrumentation, December 10 - 11, 2009, Astronomical Observatory Ramon Maria Aller, Santiago de Compostela, Spain
- IAU Symposium 266: Star Clusters - Basic Galactic Building Blocks Throughout Time And Space, August 10 - 14, 2009, IAU General Assembly, Rio de Janeiro, Brazil
- Nordic-Baltic Astronomy Summer School on Star Formation in the Milky Way and Nearby Galaxies, June 8 - 18, 2009, Tuorla Observatory, Turku, Finland
- Constellation EU RTN School on X-rays from Star Forming Regions, May 18 - 22, 2009, INAF-Oss. Astronomico di Palermo, Palermo, Italy
- TIARA Winter School on Star Formation, February 5 - 10, 2009, National Tsing Hua University, Hsinchu, Taiwan
- Formation and Evolution of Globular Clusters, January 12 - 16, 2009, Kavli Institute for Theoretical Physics, UC Santa Barbara, USA

3. Upcoming Meetings

- Protostars and Planets VI, 15 July 15 - 20, 2013, Heidelberg Convention Center, Heidelberg, Germany
- International Astronomical Union General Assembly, August 20 - 31, 2012, Beijing, China
- EPoS 2012 The Early Phase of Star Formation - Assembling Pieces of the Missing Paradigm, July 1 - 6, 2012, Max-Planck-Society Conference Center Ringberg Castle, Tegernsee, Germany
- Stellar Populations 55 years after the Vatican Conference, at the European Week of Astronomy and Space Science, July 2 - 6, 2012, Rome, Italy
- The Labyrinth of Star Formation, June 18 - 22, 2012, Crete, Greece
- From Atoms to Pebbles: Herschel's View of Star and Planet Formation, March 20 - 23, 2012, Grenoble, France

4. Publications

The topic of star clusters and associations continues to be one of the most widely followed in all of astronomy. It spans the range of interest from stellar properties, to stellar clusters, to star formation and evolution, the IMF, with considerable overlap in other commissions.

Publications in Refereed Journals (AN, AJ, A&A, ApJ, ApJS, BASI, JRASC, MN-RAS, PASP, PASA, PASJ, ...) in the period from January 2009 to September 2011 tally as follows (from SAO/NASA Astrophysical Data Service):

- Globular Clusters: ~ 700 papers
- Open Clusters: ~ 600 papers
- Stellar Associations: ~ 150 papers

Some of the issues addressed in these publications are:

- the formation and dynamical evolution of star clusters
- stellar evolution and ages
- star clusters as tracers of stellar populations

- studies of specific types of objects within clusters
- nuclear clusters
- extragalactic cluster systems

The authors utilize observations covering an increasing portion of the electromagnetic spectrum, ranging from X-rays to the far-infrared, as well as advanced N-body simulations.

A newsletter with cluster results, SCYON, is edited by H. Baumgardt and E. Paunzen: http://www.univie.ac.at/scyon/

Reviews related to star clusters that appeared in the SAO/NASA Astrophysical Data Service from 2009 to September 2011 are:

- Soderblom, D.R., The Ages of Stars, 2010, ARA&A, 48, 581
- Bastian, N., Covey, K.R., & Meyer, M.R. A Universal Stellar Initial Mass Function? A Critical Look at Variations, 2010, ARA&A, 48, 339
- de Grijs, R. (ed.), Star clusters as tracers of galactic star-formation histories, 2010, Phil. Trans. R. Soc. A 368
- Heber, U., Hot Subdwarf Stars, 2009, ARA&A, 47, 211
- Portegies Zwart, S.F., McMillan, S.L.W., & Gieles, M., Young Massive Star Clusters, 2010, ARA&A, 48, 431

Additional reviews regarding a variety of aspects of star cluster research have also appeared in the proceedings of the conferences and meetings mentioned earlier.

5. Databases

- Data on Open Clusters in the Milky Way and the Magellanic Clouds can be found in the WEBDA site (http://www.univie.ac.at/webda/), which was originally developed by Jean-Claude Mermilliod from the Laboratory of Astrophysics of the EPFL (Switzerland) and is now maintained and updated by Ernst Paunzen and Christian Stütz from the Institute of Astronomy of the University of Vienna (Austria).
- Data on Galactic Globular Clusters can be found in the "Catalog of Milky Way Globular Cluster Parameters" by W.E. Harris (http://www.physics.mcmaster.ca/Globular), as well as in "The Galactic Globular Clusters Database" at Astronomical Observatory of Rome INAF-OAR: (http://gclusters.altervista.org/).
- A Catalogue of Variable Stars in Globular Clusters is maintained by Christine Clement at http://www.astro.utoronto.ca/~cclement/read.html.
- A Catalog of "Open Clusters and Galactic Structure," by W.S. Dias, *et al.* (2002 A&A 389 871), which also contains references to other catalogs, is here: http://www.astro.iag.usp.br/~wilton/.
- A Hubble Space Telescope ACS survey of 66 Galactic Globular Clusters by Sarajedini *et al.* (2007, AJ, 133, 1658) is here: http://www.astro.ufl.edu/~ata/public_hstgc/.

Bruce G. Elmegreen and Giovanni Carraro
President and Vice President of the Commission

Transactions IAU, Volume XXVIIIA
Reports on Astronomy 2009–2012
Ian F. Corbett, ed.

© International Astronomical Union 2012
doi:10.1017/S174392131200292X

DIVISION VIII GALAXIES AND THE UNIVERSE

LES GALAXIES ET l'UNIVERS

Division VIII brings together astronomers studying a wide range of problems related to galaxies and cosmology. The objects studied include individual galaxies, groups and clusters of galaxies, large scale structure, cosmic background radiation and the universe itself. The approaches and techniques used are also diverse, based around observational data (including large, multi-wavelength data sets), theoretical studies and computer simulations and modelling.

PRESIDENT   Elaine M. Sadler
VICE-PRESIDENT   Françoise Combes
PAST PRESIDENT   Sadanori Okamura
BOARD   Roger L. Davies, John S. Gallagher III,
                   Thanu Padmanabhan, Brian P. Schmidt

DIVISION VIII COMMISSIONS

Commission 28   Galaxies
Commission 47   Cosmology

DIVISION VIII WORKING GROUPS

Division VIII WG   Supernovae

TRIENNIAL REPORT 2009-2012

1. Introduction

The fields of extragalactic research and cosmology have continued to progress rapidly over the past three years, as detailed in the reports of the Commission Presidents, and we are pleased to acknowledge the award of the 2011 Nobel Prize in Physics to Saul Perlmutter, Brian P. Schmidt and Adam G. Riess for "the discovery of the accelerating expansion of the Universe through observations of distant supernovae". The Gruber Cosmology Prize was awarded in 2009 to Wendy L. Freedman, Robert C. Kennicutt and Jeremy Mould for their leadership of the Hubble Space Telescope Key Project on the Extragalactic Distance Scale, in 2010 to Charles Steidel for the identification and study of galaxies in the very distant universe, and in 2011 to Marc Davis, George Efstathiou, Carlos Frenk and Simon D.M. White for pioneering the use of numerical simulations as a tool to model and interpret the large-scale distribution of galaxies and dark matter.

2. Observations

On the observational side, progress over the past three years has continued to be driven by a range of multi-wavelength surveys of the local and distant Universe. Some highlights from 2009-12 include the release of the first data from the *Fermi* gamma-ray space telescope (launched in 2008), the successful launch of the NASA WISE (mid-IR) and ESA Herschel (far-IR/sub-millimetre) and Planck (CMB) satellites, and the 2009 refurbishment of the Hubble Space Telescope. Data also continue to flow in from a range of large ground-based spectroscopic and imaging surveys at optical, infrared, radio and millimetre wavelengths. The need to process, store and provide rapid access to an ever-growing 'data tsunami' of information will be an important challenge for our discipline in the coming years.

3. Theory and modelling

Numerical models and simulations play an increasingly important role in the analysis and interpretation of extragalactic data, both in cosmology (e.g. through models of the growth of cosmic structure over time) and in studies of galaxy evolution (e.g. via models of AGN feedback processes and their effect on the star-formation history of galaxies). The recent rapid growth in the availability and power of high-performance computers, together with the development of increasingly sophisticated algorithms, means that the resolution and detail of hydrodynamical and semi-analytic simulations are continually improving, though the innate complexity of the baryonic physics and feedback processes involved in real galaxies remains a challenge.

4. Meetings

Supporting IAU meetings is a key activity for the Division, and in 2009-12 Division VIII members have been involved in the following IAU Symposia, as well as a wide range of other meetings worldwide (see the CADC International Astronomy Meetings List at http://www1.cadc-ccda.hia-iha.nrc-cnrc.gc.ca/meetings/ for a comprehensive list).

- IAU Symp. 271, *Astrophysical dynamics – from stars to galaxies*, Nice, France, Jun 2010
- IAU Symp. 274, *Advances in plasma astrophysics*, Catania, Italy, Sep 2010
- IAU Symp. 275, *Jets at all scales*, Buenos Aires, Argentina, Sep 2010
- IAU Symp. 277, *Tracing the ancestry of galaxies (on the land of our ancestors)*, Ouagadougou, Burkina Faso, Dec 2010
- IAU Symp. 279, *Death of Massive Stars: Supernovae and Gamma-Ray Bursts*, Nikko, Japan, Mar 2012
- IAU Symp. 280, *The Molecular Universe*, Toledo, Spain, Jun 2011
- IAU Symp. 284, *The spectral energy distribution of galaxies* (SED2011), Preston, UK, Sep 2011
- IAU Symp. 289, *Advancing the physics of cosmic distances*, Beijing, China, Aug 2012
- IAU Symp. 292, *Molecular Gas, Dust, and Star Formation in Galaxies*, Beijing, China, Aug 2012
- IAU Symp. 295, *The intriguing life of massive galaxies*, Beijing, China, Aug 2012

5. Closing Remarks

Division VIII (with 1750 members) remains the largest IAU Division, and its members continue to work on a wide range of research topics with the aim of advancing our understanding of the properties of galaxies, the formation and evolution of galaxies and large-scale structure and the physics and nature of the Universe as a whole.

Elaine M. Sadler
President of the Division

Transactions IAU, Volume XXVIIIA
Reports on Astronomy 2009–2012
Ian Corbett, ed.

© International Astronomical Union 2012
doi:10.1017/S1743921312002931

COMMISSION 28   GALAXIES

GALAXIES

PRESIDENT	Roger L. Davies
VICE-PRESIDENT	John S. Gallagher III
PAST PRESIDENT	Françoise Combes
ORGANIZING COMMITTEE	Stephane J. Courteau, Avishai Dekel, Marijn Franx, Chanda J. Jog, Shardha Jogee, Naomasa Nakai, Monica Rubio, Linda J. Tacconi, Elena Terlevich

TRIENNIAL REPORT 2009-2012

1. Highlights in extragalactic research 2009-12

The membership of Commission 28 is so large, and the spread of research interests so broad, that the research highlights presented here of necessity represent just a small subset of the work carried out over the past three years. Progress in the area of galaxy evolution continues to be particularly rapid, driven by both the availability of new multi-wavelength survey data and the development of increasingly sophisticated simulations to model the complex behaviour of stars and gas (e.g. Agertz *et al.* 2011; Guo *et al.* 2011; Keres *et al.* 2009; Schaye *et al.* 2010).

1.1. Galaxy Evolution out to z∼ 1

The physical nature of early type galaxies (ETG) has been further elucidated by the ATLAS^{3D} survey (Emsellem *et al.* 2011; Cappellari *et al.* 2011), which conclusively showed that in the local Universe the great majority (86%) of ETGs with $M_K > -21.5$ have disks or disk-like kinematics. Low angular momentum systems that might be thought of as *classical* elliptical galaxies are rare, and found predominantly in the highest density region of the survey, the Virgo cluster core, where they comprise 20% of the galaxy population.

The galaxy merger rate over time is one of the fundamental tracers of galaxy evolution. Studies based on large *HST* surveys (e.g., GEMS, GOODS, COSMOS) report a wide range of values for the observed major merger fraction and the inferred merger rate among massive galaxies over the last 7 Gyr. Among massive ($M_\star > 2.5 \times 10^{10}\ M_\odot$) galaxies, the major merger rates typically range from 2% to 10% per Gyr, and the volume-averaged major merger rate from a few $\times 10^{-3}$ to $\times 10^{-4}$ galaxies Gyr^{-1} Mpc^{-3} (e.g., Lotz *et al.* 2008; Jogee *et al.* 2009; Conselice *et al.* 2009; Bundy *et al.* 2009; de Ravel *et al.* 2009; Robaina *et al.* 2010; Lotz *et al.* 2011). The dispersion in published values can be significantly reduced by properly calibrating the visibility timescale using hydrodynamic merger simulations, and taking into account differing parent galaxy selection (Lotz *et al.* 2011). The implied minor merger rate over the last seven Gyr is at least three times the major merger rate (Jogee *et al.* 2009; Lotz *et al.* 2011).

The star formation activity of merging and non-interacting galaxies is of great astrophysical interest, since the cosmic star-formation rate (SFR) density appears to have declined by up to a factor of ten since $z \sim 1$. Recent observational work has shown that out to $z \sim 1$, the average SFR of visibly-merging massive galaxies is only modestly enhanced (by a factor of 0.5–2.0) compared to non-interacting galaxies. This result is in agreement with numerical simulations by Di Matteo et al. (2007) and Cox et al. (2008), who report a similar modest enhancement in the SFR of major mergers spanning a range of mass ratio, progenitor gas fraction and Hubble types, and orbital geometry. Furthermore, it appears that visibly-merging systems only account for a small fraction (well below 30%) of the total cosmic SFR density over the last seven Gyr (Jogee et al. 2009; Robaina et al. 2010; Lotz et al. 2008; Bell et al. 2005), implying that the decline in cosmic SFR density at $z < 1$ is predominantly shaped by changes in the star-formation rate of non-interacting galaxies.

New work using deep spectra taken with DEIMOS on Keck II, together with improved modelling techniques, shows that the stellar mass Tully-Fisher (TF) relation out to $z \sim 1.3$ can be recovered with a scatter, which is 2-3 times smaller than in earlier work and in fact, comparable to the scatter at $z \sim 0$ (Miller et al. 2011). This suggests that over the past 8 billion years there is a tight relationship between the mass in stars and the total mass (derived dynamically), and that the baryonic TF relation is firmly established at $z = 1$.

2. Galaxy Evolution at z = 1.5 to 4

The global stellar mass density reaches $\sim 50\%$ of its present-day value by $z \sim 1$, suggesting that galaxies grew much of their stellar mass at redshift $z > 1$. We still do not know, however, whether the star-formation history of galaxies at $z > 1$ is primarily driven by stellar mass, mergers, galaxy environment, or other factors.

Multiple lines of evidence suggest that massive galaxies at $z = 2 - 3$ host significant disk components, with strong implications for the processes by which galaxies grow at high redshift. The SINS survey (Förster Schreiber et al. 2009) measured ionized gas kinematics of $z \sim 2$ star-forming galaxies and found examples of clumpy, turbulent, and geometrically thick systems having high velocity dispersions ($\sigma \sim 30 - 120$ km/s). About one-third of such systems show rotating disk kinematics.

Using HST near-infrared imaging from the GOODS-NICMOS survey, Weinzirl et al. (2011) analyzed the rest-frame optical structure of one of the largest (166 with $M_\star \geqslant 5 \times 10^{10}$ $M_\odot$) samples of massive galaxies at $z = 2 - 3$. They found that 40% of these massive galaxies were ultra-compact and quiescent, compared to less than 1% at $z \sim 0$. Furthermore, the majority ($\sim 65\%$) of these massive galaxies have disky morphologies, with many of these systems being extended and associated with high star formation rates (several tens to hundreds $M_\odot$ yr^{-1}). The presence of disky massive galaxies is also reported from other, smaller samples (van Dokkum et al. 2011; Forster Schreiber et al. 2011; van der Wel et al. 2011). Weinzirl et al. (2011) suggest that the large fraction of disky systems in these high mass galaxies implies that *cold mode accretion* (e.g; Dekel & Birnboim 2006; Dekel et al. 2009a,b; Kereš et al. 2005; Kereš et al. 2009; Brooks et al. 2009; Ceverino et al. 2010) must play an important role in galaxy growth at $z > 2$, in addition to gas-rich major mergers.

Direct unambiguous observational evidence for cold mode accretion is still lacking, but multiple possibilities, both in support of and against this scenario, are under consideration (e.g., Dijkstra & Loeb 2009; Steidel et al. 2010; Goerdt et al. 2010; Faucher-Giguere &

Keres 2011; Fumagalli *et al.* 2011; Kimm *et al.* 2011; Le Tiran *et al.* 2011; Giavalisco *et al.* 2011)

The cold molecular gas content of galaxies and the star formation law of galaxies at $z = 1 - 3$ are now starting to be measured. Daddi *et al.* (2010a) report gas fractions of 50-65% in massive ($M_* \sim 4 \times 10^{10} - 1 \times 10^{11}\ M_\odot$) IR-selected BzK galaxies at $z \sim 1.5$. Tacconi *et al.* (2010) have used measurements of CO emission to determine the cold gas fraction in normal star-forming galaxies at $z = 1.1 - 2.4$. For stellar masses spanning $M_* \sim 3 \times 10^{10} - 3.4 \times 10^{11}\ M_\odot$, they find cold gas masses three to ten times higher than in today's massive spiral galaxies.

In terms of star formation laws, studies based on direct CO observations suggest that over the redshift range $z = 1 - 3$, different star formation laws may apply to starbursts and mergers than to normal non-merging systems (Daddi *et al.* 2010b; Genzel *et al.* 2010). While normal non-merging galaxies appear to follow similar molecular gas-star formation relations over three orders of magnitude in gas mass or surface density, gas-rich major mergers produce on average four to 10 times more far-infrared luminosity per unit gas mass. (Genzel *et al.* 2010). A universal SF law can however be obtained when the dynamical timescale is explicitly taken into account (Daddi *et al.* 2010b; Genzel *et al.* 2010).

3.   Galaxy Evolution at $z > 5$

The luminosity functions and star-formation rates of galaxies within the reionization epoch at redshift $z > 6$ are now starting to be mapped out for the first time. Using data from the Hubble Space Telescope (HST), Bouwens *et al.* (2011a) identified 73 candidate galaxies at $z \sim 7$ and a further 59 at $z \sim 8$. They use these to derive luminosity functions at $z \sim 7$ and $z \sim 8$ for which the faint-end slope, though somewhat uncertain, is significantly steeper than seen in the local universe, implying that lower-luminosity galaxies dominate the galaxy luminosity density during the epoch of reionization.

Long-duration gamma-ray bursts (GRBs) provide a powerful alternative tool for identifying galaxies at very high redshift. Tanvir *et al.* (2009) found that the host galaxy of GRB090423 lies at a redshift of $z \sim 8.2$, implying that massive stars were being produced and dying as GRBs only $\sim$630 million years after the Big Bang. The stellar populations and star-formation histories of $z > 6$ galaxies are also starting to be studied (e.g. Finkelstein *et al.* 2010, 2011; Bouwens *et al.* 2011b; Finlator *et al.* 2011). The chemical enrichment of the high-redshift interstellar medium has been studied by Ryan-Weber *et al.* (2009), who find evidence for a very rapid build-up of intergalactic CIV over a period of only ~ 300 Myr at $z > 5$. This could reflect the accumulation of metals associated with the rising levels of star formation activity from after $z \sim 9$ as indicated by galaxy counts, and/or an increasing degree of ionization of the intergalactic medium (IGM) over this redshift range.

Observations of quasars are pushing further into the reionization epoch, with Mortlock *et al.* (2011) reporting the discovery of a quasar at redshift $z = 7.085$, only 0.77 billion years after the Big Bang, with an estimated black hole mass of 2×10^9 solar masses. They find that the neutral gas fraction in the interstellar medium at this epoch may be as high as 10%, consistent with the results of Stark *et al.* (2010) and Schenker *et al.* (2012) which also suggest that at redshift $z \sim 7 - 8$ we are entering the era where the intergalactic medium is partially neutral. The next three years are likely to see rapid advances in the study of both galaxies and the interstellar medium at $z > 6$.

4. Closing remarks

Key developments over the past three years include the first studies of galaxies and stellar populations at $z > 7$ and the increasing recognition that 'cold-flow' accretion of gas from the cosmic web is likely to play an important role in the evolution of normal galaxies. The next three years should be an exciting time. The commissioning of KMOS and MUSE on the VLT will facilitate a huge increase in the application of integral-field techniques to the study of ever more distant galaxies. At the same time, ALMA will provide new insights into the molecular gas content and star-formation rates of distant star-forming galaxies, and the first results should start to flow from a new generation of SKA precursor and pathfinder telescopes working in HI and radio continuum. Together, these and other new facilities will enable further progress in the study of galaxies at all epochs of cosmic time.

Roger L. Davies
President of the Commission

References

Agertz, O., Teyssier, R., & Moore, B. 2011, *MNRAS*, 384, 386
Bell, E. F., *et al.* 2005, *ApJ*, 625, 23
Bouwens, R. J. *et al.* 2011a, *ApJ*, 737, 90
Bouwens, R. J. *et al.* 2011b, *ApJ*, submitted, arXiv:1109.0994
Brooks, A. M., Governato, F., Quinn, T., Brook, C. B., & Wadsley, J. 2009, *ApJ*, 694, 396
Bundy, K., *et al.* 2009, *ApJ*, 697, 1369
Cappellari *et al.* 2011, *MNRAS*, 416, 1680.
Ceverino, D., Dekel, A., & Bournaud, F. 2010, *MNRAS*, 404, 2151
Conselice, C. J. 2009, *MNRAS*, 399, L16
Cox, T. J., Jonsson, P., Somerville, R. S., Primack, J. R., & Dekel, A. 2008, *MNRAS*, 384, 386
Daddi, E., *et al.* 2010a, *ApJ*, 713, 686
Daddi, E., *et al.* 2010b, *ApJ*, 714, L118
Dekel, A. & Birnboim, Y. 2006, *MNRAS*, 368, 2
Dekel, A., Sari, R., & Ceverino, D. 2009a, *ApJ*, 703, 785
Dekel, A., *et al.* 2009b, *Nature*, 457, 451
de Ravel, L. *et al.* 2009, *A&A*, 498, 379
Dijkstra, M. & Loeb, A. 2009, *MNRAS*, 400, 1109
Di Matteo, P., Combes, F., Melchior, A.-L., & Semelin, B. 2007, *A&A*, 468, 61
Emsellem *et al.* 2011, *MNRAS*, 414, 888
Faucher-Giguère, C.-A. & Kereš, D. 2011, *MNRAS*, 412, L118
Finkelstein, S. L. *et al.* 2010, *ApJ*, 719, 1250
Finkelstein, S. L. *et al.* 2011, *ApJ*, submitted, arXiv:1110.3785
Finlator, K., Oppenheimer, B. D., & Davé, R. 2011, *MNRAS*, 410, 1703
Förster Schreiber, N. M., *et al.* 2009, *ApJ*, 706, 1364
Förster Schreiber, N. M., *et al.* 2011, *ApJ*, 731, 65
Fumagalli, M., *et al.* 2011, *MNRAS*, 418, 1796
Genzel, R., *et al.* 2010, *MNRAS*, 407, 2091
Giavalisco, M., *et al.* 2011, *ApJ*, 743, 95
Goerdt, T., *et al.* 2010, *MNRAS*, 407, 613
Guo, Q., *et al.* 2011, *MNRAS*, 413, 101
Jogee, S., *et al.* 2009, *ApJ*, 697, 1971
Kereš, D., Katz, N., Weinberg, D. H., & Davé, R. 2005, *MNRAS*, 363, 2
Kereš, D., Katz, N., Davé, R., Fardal, M., & Weinberg, D. H. 2009, *MNRAS*, 396, 2332
Kimm, T., Slyz, A., Devriendt, J., & Pichon, C. 2011, *MNRAS*, 413, L51

Le Tiran, L., Lehnert, M. D., Di Matteo, P., Nesvadba, N. P. H., & van Driel, W. 2011, *A&A*, 530, L6

Lotz, J. M., *et al.* 2008, *ApJ*, 672, 177

Lotz, J. M., *et al.* 2011, *ApJ*, 742, 103

Miller, S. H., Bundy, K., Sullivan, M., Ellis, R. S., & Treu, T. 2011, *ApJ*, 741, 115

Mortlock, D. J., *et al.*, 2011, *Nature*, 474, 616

Robaina, A. R., *et al.* 2010, *ApJ*, 719, 844

Ryan-Weber, E. V., Pettini, M., Madau, P., & Zych, B. J. 2009, *MNRAS*, 395, 1476

Schaye, J., *et al.* 2010, *MNRAS*, 402, 1536

Schenker, M. A., *et al.* 2012, *ApJ*, 744, 179

Stark, D. P., Ellis, R. S., Chiu, K., Ouchi, M., & Bunker, A. 2010, *MNRAS*, 408, 1728

Steidel, C. C., *et al.* 2010, *ApJ*, 717, 289

anvir, N. R. *et al.* 2009, *Nature*, 461, 1254

van Dokkum, P. G., *et al.* 2011, *ApJ*, 743, L15

van der Wel, A., *et al.* 2011, *ApJ*, 730, 38

Weinzirl, T., *et al.* 2011, *ApJ*, 743, 87

Transactions IAU, Volume XXVIIIA
Reports on Astronomy 2009–2012
Ian Corbett, ed.

© International Astronomical Union 2012
doi:10.1017/S1743921312002943

COMMISSION 47

COSMOLOGY
COSMOLOGIE

PRESIDENT
VICE-PRESIDENT
PAST PRESIDENT
ORGANIZING COMMITTEE

Thanu Padmanabhan
Brian Schmidt

Andrew J. Bunker, Benedetta Ciardi
Yipeng Jing, Anton M. Koekemoer
Ofer Lahav, Olivier Le Fevre
Douglas Scott

COMMISSION 5 WORKING GROUPS

TRIENNIAL REPORT 2009-2011

1. Introduction

The period 2009-2011 has seen a consolidation of our theoretical understanding around the Flat-Λ CDM Universe model, with experiments on several fronts providing observations consistent with this model, and leading to improved constraints on the values of H_0, Ω_B, Ω_M, and Ω_Λ. The recently launched Planck Satellite has started to provide a wealth of new observations of the Cosmic Microwave Background (CMB) and should lead to substantial progress in the physics of the CMB over the coming years. There has also be a steady progress in mapping the formation of structure and galaxies to higher and higher redshifts. Observations of objects at the highest redshifts are suggestive that the reionization of the Universe might not be complete at these epochs. This epoch will be probed in the near future with low-frequency radio surveys. This report, prepared from the inputs received from the committee members, concentrates on these areas.

2. Developments within the past triennium

2.1. *Progress in Understanding The Cosmological Model*

After the discovery in 1998 that the Universe is currently undergoing accelerated expansion, the field of cosmology witnessed a series of cosmological experiments which could probe this unexpected result. All these observations are consistent with a spatially flat Universe dominated (73%) by a Cosmological Constant, and pressureless matter (27%), split between baryonic material (4.5%) and an unknown form of cold, non-interacting matter usually described as Cold Dark Matter (22.5%).

This Flat-Λ CDM model is not theoretically motived, but rather is the simplest model which adequately describes observations of the Universe. In addition to the complexity of three cosmologically relevant species of matter, it would appear we live in a special time of the Universe, where the Cosmological Constant and the pressureless matter components have roughly the same density - a state that persists for only a short period in the cosmological history. A significant amount of theoretical effort has been put into describing alternative theories that might provide insights into why the Universe is either constructed as the Flat-Λ CDM model suggests, or how a different physical paradigm might mimic this model. Some of the major areas of theoretical investigation have included looking at modifications of General Relativity, the assumption of homogeneity in the cosmological model, and a range of scalar field alternatives for the Cosmological constant. It is probably fair to say that cosmological constant remains the simplest description of dark energy consistent with all observations. Observational cosmology has marched on, undertaking a series of new and/or refined measures to test the predictions of the Flat-Λ CDM model.

Type Ia supernovae still provide one of the most precise probes of the Concordance Model through relatively local measurements ($0 < z < 1.5$) of cosmological expansion. A huge dataset useful for cosmology has now been assembled by several teams. Additions since 2009 include nearby SN Ia ($z < 0.1$) from the CfA (Hicken *et al.* 2009), Berkeley (Ganeshalingam, *et al.* 2010), and Carnegie groups (Folatelli, *et al.* 2010), moderate redshift ($0.1 < z < 0.4$) from the SDSS-II consortium (Kessler *et al.* 2009), high-redshift ($z > 0.3$) from the Supernova Legacy Search (Guy *et al.* 2010) and *HST* program of Amanullah *et al.* (2010). These programs show the increasing difficultly of eliminating systematic biases in the samples, including the effects of dust, supernova intrinsic color, and environmental effects, but in total, the experiments appear consistent with the Flat-Λ CDM model's prediction of luminosity distance evolution over the entire redshift range to a level of a few percent.

The technique of measuring the baryon acoustic oscillations (BAO), which trace the anisotropies seen in the CMB as imprinted in large scale structure, was first undertaken through the SDSS and 2dF redshift surveys. More targeted experiments are now completed (WiggleZ) or well underway (BOSS; www.sdss-3.org), with the combination of all measurements able to directly measure the acceleration of the Universe (Blake *et al.* 2011a), which have yielded similar results to the supernova experiments. These surveys have also enabled the measurement of the growth of structure to $z > 0.6$ (Blake *et al.* 2011b), finding a result in accord with the Flat-Λ CDM model. The *BOSS* project will provide a substantial improvement in precision of the BAO scale in the redshift range $0.3 < z < 0.7$ over the next few years. In addition, *BOSS* and the innovative *HETDEX* (http://hetdex.org/) project will probe the BAO scale at $z > 2$

Measurements of the anisotropies in the cosmic microwave background (CMB) have reached a new level of refinement through the release of the *WMAP* 7-year maps (Komatsu *et al.* 2011). Coupled with a better measurement of the Hubble constant from the geometrically derived radio maser distance to NGC4258, *HST* Cepheid variable stars, and optical observations of SN Ia (Riess *et al.* 2011), *WMAP* provides remarkably tight constraints on the curvature of the Universe, and additional useful constraints which, when combined with other methods such as baryon acoustic oscillations and supernovae, can simultaneously constrain curvature, matter density, dark energy density, and the equation of state of the dark energy (Komatsu *et al.* 2011, Sullivan *et al.* 2011, Blake *et al.* 2011a). These comparisons find that the Flat-Λ CDM model continue to withstand this intense scrutiny.

As to be expected, future experiments are becoming more ambitious in scale and in budget. BigBoss (http://bigboss.lbl.gov) aims to equip the Mayall 4-m telescope with a huge purpose-built spectrograph, and undertake a 20 million galaxy BAO survey. BAOs are also targets of the proposed *EUCLID* (http://sci.esa.int/euclid) and *WFIRST* (http:// wfirst.gsfc.nasa.gov) satellites. These satellites also aim to use weak gravitational lensing to measure the properties of dark energy. The Large Synoptic Survey Telescope (http://www.lsst.org), aims to also exploit gravitational lensing through a deep map of the entire southern sky. These future experiments will push the current tests of the Concordance Model an order of magnitude higher than currently available, and will hopefully help us to understand two of cosmology's greatest mysteries - nature of the dark matter and dark energy.

2.2. *Progress in Cosmic Microwave Background Observations*

A key event for cosmology in this period was the dual launch on 14th May 2009 of the *Planck* satellite and the *Herschel Space Observatory*. This successful launch ushered in a new era of precision mapping of the cosmic microwave background (CMB), as well as opening up the far-infrared and submillimeter wavebands for a wide range of studies, including cosmological ones.

Planck (Tauber *et al.* 2010) is the third generation CMB satellite, following on from *COBE* and *WMAP*. Its two instruments image the entire sky over frequencies from 30 GHz to 857 GHz, with improved sensitivity and angular resolution. Press releases in 2010 demonstrated that *Planck* worked according to design, and in January 2011 a series of 25 papers was submitted. These focused mainly on 'foreground' science, but included studies of the first all-sky sample of clusters found using the Sunyaev-Zeldovich effect (Planck Collaboration VII 2011), as well as an investigation of clustering in the cosmic infrared background (Planck Collaboration XVIII 2011). The release of the first primary CMB cosmology results is planned for January 2013.

Herschel has been carrying out a number of surveys which are important for the study of galaxy formation and evolution, particularly (in order of shallow and wide to deep and narrow) H-ATLAS (Eales *et al.* 2010), HerMES (Oliver *et al.* 2010) and H-GOODS (Elbaz *et al.* 2011), using the SPIRE and PACS instruments. Results have included counts of galaxies at wavelengths from 70 to 500 μm, constraints on the evolution of the bolometric luminosity function, studies of correlations in the submillimeter background, investigation of the relationship between *Herschel*-detected galaxies and a wide range of other wavelengths, and the discovery of strongly lensed systems at high redshift.

WMAP has continued to release updates of results for the years 5 (Komatsu *et al.* 2009) and 7 (Komatsu *et al.* 2011) of the satellite's data-set. These have firmed up the values of cosmological parameters, and provided a range of tests of the Flat-Λ CDM paradigm.

On smaller angular scales, the Atacama Cosmology Telescope (Swetz *et al.* 2011) and the South Pole Telescope (Carlstrom *et al.* 2011) have provided exquisite new measurements of the CMB anisotropy power spectrum, showing several additional acoustic peaks. These experiments have also demonstrated the ability to find galaxy clusters, as well as to measure the effects of gravitational lensing on the CMB anisotropies.

2.3. *Progress in understanding the Formation and Evolution of galaxies from large deep surveys*

Deep imaging and spectroscopic surveys constitute a workhorse resource to test galaxy formation and evolution. During this period, multi-wavelength surveys with large visible and near-IR arrays have developed towards greater depth and larger areas. One noticeable evolution has been the broader wavelength coverage including the near-IR (*JHK*) of the most recent surveys, offering a better constraint onto the spectral energy distributions (SEDs) and the morphological properties. One can note the important impact of the mid-IR space observatory Spitzer, and the far-IR space observatory Herschel at the end of the period, further extending the multi-wavelength data to include the radiation of cold components.

Existing and new large imaging cameras (the largest have about $0.5°$ on a side) have produced large quantities of data, mostly in the visible and near-IR domains. Deep galaxy imaging surveys in the visible domain (*ugriz*) now reach several hundreds of square degrees (CFHTLS), very faint magnitudes AB $\sim 28-30$ at visible wavelengths (CFHTLS, COSMOS, GOODS), $AB \sim 25$ at near infrared (UKIDSS, Lawrence et al., 2007; Ultra-Vista, on going), or $AB \sim 23$ at mid/far infrared (Sanders et al., 2007). The HST-ACS, followed by WFC3 have opened the way to extremely deep surveys under high spatial resolution, like the COSMOS $2\deg^2$ survey (Scoville et al., 2007) or the CANDELS survey (Grogin et al., 2011).

Narrow-band imaging surveys are now deep enough to identify relatively large samples of emission line objects, particularly focusing on the Ly-α 1215 Å Hydrogen line used to track the highest redshift objects. Atmospheric windows are exploited in narrow trans-mission ranges at redshifts $z \sim 5.7, 6.5, 7.7$, (Ouchi et al., 2008); but attempts at $z = 8.8$ have been unsuccessful up to now (e.g. Cuby et al., 2007).

Using the large multi-wavelength datasets, the technique of photometric redshifts de-termined from template fitting to the SED has considerably matured. The best surveys now reach a redshift accuracy of $\sim 0.03(1 + z)$ with catastrophic failures less than 2% (e.g. Ilbert et al., 2006). As this applies to all sources measured on images, this opens up statistically robust analysis from very large samples reaching close to one million galaxies (COSMOS, Ilbert et al., 2009).

As deep photometry has been gaining in area and depth, so have the large spectro-scopic surveys. Large and efficient multi-object spectrographs on 8–10 m telescopes have continued to deliver large quantities of data. The exploitation of the DEEP2 (Faber et al., 2007) and VIMOS VLT Deep Survey (Garilli et al., 2008), with about 50,000 spectro-scopic redshifts each in $0.5 < z < 1.4$ and $0 < z < 5$ respectively, has continued, and new surveys like zCOSMOS bring a new insight with 20,000 redshifts in the COSMOS field with $0 < z < 1.2$ (Lilly et al., 2009). The larger volumes sampled to $z \sim 1$ have opened-up the possibility to study the properties of galaxies in relation to different types of environment.

An important development has been the demonstration of the evolution of galaxy type vs. local density relation, well known locally. At $z \sim 1$ this relation is almost flat with galaxies of different types no longer segregated in relation to environment (Cucciati et al., 2006; Cooper et al., 2006; Tasca et al., 2009). The picture of galaxy evolution connected to hierarchical growth of the dark matter halos has received support from the evidence of the major role of mergers since $z \sim 1$ (Lotz et al., 2008; de Ravel et al., 2009), with a strong dependency of the merger rate on the mass.

The star formation rate density evolution and the stellar mass density, two major statistical indicators of evolution have received considerable attention, with new mea-surements available out to early epochs. The star formation rate density has evolved in several steps, with a strong increase from redshift 7 to 4 (to be confirmed from larger samples, Bouwens et al. 2007), a constant high rate from $z \sim 3$ to $z \sim 1$ (Tresse et al., 2007; Cucciati et al., 2011), and a strong decrease by about one order of magnitude from $z \sim 1$ to the present, as observed by earlier surveys like CFRS. The stellar mass density evolution has been traced robustly out to $z \sim 2$ (Arnouts et al., 2007; Ilbert et al., 2010). It is found that between $z \sim 2$ and $z \sim 1$ there is a strong build-up of mass in early-type galaxies by almost a factor of 10, pointing to this epoch as an important phase in early-type galaxy evolution.

The question of the way that mass has assembled into galaxies is still hotly debated. While major mergers are readily demonstrated to contribute 25% to 40% of the mass growth in galaxies since $z \sim 1$ (de Ravel et al., 2009), cold accretion along the filaments of the cosmic web has been claimed to be mostly responsible for the mass accretion onto galaxies at $z \sim 2$ (Genzel et al., 2008; Dekel et al., 2009). This is challenged by the observation of a large fraction of about (1/3) of galaxies in close pairs at $z \sim 1.5 - 2$

(Forster-Schreiber *et al.*, 2009; Epinat *et al.*, 2009). This issue should become clarified once larger samples are available.

The question of the complete census of galaxies, necessary to properly account for all star formation and mass content, is still open. To help find the needle in the haystack, various a priori selection criteria have been used to detect the highest redshift populations based on photometry, like Lyman Break Galaxy or BzK ($z \sim 2$) selection. However, these techniques may miss out a fraction of the population, as the counts of galaxies from these seem to be significantly lower than from direct flux selected samples (Cucciati *et al.*, 2011). Compounded with the difficulty of constraining the faint end (low mass) slope of the luminosity (mass) function beyond redshift $z \sim 2$, this uncertainty in counts results in significant uncertainties on the star formation or mass density (among other things) at $z > 2$. New surveys are addressing this issue using near-IR flux selection (van Dokkum *et al.*, 2006), or a combination of criteria to minimize the possible biases.

Despite this enormous progress, it is clear that deep surveys will need to expand further towards the classical avenues of wider, deeper, and redder observations. The next years will see a consolidation of the results from Herschel, and the start of operations of ALMA to look at high redshift galaxies from their cold components, to estimate the total gas and dust content and obtain a complete star formation census. New generation facilities will start operations or be developed in the next few years to enable these new generation surveys, including new near-IR multi-object spectrographs. All-sky surveys at $z \sim 1 - 2$ are now within reach of technical capabilities, and facilities are being proposed to observe several millions or tens of millions of distant galaxy spectra (ESA-*Euclid*, *BigBoss*, *Subaru-PFS*). On the extremely distant front, *JWST* and the *ELT*s will follow-up the candidates identified in large surveys with imaging cameras, hopefully before the end of this decade. One major goal will be to track the reionization epoch, finding first light.

2.4. *Progress in Understanding the Epoch of Reionization*

The epoch of reionization (EoR) sets a fundamental benchmark in cosmic structure formation, corresponding to the birth of the first luminous objects that act to ionize the neutral intergalactic medium (IGM). The H-ionizing photons emitted by such sources in fact produce HII bubbles in the surrounding medium on a range of size scales. With the passage of time, the bubbles finally punch through the walls separating each other, to leave behind an almost fully ionized IGM, which remains such to the present. Despite the agreement reached by the scientific community on the general characteristics of the EoR (for example it is thought to be a gradual process mainly driven by stellar-type sources), its details are still being debated.

The progress made in the development of the sophisticated numerical techniques necessary to model the EoR has recently made possible a more accurate treatment of the physical processes involved, by considering e.g. a larger mass range, the contribution of x-ray photons from sources more energetic than stars and the impact of He physics on H reionization, among others (e.g. Baek *et al.* 2010; Iliev *et al.* 2011; Ciardi *et al.* 2011). But our persisting ignorance in particular of the efficiency of star formation, the escape fraction of ionizing photons and the stellar properties of high-z galaxies makes it difficult to improve substantially such modeling until more stringent observational constraints become available.

In fact, present observations offer information on the final stages of reionization and on the global amount of electrons produced during the process, but a crucial observational insight on the evolution of the EoR and the properties of its sources is still lacking. The latest release from the WMAP satellite (Larson *et al.* 2011) gives a Thomson scattering optical depth of $\tau_e = 0.088 \pm 0.015$, which provides an estimate of the global amount of electrons produced during the reionization process. On the other hand, this does not provide any constraint on the evolution of the electron number density, i.e. on the reionization history. If reionization were an instantaneous process, the above value would correspond to a reionization redshift of $z_{reion} = 10.5 \pm 1.2$. But it is very likely that the process is gradual and hence is expected to start at higher redshift.

Information on the latest stages of reionization has been gathered through the years from observations at near-IR wavelengths of absorption by the IGM in the spectra of the highest redshift quasars (e.g. Becker $et\ al.$ 2001; Fan $et\ al.$ 2006), suggesting that the IGM is mostly ionized by $z \sim 6$. Recently though, a quasar at $z = 7.09$ has been detected (Mortlock $et\ al.$ 2011) with a highly ionized near zone which is smaller than those around quasars of similar luminosity at $z \sim 6$, indicating that the volume averaged HI fraction is larger than 0.1 (Bolton $et\ al.$ 2011), i.e. the IGM is still significantly neutral. This might be compatible with the possibility that reionization was yet to fully complete by $z \sim 6$ (Mesinger 2010). The possibility of a significant neutral fraction in the IGM surrounding such a quasar makes it an excellent target for studies of the IGM ionization state using the redshifted 21-cm transition.

Indeed, it has long been known (e.g. Field 1959) that neutral hydrogen in the IGM and gravitationally collapsed systems may be directly detectable in emission or absorption against the CMB at the frequency corresponding to the redshifted HI 21 cm line. In general, 21-cm spectral features will display angular structure as well as structure in redshift space, due to inhomogeneities in the gas density field, hydrogen ionized fraction, and spin temperature. Several different signatures have been investigated in the recent literature (see the review by Morales & Wyithe 2010): (i) fluctuations in the 21-cm line emission induced by the 'cosmic web', by the neutral hydrogen surviving reionization and by minihalos; (ii) a global feature ('reionization step') in the continuum spectrum of the radio sky that may mark the abrupt overlapping phase of individual intergalactic HII regions; (iii) and the 21-cm narrow lines generated in absorption against very high-redshift radio sources by the neutral IGM and by intervening minihalos and protogalactic disks.

A number of radio observational facilities are presently being built (e.g. LOFAR, MWA, PAPER) and planned (SKA) with the aim of providing critical insight into the EoR. Among these, LOFAR has started its commissioning phase in early 2011 and, together with the other radio interferometers, will hopefully, in the next 5–10 years, provide more stringent constraints on the reionization history.

3. Closing remarks

As part of the lead up to the 2012 IAU General Assembly, Commission 47 will create a strategic plan for the next 3 years. This plan will be developed by the organising committee with inputs from commission membership. The Vice-President, Brian Schmidt, will take the lead in preparing and finalising it for adoption at the General Assembly in Beijing.

Thanu Padmanabhan
President of the Commission

References

Amanullah, R., Lidman, C., Rubin, D., $et\ al.$ 2010, ApJ, 716, 712

Arnouts, S., Walcher, C. J., Le Fèvre, O., *et al.* 2007, *A&A*, 476, 137

Baek, S., Semelin, B., Di Matteo, P., Revaz, Y., & Combes, F. 2010, *A&A*, 523, A4

Becker, R. H., Fan, X., White, R. L., *et al.* 2001, *AJ*, 122, 2850

Blake, C., Brough, S., Colless, M., *et al.* 2011, *MNRAS*, 415, 2876

Blake, C., Davis, T., Poole, G. B., *et al.* 2011, *MNRAS*, 415, 2892

Bolton, J. S., Becker, G. D., Raskutti, S., *et al.* 2011, *MNRAS*, 1962

Bouwens, R. J., Illingworth, G. D., Franx, M., & Ford, H. 2007, *ApJ*, 670, 928

Carlstrom, J. E., Ade, P. A. R., Aird, K. A., *et al.* 2011, *PASP*, 123, 568

Cooper, M. C., Newman, J. A., Croton, D. J., *et al.* 2006, *MNRAS*, 370, 198

Cuby, J.-G., Hibon, P., Lidman, C., *et al.* 2007, *A&A*, 461, 911

Cucciati, O., Iovino, A., Marinoni, C., *et al.* 2006, *A&A*, 458, 39

Cucciati, O., Tresse, L., Ilbert, O., & Le Fèvre, O. 2011, SF2A-2011: Proceedings of the Annual
    meeting of the French Society of Astronomy and Astrophysics Eds.: G. Alecian, K. Belka-
    cem, R. Samadi and D. Valls-Gabaud, pp. 85-89, 85

de Ravel, L., Le Fèvre, O., Tresse, L., *et al.* 2009, *A&A*, 498, 379

Dekel, A., Birnboim, Y., Engel, G., *et al.* 2009, *Nature*, 457, 451

Eales, S., Dunne, L., Clements, D., *et al.* 2010, *PASP*, 122, 499

Elbaz, D., Dickinson, M., Hwang, H. S., *et al.* 2011, *A&A*, 533, A119

Epinat, B., Contini, T., Le Fèvre, O., *et al.* 2009, *A&A*, 504, 789

Faber, S. M., Willmer, C. N. A., Wolf, C., *et al.* 2007, *ApJ*, 665, 265

Förster Schreiber, N. M., Genzel, R., Bouché, N., *et al.* 2009, *ApJ*, 706, 1364

Fan, X., Strauss, M. A., Becker, R. H., *et al.* 2006, *AJ*, 132, 117

Field, G. B. 1959, *ApJ*, 129, 551

Folatelli, G., Phillips, M. M., Burns, C. R., *et al.* 2010, *AJ*, 139, 120

Ganeshalingam, M., Li, W., Filippenko, A. V., *et al.* 2010, *ApJs*, 190, 418

Garilli, B., Le Fèvre, O., Guzzo, L., *et al.* 2008, *A&A*, 486, 683

Genzel, R., Burkert, A., Bouché, N., *et al.* 2008, *ApJ*, 687, 59

Grogin, N. A., Kocevski, D. D., Faber, S. M., *et al.* 2011, arXiv:1105.3753

Guy, J., Sullivan, M., Conley, A., *et al.* 2010, *A&A*, 523, A7

Hicken, M., Wood-Vasey, W. M., Blondin, S., *et al.* 2009, *ApJ*, 700, 1097

Ilbert, O., Arnouts, S., McCracken, H. J., *et al.* 2006, *A&A*, 457, 841

Ilbert, O., Capak, P., Salvato, M., *et al.* 2009, *ApJ*, 690, 1236

Ilbert, O., Salvato, M., Le Floc'h, E., *et al.* 2010, *ApJ*, 709, 644

Iliev, I. T., Mellema, G., Shapiro, P. R., *et al.* 2011, arXiv:1107.4772

Kessler, R., Becker, A. C., Cinabro, D., *et al.* 2009, *ApJs*, 185, 32

Komatsu, E., Dunkley, J., Nolta, M. R., *et al.* 2009, *ApJs*, 180, 330

Komatsu, E., Smith, K. M., Dunkley, J., *et al.* 2011, *ApJs*, 192, 18

Larson, D., Dunkley, J., Hinshaw, G., *et al.* 2011, *ApJs*, 192, 16

Lawrence, A., Warren, S. J., Almaini, O., *et al.* 2007, *MNRAS*, 379, 1599

Lilly, S. J., Le Brun, V., Maier, C., *et al.* 2009, *ApJs*, 184, 218

Lotz, J. M., Davis, M., Faber, S. M., *et al.* 2008, *ApJ*, 672, 177

Mesinger, A. 2010, *MNRAS*, 407, 1328

Morales, M. F. & Wyithe, J. S. B. 2010, ARAA, 48, 127

Mortlock, D. J., Warren, S. J., Venemans, B. P., *et al.* 2011, *Nature*, 474, 616

Oliver, S. J., Wang, L., Smith, A. J., *et al.* 2010, *A&A*, 518, L21

Ouchi, M., Shimasaku, K., Akiyama, M., *et al.* 2008, *ApJs*, 176, 301

Planck Collaboration, Ade, P. A. R., Aghanim, N., *et al.* 2011,
    arXiv:1101.2028

Planck Collaboration, Ade, P. A. R., Aghanim, N., *et al.* 2011,
    arXiv:1101.2041

Riess, A. G., Macri, L., Casertano, S., *et al.* 2011, *ApJ*, 730, 119

Sanders, D. B., Salvato, M., Aussel, H., *et al.* 2007, *ApJs*, 172, 86

Scoville, N., Aussel, H., Brusa, M., *et al.* 2007, *ApJs*, 172, 1

Sullivan, M., Guy, J., Conley, A., *et al.* 2011, *ApJ*, 737, 102

Swetz, D. S., Ade, P. A. R., Amiri, M., *et al.* 2011, *ApJs*, 194, 41
Tasca, L. A. M., Kneib, J.-P., Iovino, A., *et al.* 2009, *A&A*, 503, 379
Tauber, J. A., Mandolesi, N., Puget, J.-L., *et al.* 2010, *A&A*, 520, A1
Tresse, L., Ilbert, O., Zucca, E., *et al.* 2007, *A&A*, 472, 403
van Dokkum, P. G., Quadri, R., Marchesini, D., *et al.* 2006, *ApJl*, 638, L59

Transactions IAU, Volume XXVIIIA
Reports on Astronomy 2009–2012
Ian F. Corbett, ed.

© International Astronomical Union 2012
doi:10.1017/S1743921312002955

DIVISION VIII   OPTICAL and INFRARED TECHNIQUES
TECHNIQUES OPTIQUES ET INFRAROUGE

PRESIDENT   Andreas Quirrenbach
VICE-PRESIDENT   David Silva
PAST PRESIDENT   Rolf-Peter Kudritski
Organising Committee   Michael G. Burton
                   Ian S. McLean
                   Eugene F. Milone
                   Jayant Murthy
                   Stephen T. Ridgway
                   Gražina Tautvaišiene
                   Andrei A. Tokovinin
                   Guillermo Torres
                   Xiaonian ZHENG

DIVISION IX COMMISSIONS

Division IX Commission 21 Galactic and Extragalactic Background Radiation
Division IX Commission 25 Astronomical Photometry and Polarimetry
Division IX Commission 30 Radial Velocities
Division IX Commission 54 Optical & Infrared Interferometry

DIVISION IX WORKING GROUPS

Division IX WG Adaptive Optics
Division IX WG Infrared Astronomy
Division IX WG Large Telescope Projects
Division IX WG Site Testing Instruments
Division IX WG Sky Surveys

INTER-DIVISION WORKING GROUPS

Inter-Division IX-X-XI WG Astronomy from the Moon
Inter-Division IX-X WG Encouraging the International Development of Antarctic Astronomy

Andreas Quirrenbach
President of the Division

Transactions IAU, Volume XXVIIIA
Reports on Astronomy 2009–2012
Ian Corbett, ed.

© International Astronomical Union 2012
doi:10.1017/S1743921312002967

COMMISSION 25

STELLAR PHOTOMETRY
AND POLARIMETRY

STELLAIRE PHOTOMÉTRIE ET
POLARIMÉTRIE

PRESIDENT
VICE-PRESIDENT
PAST PRESIDENT
ORGANIZING COMMITTEE

Eugene F. Milone
Alistair Walker
Peter Martinez
Barbara Anthony-Twarog, Pierre Bastien,
Jens Knude, Donald Kurtz,
John Menzies, Aleksey V. Mironov,
Shengbang Qian

COMMISSION 25 WORKING GROUPS

Div. IX / Commission 25 WG   Infrared Working Group

TRIENNIAL REPORT 2009-2011

1. Introduction

The Commission on Photometry and Polarimetry has a long and distinguished history in the IAU and its contributions to astronomy have been extensive and profound. Its efforts are centered on the issues of atmospheric extinction, photometric passbands, transformations among photometric systems, and calibration. Photometric and polarimetric techniques and standardization are essential tools in our exploration and investigation of astronomical objects and astrophysical quantities.

The volume *Astronomical Photometry: Past, Present, and Future* edited by Milone & Sterken (2011), and published in Springer's Astrophysics and Space Science Library series, summarizes the march to increased precision and accuracy, and in so doing also illustrates the role that Commission 25 has played in that development. Sterken *et al.* (2011) describe efforts to improve photometry over time while Milone & Pel (2011) focus on differential photometry as the historically most successful way to achieve this from the ground. This article is basically a tribute to Theodore Walraven, who passed away during the previous triennium, and his legacy. Howell (2011) carries the discussion on differential photometry into the most recent period, discussing CCD instruments and methods. Ambruster *et al.* (2011) describe the highly successful Pierce-Blitzstein pulse-counting differential photometer in use at the Flower and Cook Observatory for more than half a century. Landolt (2011) discusses the rise of "Johnson Photometry and its descendants." Wing (2011) discusses the use of photometry in and for spectral classification. Cohen (2011) and Adelman (2011) discuss absolute calibration in the optical & infrared photometry and in optical spectrophotometry, respectively. Bastien (2011) reviews the development of polarization studies, and Milone & Young (2011a) do the same for the infrared.

Thus the work by members of this commission is essential for photometric, spectrophotometric, and polarimetric surveys, and for both ground-based and space mission operations.

According to the IAU membership database, Commission 25 currently has 252 members. This represents about an 8% increase over the previous triennium. Because photometry and polarimetry are active and vital areas of astronomy, we can expect further growth.

In 2007, the Commission's website was moved from its previous host site at the Vrije Universiteit Brussel to the South African Astronomical Observatory. The URL of the current Commission 25 website is http://iau_c25.saao.ac.za/. We thank Dr Christiaan Sterken for having established the Commission's website and for having maintained it for a number of years. Its present administrator is Dr. Peter Martinez, past president of Comm. 25. In 2010 the Commission updated its website to bring it in line with the "look and feel" of the official IAU website. The Commission 25 website now contains a link to a very useful paper by G. Torres (2010), entitled "On the Use of Empirical Bolometric Corrections for Stars," that is essential reading for all photometrists. Finally, regarding IR sources, links are also provided to the key papers discussing the Infrared Working Group (IRWG) system. The website has a list of faint standard stars in the near-IR passbands Mauna Kea photometry system. Because the IRWG standard star magnitudes listed in Milone & Young (2005) have had zero points added to them to provide "familiar" values, this list can probably be used to standardize observations made with the IRWG iJ, iH, and iK filters at Mauna Kea. However, a better strategy would be to extend the list of IRWG standard stars found in Milone & Young (2005), to cover the entire sky, by determining secondary standards.

Now we discuss specific developments during the 2009–2011 triennium.

2. Photometry

2.1. *Photometric Standards and Calibration (B. J. Anthony-Twarog)*

Work on photometric standards and calibration, which underpins all photometric applications, proceeded during the triennium under review. Discussions in the earliest IAU General Assemblies dealt with such issues, namely the establishment of a widely accessible set of standards (e.g., the North Polar Sequence), as well as improved techniques, magnitude scales and zero points, and improved precision in measurement.

Some issues, past and present, of photometric and spectrophotometric calibration were discussed by Martin Cohen and Saul Adelman in two Historical Astronomy Division sessions of the Long Beach AAS meeting held in January, 2009, and reviewed in more detail in Sections 4 and 5, below.

From traditional mainstays to innovative space-based calibration solutions, interest in photometric calibration strategies is high. Tying ground based calibration efforts to space missions is the Absolute Color Calibration Experiment for Standard Stars (Kaiser *et al.* 2009). The goal of ACCESS is to transfer absolute laboratory standards to the stars by observing a select set of spectrophotometric standards from space. If successful, the precise calibration of flux from 0.35 to 1.7 microns will be achieved.

The already rich contributions by Arlo Landolt have been extended further by new $UBVRI$ (Johnson-Kron-Cousins) photoelectric observations for 202 stars centered on the celestial equator and at magnitude ranges suitable for faint CCD calibration (to $V \sim 16.3$). A catalog of VR (Johnson-Cousins) photometry developed for the Deep Ecliptic Survey has been published by Buie *et al.* (2011). Over 200,000 stars within

6 degrees of the ecliptic plane are included, extending to $R \sim 16$ and $V \sim 17$. The calibration builds on the equatorial standard system of Landolt.

In this context, we note that Peter Stetson provides standard stars on the Landolt system for many fields, mostly centered on the Landolt fields and on globular clusters. This is an on-going project, 89360 stars in total as of 2010 December. Magnitudes and positions can be conveniently downloaded†.

As space-based astronomy explores wavelength regimes previously inaccessible from the ground, new and extended lists of calibrated standards are called for. Siegel *et al.* (2010) present a small but crucially uniform catalog of eleven new faint ultraviolet standards. This may sound like a small catalog — the authors point out that **four** white dwarfs, all brighter than $V = 13.4$, have comprised the ultraviolet calibration standard set until now. Using data from SDSS, GALEX and Swift, the new standards have consistently calibrated observations in 11 passbands, SDSS spectra and well-constrained model spectra for future applications.

The calibration of the photometric system for the wide-field camera of the UKIRT telescope ($ZYJHK$) has been described by Hodgkin *et al.* (2009). Among other uses, the WFCAM hosts the UKIRT Infrared Deep Sky Survey (UKIDSS). The $ZYJHK$ passbands extend from 0.84 to 2.37 mum. The 2MASS point source catalog provided the primary foundation for the calibration.

Photometry of resolved populations depends on precisely calibrated standard relations for determination of reddenings and chemical abundances. An update of the critical intrinsic Strömgren color calibration was published by Karataş & Schuster (2010). The calibration relation for $(b - y)_0$ includes 11 terms and is based on nearly 400 stars determined to be within 70 pc of Earth and therefore essentially unreddened. Differences between this and previously calibrated mean relations are gratifyingly minor.

What if every CCD exposure included a standard or two, tied to several standard photometric systems? Towards this ambitious goal, Pickles & Depagne (2011) present on-line catalogs of synthetically calibrated magnitudes in several bandpass systems (e.g., $UBVRI$–ZY and $ugriz'$) for several large-scale surveys, including the Tycho2, NOMAD and 2MASS surveys. Photometry for millions of stars has been synthesized by interpolating within a set of 20 CALSPEC spectrophotometric standards.

An entirely new photometric system has been characterized by Aparicio Villegas *et al.* (2010). The ALHAMBRA system uses 20 contiguous, equal-width bandpasses from 350 to 970 nm. The system is defined by a set of standard stars including several classic spectrophotometric standards as well as nearly 300 additional stars.

The detailed process of establishing the input catalog for stars studied by the Kepler mission is described by Brown *et al.* (2011). An important discriminant for each of the 160,000 potential Kepler targets was distinction between cool giants and cool dwarfs; four of the five SDSS bandpasses, plus a bandpass patterned after a DDO filter at 510 nm, were calibrated by ground-based observations and tied to model atmosphere fluxes. In addition to data for 284 primary standards, information for the Kepler Input Catalog is publically available through MAST.

While most astronomers strive for in-focus imaging and spectroscopy, Southworth *et al.* (2009) demonstrate that extreme de-focusing can be used to generate light curves with 0.0005 mag precision. Their paper discusses transit events for the extrasolar planetary system WASP-5, observed with the ESO Danish 1.5-m telescope. Extra-focal images were sometimes employed to achieve higher precision in the photographic era (cf. p. 6 of Sterken *et al.* (2011), for examples).

† http://www3.cadc-ccda.hia-iha.nrc-cnrc.gc.ca/community/STETSON/

Finally, the calibration of the photon detectors of the UVIT instruments to be flown on the Indian Space Research Organization's Astrosat mission has been reported by Postma, Hutchings, & Leahy (2011). This mission is a collection of three x-ray and two UV-optical telescopes on the same platform aligned to allow simultaneous detection of each pointed target.

2.2. *Large-Scale Photometric surveys (J. Knude)*

Among large-scale surveys' major data releases announced during the present triennium, was the final Sloan Digital Sky Survey II data release (DR7). Data releases for SDSS III have started in 2011 with DR8.

This may be the appropriate place to discuss photometry progress of the Gaia ESA keystone space mission project. The Gaia broad band photometric system (white light G, blue G_BP, red G_RP and G_RVS bands) was characterized and color-color transformations to other commonly used photometric systems (Johnson-Cousins, Sloan Digital Sky Survey, Hipparcos and Tycho) were established. The data and tools provided in Jordi *et al.* (2010) allow planning scientific exploitation of Gaia data, performing simulations of the Gaia-like sky, planning ground-based complementary observations and for building catalogues with auxiliary data for the Gaia data processing and validation.

Huge efforts are being devoted to the calibration of Gaia data acquired throughout the planned five years mission with more than 100 CCDs in the focal plane. The calibration is performed in two stages:

1. A first calibration step based on a large number of flux-constant sources defining an internal mean instrumental system; and

2. A second step tying the internal system to the real world and to physical units through a set of well defined Spectrophotometric Standard Stars (SPSS).

The results up to now for the first step, using simulated data, are detailed in Jordi *et al.* (2010). Spectrophotometric and photometric observational campaigns with the TNG telescope at La Palma, the NTT and REM at la Silla (ESO), the 2.2-m telescope at CAHA, the Cassini telescope at Loiano and the 1.5-m telescope at San Pedro Mártir, have been conducted semiannually or annually.

Other space projects are described more fully below.

2.3. *Space Photometry of Stars (D. W. Kurtz)*

Over the past decade the precision of photometric measurements of stellar brightness has improved to the parts per million level as a result of space-based photometric telescopes, such as the star-tracker on the WIRE mission (e.g., Bruntt *et al.* 2009), the Canadian MOST mission (e.g., Gruberbauer *et al.* 2011), the French-led CoRoT mission (Auvergne *et al.* 2009), and the NASA *Kepler* mission (Gilliland *et al.* 2011). This precision exceeds the best obtained from the ground (e.g., Kurtz *et al.* 2005) by a factor of 10, and it is in general an improvement of $100 - 1000$ times better than traditional ground-based photometric studies. This is coupled with duty cycles that can exceed 90% as a consequence of being above the weather and free of the Earth's day-night cycle. This combination has led to a revolutionary view of stellar variability, with deeper understanding of stellar astrophysics and the discovery of many new phenomena. The dominant missions in this triennium of the IAU are the CoRoT and *Kepler* Missions; *Kepler* will be used here to illustrate the new regime.

A consequence of this new era in photometric precision is that the activities of C25 and the methods of stellar photometry are fundamentally changed. When the IAU was founded in 1919 the measurement of stellar brightness was done by eye and by densitometry measurements of images on photographic plates with a quantum efficiency less than

10%. The growth of the electronic photometer through the 1930s led to its dominance until the advent of CCD detectors. Now we have >90% quantum efficiency with only optical losses of the starlight in the photometric space missions. In past decades, studies were conducted on individual stars, often by single astronomers, or small groups. Now the *Kepler* mission is observing 160 000 stars nearly continuously for a minimum of 3.5 y with precision of $1 - 10\,\mu$mag, depending on the brightnesses of the targets. This means that many hundreds of astronomers are working in large teams to discover transiting extra-solar planets and study stellar activity, rotation, flares, eclipsing binary stars, asteroseismology of many types of pulsating stars, stellar structure and evolution. There are also many people working on further improvements to photometric precision, both in the hardware and in data processing.

This trend will continue with the upcoming ground-based surveys of PanStarrs[†] (e.g., Stubbs *et al.* 2007), DECam[‡], scheduled for installation at CTIO as we write this, and the Large Synoptic Survey Telescope (LSST)[¶] (e.g., Ivezić *et al.* 2008), as well as the ESA Gaia mission[||] (e.g., Eyer 2006). PanStarrs is primarily searching for near-Earth asteroids, the LSST will survey for all time variable objects from Supernovae to asteroids, and Gaia will measure the parallaxes of a billion stars to micro-arcsecond precision, as well as obtain light curves for over a million variable stars, half a million quasars, and half a million solar system objects. The cameras for these projects will exceed the 95 megapixel, 42-CCD camera of *Kepler* by more than a factor of 10: 1.4 gigapixels for PanStarrs and 3.2 gigapixels for LSST. Gaia has many instruments; it will have 180 CCDs, 110 of which are devoted to astrometry. Although the photometric precision and observing strategies differ for these projects, what they all share is data acquisition on an unprecedented scale in astronomy.

This flood of data will take us from the current working method of large teams to a situation where humans must be taken out of the data acquisition and initial analysis. LSST and Gaia will produce millions of light curves for many kinds of variable astronomical objects. Machines will have to make the first classifications of these and select subsets of the astrophysically most interesting targets to be studied by humans, hence machine learning studies will play a large part in future astrophysics. This part of the work of C25 will then be similar to that of the large particle physics experiments now, bringing changes in the way teams are managed, how students are trained, and how results are published.

At the time of this writing the problem of photometric *accuracy* (as opposed to *precision*) with the space data has not been studied. All four of the missions — WIRE, MOST, CoRoT and *Kepler* — use a single, broad white light bandpass with no calibration to standard star fluxes. WIRE, MOST and CoRoT are all in low Earth orbits, thus have to contend to varying degree — depending on their pointing — with scattered Earth-light and the radiation environment, particularly the South Atlantic Anomaly. They also have regular data gaps for targets that are not in the continuous viewing zone. The *Kepler* mission in its Earth-trailing solar orbit does not suffer from those problems, hence illustrates our new understanding of the limitations of CCD photometric precision. At μmag levels we not only see the stars in unprecedented detail, we also see the limitations of our instruments in new ways.

---

† http://pan-starrs.ifa.hawaii.edu/public/
‡ http://www.darkenergysurvey.org/
¶ http://www.lsst.org/lsst/
|| http://www.esa.int/science/gaia

With ground-based CCD photometry flat-fielding and bias subtraction are standard image processing steps, as they are for spaced-based data, although in this case the flat field can be determined only in the laboratory, long before launch, and there is no way to monitor changes with aging in the instrument. The *Kepler* stellar images spread over many pixels (for the brightest stars this may be of the order of 1000 pixels) and the telescope pointing accuracy is to a small fraction of a pixel. Pointings last for one month between data downloads. Sources of noise are *intra*-pixel sensitivity variation (which usually is not considered in ground-based studies), differential velocity aberration, vibrations caused by reaction wheels, temperature variations caused by pointing changes for data downloading and solar panel and radiator positioning, cosmic rays, and, very importantly, changes in contamination by background stars with tiny pointing drift changes. At μmag precision nearly all stars are variable, so that variability in background stars is noise for target stars.

It might be thought at first that with 160 000 target stars differential photometry could remove instrumental effects. But many of those listed above are pixel specific, and all stars are variable. By using ensembles of thousands of stars it is possible to produce co-trending vectors that do allow some significant improvement in results from differential photometry. These are discussed in the *Kepler* data release notes †, which are essential reading for those making use of the data. It is interesting to note that the ultimate limit to our photometric knowledge of stellar brightness is now limited by the stars themselves for periods of hours and longer. Stellar activity is the major source of noise at μmag precision on this time scale.

Remarkably and unexpectedly, one lesson learned from the *Kepler* mission is that saturation is not a barrier to high precision CCD photometry. RR Lyrae itself is in the *Kepler* field-of-view and is extravagantly saturated. But once a potential well of a pixel is filled with electrons, new photons are still detected and the electrons flow to neighbouring pixels. This is well-known. The surprise is that by designing a mask that uses all the overflow pixels, all the photons can be counted precisely. That has been exploited for RR Lyrae with novel insight into the Blazhko effect and the discovery of period-doubling (Kolenberg *et al.* 2011). Thus it must also be true that ground-based CCD studies can be performed on bright targets, so long as all of the overflow electrons are captured by pixels on the chip.

We are in the midst of a revolution of new discovery of exoplanets and stellar astrophysics as a consequence of the vast improvement in precision of stellar photometry wrought by the photometric space missions. There will be no new grand improvement in precision as a result of improvements to CCD technology, or even a new photon detection technology: we now are close to being 100% efficient in capturing the photons gathered by our telescopes; the limitations are in the optics, not the detectors, and ultimately in the variability of the stars themselves. The great new instrumental developments are in the scale of the projects, such as PanStarrs, LSST and Gaia. The challenges for C25 in this field are in data management and exploitation with machine learning. Teams of astronomers will then study the most interesting objects, but there will still be space for the lone brilliant individual to discover what no else has seen or understood in the deluge of data. That has not changed since the inception of the IAU in 1919.

2.4. *Other Photometric Developments*

Neugent & Massey (2010) discuss the night sky spectrum at Kitt Peak over the past twenty years and report that the Kitt Peak sky brightness has changed little over that

† http://archive.stsci.edu/kepler/data_release.html

interval, with an increase of 0.1 mag. brighter at zenith and 0.3 mag brighter in the direction of Tucson. In addition, however, they found that the zenith brightness remained the same and the brightness in the Tucson direction decreased, compared with values ten years earlier. This is important for photometry, because it suggests that Tucson's light abatement strategies have been successful, despite increasing populations around Kitt Peak.

Finally, Stan Walker reports that members of Variable Stars South, RASNZ, are using standard DSLR cameras to observe Cepheids and eclipsing binaries to $V = 8.5$ with transformations to the standard $BVRc$ system accurate to 1% - 2% †.

3. Polarimetry (Pierre Bastien)

3.1. *Polarimetric Studies – Some Highlights*

With more than 14 000 hits on ADS during the period 2008 - 2011 (October), the subject of polarimetry is very active. Some particularly active areas are mentioned here. Stellar magnetism studies of more than 1 000 stars have been obtained with FORS1 at the VLT, HARPS at ESO and ESPaDOnS at the CFHT: chemically peculiar stars, hot stars [the MiMeS project; Wade *et al.* (2011)], Herbig Ae/Be stars and stars in clusters attract a significant attention. For example, it was found from the study of Ap/Bp stars in open clusters with known ages that the magnetic field strength decreases significantly during their Main Sequence lifetime [Landstreet *et al.* (2007), Landstreet *et al.* (2008)].

Circumstellar disks of young stars and exoplanets are under intense scrutiny. Hashimoto *et al.* (2011) reported high-resolution imaging polarimetry and detection of fine structures in the protoplanetary disk around the Herbig Ae star AB Aur with the high-contrast instrument HiCIAO on the Subaru thelescope. Their polarized intensity image in the H-band has a spatial resolution of only 9 AU. Quanz *et al.* (2011) resolved the disk around the Herbig Be star HD 100546 with polarimetric differential imaging using the AO assisted high-resolution camera NACO at the VLT in the H and K_s filters.

The Sun, pulsars, supernovae, active galaxies and Gamma-ray bursters continue to be favored targets for polarimetry.

3.2. *New Polarimetric Instruments*

A fast-switching spectropolarimeter has been used on the 1.8-m Plaskett telescope at the Dominion Astrophysical Observatory since 2007 mostly in support of the MiMeS project (see highlights above). A polarimetric unit, HARPSpol, has been installed for use with the HARPS spectrograph at the Cassegrain focus of the ESO 3.6-m telescope. See Piskunov *et al.* (2011).

High-contrast imaging polarimetry has been tested with ExPo at the William-Herschel Telescope for improving data reduction techniques in order to reach contrast ratios of 10^5. GPI for Gemini [McBride *et al.* (2011)] and SPHERE for the VLT [see Beuzit *et al.* (2010) and Schmid *et al.* (2010) for ZIMPOL, the polarimeter for SPHERE], both with polarimetric modes, are scheduled to be delivered to their respective telescopes some time in 2012.

The submillimeter polarimeter, POL-2, to be used with the SCUBA-2 detector on the James-Clerk-Maxwell Telescope has been installed on the SCUBA-2 cryostat in July 2010. Further work, i.e., early observations, completion of the data reduction software and commissioning of the instrument is pending the optimization of SCUBA-2 and its release to the astronomy community.

† http://www.variablestarssouth.org/index.php/research-projects/dslr-projects

4. Conferences

Conferences specifically on photometry and/or polarimetry were held during the triennium. Because of limited space, we do not attempt to report on the many other conferences on various astronomical topics in which photometric and polarimetric work was described and discussed!

The proceedings for the conference *Astronomical Polarimetry 2008 – Science from Small to Large Telescopes* which took place at the Fairmont Le Manoir Richelieu in La Malbaie, Québec, Canada are almost ready for publication [Bastien *et al.* (2011)].

Two sessions on the topic "Photometry: Past and Present" were organized by E. F. Milone and held at the Long Beach, California, meeting of the American Astronomical Society in January, 2009. Papers were presented on the development of: precise optical and infrared photometry (Milone, Sterken, and Young); CCD photometry (Steve Howell); the *UBVRI* system (Arlo Landolt); calibration in visual and infrared passbands (Martin Cohen); spectrophotometry (Saul Adelman); and polarimetry (Pierre Bastien). A number of relevant poster papers (e.g., on photometric systems by Robert Wing) and another oral paper on the Pierce-Blitzstein photometer of the Flower & Cook Observatory by Carol Ambruster *et al.*), were also noted. All of these authors contributed to a volume of papers on these topics, which has now been published (Milone & Sterken, 2011).

The conference, *Stellar Polarimetry: From Birth to Death* was held 27 – 30 June 2011 in Madison, WI, with 81 registered participants. The focus was on current problems and future opportunities in the area of stellar polarimetry, with applications for forming stars (T Tauri stars, Herbig Ae/Be stars, UCHII, etc.), main sequence stars (Ap and Bp stars, Zeeman Doppler Imaging, etc.), post-main sequence stars (red and blue supergiants, WR stars, LBVs, etc.), stellar deaths (SNe, GRBs) and circumstellar media (disks, stellar winds, interacting binaries, colliding winds). Theorists and observers met with the goal of improving our understanding the physical processes operating in and around stars. More information is available on the web site: http://arwen.etsu.edu/starpol/.

We note that several conferences are planned for 2012, as reported by Russ Genet and by J. Allyn Smith, thus guaranteeing further discussion of devlopments in photometry, photometric telescopes, and calibrations, in the next triennial report.

5. Infrared Astronomy Working Group (E. F. Milone)

A separate report for the IRWG may be found elsewhere in this volume, so the report of the working group here will be brief. The IRWG maintains a separate website†, but other, updated information, and references, may be found at another site, as well ‡. The IRWG continues to champion the use of the optimized IRWG passband set described initially in Young *et al.* (1994), for which a set of preliminary standards were provided in Milone & Young (2005). The optimization of passband width and central wavelength (assuming triangular spectral profiles) created a set of passbands in which the water vapour absorptions are minimized. In the current triennium, emphasis was placed on the usefulness of the near-IR set, and the iN passband, for sites at *all* elevations where photometry can be carried out. This includes observatories that have traditionally carried out only optical photometry, and it includes amateur astronomers who have acquired experience in infrared photometry, perhaps at a local university astronomical observatory. The AAVSO has begun to encourage amateur astronomers to observe with a relatively

† http://people.ucalgary.ca/~milone/IRWG/
‡ http://people.ucalgary.ca/~milone/oip.html#IRPP

inexpensive if uncooled system making use of the widely available Mauna Kea K_s filter, one of the better unoptimized filters.

Presentations emphasizing these points were made in several venues over the past few years. In the interval 2009-2011, a note was placed in the IAU daily newspaper during the IAU General Assembly in Rio de Janeiro in August, 2009, describing the suitability of these passbands for lower elevation sites, as well as for superior transformation properties and SN at the traditional high-elevation sites. Oral presentations on this theme were made also at the Telescopes from AFAR meeting in Waikalua, Hawaii, in February, 2011, and to the meeting of the Calgary Centre of the Royal Astronomical Society of Canada in September, 2011. The Waikalua presentation and the paper based on it (Milone & Young, 2011b) can be found on-line†.

We noted in these presentations that the organization of a "mass buy" of the Mauna Kea Near Infrared passbands in the mid 2000's, although inferior for photometric transformations of data obtained at traditional, high-elevation observatories, essentially precluded a mass purchase of the IRWG near-IR set. Of course, observatories will have to obtain new filters to replace old interference filters within a few years, because multilayer coatings can scarcely be expected to remain intact for as long as a decade, even if kept under vacuum conditions. Therefore it is our hope and expectation that the IRWG filters (thus far produced only by Custom Scientific, Inc., of Phoenix, AZ) ‡ will become the filters of choice for future infrared filter purchases, and so become available for precise IR photometry. See the IRWG report for more details and IR work carried out during the 2009-2011 triennium.

6. Ongoing and Future Work of Commission 25

The various large-scale surveys are generating a flood of new standard stars and standard star observations. The Commission believes that some form of coordination among the various initiatives would be helpful to the astronomical community at large. The Commission is thus planning to develop an IAU standard star data portal. This will be a one-stop location for all standard star data access. It would be an internet gateway to various recommended data servers. The information on the portal would be quality-controlled by a group of experts, so that observers could easily choose suitable and reliable standard stars that they need to use for their observations.

Apart from providing reliable standard star data, the envisaged data base could provide information on how to use those standards properly. It has been suggested that the Commission could compile educational material in the form of a "cookbook" on methods of doing and using photometry. In this regard the proceedings of the conference on *The Future of Photometric, Spectrophotometric and Polarimetric Standardization*, held in Blankenberge, Belgium in May 2006, would be particularly useful. A working group could take up these ideas at the next General Assembly. In keeping with this, it has been suggested that the Commission could consider organizing a symposium on standardization topics across the electromagnetic spectrum.

With regard to polarimetry, the need for faint polarized standard stars for large telescopes was raised at the Astronomical Polarimetry 2008 meeting, described in the report of the previous triennium (Martinez *et al.*, 2009). It was suggested that Commission 25 would be an appropriate forum to collate all the results in one place. A list of papers

---

† http://tfa.cfht.hawaii.edu/papers/milone_tfa_paper.pdf
‡ http://www.customscientific.com/

containing polarized standards, bright and faint, should be in place on the Commission 25 web site in 2012.

Finally, a perennial problem is the proliferation of names of passbands. For example, the "Z" and "Y" designations of passbands in the near infrared by Warren and Hewett (2002), or the iz, iJ, iH, and iK designations for near Infrared passbands by Young et al. (1994). The "Y" is similar although not identical to the "iz" passband, and both are similar to a passband designated "Y" by Hillenbrand et al. (2002). Unfortunately, the "Y" designation was already given to a Vilnius passband, more than 40 years ago. Further discussion of passband designations for this spectral region can be found in Milone & Young (2008), and a discussion of other designation conflicts regarding the Johnson infrared passband designations can be found in Milone & Young (2011a). In addition, a new "y", not related the Strömgren "y" has recently been appropriated for an infrared passband with a central wavelength of 990 nm by High et al. (2010). The new filter is to be used with a red-enhamced CCD. The authors, themselves, decry such "degeneracy in terminology" but note that it "seems likely to persist." If something is to be done about providing guidelines to avoid such confusion, Commission 25 should do it. Consequently the Organizing Committee is reviewing the problem and may be making recommendations to the IAU at the Beijing General Assembly.

6.1. Acknowledgements

Milone thanks the members of the Organizing Committee of Commission 25 who provided helpful suggestions, comments, and summaries during the triennium 2009 – 2011. We thank also members of the Commission who apprised us of their activities. Carme Jordi kindly supplied an update on Gaia project photometry.

References

Adelman, S. 2011, in E. F. Milone and C. Sterken, eds., *Astronomical Photometry: Past, Present and Future*, Springer Science+Business Media, ASSL 373, 187

Ambruster, C. W., Hull, A. B., Koch, R. H., & Mitchell, R. J. 2011, in E. F. Milone & C. Sterken, eds., *Astronomical Photometry: Past, Present and Future*, Springer Science+Business Media, ASSL 373, 83

Aparicio Villegas, T., et al. 2010, *AJ*, 139, 1242

Auvergne M., et al., 2009, *A&A*, 506, 411

Bastien, P. 2011, in E. F. Milone & C. Sterken, eds., *Astronomical Photometry: Past, Present and Future*, Springer Science+Business Media, ASSL 373, 199

Bastien, P., Manset, N., Clemens, D. P., & St-Louis, N., eds., 2011, *Astronomical Polarimetry 2008 – Science from Small to Large Telescopes*, Proc. Intern. Conf., La Malbaie, Québec, Canada, *ASP-CS*, 499

Beuzit, J.-L., et al. 2010, in: A. Bocaletti (ed.), *In the Spirit of Lyot 2010: Direct Detection of Exoplanets and Circumstellar Disks*, Proc. conf. Univ. de Paris Diderot, Paris, France, 25-29 October 2010.

Brown, T. M., Latham, D. W., Everett, M. E., & Esquerdo, G. A. 2011, *AJ*, 142, 112

Bruntt H., et al. 2009, *MNRAS*, 396, 1189

Buie, M. W., Trilling, D. E., Wasserman, L. H., & Crudo, R. A. 2011, *ApJS*, 194, 40

Cohen, M. 2011, in E. F. Milone and C. Sterken, eds., *Astronomical Photometry: Past, Present and Future*, Springer Science+Business Media, ASSL 373, 177

Eyer L. 2006, *MmSAI*, 77, 549

Gilliland R. L., et al. 2011, arXiv:1107.5207

Gruberbauer M., et al. 2011, *A&A*, 530, A135

Hashimoto, J., et al. 2011, *ApJL*, 729, L17

High, F. W., et al. 2010, *PASP*, 122, 722

Hillenbrand, L. A., Foster, J. B., Persson, S. E., & Matthews, K. 2002, *PASP*, 114, 708

Hodgkin, S. T., Irwin, M. J., Hewett, P. C., & Warren, S. J. 2009, *MNRAS*, 394, 675

Howell, S. 2011, in E. F. Milone and C. Sterken, eds., *Astronomical Photometry: Past, Present and Future*, Springer Science+Business Media, ASSL 373, 69

Ivezić, Ž., *et al.* 2008, AIPC, 1082, 359

Jordi, C., *et al.* 2010, *A&A*, 523, 48

Kaiser, M. E., *et al.* 2009, *BAAS*, 41, 437

Karate, Y. & Schuster, W. J. 2010, *New A*, 15, 444

Kolenberg, K., *et al.*, 2011, *MNRAS*, 411, 878

Kurtz, D. W., *et al.* 2005, *MNRAS*, 358, 651

Landolt, A. U. *AJ*, 137, 4186

Landolt, A. U. 2011, in E. F. Milone & C. Sterken, eds., *Astronomical Photometry: Past, Present and Future*, Springer Science+Business Media, ASSL 373, 107

Landstreet, J. D., *et al.* 2007, *A&A*, 470, 685

Landstreet, J. D., *et al.* 2008, *A&A*, 481, 465

Martinez, P., *et al.* 2009, in *Reports on Astronomy 2006-2009*, Karel A. van der Hucht, ed., *IAU Transactions*, /bf 4, Issue XXVIIA, 304

McBride, J., *et al.* 2011, *PASP*, 123, 692

Milone, E. F. & Pel, J. W. 2011, E. F. Milone, and C. Sterken, eds., *Astronomical Photometry: Past, Present and Future*, Springer Science+Business Media, ASSL 373, 33

Milone, E. F. & Sterken C, eds. 2011, *Astronomical Photometry: Past, Present and Future*, Springer Science+Business Media, *ASSL* 373

Milone, E. F. & Young, A. T. 2005, *PASP*, 117, 485

Milone, E. F. & Young, A. T., 2008, *JAAVSO*, 36, 110

Milone, E. F. & Young, A. T., 2011a, in E. F. Milone & C. Sterken, eds., *Astronomical Photometry: Past, Present, and Future*, (New York: Springer) ASSL 373, 125

Milone, E. F. & Young, A. T., 2011b, http://tfa.cfht.hawaii.edu/papers/milone_tfa_paper.pdf

Neugent, K. F. & Massey, P., 2010, *PASP*, 122, 1246

Pickles, A. & Depagne, É. 2011, *PASP*, 122, 1437

Piskunov, N., *et al.* 2011, *Msngr*, 143, 7

Postma, J., Hutchings, J. B., & Leahy, D. 2011, *PASP*, 123, 833

Quanz, S. P., *et al.* 2011, *ApJ*, 738, 23

Schmid, H. M., *et al.* 2010, in: A. Bocaletti (ed.), *In the Spirit of Lyot 2010: Direct Detection of Exoplanets and Circumstellar Disks*, Proc. conf. Univ. de Paris Diderot, Paris, France, 25-29 October 2010.

Siegel, M. H. *et al.* 2010, *ApJ* 725, 1215

Southworth, J. *et al.* 2009, *MNRAS*, 396, 1023

Sterken, C., Milone, E. F., & Young, A. T. (2011) in E. F. Milone, and C. Sterken, eds., *Astronomical Photometry: Past, Present and Future*, Springer Science+Business Media, ASSL 373, 1

Stubbs, C. W., *et al.* 2007, *PASP*, 119, 1163

Torres, G. 2010, *AJ*, 140, 1158

Wade, G. A., *et al.* 2011, in: *Active OB stars: structure, evolution, mass loss, and critical limits*, Proc. IAU Symp., 272, 118

Warren, S. & Hewett, P. (2002), in N. Metcalfe and T. Shanks, eds., ASP Conf. Ser. Vol. 283, *A New Era in Cosmology*, Astron. Soc. Pac., San Francisco, p. 369

Wing, R. F. 2011, in E. F. Milone & C. Sterken, eds., *Astronomical Photometry: Past, Present and Future*, Springer Science+Business Media, ASSL 373, 143

Young, A. T., Milone, E. F., & Stagg, C. R. 1994, *A&PS*, 105, 259

Transactions IAU, Volume XXVIIIA
Reports on Astronomy 2009–2012
Ian Corbett, ed.

© International Astronomical Union 2012
doi:10.1017/S1743921312002979

COMMISSION 30

RADIAL VELOCITIES
VITESSES RADIALES

PRESIDENT	Guillermo Torres
VICE-PRESIDENT	Dimitri Pourbaix
PAST PRESIDENT	Stephane Udry
ORGANIZING COMMITTEE	Geoffrey W. Marcy,
	Robert D. Mathieu, Tsevi Mazeh,
	Dante Minniti, Claire Moutou,
	Francesco Pepe, Catherine Turon,
	Tomaz Zwitter

COMMISSION 30 WORKING GROUPS

Div. IX / Commission 30 WG	Stellar Radial Velocity Bibliography
Div. IX / Commission 30 WG	Radial Velocity Standards
Div. IX / Commission 30 WG	Catalogue of Orbital Elements of
	Spectroscopic Binary Systems (SB9)

TRIENNIAL REPORT 2009-2012

1. Introduction

The past three-year period has seen steady efforts to collect large numbers of radial-velocity (RV) measurements, as well as important applications of radial velocities to astrophysics. Improvements in precision continue to be driven largely by exoplanet research. A workshop entitled "Astronomy of Exoplanets with Precise Radial Velocities" took place in August of 2010 at Penn State University (USA), and was attended by some 100 researchers from around the world. The meeting included thorough discussions of the current capabilities and future potential of the radial velocity technique, as well as data analysis algorithms to improve precision at visible and near-infrared wavelengths.

While most of these discussions were focused on the search for and characterization of exoplanets, it is clear that more classical applications of radial velocities are also benefiting from the improvements, as evidenced by recent work on binary stars described herein. Below is a summary of other activity in the field of radial velocities during this triennium. Due to space limitations, we include only a selection of efforts and results in this area.

2. Large-scale radial-velocity surveys (T. Zwitter and G. Torres)

The RAdial Velocity Experiment (RAVE; http://www.rave-survey.org) is an ongoing international collaboration of ~60 scientists from nine countries, led by M. Steinmetz from the AIP in Potsdam. It is continuing to use the UK Schmidt telescope at the Australian Astronomical Observatory to record a large unbiased sample of stellar spectra selected only by their I-band magnitude. During this triennium RAVE publicly released its full pilot survey Siebert *et al.* 2011a, which contains 86,223 RV measurements for 81,206 stars in the southern hemisphere. In addition, stellar parameters for 42,867 of the stars were published. Altogether RAVE has already collected over 500,000 spectra, with approximately 10% of the observing time being devoted to repeat observations. The mean radial velocity error is $\sim$2 km s^{-1}, and 95% of the measurements have an internal error better than 5 km s^{-1}. This can be combined with distances based on 2MASS photometry, spectroscopically determined values of the stellar parameters, and stellar isochrone fitting (see Breddels *et al.* 2010, Zwitter *et al.* 2010, Burnett *et al.* 2011). Such distances are accurate to $\sim$20% and cluster around 300 pc for dwarfs and 1 or 2 kpc for giants.

This collection of information allows comprehensive studies of the kinematics of our part of the Galaxy, as well as its structure and formation history. RAVE is suited to searching for stellar streams, some of which are remnants of dwarf galaxies that merged with the Milky Way during galaxy formation. A new stream, dubbed the Aquarius stream, is an example of such remnants that can be found with RAVE (Williams *et al.* 2011). Another kind of stellar streams, known as moving groups, are born inside our Galaxy. New members of nearby moving groups have been found in RAVE (Kiss *et al.* 2011), and the survey promises to reveal more in the future. RAVE allows for efficient searches for the very first stars (Fulbright *et al.* 2010), and enables the detection of interesting trends in the motions of the stars in the vicinity of the Sun (Siebert *et al.* 2011b). The survey is well suited to study our Galaxy's thick disk. Two recent studies from RAVE (Wilson *et al.* 2011, Ruchti *et al.* 2011) have focused on uncovering its origin. The survey will continue in 2012.

Another large-scale survey released during this triennium that includes radial-velocity measurements (albeit of low precision) for vast numbers of stars is SEGUE (Sloan Extension for Galactic Understanding and Exploration). A paper by Yanny *et al.* (2009) describes these spectroscopic results, which are based on some 240,000 low-resolution ($R \sim 1800$) spectra of fainter Milky Way stars down to a magnitude limit of $g \approx 20.3$. One of the goals is to enable studies of the kinematics and populations of our Galaxy and its halo. The RV precision varies from 4 km s^{-1} at the bright end ($g \approx 18$) to 15 km s^{-1} at the faint end. In addition to the velocities, atmospheric parameters including effective temperature, surface gravity, and metallicity were derived for the stars with suitable signal-to-noise ratios. The individual spectra along with associated parameters are publicly available as part of the Sloan Digital Sky Survey Data Release 7.

3. The role of radial-velocity measurements in studies of stellar angular momentum evolution and stellar age (S. Meibom)

Radial-velocity measurements with multi-object spectrographs have played a critical role in defining the mile-posts that are the foundation for much of our understanding of the time-evolution of stars. These mile-posts are star clusters — coeval, cospatial, and chemically homogeneous populations of stars over a range of masses for which the age can be determined well by fitting model isochrones to single cluster members in the color-magnitude diagram (CMD). However, the inherent qualities of clusters can only be fully exploited if pure samples of kinematic members are identified and characterized. This can be accomplished most securely and effectively with radial-velocity measurements (e.g., Geller *et al.* 2008, Hole *et al.* 2009).

Recent dedicated photometric surveys for stellar rotation periods in young clusters have begun to see dependencies of stellar rotation on stellar age and mass. These dependencies guide our understanding of the angular momentum evolution of FGK dwarfs by determining the mass- and time-dependence of their rotation periods. Over the past three years such surveys have been combined with radial-velocity surveys for cluster membership and binarity in open clusters with different ages, revealing well-defined relations between stellar rotation period, color (mass), and age not previously discernible (Meibom *et al.* 2009a, Meibom *et al.* 2011a, Meibom *et al.* 2011b). These relations offer crucial new constraints on internal and external angular momentum transport and on the evolution of stellar dynamos in late-type stars of different masses.

Furthermore, stellar rotation has emerged as a promising and distance-independent indicator of age ("gyrochronology"; Kawaler 1989, Barnes 2003, Barnes 2007), and open clusters fulfill an important role in calibrating the relation between age, rotation, and mass. Indeed, open clusters can define a surface in the three-dimensional space of stellar rotation period, mass, and age, from which the latter can be determined from measurements of the former two. It is critical, however, to establish the cluster ages from CMDs in which non-members have been removed and single members identified. It is also important to identify short-period binaries where tidal mechanisms may have modified the stellar rotation. Radial velocities are an efficient and proven technique to identify both single and short-period binary members. The tight mass-rotation relations seen in clusters over the past three years reflect the powerful combination of time-series spectroscopy for cluster membership and time-series photometry for rotation periods.

4. Radial velocities in open clusters (R. Mathieu)

Studies of kinematic membership and binarity in open clusters based on radial-velocity measurements have a long history. During this triennium the WIYN Open Cluster Study (WOCS; Mathieu 2000) has continued to acquire intermediate-precision ($\sigma_{RV} = 0.4 \, \mathrm{km \, s^{-1}}$) radial-velocity measurements on its core open clusters. Currently the project has in hand a total of more than 60,000 measurements of some 11,800 stars in the open clusters M34, M35, M37, M67, NGC 188, NGC 2506, NGC 6633, NGC 6819, and NGC 7789. Some of these data and associated results have already appeared in the literature (Geller *et al.* 2008, Geller *et al.* 2009, Geller *et al.* 2010, Hole *et al.* 2009, Meibom *et al.* 2009a, Meibom *et al.* 2009b, Meibom *et al.* 2011b).

Particularly notable progress has been made on understanding the nature of blue stragglers in the open cluster NGC 188 (Mathieu & Geller 2009, Geller & Mathieu 2011). Sixteen of the 21 blue stragglers are spectroscopic binaries. These binaries have a remarkable eccentricity versus log period distribution, with all but two having periods within a decade of 1000 days. The two short-period binaries are double-lined, one of which comprises *two* blue stragglers. A statistical analysis of the single-lined binary mass functions shows the secondary mass distribution to be narrowly confined around a mass of 0.5 $M_\odot$. The combination of these results strongly suggests a mass-transfer origin for the blue stragglers, leaving behind white dwarf companions. However, the shortest period binaries are certainly the product of dynamical encounters, leaving open the possibility of collisional origins for those blue stragglers.

5. Toward higher radial-velocity precision (F. Pepe, C. Moutou, C. Lovis)

Broadly speaking, this period has been characterized by three general trends regarding precise radial-velocity measurements as applied to exoplanet research. Firstly, the precision has been pushed to its limits such that very small RV signals even below 1 $\mathrm{m \, s^{-1}}$

have now been detected. This capability has revealed a large population of super-Earths and Neptunes, demonstrating that they are common around solar-type stars in the Milky Way (Howard *et al.* 2011, Mayor *et al.* 2011). Secondly, Doppler shift measurements have become a tool complementary to other techniques, and in particular to the transit method of detecting exoplanets. And thirdly, RVs are moving into the near-infrared domain. The combination of red wavelengths and very high spectral resolution has only become available in recent years, but has already brought on previously unavailable opportunities for the observation of stars that are very young, very active, or of very late spectral type, and opened up possibilities for the detection of planetary signatures among those stars.

Recent developments have demonstrated that at the few $m\,s^{-1}$ level the star is not necessarily the limiting factor, and that there is good reason to aim for *sub*-$m\,s^{-1}$ instrumental precision (see, e.g., Pepe *et al.* 2011) provided the star is chromospherically quiet and that photon noise is not the limit. Considerable progress has been made on instrumental issues. One of the limiting factors has been the non-uniform illumination of the spectrograph, where even the use of (circular) fibers does not remove this problem entirely. Non-circular fibers have shown great promise for their scrambling properties, although much of this work has not yet appeared in the literature. One exception is the study by Perruchot *et al.* (2011) with octagonal-section fibers. Using these devices it has been possible to improve the RV precision on the SOPHIE spectrograph mounted on the 1.93 m OHP telescope from about 8 $m\,s^{-1}$ to 1.5 $m\,s^{-1}$. The other important factor limiting instrumental precision is the wavelength calibration. The two main techniques used for this (thorium-argon lamps, and the iodine cell method) have a limited wavelength coverage, suffer from line blending, and have other drawbacks (including large dynamic range for the lines and limited lifetime of hollow-cathode lamps, and light absorption as well as sensitivity to ambient conditions for the iodine cell). The use of laser frequency combs as a path to achieving $cm\,s^{-1}$ precision has been explored for several years, and a number of these systems are now under development for both visible (Osterman *et al.* 2007, Steinmetz *et al.* 2008, Li *et al.* 2009) and infrared wavelengths (Osterman *et al.* 2011, Schettino *et al.* 2011). Challenges still remain, but the expectation is that these devices will be available on several telescopes around the world on a timescale of a few more years.

The problems posed by the spectrograph illumination and wavelength calibration will likely be solved soon. Present-generation spectrographs are already implementing solutions to those challenges based on the technologies mentioned above. At the $cm\,s^{-1}$ level, however, stellar "jitter" will still be an important source of error. Current efforts to overcome this have focused on filtering the stellar noise contribution (*p* modes, granulation, activity) by applying optimal observation strategies (see, e.g., Dumesque *et al.* 2011). Future planet search programs requiring extremely high precision will likely have to pre-select targets with very low or very well-known stellar jitter, so that these effects either have minimal impact on the RVs, or can be modeled and removed. And of course, beating down photon noise in the search for Earth-like planets will require ever larger telescopes, or restricting the searches to relatively bright stars.

Achieving very high velocity precision in the near-infrared has so far lagged behind the optical regime. Performance at the $m\,s^{-1}$ has not yet been achieved, although 5 $m\,s^{-1}$ has been demonstrated in a few cases (e.g., Bean *et al.* 2010, Figueira *et al.* 2010). The problems to be overcome include the treatment of telluric lines, detector technology, and cryogenic optics.

Several new radial-velocity instruments are presently under construction that should come online in the next few years. A non-exhaustive list with an indication of the wavelength regime, telescope on which they will be mounted, and expected first-light date includes HARPS-N (visible, TNG, 2012), PEPSI (visible, LBT, 2012), GIANO (IR, TNG, 2012), HZPF (IR, HET, 2013), CARMENES (visible-IR, 3.6 m Calar Alto, 2014), SPIROU (IR, CFHT, 2015), and ESPRESSO (visible, VLT, 2016).

6. High-precision radial velocities applied to studies of binary stars (G. Torres)

As indicated above, one of the procedures used in exoplanet research for ensuring high precision in the radial-velocity measurements relies on an iodine cell in front of the spectrograph slit to track instrumental drifts and changes in the point-spread function that normally lead to systematic errors (see, e.g., Marcy & Butler 1992, Butler *et al.* 1996). Some years ago Konacki (2005) extended the iodine technique to composite spectra, showing that precisions of a few tens of $m s^{-1}$ can be reached in selected double-lined spectroscopic binaries. This enables considerably higher precision to be obtained for the masses of binary stars than has usually been achieved (see also earlier work by Lacy 1992).

A recent study by Konacki *et al.* (2010) focused on a handful of favorable (nearly edge-on) binaries, and combined spectroscopy with long-baseline interferometric observations, which yield the inclination angle of the orbit, to achieve record precision for one of their systems, HD 210027. Relative errors in the masses are as low as 0.066%, the smallest obtained for any normal star. Other studies by the same group have also reached very small uncertainties (Hełminiak & Konacki 2011, Hełminiak *et al.* 2011), made possible by the much improved velocities using their technique. The precision of the masses of HD 210027 rivals that of the best known determinations in double neutron star systems, measured by radio pulsar timing.

7. Doppler boosting effect (T. Mazeh)

In the last two years a new type of stellar radial-velocity measurement has emerged, based on the photometric beaming (aka Doppler boosting) effect. This causes the bolometric flux of a star to increase or decrease as it moves toward or away from the observer, respectively. The magnitude of the beaming effect is approximately $4V_r/c$, where V_r is the stellar radial velocity and c is the speed of light, and is therefore on the order of 10^{-3} to 10^{-4} of the stellar intensity for a solar-type star with a stellar secondary and a period of 10 days or so.

While the beaming effect had been observed previously from the ground in one or two very favorable cases (e.g., Maxted *et al.* 2000), the availability of a quarter of a million very precise, continuous light curves produced by the *CoRoT* and *Kepler* missions has opened the door to the detection of new binary systems by this method (see also Loeb & Gaudi 2003, Zucker *et al.* 2007, Faigler & Mazeh 2011). Seven new *non-eclipsing* binaries with orbital periods between 2 and 6 days have already discovered by this effect in the *Kepler* data, and were confirmed by classical spectroscopic radial-velocity measurements (Faigler *et al.* 2011). The effect has now also been seen in ground-based photometry of two extremely short period double white dwarf eclipsing binaries with periods of 5.6 and 0.2 hours (Shporer *et al.* 2010, Brown *et al.* 2011), as well as in other eclipsing systems observed by *Kepler* that also contain white dwarfs (van Kerkwijk *et al.* 2010, Carter *et al.* 2011, Breton *et al.* 2011).

8. Working Groups (H. Levato, G. Marcy, D. Pourbaix)

Below are the reports of the three active working groups of Commission 30. Their efforts are focused on providing a service to the astronomical community at large through the compilation of a variety of information related to radial velocities.

8.1. *WG on Stellar Radial Velocity Bibliography (Chair: H. Levato)*

This WG is a very small one that was created with the purpose of continuing the cataloging of the bibliography of radial velocities of stars made by Mme Barbier in successive catalogues until her retirement in 1990 (see Barbier-Brossat *et al.* 1990).

The new compilation was started late in 1990. The first version of the catalogue after the retirement of Mme Barbier was published for the 1991–1994 triennium. The catalogue is updated every six months at the following web page:

http://www.icate-conicet.gob.ar/basededatos.html

During the 2009–2011 period the WG searched 33 journals for papers containing measurements of the radial velocities of stars. As of December 2010 a total of 198,063 entries had been cataloged. By the end of 2011 this is expected to increase to about 285,000 records. It is worth mentioning that at the end of 1996 the number of entries was 23,358, so that in 15 years the catalogue has grown by more than an order of magnitude. The main body of the catalogue includes information about the technical characteristics of the instrumentation used for the radial velocity measurements, and comments about the nature of the objects.

The future of radial velocities is becoming very attractive and the same time more complex. Large numbers of new radial velocity measurements are expected to be published, and it may be necessary to discuss if the present approach is the best way to keeping a record of the bibliography of radial velocity measurements.

8.2. *WG on Radial Velocity Standards (Chair: S. Udry)*

During this triennium significant progress has been made towards establishing lists of stars that can serve as radial-velocity standards, to a much higher level of precision than lists that have been used in the past. Two main efforts have taken place.

One was summarized by Crifo *et al.* (2010), who report the compilation of an all-sky list of 1420 relatively bright (mostly $V \approx 6$–10) stars developed specifically for use by the Gaia project, but which is of course very useful to the broader community. The list is based largely on measurements published by Nidever *et al.* (2002), Nordström *et al.* 2004, and Famaey *et al.* (2005). The radial velocities of most these stars are believed to be accurate at the ~ 300 m s^{-1} level, and a large fraction of them are being re-observed at higher precision with modern instruments (SOPHIE, NARVAL, CORALIE; see Chemin *et al.* 2011). It is expected that the accuracy will be improved to 100 or possibly 50 m s^{-1} when this task is concluded. A link to this list of potential standards is available on the Commission web page.

A parallel effort reported by Chubak & Marcy (2011) has been carried out by the California Planet Search group using the HIRES spectrometer on the Keck I telescope. They present radial velocities with an accuracy (RMS compared to present IAU standards) of 100 m s^{-1} for 2086 stars of spectral type F, G, K, and M based on some 29,000 spectra. Additional velocities are presented for 132 RV standard stars, all of which exhibit constant radial velocity for at least 10 years, with an RMS less than 10 m s^{-1}. All velocities were measured relative to the solar system barycenter and are placed on the velocity zero-point scale of Nidever *et al.* (2002). They contain no corrections for convective blueshift or gravitational redshift. An innovation was to determine a secure wavelength zero-point for each spectrum by following the suggestion of Roger Griffin in using telluric lines (the origin of the iodine cell concept). Specifically, they used the telluric A and B bands at 7594–7621 Å and 6867–6884 Å, respectively, which were present in all of the spectra. This allows to correct for small changes in the CCD position, the spectrometer optics, and guiding errors for the specific observation of the program star.

There is a significant overlap between the lists of Crifo *et al.* (2010) and Chubak & Marcy (2011), providing excellent radial velocity integrity for the stars in common. It is expected that the combination of these lists will serve as standards for studies of long-period binary stars, star cluster dynamics, and for surveys of the chemical and dynamical structure of the Galaxy such as SDSS, RAVE, Gaia, APOGEE, SkyMapper, HERMES, and LSST.

8.3. *WG on the Catalogue of Orbital Elements of Spectroscopic Binaries (SB9) (Chair: D. Pourbaix)*

At the 2000 General Assembly in Manchester, a WG was set up to work on the implementation of the 9th Catalogue of Orbits of Spectroscopic Binaries (SB9), superseding the 8th release of Batten *et al.* (1989) (SB8). SB9 exists in electronic format only. The web site (http://sb9.astro.ulb.ac.be) was officially released during the summer of 2001. This site is directly accessible from the Commission 26 web site, from BDB (in Besançon), and from the CDS, among others.

Substantial progress have been made since the last report, in particular in the way complex multiple systems can be uploaded together with their radial velocities. The way data weights can be supplied has also been improved.

As of this writing the SB9 contains 3039 systems (SB8 had 1469) and 3784 orbits (SB8 had 1469). A total of 623 papers were added since August 2000, with most of them coming from *outside* the WG. A significant number of papers with orbits still await uploading into the catalogue. According to the ADS, the release paper (Pourbaix *et al.* 2004) has received 152 citations since 2005. This is about three times more than the old Batten *et al.* catalogue over the same period, with the SB8 still being cited in the current literature.

The important work of cross-checking the identification of systems is carried out by the CDS (Strasbourg). Indeed, with the SBC9 identifier now added to SIMBAD, each new release of the SB9 tar ball is cross checked for typos prior to integration at the CDS. Whereas some of these mistakes are ours, some authors share the responsibility as well. Users have also helped in pinning down some problems.

Although this work is very welcome by the community (about 500–1000 successful queries received every month, with 50 distinct IP addresses over the past month) and some tools have been designed to make the job of entering new orbits easier (input file checker, plot generator, etc.), the WG still suffers from a serious lack of manpower. Few colleagues outside the WG spontaneously send their orbits (though they are usually happy to send their data when we asked). Any help from authors, journal editors, etc., is therefore very welcome. Uploading an orbit into SB9 also means checking it against typographical errors. In this way we have found a number of mistakes in published solutions. Sending orbits to SB9 prior to publication (e.g., at the proof stage) would therefore be a way to prevent some mistakes from making their way into the literature.

Guillermo Torres
president of the Commission

References

Barbier-Brossat, M., Petit, M., & Figon, P. 1990, *A&AS*, 85, 885

Barnes, S. A. 2003, *ApJ*, 586, 464

Barnes, S. A. 2007, *ApJ*, 669, 1167

Batten, A. H., Fletcher, J. M., & MacCarthy, D. G. 1989, Eighth catalogue of the orbital elements of spectroscopic binary systems, *Publ. Dom. Astr. Obs.*, 17, 1

Bean, Jacob L., Seifahrt, A., Hartman, H. *et al.* 2010, *ApJ*, 711, L19

Breddels, M. A., Smith, M. C., Helmi, A. *et al.* 2010, *A&A*, 511, 90

Breton, R. P., Rappaport, S. A., van Kerkwijk, M. H., & Carter, J. A. 2011, *ApJ*, in press (arXiv:1109.6847)

Brown, W. R., Kilic, M., Hermes, J. J. *et al.* 2011, *ApJ*, 737, L23

Butler, R. P., Marcy, G. W., Williams, E., McCarthy, C., Dosanjh, P., & Vogt, S. S. 1996, *PASP*, 108, 500

Burnett, B., Binney, J., Sharma, S. *et al.* 2011, *A&A*, 532, A113

Carter, J. A., Rappaport, S., & Fabrycky, D. 2011, *ApJ*, 728, 139

Chemin, L., Soubiran, C., Crifo, F., Jasniewicz, G., Katz, D., Hestroffer, D., & Udry, S. 2011, to appear in *SF2A2011 conference proceedings*, eds. G. Alecian, K. Belkacem, S. Collin, R. Samadi & D. Valls-Gabaud, arXiv:1110.2944

Chubak, C. & Marcy, G. W. 2011, *ApJ*, submitted

Crifo, F., Jasniewicz, G., Soubiran, C., Katz, D., Siebert, S., Veltz, L., & Udry, S. 2010, *A&A*, 524, A10

Dumusque, X., Udry, S., Lovis, C., Santos, N. C., & Monteiro, M. J. P. F. G. 2011, *A&A*, 525, 140

Faigler, S. & Mazeh, T. 2011, *MNRAS*, 415, 3921

Faigler, S., Mazeh, T., Quinn, S. N., Latham, D. W., & Tal-Or, L. 2011, *ApJ*, in press (arXiv:1110.2133)

Famaey, B., Jorissen, A., Luri, X., Mayor, M., Udry, S., Dejhonghe, H. l, & Turon, C. 2005, *A&A*, 430, 165

Figueira, P., Pepe, F., Melo, C. H. F. *et al.* 2010, *A&A*, 511, A55

Fulbright, J. P., Wyse, R. F. G., Ruchti, G. R. *et al.* 2010, *ApJ*, 724, 104

Geller, A. M. & Mathieu, R. D. 2011, *Nature*, 478, 356

Geller, A. M., Mathieu, R. D., Braden, E. K. *et al.* 2010, *AJ*, 139, 1383

Geller, A. M., Mathieu, R. D., Harris, H. C., & McClure, R. D. 2008, *AJ*, 135, 2264

Geller, A. M., Mathieu, R. D., Harris, H. C., & McClure, R. D. 2009, *AJ*, 137, 3743

Hełminiak, K. G. & Konacki, M. 2011, *A&A*, 526, A29

Hełminiak, K. G., Konacki, M., Muterspaugh, M. W. *et al.* 2011, *MNRAS*, in press (arXiv:1109.5059)

Hole, K. T., Geller, A. M., Mathieu, R. D. *et al.* 2009, *AJ*, 138, 159

Howard, A., Marcy, G. W., Johnson, J. A. *et al.* 2011, *BAAS*, 217, #415.06

Kawaler, S. D. 1989, *ApJ*, 343, L65

Kiss, L. L., Moór, A., Szalai, T. *et al.* 2011, *MNRAS*, 411, 117

Konacki, M. 2005, *ApJ*, 626, 431

Konacki, M., Muterspaugh, M. W., Kulkarni, S. R., & Hełminiak, K. G. 2010, *ApJ*, 719, 1293

Lacy, C. H. 1992, in *Complementary Approaches to Double and Multiple Star Research*, IAU Colloq. 135, eds. H. A. McAlister & W. I. Hartkopf, ASP Conf. Ser. Vol 32 (San Francisco: ASP), 152

Li, C.-H., Benedick, A., Blake, C. *et al.* 2009, *BAAS* 213, #601.01

Loeb, A. & Gaudi, B. S. 2003, *ApJ*, 588, L117

Marcy, G. W. & Butler, R. P. 1992, *PASP*, 104, 270

Mathieu, R. D. 2000, in *Stellar Clusters and Associations: Convection, Rotation, and Dynamos*, ASP Conf. Ser. Vol. 198, eds. R. Pallavicini, G. Micela & S. Sciortino (San Francisco: ASP), 517

Mathieu, R. D. & Geller, A. M. 2009, *Nature*, 462, 1032

Mayor M., Marmier M., Lovis, Ch. *et al.* 2011, *A&A*, submitted (arXiv1109.2497)

Maxted, P. F. L., Marsh, T. R., & North, R. C. 2000, *MNRAS*, 317, L41

Meibom, S., Barnes, S. A., Latham, D. W. *et al.* 2011a, *ApJ*, 733, L9

Meibom, S., Grundahl, F., Clausen, J. V. *et al.* 2009b, *AJ*, 137, 5086

Meibom, S., Mathieu, R. D., & Stassun, K. G. 2009a, *ApJ*, 695, 679

Meibom, S., Mathieu, R. D., Stassun, K. G., Liebesny, P., & Saar, S. H. 2011b, *ApJ*, 733, 115

Nidever, D. L., Marcy, G. W., Butler, R. P., Fischer, D. A., & Vogt, S. S. 2002, *ApJS*, 141, 503

Nordström, B., Mayor, M., Andersen, J. *et al.* 2004, *A&A*, 418, 989

Osterman, S., Diddams, S., Beasley, M. *et al.* 2007, *SPIE*, 6693, 44

Osterman, S., Diddams, S., Quinlan, F. *et al.* 2011, *Research, Science and Technology of Brown Dwarfs and Exoplanets: Proceedings of an International Conference held in Shangai on Occasion of a Total Eclipse of the Sun, Shangai, China*, eds. E. L. Martin, J. Ge & W. Lin, EPJ Web of Conferences, Vol. 16, id.02002, 160, 2002

Pepe, F., Lovis, C., Ségransan, D. *et al.* 2011, *A&A*, 534, A58

Perruchot, S., Bouchy, F., Chazelas, B. *et al.* 2011, *SPIE*, 8151, 37

Pourbaix, D., Tokovinin, A. A., Batten, A. H. *et al.* 2004, *A&A*, 424, 727

Ruchti, G. R., Fulbright, J. P., Wyse, R. F. G. *et al.* 2011, *ApJ*, 737, 9

Schettino, G., Baffa, C., Giani, E., Inguscio, M., Oliva, E., Tozzi, A., & Cancio Pastor, P. 2011, *SPIE*, 7808, 44

Shporer, A., Kaplan, D. L., Steinfadt, J. D. R. *et al.* 2010, *ApJ*, 725, L200

Siebert, A., Famaey, B., Minchev, I. *et al.* 2011b, *MNRAS*, 412, 202

Siebert, A., Williams, M. E. K., Siviero, A. *et al.* 2011a, *AJ*, 141, 187

Steinmetz, T., Wilken, T., & Araujo-Hauck, C. 2008, *Science*, 321, 1335

van Kerkwijk, M. H., Rappaport, S. A., Breton, R. P. *et al.* 2010, *ApJ*, 715, 5

Williams, M. E. K., Steinmetz, M., Sharma, S. *et al.* 2011, *ApJ*, 728, 102

Wilson, M. L., Helmi, A., Morrison, H. L. *et al.* 2011, *MNRAS*, 413, 2235

Yanny, B., Rockosi, C., Newbert, H. J. *et al.* 2009, *AJ*, 137, 4377

Zucker, S., Mazeh, T., & Alexander, T. 2007, *ApJ*, 670, 1326

Zwitter, T., Matijevič, G., Breddels, M. A. *et al.* 2010, *A&A*, 522, 54

Transactions IAU, Volume XXVIIIA
Reports on Astronomy 2009–2012
Ian Corbett, ed.

© International Astronomical Union 2012
doi:10.1017/S1743921312002980

DIVISION IX / COMMISSION 30 / WORKING GROUP
SB9 - Spectroscopic Binary Systems

CHAIR	**Dimitri Pourbaix**
VICE-CHAIR	...
PAST CHAIR	...
BOARD	A. Batten
	F. Fekel
	W. Hartkopf
	H. Levato
	N. Morrell
	A. Tokovinin
	W. Torres
	S. Udry

TRIENNIAL REPORT 2009-2012

1. Introduction

In Manchester, a WG was set up to work on the implementation of the 9th catalogue of orbits of spectroscopic binaries (SB9), superseding the 8th release of Batten *et al.* (1989) (SB8). SB9 exists in electronic format only. The web site http://sb9.astro.ulb.ac.be was officially released during Summer 2001. This site is directly accessible from the Commission 26 web site, from BDB (in Besancon) and from the CDS (at least).

2. Developments within the past triennium

Since the last report, some substantial progress have been accomplished, in particular in the way complex systems can be uploaded together with their radial velocities. The way data weights can be supplied has also been improved.

For the time being, SB9 contains 3039 systems (1469 in SB8) and 3784 orbits (1469 in SB8). A total of 623 papers were added since August 2000 but most of them come from OUTSIDE the WG. We still have a lot of papers with orbits which are waiting for being uploaded. According to ADS, the replease paper (Pourbaix *et al.* 2004, A&A) has been cited by 152 references since 2005. This is only three times more than the old Batten *et al.* paper over the same period even though that one is still cited nowadays.

An important work of cross checking the identification of systems is carried out by the CDS (Strasbourg). Indeed, with the SBC9 identifier added to Simbad, each new release of the SB9 tar ball is cross checked for typos prior to integration at the CDS. Whereas some of these mistakes are ours, some authors share the responsibility as well. Users have also helped pinning down some problems.

Although this work is very welcome by the community (about 500-1000 success- ful queries every month, with 50 distinct IP addresses over the past month) and some tools have been designed to make the job of entering new orbits easier (input file checker, plot generator, ...), the WG still suffers from a serious lack of manpower. Few colleagues outside the WG spontaneously send their orbits (but they are usually pleased to send their data when we ask for them). Any help (from authors, journal editors, ...) is therefore very welcome. Uploading an orbit in SB9 also means checking it against typos. We thus found some mistakes in the published solutions. Sending orbits to SB9 prior to publication (e.g. at the proof stage) would therefore be a way to prevent some mistakes from going to the litterature.

Transactions IAU, Volume XXVIIIA
Reports on Astronomy 2009–2012
Ian Corbett, ed.

© International Astronomical Union 2012
doi:10.1017/S1743921312002992

COMMISSION 54

OPTICAL/INFRARED INTERFEROMETRY

L'INTERFÈROMÈTRIE
OPTIQUE/INFRAROUGE

PRESIDENT	Stephen T. Ridgway
VICE-PRESIDENT	Gerard van Belle
SECRETARY	Denis Mourard
PAST PRESIDENT	Guy Perrin
ORGANIZING COMMITTEE	Gilles Duvert, Reinhard Genzel, Christopher Haniff, Christian Hummel, Peter Lawson, John Monnier, Peter Tuthill, Farrokh Vakili

COMMISSION 54 WORKING GROUPS

Div. XII / Commission 54 WG	Interferometry Data standards
Div. XII / Commission 54 WG	Imaging Algorithms
Div. XII / Commission 54 WG	Calibrator Stars

TRIENNIAL REPORT 2009–2012

1. Introduction

The Commission was created in 2006, in response to an initiative by members of the international interferometry community, and as a natural expansion of the work of the earlier Working Group on Optical/Infrared Interferometry. At that time, optical interferometry had been in regular use in modern astronomy for approximately 20 years, primarily with first and second generation prototype and experimental facilities. Also at this time, the first observatory-scale user facilities were coming into operation at ESO, Keck, and CHARA.

The focus of the Commission is to establish scientific and technical standards that facilitate the future growth of the field. It has also found a natural function in supporting communication within the community, and recognizing and publicizing accomplishments.

2. The Working Group on Interferometry Data Standards

The Working Group on Interferometry Data Standards, chaired by J. Young, continues the work of the earlier Working Group on Optical/Infrared Interferometry, which provided an important service to the community by defining the Optical Interferometry Data Exchange Format.

During 2011, members of the Working Group participated in discussions about possible enhancements to the OIFITS optical interferometry data exchange standard. These discussions started at the instigation of the Jean-Marie Mariotti Center (JMMC) and the

ESO Very Large Telescope Interferometer (VLTI) instrument teams, which had identified several areas where the existing format was inadequate for describing the final and intermediate outputs of the VLTI/AMBER and VLTI/MIDI data reduction pipelines. The desired enhancements can be summarized as follows:

1. A more prescriptive and precise definition of the existing OIFITS standard (while retaining backwards-compatibility with existing files);

2. Specific support for correlated flux and differential phase data at various stages of calibration (alternative to or augmentation of the existing $OI_V IS$ complex visibility tables);

3. Definition of calibrator tables which record information about the model used to calculate the visibilities for each calibrator star. Such tables would allow identification of cases where re-calibration is required, and perhaps (re-)calibration using an alternative model.

Discussions related to [2] and [3] had previously taken place within the Group in late 2008 (in the case of [3] also involving C. Hummel from the Working Group on Calibrator Stars), but the Group had been unable to agree on specific proposals for a new version of the standard.

The chairman reports: Addressing [1] should be straightforward. The Working Group should also aim to collaborate with VLTI in writing a proposal for [2]. The remaining enhancement [3] is more problematic; the root cause of the ongoing disagreement is probably the variety of use cases that stakeholders envisage for the calibrator information, which leads to differing ideas about the most appropriate implementation. We should either define something very simple which supports only the most limited set of use cases, or decide not to alter the current standard in this respect.

3. Working Group on Imaging Algorithms

The Working Group on Imaging Algorithms was formed at a time when the relative performance of different algorithmic approaches to imaging in optical interferometry was not at all clear. The IAU Imaging Beauty Contest is a competition aimed at showcasing the performance of image reconstruction software in optical interferometry. Science cases are selected by the organizers, then synthetic data sets in the OIFITS format are generated from model images (which remain secret during the competition). Images are reconstructed from the data sets, and the reconstruction closest to the model is declared the winner.

The 2010 Imaging Beauty Contest consisted of blind imaging of simulated interferometric datasets of a scaled infrared model of Betelgeuse plus faint companion. Both low resolution (R = 35), broadband, and moderate resolution (R=1500) datasets were used in the judging. The simulated observations were those possible with the ESO AMBER instrument on three configurations of four telescopes of the VLTI.

The contest was organized by F. Malbet of the Laboratoire d'Astrophysique de l'Observatoire de Grenoble (LAOG), the dataset produced by G. Duvert, also of LAOG and the judging by W. Cotton of the National Radio Astronomy Observatory (NRAO). Entries were submitted by J. Young, F. Baron and D. Buscher of the University of Cambridge using the BSMEM package, by S. Rengaswamy of ESO, using the RPR software, by F. Baron, B. Kloppenberg and J. Monnier of the Universities of Michigan and Denver using the SQUEEZE software, and by M. Vannier and L. Mugnier of the Université of Nice and ONERA using the Wisard package. The results were announced at the 2010 SPIE meeting in San Diego, California, USA. A description of the contest and the entries is published in Malbet *et al.* (2010). The BSMEM entry of Young, Baron and Buscher was the winning entry.

The 5th Beauty Contest will take place in early July 2012 during the SPIE Astronomical Telescopes and Instrumentation conference in Amsterdam under the leadership of F. Baron. In this edition of the contest, two imaging scenarios will be offered for reconstruction: imaging a Young Stellar Object (YSO) with VLTI, and imaging a spotted star with the CHARA Array. In both case the data will have realistic signal-to-noise and UV coverages, simulating actual combiners currently installed at VLTI and CHARA. As in the previous contest, the YSO data set will be based on a polychromatic model and will include differential visibility data. The spotted star data set will be broadband data but its reconstruction will constitute a challenging imaging exercise inspired by actual CHARA data. Due to the higher difficulty, the organizers expect the submissions to be much more varied than they were the previous years, and thus to reflect more the capabilities of individual algorithms.

The Commission would like to acknowledge the essential work by P. Lawson in developing and carrying forward the contest, which has proven to be a popular and successful mechanism for advancing the Group's objectives.

4. Other Working Groups

Most optical interferometry observing techniques require calibrator stars, small enough to be near-point-like, and preferably with known diameter. It was recognized early that it can be difficult to ensure that a candidate calibrator is not a binary or otherwise poorly suited. The Working Group on Calibrator Stars was formed to systematize the exchange of information on methods and actual calibrator lists. This group is currently chaired by Christian Hummel. The Bad Calibrator Registry, originally operated by John Monnier at Michigan, has moved to an ESO server, at

http://www.eso.org/sci/observing/tools/catalogues/bcr.html.

It has been updated to work with Simbad 4.

A planned Working Group on future large arrays proved to be premature, as the consensus in interferometry turned from planning larger and more ambitious facilities to fully exploiting existing facilities. However, work in this area may be ramping up again, with planned discussion in 2012 and the possibility of considering anew an IAU Working Group as a locus of activity.

Intensity interferometry involves very different technology than the now widely used amplitude interferometry, but the two techniques have much in common in data interpretation and science goals. Intensity interferometry has excited new interest in recent years, owing to developments in high speed electronics and to the increasing deployment of large light collectors for Cherenkov arrays, which might be used for intensity interferometry. Also it was demonstrated that phase reconstruction algorithms make imaging possible with intensity interferometry provided a dense enough coverage of the reciprocal plane is available. A webpage with overview and links to other information is maintained at http://www.cta-observatory.org/?q=node/92. A number of Commission members work in this area and the possibility of forming a Working Group on Intensity Interferometry is under consideration.

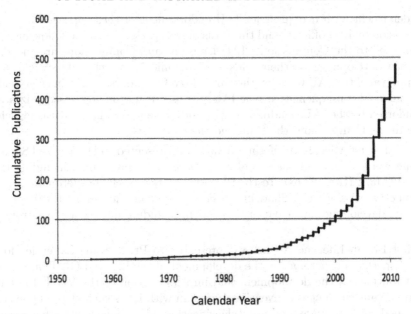

Figure 1. Refereed science publications from optical interferometry, as compiled on the OLBIN web page, on November 1, 2011

5. Developments within the past triennium

The Commission has a web site hosted by the Optical Long Baseline Interferometry News (OLBIN) and maintained by P. Lawson (at http://olbin.jpl.nasa.gov/iau/index. html). The email exploder associated with OLBIN communicates with a larger community of 419 subscribers, many of whom participate in spite of having not chosen to formally join the Commission. The Commission makes liberal use of this mailing list for communicating with the broader interferometry community. The OLBIN site also offers a very complete bibliography of publications in the main scientific and technical areas of interest to the Commission. It also offers tools develped by F. Malbet and colleagues at the Jean-Marie Mariotti Center for analyzing the bibliography content.

With continuing full-time operation of the VLTI, CHARA Array, and NOI, the OLBIN database of optical interferometry publications has been recently growing at nearly 100 per year. Figure 1, prepared with data and tools at the OLBIN web site, shows the cumulative publication count just for refereed science papers.

The main venue for interferometry community meetings has proven, in recent years, to be the biennial SPIE conference, Astronomical Telescopes and Instrumentation. In 2008 and 2010, the IAU collaborated with conference organizers to host special discussion sessions, at which IAU activities were included on the agenda. One idea that originated in these sessions was the possibility of offering a prize in the area of the Optical/Infrared Interferometry.

In 2009, the Commission officers initiated a prize program, with two prizes to be offered at intervals of no more frequently than every 2 years. The Fizeau Prize, sponsored by the Observatoire de la Côte d'Azur, recognizes either lifetime or recent accomplishments in the technical areas of interferometry. The Michelson Prize, hosted by the Mt Wilson Institute, recognizes either lifetime or recent accomplishments in the astrophysical sciences applications of interferometry. Both prizes also may recognize related efforts in education and public outreach. The Commission approached IAU then-General Secretary

K. van der Hucht with a request for IAU endorsement of these prizes. This required a policy decision by IAU officers, and this decision was presented in the following statement communicated to the Commission: "IAU Divisions and Commissions are encouraged to support new developments in their fields of astronomical research and technology. Subject to the approval of the IAU Officers, they are allowed to endorse prizes in their fields which have no financial consequences for the IAU." Subsequently our proposed prizes were formally endorsed by the IAU, lending them considerable additional gravitas, which will be valuable to the Commission, the field, and the recipients.

In 2010, the first Michelson Lifetime Award was presented to Dr. Michael Shao "for his pioneering work on ground-based and space-based interferometers, including the Mark I, Mark II, Mark III, Palomar Testbed Interferometer, Keck Interferometer, and Space Interferometry Mission. Dr. Shao has been a prominent leader in the interferometry community, developing new avenues of research, including narrow-angle astrometry and nulling."

The first Fizeau Lifetime Award was presented to Pr. Antoine Labeyrie "for his invention of speckle interferometry, the development of the I2T and GI2T interferometers, and contributions to the development and implementation of the VLTI. Pr. Labeyrie's innovative genius challenges conventional wisdom with ideas such as the hypertelescope, laser-trapped space mirrors, and pupil densification; his visionary work has meant much to the community, and has been a forceful reminder that our scientific ambitions are limited only by our imaginations."

It is currently expected that nominations for the prizes will be invited in conjunction with the 2012 General Assembly.

6. Closing remarks

From an initial 14 members, scarcely more than the organizing committee, the Commission has grown to, at this writing, 109 members and 2 consultants, with 10 additional member and consultant applicants in process. In growing, the Commission has been partially responsible for some dozens of astronomers choosing to join the IAU in order to enable Commission membership. Since Commission 54 is very young, and many of the members are new to the IAU, the membership is still developing consensus on objectives and paths toward them. The officers will meet with the community at the SPIE meeting in Amsterdam in July, 2012, and at the General Assembly in Beijing in August, 2012, to discuss plans for the next three years.

Stephen T. Ridgway
president of the Commission

Reference

Malbet, F. *et al.* 2010, *SPIE* 7734, 83

Transactions IAU, Volume XXVIIIA
Reports on Astronomy 2009–2012
Ian F. Corbett, ed.

DIVISION IX / COMMISSION 25 / WORKING GROUP
INFRARED ASTRONOMY

CHAIR	Eugene F. Milone
VICE-CHAIR	Andrew T. Young
BOARD	Roger A. Bell, Michael Bessell,
	Richard P. Boyl, Brian Carter,
	T. Alan Clark, Martin Cohen,
	David J.I. Fry, Robert Garrison,
	Ian S. Glass, John Graham,
	Anahi Granada, Lynn Hillenbrand,
	Robert L. Kurucz, Ian McLean,
	Matthew Mountain, George Riecke,
	Rogerio Riffel, Ronald G. Samec,
	Stephen J. Schiller, Douglas Simons,
	Michael Skrutskie, C. Russell Stagg,
	Christiaan L. Sterken, Roger I. Thompson,
	Alan Tokunaga, Kevin Volk, Robert Wing.

TRIENNIAL REPORT 2009-2011

1. Introductory Background

The formal commissioning of the IRWG occurred at the 1991 Buenos Aires General Assembly, following a Joint Commission meeting at the IAU GA in Baltimore in 1988 that identified the problems with ground-based infrared photometry. The meeting justification, papers, and conclusions, can be found in Milone (1989). In summary, the challenges involved how to explain the failure to achieve the milli-magnitude precision expected of infrared photometry and an apparent 3% limit on system transformability. The proposed solution was to redefine the broadband Johnson system, the passbands of which had proven so unsatisfactory that over time effectively different systems proliferated, although bearing the same "*JHKLMNQ*" designations; the new system needed to be better positioned and centered in the spectral windows of the Earth's atmosphere, and the variable water vapour content of the atmosphere needed to be measured in real time to better correct for atmospheric extinction.

The IRWG then established criteria for judging the performance of existing infrared passbands and experimented with passband shapes, widths, and placements within the atmospheric spectral windows. The method and coding were initiated and largely carried out by A. T. Young, and, aided by C. R. Stagg, Milone ran the simulations. The full details of the criteria and results of the numerical simulations were presented by Young *et al.* (1994). Subsequent work, described in WG-IR and/or Commission 25 reports, included the use of a newer MODTRAN version (3.7) to check and extend previous work. This part of the program proved so successful in minimizing the effects of water vapour on the source flux transmitted through the passband that the second stage, real-time monitoring of IR extinction, was not pursued, although this procedure remains desirable for unoptimized passbands designed for specific astrophysical purposes. Considerable work has now been done in measuring the emission of precipitable water-vapor (as we note below), so this goal may be nearing achievement.

During subsequent triennia, the WG concentrated on gathering and presenting evidence of the usefulness of the near-IR IRWG infrared passband set, viz., the iz, iJ, iH, iK passbands. A series of field trials of this suite was conducted over the years 1999–2003 with an InSb detector in a double-well Dewar mounted on the 1.8-m telescope at the Rothney Astrophysical Observatory of the University of Calgary in the foothills of the Canadian Rockies. The results of those trials and the details of further work were presented in Milone & Young (2005). This paper showed that not only were the near-IR suite of the IRWG passbands more useful to secure precise transformations than all previous near-IR passbands, but that they were also superior in at least one measure of the signal-to-noise ratio. This evidence was further discussed and refined in Milone & Young (2007). As a consequence, the original purpose of the IRWG has been achieved, but resistance to the widespread adoption of the new passband system is, nevertheless, still strong, and passbands that somewhat compromise the IRWG recommendations have been advanced in order to provide more throughput, at the cost of precision and standardization. Thus, nonoptimized passbands are still in use at infrared observatories around the world. The situation is described in Milone & Young (2011a), and below.

It would be incorrect to conclude, however, that the work of the IRWG has had no impact on the IR community. As noted in previous IRWG reports, there is now a general acceptance of the principles enunciated in Young, Milone, & Stagg (1994).

2. Membership in the IRWG

The WG-IR has had the policy of being open to input from its members at all times following the initial consultations with all segments of the infrared community. We maintain this open policy for membership, and consider all members interested in improving the precision and accuracy of infrared photometry to be members. The above list, therefore, is a subset of the full membership.

3. Developments in the 2009-2011 Triennium

The IRWG met during the Rio de Janeiro GA in Session 2 on Aug. 7, 2009, and was chaired by E. F. Milone. Recent highlights of the work of the IRWG were described in Milone (2009). In Milone & Young (2005), and further in Milone & Young (2007), correlations were shown among: our figure of (de)merit, θ, a measure of the distortion of the spectral irradiance of starlight as it descends the Earth's atmosphere; a measure of the Forbes effect (the rapid change in slope of the extinction curve with decreasing airmass); the extinction coefficient between 1 and higher airmasses; and a measure of the signal to noise ratio. Milone & Young (2008) argued for the suitability of the IRWG passbands to provide millimagnitude precision for variable star infrared photometry at *any* photometric site, irrespective of its elevation. Directed to both professional and amateur astronomers, this paper compares extinction coefficients obtained using a sample of old IR filters with those using the IRWG passbands determined from the same night at the RAO. The coefficients for the old passbands are seen to be greater by factors ~ 2 or more, as predicted by the simulations and numerical experiments. The very small Forbes effect seen with the IRWG passbands permits the use of the Bouguer extinction coefficients to obtain more accurate outside-atmosphere magnitudes than is possible with others, for which the Forbes effect can be debilitatingly large. A similar theme was emphasized in a column in the General Assembly's daily newspaper, *Estrela D'alva*, Day 10, p.4. Thanks to bulk prices (for lots of 9 or greater) by Custom Scientific, Inc. of Phoenix, Arizona, it may be possible to achieve substantial cost savings in the manufacture of the IRWG filters. Hopefully, photometer manufacturers will be induced to offer installation of the IRWG iz and iH filters in their IR instruments.

Subsequent papers supporting the IRWG passbands were read at three venues, in the interval 2009–2011. A paper on IR as well as optical precision was presented at the AAS meeting in Long Beach, California, in Jan., 2009, in the first of two sessions entitled Photometry: Past and Present. The sections were organized by Milone for the Historical Astronomy Division of the AAS. The paper was expanded into three articles in the volume, *Astronomical Photometry: Past, Present, and Future* (Milone & Sterken (2011)). The article on the development of IR photometry is cited as Milone & Young (2011a). At the *Telescopes from AFAR* conference, held in Waikaloa, HI in Feb., 2011, Milone & Young (2011b) summarized the development of the IRWG passband recommendations. This presentation and the paper based on it are, at present writing, posted on the conference website †.

For anyone interested in observing with the new IRWG passbands, a list of standard stars is available in Milone & Young (2005). Although this list is not extensive, zero points were adopted to approximately match those of the Mauna Kea near-IR set (for the JHK passbands); a link to the latter set is available at the Commission 25 website‡.

Because the IRWG standard star magnitudes listed in Milone & Young (2005) have had zero points added to them to provide "familiar" values, this list can also be used to standardize observations made with the IRWG filters, at sites such as Mauna Kea. Moreover, they may be useful at lower altitude observatory sites, but ONLY if the IRWG passband are used. This is because the Forbes effect is significant for any passband system than the IRWG, but will be negligible for much of the photometry carried out with the IRWG near-IR set, and in the iN passband.

4. Other New Developments in IR Astronomy

The following sample of work reported in the 2009–2011 interval may be of interest to IR photometrists.

- van Dokkum *et al.* (2009) used three overlapping "J" passbands and two "H" passbands to locate the Balmer jump & 400nm break in galaxies over the z red shift range 1.5 to 3.5 and compare their passbands to the atmospheric transmission curve. The breaking of the conventional passbands into shorter segments, at least in part to improve transmission through the atmosphere, is a welcome step, even if it is not the main purpose of the work.

- Hodgkin *et al.* (2009) described the calibration of the photometric system for the wide-field camera of the UKIRT telescope ("$ZYJHK$"). Among other uses, the WFCAM hosts the UKIRT Infrared Deep Sky Survey (UKIDSS). The "$ZYJHK$" passbands extend from 0.84 to 2.37 μm. The 2MASS point source catalog provided the primary foundation for the calibration. Neither the 2MASS JHK_2 nor the Near-IR Mauna Kea JHK passbands were optimized to block water-vapor absorptions.

- Kanneganti *et al.* (2009) discuss a new infrared camera for the University of Virginia's Fan Mountain Observatory. The camera employs an IGC Polycold closed-cycle refrigeration system. "FanCam" employs a HAWAII-I 18.5 μm pixel detector array, and

† http://tfa.cfht.hawaii.edu/papers/milone_tfa_paper.pdf
‡ iauc_25.saao.ac.za

is used on a 31-in telescope. The camera appears to work well, although the background is very high, a condition to which the "JHK" passbands probably contribute strongly. The paper contains the statement that despite the wet, low-elevation characteristics of the site, "the J, H, and K_s bands are virtually unaffected by water-vapor, ... permitting high precision photometric observations." This phrase alone indicates that the work of the IRWG is far from complete. Of course, if the IRWG passband designations iz,iJ,iH,iK were inserted instead, the statement would be more nearly correct.

- Meixner *et al.* (2010) describe the WHIRC high-resolution infrared camera installed on the 3.5-m WIYN telescope at Kitt Peak. The detector is a Raytheon VIRGO 2048 x 2048 HgCdTe array. A tip-tilt module provides diffraction–limited imaging from 0.4 to 2.4 μm. It is currently equipped with 1.061 μm, He (1.083 μm), Pa β, [Fe II], Br γ (the latter three repeated with a red-shift equivalent to 4500 km^{-1}), H_2, and CO. They report internal precision among a 5 star sample of ± 0.02 mag.

- Monson & Pierce (2009) discuss the performance of the BIRCAM array camera of the Wyoming Red Buttes Observatory, where it is mounted on a 24-in telescope. They equip this instrument with "JHK" filters.

- Pickels & Depagne (2010) synthesize "ZY" passband magnitudes, among many others, for 2.7 million stars in the Tycho2 Catalog, by integrating spectral atlases. The "Z" passband does not resemble the "iz" or "yz" (as it was referred to in Young, Milone, & Stagg (1994)) passband of the IRWG set, but "Y" resembles it, as does the "Y" passband proposed by Hillenbrand *et al.* (2002).

- Querel, Naylor, & Ferber (2011) present an algorithm with which to calculate precipitable water vapor from infrared water-vapor emission measured with their IRMA instrument, and others, by matching the observations with their Blue Sky Transmission and Radiance Atmospheric Model (BTRAM), which makes use of the HITRAN 2008 molecular parameters data base. This is the sort of achievement which can lead to the complete solving of the infrared extinction problem. The real-time modeling of water-vapor content is what is needed to obtain the remaining corrections to observations made through passbands already optimized to block water-vapor.

5. Discussion

Clearly, a sharp break from the nomenclature of the Johnson passbands has not occurred, as the above papers illustrate. Many Infrared astronomers continue to use "$JHKL$" designations (now more commonly designated with a subscript "s" on "K" or by primes, to indicate a slight change in profile to lessen the effect of the water-vapor bands' absorption and thermal emission) even though there has been, at least until recently, no single passband system in use with those designations.

The Mauna Kea near-IR suite of passbands described by Tokunaga & Vacca (2007), is one of the best of the incrementally improved systems, but it is, nevertheless, *not* optimized for extinction and standardization, nor for a measure of the signal-to-noise ratio, and falls short of the IRWG specifications. This will result in a significant Forbes effect at sites other than the highest elevation sites (Mauna Kea, Mt. Evans).

That observers are applying the Johnson designations to the Mauna Kea near-IR set is not surprising because that is what Tokunaga, Simons, & Vacca (2002) called them, thus resulting in confusion in nomenclature with earlier passbands.

The IRWG suggests that the designations be assigned, instead, only to the atmospheric windows most prominently associated with the original Johnson passbands — see Milone & Young (2005).

The Mauna Kea near-infrared suite, an incremental improvement over older passbands sets, although falling short of the IRWG prescription, typifies the movement of infrared astronomy toward the goals of the IRWG.

In addition to promotion of the IRWG passbands, the following tasks still need to be carried out by future Infrared Astronomy Working Groups:

• Acquisition and testing of several sets of near IRWG passbands for distribution to several observing sites†.

• Establishment of standard star observations across the entire sky, and their dissemination, including placement on the IRWG‡ or Commission 25 website.

• Encouragement of instrument suppliers to make use of the iz and iH passbands in low-cost photometers.

• Fabrication of the iL, iN, and possibly the in filters. The simulations consistently showed the iN passband to result in quite modest Forbes effects even at low-elevation sites.

6. Closing remarks

The IRWG has now been in existence for two decades. Even with incremental improvements in IR passbands in use at the major IR observing sites around the world, other passbands suffer from the penchant of many IR observers to use the spectrally widest passbands they can fit into the atmospheric windows. Such passbands are, de facto, less desirable for use at lower sites, such as KPNO, CTIO, and all other observatory sites below 4 km elevation.

We do not agree with the assertion of Tokunaga & Vacca (2007) that the large Forbes effect from nonoptimized passbands is important only in photometry of very cool objects or that it is removably generally by application of simple color terms. However, the issue can be resolved by simultaneous observation of the same objects through the IRWG and the Mauna Kea passbands at 2-km or lower elevation sites.

More work clearly needs to be done to promote the IRWG passbands, for all applications, and especially time–variable observations, because of the continuous variation in water-vapor content, and thus, in the Forbes effect.

We recognize that astronomers are a conservative lot, and even though there is no standard $JHKLMNQ$ system to which they need to be loyal, infrared astronomers have been particularly reluctant to adopt and try new broadband filters, except to isolate particular spectral features. In most cases they may not want to sacrifice white-light filters (which is what conventional infrared filters have been, effectively) for narrower ones that provide less overall throughput, but are *not* defined by the edges of the atmospheric windows.

Such an attitude is understandable in observers operating in what can be called a "discovery" mode, where high precision and accuracy are not critical, but photometrists have basically different aims. One legitimate concern is that data taken in other IR passbands may not transform to the IRWG system, with the possible exception of data in the newer Mauna Kea set used at one of the few sites in the world where it is truly suitable, and when conditions there are dry. Consequently, it seems that only demonstrations of the superiority of the IRWG passbands set when used on specific targets will convince many to use them. Therefore, we urge photometrists with a strong interest in precise

† IRWG filters are now available at bulk prices from Custom Scientific, Inc., of Phoenix, AZ. Current prices may be obtained from that company.

‡ http://people.ucalgary.ca/~milone/IRWG/

photometry to give these passbands a try at observatories where photometry is done, either with a chopping secondary and LIA system, as at the Rothney Astrophysical Observatory, or with array cameras.

The widespread adoption of the Mauna Kea set as the "JHK" passbands of the day indicates that evolution, if not revolution, is possible in this field. The fact that generations of broadband filters, although widely different in profile, bear the same names, is, of course, bizarre. Commission 25 has added the issue of passband designation recommendations to its agenda.

Eugene F. Milone
Chair of the Working Group

References

van Dokkum, P. G., *et al.* 2009, *PASP*, 121, 2

Hillenbrand, L. A., Foster, J. B., Persson, S. E., & Matthews, K. 2002, *PASP*, 114, 708

Hodgkin, S. T., Irwin, M. J., Hewett, P. C., & Warren, S. J. 2009, *MNRAS*, 394, 675

Kanneganti, S., *et al.* 2009, *PASP*, 121, 885

Meixner, M., *et al.* 2010, *PASP*, 122, 451

Milone 1989, in: E. F. Milone (ed.), *Infrared Extinction and Standardization*, Proc., Two Sessions of IAU Commissions 25 and 9, Baltimore, MD, USA, 4 August 1988, *Lecture Notes in Physics*, Vol. 341 (Heidelberg: Springer), 1

in: Karel A. van der Hucht, ed., *Reports on Astronomy 2006-2009*, *IAU Transactions*, XXVIIA, 313

Milone, E. F. & Sterken, C. 2011 *Astronomical Photometry: Past, Present, and Future*,

Milone, E. F. & Young, A. T. 2005, *PASP*, 117, 485

Milone, E. F. & Young, A. T. 2007, in: C. Sterken (ed.), *The Future of Photometric, Spectrophotometric, and Polarimetric Standardization*, Proc. Intern. Workshop, Blankenberge, Belgium, 8-11 May 2006, *ASP-CS*, 364, 387

Milone, E. F. & Young, A. T. 2008, *JRASC*, 36, 110

Milone, E. F. & Young, A. T., 2011a in: E. F. Milone & C. Sterken, eds., *Astronomical Photometry: Past, Present, and Future*, (New York: Springer) ASSL 373, 125

Milone, E. F. & Young, A. T., 2011b in: *Telescopes from AFAR*, on-line publication, http://tfa.cfht.hawaii.edu/papers/milone_tfa_paper.pdf

Monson, A. J. & Pierce, M. J. 2009, *PASP*, 121, 728

Pickles, A. & Depagne, É. 2010, *PASP*, 122, 1437

Querel, R. R., Naylor, D. A., & Kerber, F. 2011, *PASP*, 123, 222

Tokunaga, A. T. & Vacca, W. D. 2007, in: C. Sterken (ed.), *The Future of Photometric, Spectrophotometric, and Polarimetric Standardization*, Proc. Intern. Workshop, Blankenberge, Belgium, 8-11 May 2006, *ASP-CS*, 364, 409

Tokunaga, A. T., Simons, D. A., & Vacca, W. D. 2002, *PASP*, 114, 180

Young, A. T., Milone, E. F., & Stagg, C. R. 1994, *A&AS*, 105, 259

Transactions IAU, Volume XXVIIIA
Reports on Astronomy 2009-2012   © International Astronomical Union 2012
Ian F. Corbett, ed.   doi:10.1017/S1743921312003018

DIVISION X   RADIO ASTRONOMY

RADIOASTRONOMIE

Division X coordinates, within the structure of IAU, activities of observational and theoretical astronomers interested in phenomena that are detectable at radio wavelengths and in instrumentation for such observation.

PRESIDENT	Russ Taylor
VICE-PRESIDENT	Jessica Chapman
PAST PRESIDENT	Nan Rendong
BOARD	Christopher Carilli, Gabriele Giovannini,
	Richard Hills, Hisashi Hirabayashi,
	Justin Jonas, Joseph Lazio,
	Raffaella Morganti, Monica Rubio
	Prajval Shastri

DIVISION X COMMISSIONS

Commission 40	Radio Astronomy

DIVISION X WORKING GROUPS

Division X WG	Astrophysically Important Spectral Lines
Division X WG	Interference Mitigation

INTER-DIVISION WORKING GROUPS

Division IX-X-XI WG	Astronomy from the Moon
Division IX-X WG	Encouraging the International Development of Antarctic Astronomy
Division X-XII WG	Historical Radio Astronomy

TRIENNIAL REPORT 2009-2011

1. Introduction

This triennium has seen a phenomenal investment in development of observational radio astronomy facilities in all parts of the globe at a scale that significantly impacts the international community. This includes both major enhancements such as the transition from the VLA to the EVLA in North America, and the development of new facilities such as LOFAR, ALMA, FAST, and Square Kilometre Array precursor telescopes in Australia and South Africa. These developments are driven by advances in radio-frequency, digital and information technologies that tremendously enhance the capabilities in radio astronomy. These new developments foreshadow major scientific advances driven by radio observations in the next triennium. We highlight these facility developments in section 3 of this report. A selection of science highlight from this triennium are summarized in section 2.

This report also includes updates from Division X working groups on Astrophysically Important Spectral Lines, and on Interference Mitigation. Both are areas of growing importance as instantaneous bandwidths, sensitivities and the coverage of the electromagnetic spectrum at radio wavelengths increase with the advance of new technologies. A report from the WG on Astronomy from the Moon is also included. Ultra low frequency radio astronomy promises to be one of the first astronomical uses of the lunar platform.

The former DX Working Group on Global VLBI was terminated during this triennium. This working group was formed during the early days of the expansion of VLBI to global baselines, and provided a forum for coordination of programs and standardization of data and technologies from diverse regions of the globe. Intercontinental VLBI is now well served by dedicated arrays, correlation facilities and programs that are managed by national and international agencies. We take this opportunity to thank the members of the GVWG who have played a strong role in the successful development of this discipline as a global enterprise.

2. Science Highlights

2.1. *Pulsars*

Efforts to detect a stochastic gravitational wave background via high precision pulsar timing observations continue world-wide. The latest results have been reported two groups, one combining the Parkes radio telescope and Arecibo Observatory data, the other combining data from various European radio telescopes. Upper limits continue to improve—steadily approaching the level expected from the ensemble population of supermassive black hole binaries expected from mergers of galaxies—helped by a combination of improved understanding of instrumental effects and new analysis techniques of the data (Yardley *et al.* 2011; van Haasteren *et al.* 2011).

In conjunction with NASA's Fermi Gamma-ray Space Telescope, a world-wide network of radio telescopes has been observing un-identified Fermi sources for pulsars. The result has been a dramatic increase in the number of millisecond pulsars, with radio observations of Fermi sources likely to discover many more millisecond pulsars than were discovered since the original discovery of millisecond pulsars (Abdo, Ackermann & Ajello 2010).

Analysis of a pulsar survey with the Parkes radio telescope has revealed a millisecond pulsar with an extremely unusual companion. The companion's mass is comparable to that of Jupiter but its density suggests that it is an ultralow-mass carbon white dwarf. Such a system may help reveal how millisecond pulsars are produced as well as elucidate various binary evolution channels. (Bailes, *et al.* 2011).

A new mass determination for PSG J1614-2230 makes it the most massive pulsar known and rules out a number of equations of state for nuclear matter, including many exotic hyperon, kaon models. The derived companion mass is $0.500 \pm 0.006\ M_\odot$, and the pulsar mass is $1.97 \pm 0.04\ M_\odot$ (Demorest *et al.* 2010).

2.2. *Galaxies*

The Arecibo Legacy Fast ALFA Survey (ALFALFA) continued with its blind search of the Arecibo sky for neutral hydrogen 21 cm emission. When complete, ALFALFA will begin to approach the volumes needed for a representative cosmological survey. Currently, the ALFALFA survey is 40% completed and already is yielding significant improvements in the source density compared to previous HI surveys. Already it is clear that there are few gas-rich, optically dark systems in the local Universe (Haynes *et al.* 2011)

A variety of instruments, including the Very Large Array (VLA) and EVLA, Multi-Element Radio-Linked Interferometer Network (MERLIN), and the European Very Long

Baseline Network (EVN), have been used to probe sub-millimeter galaxies (SMGs) to assess the relative contributions between accretion onto supermassive black holes and star formation in producing their luminosity (e.g. Briggs *et al.* 2011),

A suite of radio instruments is being used to probe the innermost regions of active galactic nuclei (AGN) where the most energetic processes happen. The observations are being conducted in coordination with NASA's Fermi Gamma-ray Space Telescope and include a world-wide network of both single-dish and imaging instruments in both hemispheres.

Observations of the CO line over cosmological time scales are beginning to illustrate how galaxies turn their gas into stars. In particular, existing instrumentation is beginning to illuminate processes at redshifts larger than 1, when the star formation rate density of the Universe reached its peak. (e.g. Genzel *et al.* 2010). CO emission has been detected from the most distant known submm galaxy at z=5.3. These observations show the cold gas that fuels the star formation, implying extreme amounts ($> 10^{10}$ $M_\odot$) of dense gas in this forming elliptical galaxy. (Riechers *et al.* 2010)

2.3. *Galaxy Clusters*

A bright, giant radio halo was detected in MACS J0717.5+3745; the most distant cluster currently known to host a radio halo. This radio halo is also the most powerful ever observed, and the second case for which polarized radio emission has been detected, indicating that the magnetic field is ordered on Mpc scales (Bonafede *et al.* 2009).

It has been confirmed that the radio spectra of halos are related to the cluster temperature, being flatter in hotter clusters (Giovannini *et al.* 2009). Since the cluster temperature is a good indication of the turbulence present in the ICM, this correlation favours the interpretation that turbulence is the mechanism responsible for supplying energy to relativistic electrons.

The correlation between cluster X-ray luminosity and radio power confirms the dichotomy between merging clusters and relaxed clusters. Models of diffusive shock acceleration suggest that in shocks that occur during cluster mergers, particles are accelerated to relativistic energies. In the presence of magnetic fields, these particles emit synchrotron radiation and may form radio relics (van Weeren *et al.* 2010). The new detections of many radio relics that display highly aligned magnetic fields, a strong spectral index gradient, and a narrow relic width, confirm this model and give measures of the magnetic field in a previously unexplored site of the universe.

2.4. *Active Galactic Nuclei*

In 2009 for the first time e-VLBI science observations were carried out with a global array, reaching a maximum baseline length of 12,458 km, including telescopes in Europe, East Asia, and Australia. The γ-ray narrow line Seyfert 1 PMN J0948+0022 observed at 22 GHz, showed a structure dominated by a bright component, more compact than 55 μarcsec, with a fainter component (Giroletti *et al.* 2011). Relativistic beaming is required by the observed brightness temperature. The results show that global e-VLBI is a reliable and promising technique for future studies.

M87 was observed at six frequencies with VLBA, allowing a positional accuracy of the location of the central black hole relative to the jet base (radio core) of about 20 μarcsec (Hada *et al.* 2011). As the jet base becomes more transparent at higher frequencies, the multifrequency position measurements of the radio core enable a determination of the upstream end of the jet. The data reveal that the central engine of M87 is located within 14-23 R_s of the radio core.

3. International Radio Astronomy Facilities Development

The *Atacama Large Millimetre Array* (ALMA), which is a partnership involving institutes in Europe, North America and East Asia, in cooperation with the Republic of Chile, is making rapid progress towards it goal of providing dramatic improvements in sensitivity and resolution for astronomy at millimeter and sub-millimeter wavelengths. The initial commissioning results have confirmed that the accuracy, sensitivity and stability of the key components are close to the ambitious goals set for them. The tests have also demonstrated that all parts of the system can be made to work together to produce high-quality images. A series of observations have been made for the Scientific Verification program - aimed at producing data that can be compared directly with observations made with other (sub-)millimeter arrays - and the results are being released publicly as they become available. The first call for proposals ("Cycle 0") was sent out in March 2011 and scheduled observations with 16 antennas, 4 receiver bands and baselines of up to 400m started on 30th Sept.

As of November 2011, 24 antennas were in operation. A further eight antennas, with full complements of receivers, are being tested. The antennas in operation include examples of all three of the 12m designs, as well as four of the 7-m diameter antennas for the ALMA Compact Array. Final versions of the local oscillator system, the correlators and all the other electronics and control and monitor systems are in place, and the infrastructure (roads, power and communications) is expected to be completed early in 2012. ALMA's capabilities will continue to grow during 2012 and there will be further calls for proposals from the user community. It remains the target to have the full system with sixty-six antennas and baselines of up to 16km in operation during 2013.

The *Expanded Very Large Array* construction project continues to proceed on schedule for delivery by December, 2012. Implementation of all major systems is complete or approaching completion, and the new instrument is well into standard operations. Remaining tasks include the completion of Front End receiver hardware, the installation of the 4 Gs/s 3-bit samplers, and the implementation of specific software support. Antennas continue to be outfitted with new wide band receivers. Well over half of these are now in service. The schedule to deploy these remaining receiver bands has held steady for the past two years. The retrofit of existing Data Transmitter modules with wideband samplers has begun with the installation of the first twelve. The 3-bit samplers, required for the full 8-GHz operation of the EVLA, are schedule to be fully deployed by September, 2012. Observing capability with 2 GHz bandwidth is complete, monitor and control of antenna electronics is fully realized, and 3-bit operation is supported by the correlator and EVLA. Notable work which remains includes providing support for sub arrays and data pipeline processing.

LOFAR is a low-frequency radio interferometer nearing completion in Europe. The antenna array consists of two distinct antenna types: the Low Band Antenna (LBA) operates between 10 and 90 MHz and the High Band Antenna (HBA) between 110 and 250 MHz. Within the Netherlands 33 stations are currently online and are distributed over an area about one hundred kilometres in diameter in the North-East of the Netherlands. The remaining seven stations in the Netherlands are expected to be completed in early 20120. Eight international stations are completed inGermany (5), Sweden (1), the UK (1) and France (1). The array is currently undergoing intensive commissioning with an initial all-sky calibration survey underway. In addition to the station roll-out, the final deployment of computing hardware at the LOFAR central processing facility is complete and development of the science processing pipelines is making steady progress. An initial release of the full system is slated for beginning of 2012. These capabilities will be offered to the community as part of an initial Announcement of Opportunity in early 2012.

At the *Westerbork Synthesis Radio Telescope* a Square Kilometre Array pathfinder project, APERTIF, aims to increase survey speed by installing phase-array feed systems operating at 1.4 GHz. The project successfully passed its detailed design phase and is expected to roll out in 2013. ASTRON has started the process that will define the science to be done with APERTIF. Most of the observing time with APERTIF will be devoted to large surveys, the data products of which will be made available to the entire community through an open-access archive. A call for expressions of interest for APERTIF was issued in July 2010. The response was very large, and efforts are underway to combine the proposed surveys into a coordinated campaign of duration about five years.

The *Giant Meter Wave Telescope* is undergoing a major upgrade whose main aims are to provide seamless frequency coverage from 50 to 1450 MHz, improved sensitivity with better feed and receivers, increase instantaneous bandwidth to 400 MHz, and improve computing and infrastructure facilities. As part of these developments, a series of significant upgrades have been completed during 2009-11, notable amongst which are completion and release of a 32 MHz software backend for correlation, beam forming and pulsar processing, development of a wide-band optical fibre link, and development of automated software systems for scheduling, data archiving and retrieval.

The *Five-hundred-meter Aperture Spherical Telescope* (FAST) will be the world's largest single dish sited in a karst depressions called Dawodang in south Guizhou province in China. Its active 500-metre diameter reflector directly corrects for spherical aberration. The light-weight focus cabin is driven by steel cables and has a robotic secondary system to precisely position the receivers. Working at frequency range of 70MHz - 3 GHz, the telescope will provide a powerful tool for HI surveys, pulsar science, radio spectra and international VLBI. The report on starting of construction was approved by the Chinese government in March 2011 and early science is expected to start in 2016.

The *Sardinia Radio Telescope* is a 64-m dish which is being completed in the Sardinia Island (south Italy) and will be used as a single dish, for VLBI, and for satellite tracking. It will be the second largest dish in Europe, and the largest one with an active surface; the primary mirror is made of a mosaic of more than 1000 actuated adjustable panels. It will represent a major addition to the European VLBI Network and will be used for space VLBI observations in conjunction with the *RadioAstron* antenna. First light receivers are in the range 0.3 - 22 GHz. The high frequency receiver is a 7-horns array, supplying a larger field of view and high imaging speed. Early science is expected in 2012.

The *RadioAstron* space radio telescope was launched from Baikonur on July 18, 2011, and its 10-meter antenna was successfully unfolded on July 23. Since then, the telescope has been undergoing in-orbit checkout. In particular, the on-board hydrogen maser has been switched on and is working properly; the radiometers have been turned on and the system temperature in all four bands (P, L, C, and K) has been found to be close to the specifications. First light from Cassiopeia A was detected at 92 and 18 cm and the effective areas are close to the expected values. Fringe searches are expected to start on November 15 2011 and the Early Science Program at the beginning of 2012.

The *Square Kilometre Array* (SKA) will be the next-generation global radio telescope operating from metre to centimetre wavelengths. The early stage planning and design of the SKA has been undertaken as a collaboration of institutes from over 20 countries. A crucial steps for the SKA project were taken in the last week of March 2011. A Founding Board was created with the aim of establishing a legal entity for the project by end of 2011, and agreeing the resourcing of the pre-construction phase from 2012 to 2015. Nine countries signed the initial agreement, and a number of additional countries are expected to join the Board upon establishment of the SKA legal entity. At its first meeting on 2 April, the Founding Board decided that the location of the SKA Project Office during the pre-construction phase will be at the Jodrell Bank Observatory in the UK.

The SKA System Conceptual Design Review was passed in February 2011. Several of the SKA sub-system CoDRs (signal processing, aperture arrays, antenna) have now also passed review and all reviews are scheduled for completion by January 2012. The overall system will reach Preliminary Design Review stage at the end of 2012; construction is currently planned to commence in 2016.

The site characterization for the SKA location is well advanced. Information and reports from both of the candidate sites in Australia/New Zealand and southern Africa are under review. The decision on the SKA site is expected in early 2012. SKA precursor telescopes that demonstrate technologies under development for the SKA are under construction on both proposed sites.

The *Australian Square Kilometre Array Pathfinder* (ASKAP) telescope is a wide-field (30 sq deg) radio survey telescope currently being constructed at the Murchison Radioastronomy Observatory in Mid-West region of Western Australia. The telescope will comprise 36 12m-diameter antennas equipped with 192-element phase array feeds providing up to 30 simultaneous observing beams. The telescope will be operational over the range 0.7-1.8 GHz and a maximum angular resolution of 10 arcsec. It will have processed bandwidth of 300M Hz over 16K channels. As of October 2011, 10 antennas have been constructed, will all 36 to be built on site by March 2011. The first six antennas are expected to be conducting science observations with the phased array feed system by mid-2012. The wide field-of-view provided by the focal plane array receiver systems optimizes ASKAP for rapid imaging surveys. Ten large Survey Science Projects have been awarded survey status; comprising 350 scientists from over 150 institutions internationally. ASKAP operates an open access model, and all data will be available to the international community with no proprietary period.

The *MeerKAT* array, being constructed on the South African SKA candidate site in the remote and arid Karoo region, will consist of 64 Gregorian offset 13.5-m antennas. The array will initially extend out to 8 km baselines. MeerKAT is a precursor for the mid-frequency dish component of the SKA. Novel reflector manufacturing techniques are being prototyped to ensure good performance across the operating bandwidth at an affordable cost for both MeerKAT and the SKA. The antennas will be equipped with a suite of cryogenically cooled single-pixel receivers covering a frequency range of 580 MHz to 15 GHz. The radio frequency outputs from the receivers will be digitized directly to ensure signal fidelity, and the array signal processor will be based on FPGA processors (developed as part of the CASPER collaboration) and commodity computing elements (such as GPUs and other multi-core devices). MeerKAT will be the most sensitive centimetre wavelength instrument in the southern hemisphere and will be completed in 2016. Ten large-scale science surveys involving more than 500 scientists have been allocated 70% of the initial observing time on the MeerKAT, amounting to five years in total. A seven-antenna prototype array, named KAT-7, has been constructed on the Karoo site, and is currently conducting commissioning science observations.

4. WG Reports

4.1. *Astronomy from the Moon - chair: Heino Falke*

The European Space Agency (ESA) plans its first lunar lander in the region of lunar south pole in 2018. A primary goal is to demonstrate a soft precision landing.

Scientifically the following topics are being studied by ESA as part of the mission:

(*a*) Investigating the lunar conditions (dust, plasma, ionosphere, availability of natural resources) in preparation for human settlements on the moon.

(*b*) Performing Moon-based Ultra-Long-Wavelength Astronomy (ULWA). Using a single radio antenna covering kHz-100 MHz a variety of scientific issues can be addressed, including measurement of Solar bursts and radio emissions from the planets, the global signal of the Epoch of Reionization of the universe (Jester & Falcke, 2009), and emission of ultra high-energy cosmic rays and neutrinos from the moon regolith (Scholten 2010).

A Dutch-Chinese collaboration has funded a joint PhD study aiming at engineering of Moon-based instrumentation for ULWA. An Indian - Russian collaboration is considering an ULWA experiment in the framework of the joint Moon exploration program with possible implementation after 2018. Using the results of UK-SE FIRST and ESA-NL DARIS studies, a concept of a free-flying ULWA mission SURO (a low-cost array of 9 satellites near L2) was proposed for a cosmic vision call. Although the proposal did not succeed, new submissions are planned.

In the US NASA's Lunar Science Institute has been very active with various projects.

• Dark Ages Radio Explorer. A concept for a lunar-orbiting spacecraft carrying a dipoleantenna for the purposes of making measurements of the global highly-red shifted 21-cm signal from the end of the Dark Ages and Cosmic Dawn. A proposal was submitted to the NASA Explorer program. While it received favorable reviews, it was not funded.

• Technology development for a future lunar radio array based on polyimide film. An antenna mated to a commercial off-the-shelf receiver has been fielded. Absorption due to the terrestrial ionosphere was detected, though the original target was a diurnal variation in power due to the Galactic background, which was not achieved. The test results are being assessed.

• Technology development for the deployment of the antennas of a future lunar radio array using a spring-loaded anchor system. Initial tests of a proof-of-concept system have been conducted, and more detailed engineering designs are being developed.

4.2. *Astrophysically Important Spectral Lines - chair: Masatoshi Ohishi*

The frequency range, 275 -1000 GHz, is used for radio astronomy observations of important spectral lines and continuum bands. New receiver technology and new instruments (both ground-based and space based) being used in the 275 - 1000 GHz region are helping to refine the results of radio astronomy observations in this spectrum range, while similar developments in the 1000-3000 GHz range are leading to a better understanding of specific spectral lines and atmospheric windows that are of interest to radio astronomers. Significant infrastructure investments are being made under international collaboration for the use of these bands between 275 and 3000 GHz.

Frequency allocations for the use of this frequency range are not available, but the radio astronomy community is requested to identify a list of specific bands of interest between 275 and 3000 GHz towards World Radio Communication Conference 2012 (WRC2012) held by the International Telecommunication Union. A new ITU-R Recommendation RA.1860 (Preferred frequency bands for radio astronomical measurements in the range 1-3 THz) was published on February 2010, by including a list of astrophysically most important spectral lines in the frequency range between 275 and 3000 GHz that was established by the IAU Working Group on Important Spectral Lines.

Proposals for WRC2012 are being submitted (as of the end of October 2011) by governments and regional telecommunication bodies, all of which are identical and incorporate the outcome of this Working Group. To the best of our knowledge, there have been no objections to provide science communities with better electromagnetic environment. It is thus expected that the importance of radio astronomical observations in the frequency range between 275 and 3000 GHz will be better recognized after February 2012.

4.3. *Interference Mitigation - chair: Willem Baan*

The issue of RFI Mitigation has gained more visibility within the radio astronomy community in recent years. On one side, the computing capabilities available at observatories have been augmented dramatically, while on the other side the presence of RFI has become more damaging because of changing observing capabilities. The vulnerability of radio astronomy observing systems to RFI has increased because of routinely lower system temperatures and steadily increasing observing bandwidths. Observing bandwidths now routinely cover large chunks of frequency space outside bands allocated to the Radio Astronomy Service (RAS), including bands allocated to other communication services for transmissions.

The use of these large observing bandwidths also creates the urgent need of mitigating the signals in bands outside those allocated to the RAS. Mitigation techniques implemented in new (and existing) observing systems remove the effects of RFI using automated implementations of the familiar (time-consuming) data flagging procedures. As long as the percentages data loss is small, such applications will be adequate. However, when data loss becomes significant, other techniques are needed that subtract the RFI component from the data and leave the astronomical data mostly intact. Certain forms of waveform estimation and higher-order statistics can do this but are computationally complicated and difficult to implement. RFI mitigation techniques are being introduced in modern telescopes, such as LOFAR, WMA, EVLA, eMERLIN, GMRT, and others.

The third in a series of RFI Mitigation workshop was held in Groningen, The Netherlands on 29-31 March 2010, where the most recent achievements in algorithms and implementations were presented. The presentations of this meeting can be found at http://www.astron.nl/rfi. All papers are available in e-literature.

Working Party 7D on Radio Astronomy of the Radiocommunication Sector of the International Telecommunication Union is preparing a Report on RFI Mitigation techniques.

A. Russell Taylor
President of the Division

References

Abdo, A. A., Ackermann, M., & Ajello, M. 2010, *ApJS*, 187, 460
Bailes, M., Bates, S. D., Bhalerao, V., *et al.* 2011, *Science*, 333, 1717)
Bonafede, A., Feretti, L., Giovannini, G. *et al.* 2009, *A&A*, 503, 707.
Biggs, A. D., Younger, J. D., & Ivison, R. J. 2011, *MNRAS*, 408, 342)
Demorest, P. B., Pennucii, T., Ransom, S. M. *et al.* 2010, Nature, 467, 1081.
Genzel, R., Tacconi, L. J., Gracia-Carpio, J., *et al.* 2010, *MNRAS*, 407, 2091
Giovannini, G., Bonafede, A., Feretti, L. *et al.*. 2009, *A&A*, 507, 1257.
Giroletti, M., Paragi, Z., Bignall, H., *et al.* 2011, *A&A*, 528, 1.
Hada, K., Doi, A., Kino, M. *et al.* 2011, *Nature*, 477, 185.
Haynes, M. P., Giovanelli, R., Martin, A. M., *et al.* 2011, *AJ*, 142, 170
Jester, S. & Falcke, H. 2009, *New Astronomy Reviews*, 53, 1.
Reicher, D. A., Capak, P. L., & Carilli, C. L. 2010, *ApJ*, 720, L131.
Scholten, O. 2010, *Journal of Physics: Conference Series*, 239, 012003
Yardley, D. R. B., Coles, W. A., Hobbs, G. B., *et al.* 2011, *MNRAS*, 414, 1777
van Haasteren, R., Levin, Y., Janssen, G. H., *et al.* 2011, *MNRAS*, 414, 3117)
van Weeren, R. J., Reinout, J., Röttgering, H. J. A., *et al.* 2010, *Science*, 330, 347.

Transactions IAU, Volume XXVIIIA
Reports on Astronomy 2009–2012
Ian Corbett, ed.

© International Astronomical Union 2012
doi:10.1017/S174392131200302X

DIVISION X, XII / COMMISSION 40, 41 / WORKING GROUP RADIO ASTRONOMY

CHAIR	**Kenneth Kellermann**
VICE-CHAIR	**Wayne Orchiston**
BOARD	**Rod Davies**
	Leonid Gurvits
	Masato Ishiguro
	James Lequeux
	Govind Swarup
	Jasper Wall
	Richard Wielebinski
	Hugo van Woerden

TRIENNIAL REPORT 2006-2009

1. Introduction

The IAU Working Group on Historical Radio Astronomy (WGHRA) was formed at the 2003 General Assembly of the IAU as a Joint Working Group of Commissions 40 (Radio Astronomy) and 41 (History of Astronomy), in order to: a) assemble a master list of surviving historically-significant radio telescopes and associated instrumentation found worldwide; b) document the technical specifications and scientific achievements of these instruments; c) maintain an on-going bibliography of publications on the history of radio astronomy; and d) monitor other developments relating to the history of radio astronomy (including the deaths of pioneering radio astronomers).

The WGHRA is now an Inter Division (DX and DXII) Working Group.

2. WG Web site

The IAU WGHRA maintains a web site at http://rahist.nrao.edu/ which includes past WG reports, brief biographical notes on Grote Reber Gold Medalists for Innovative Contributions to Radio Astronomy, brief memorial articles on recently deceased radio astronomers, and links to various sources of material on the history of radio astronomy.

3. Preservation

The WG noted with satisfaction that the reported deterioration of the Bell Labs horn reflector used by Penzias and Wilson to detect the CMB has been addressed by Lucent Technologies, and that the horn has been refurbished.

In the Netherlands, the 25-meter Dwingeloo dish, inaugurated in 1956, and used for major research programs up to 1998, has been repaired and modernized by CAMRAS, a foundation run by radio amateurs, since 2006. The Dutch Ministry of Education, Culture and Science has granted a major subsidy for the full restoration of the telescope, to be started in 2012. The telescope will be made available for education and research projects by high-school students. The 60th anniversary of the first 21 cm mapping of the Milky Way with the 7.5 meter dish at Kootwijk was celebrated at the original site on 11 May 2011.

In 2003, the National Radio Astronomy Observatory initiated the first Archives devoted exclusively to radio astronomy. The NRAO Archives seeks out, collects, organizes, and preserves institutional records, personal papers, audio-visual materials, and oral histories of enduring value documenting NRAO's development, institutional history, instrument construction, and ongoing activities, including its participation in multi-institutional collaborations. As the national facility for radio astronomy, the Archives also includes an increasing collection of materials on the history and development of radio astronomy and the work of individual astronomers especially in the United States. See http://www.nrao.edu/archives/.

In addition to the institutional records of NRAO, the NRAO Archives includes Web resources on early radio astronomy courses and on Nan Dieter Conklin and Harold "Doc" Ewen, as well as personal papers of Ronald Bracewell, John Findlay, David Heeschen, John Kraus, Grote Reber, Richard Thompson, and James Ulvestad. Acquisitions since 2009 include small collections of papers from Marshall Cohen, Mark Gordon, David Hogg, Kenneth Kellermann, and Paul Vanden Bout. Major acquisitions in 2011, on which processing has just begun, are the papers of the late Donald Backer and papers received from Bernard Burke.

During 2010 and 2011, Woodruff Sullivan III donated research materials gathered over 30 years in writing his book, *Cosmic Noise: A History of Early Radio Astronomy* Sullivan (2009) including 255 interviews with radio astronomers audio-taped between 1971 and 1988. The book covers the period up to 1953, and a significant portion of his interviews and his other materials illuminates post-1953 radio astronomy history. The 2011 Pollock Award from Dudley Observatory funded the digitization of the taped interviews and the preparation of detailed finding aids for the Sullivan collection. See http://www.nrao.edu/archives/Sullivan/sullivan.shtml/.

Additional material on the history of radio astronomy can be found at: http://www.astro.washington.edu/users/woody/hra.html

4. Conferences

Celebrations of the 50th anniversary of NRAO, Bridle *et al.*(2008) and Parkes in 2011 (http://www.atnf.csiro.au/research/conferences/Parkes50th/index.html) and the 40th anniversary of Westerbork (http://www.astron.nl/wsrt40/) each contained historical reviews of the development of radio astronomy. In November 2009, Kellermann and Ekers organized a session on *Discoveries in Astronomy* at the American Philosophical Society with an emphasis on radio astronomy in papers by Ekers and Kellermann (2011), Schmidt (2011), Longair (2011) and by R.W. Wilson on the *Discovery of the CMB* (unpublished). All of the presentations can be viewed on-line at http://www.amphilsoc.org/meetings/webcast/archive/y/2009/m/11.

At the 2011 General Assembly of URSI Commission J, Kellermann reviewed the careers of recently deceased radio astronomers.

5. Other Major Publications

Wielebinski and Wilson (2010) have reviewed the history of radio astronomy. Goss & McGee (2009) have published a biography of Ruby Payne-Scott which conveys her personal challenges trying to do radio astronomy in post-war Australia. In 2012, a new edition of this book for a non science audience *Making Waves: The Story of Ruby Payne-Scott, Australian Pioneer Radio Astronomer* will be published by Goss as part of the Springer *Astronomers' Universe* popular astronomy series. Several papers reviewing the history of radio astronomy in France have been published by Orchiston *et al.* (2009), Lequeux *et al.* (2009), Pick *et al.* (2011), Encrenaz *et al.* (2011). Papers on the history the Stockert radio telescope by Wielebinski, R. (2010) and the Effelsberg radio telescope by Wielebinski *et al.* (2011) document the development of radio astronomy in Germany. Kellermann has edited a translation of the 1986 book in Russian on *A Brief History of the Development of Radio Astronomy in the USSR* due to be published by Springer in 2012.

Ken Kellermann

Chair of Working Group on Historical Radio Astronomy

References

Bridle, A. H., Condon, J. J., & Hunt, G. C. 2008, *Frontiers of Astrophysics: A Celebration of NRAO's 50th Anniversary*, Astronomical Society of the Pacific confrence Series, Number 395.

kers, R. D. & Kellermann, K. I. 2011, *Discoveries in Astronomy* Publications of the American Philosophical Society, 115, 2, 129-133

Encrenaz, P., Gómez-González, Jesús, Lequeux, J., & Orchiston, W. 2011, *Highlighting the History of French Radio Astronomy 7: The Genesis of the Institute of Radioastronomy at Millimeter Wavelengths (IRAM)*, Journal of Astronomical History and Heritage 14 (2) 83-92.

Goss, W. M. & McGee, R. X. 2009, *Under the Radar, The First Woman in Radio Astronomy*, Ruby Payne-Scott (Springer ASSL series)

Lequeux, J., Steinberg, J.-L., & Orchiston, W. 2009 *Highlighting the History of French Radio Astronomy 5: The Nancay Large Radio Telescope*, Journal of Astronomical History and Heritage 13(1) 29-42.

Longair, Malcolm, 2011, *The Discovery of Pulsars and the Aftermath*, Publications of the American Philosophical Society, 115, 2, 147-157

Orchiston, W., Steinberg, J.-L., Kundu, M., Arsac, J., & Blum, E. J. 2009, *Highlighting the History of French Radio Astronomy 4: Early Solar Research at the École Normale Superieure, Marcoussis and Nancay*, Journal of Astronomical History and Heritage 12(3), 175-188.

Pick, M., Steinberg, J.-L., & Boischot, A. 2011, *Highlighting the History of French Radio Astronomy 6: The Multi-Element Grating Arrays at Nancay*, Journal of Astronomical History and Heritage 14 (1) 57-77.

Schmidt, M. 2011, *The Discovery of Quasars*, Publications of the American Philosophical Society, 115, 2, 142-146

Sullivan, W. T. III, *Cosmic Noise: A History of Early Radio Astronomy* 2009, (Cambridge U. Press)

Wielebinski, R. & Wilson, T. 2010, *The Development of Radio Astronomy*, in Heritage Sites of Astronomy and Archaeoastronomy, 213–220

Wielebinski, R. 2010, *The Stockert Radio Telescope*, in: Heritage Sites of Astronomy and Archaeoastronomy, pp. 221–222

Wielebinski, R., Junkes, N., & Grahl, B., 2011, *The Effelsberg 100-m Radio Telescope: Construction and Forty Years of Radio Astronomy*, in Journal of Astronomical History and Heritage, 14 (1) pp. 3-21

Transactions IAU, Volume XXVIIIA
Reports on Astronomy 2009–2012   © International Astronomical Union 2012
Ian F. Corbett, ed.   doi:10.1017/S1743921312003031

DIVISION XI   **SPACE AND HIGH-ENERGY**
                        ASTROPHYSICS
                        ASTROPHYSIQUE SPATIALE & DES HAUTES
                        ENERGIES

PRESIDENT   Christine Jones
VICE-PRESIDENT   Noah Brosch
PAST PRESIDENT   Günter Hasinger
BOARD   Matthew G. Baring, Martin Adrian Barstow,
                        João Braga, Evgenij M. Churazov,
                        Jean Eilek, Hideyo Kunieda,
                        Jayant Murthy, Isabella Pagano,
                        Hernan Quintana, Marco Salvati,
                        Kulinder Pal Singh, Diana Mary Worrall

DIVISION XI COMMISSIONS

Commission 44   Space & High Energy Astrophysics

DIVISION XI WORKING GROUPS

Division XI WG   Particle Astrophysics

INTER-DIVISION WORKING GROUPS

Divisions IX-X-XI WG Astronomy from the Moon

Division XI is organized by astronomers and astrophysicists who are mainly involved in space astronomy and their relevant research fields. Thus the Division XI members represent a very broad community, including radio, infrared, optical, ultraviolet, X-ray, and gamma ray, as well as cosmic ray observers and theorists. The topics of interest to the Division were extended to the study neutrino, astrophysical particles and gravitational waves, but these are currently under-represented in the Divisional membership. The relevant investigations cover almost all astronomical topics from our Solar System, stellar, Galactic and extragalactic research to studies of the deep space Universe and cosmology. This implies that communication and cooperation among the Division members, and cross-fertilization with members of other Divisions, are important and helpful to promote new space and ground based observatories and to enhance their scientific value.

TRIENNIAL REPORT 2009-2011

1. Organizational issues

Originally, Division XI concerned itself only with high-energy astrophysics (in particular UV, X-ray and gamma rays), to which was later added the domain of lower-energy astrophysics where observations are generally performed from space (optical, infrared, submillimeter and parts of the radio spectrum). The Division also includes ground-based high energy gamma ray and cosmic ray experiments, gravitational wave, and Moon-based astronomical observations. The individual expertise of the present OC reflects primarily the UV and higher energy domains. However, since there are plans within the IAU to restructure divisions, we propose that, following the changes in the Divisional structure and renewal of the OC, the new members will be recruited to broaden the spectral range of research covered by the Division.

Division XI subsums now a single Commission (#44, having the same name as the Division and the same OC as the Division) and two Working Groups ("Astronomy from the Moon", which is an inter-divisional WG together with Divisions IX and X, and "Particle Astrophysics"). While the Astronomy from the Moon WGs is active, the Particle Astrophysics WG should be substantially changed under the new IAU structure. We propose to restructure the Particle Astrophysics WG, rename it "Astroparticle WG" and invigorate it by inviting new members from the high energy physics community to join the IAU and this working group.

In the proposed restructuring of the Divisional structure of the IAU, some areas of fundamental physics, including astroparticle physics, would come under the new division that would succeed the present Division XI. We propose to initiate a strong recruiting effort among high-energy physicists to join the new Division and initiate more cross-fertilization between particle physics and astrophysics within the IAU. With this change, since the IAU Commissions will be retained after the IAU restructuring effort, it may be beneficial to create a number of new Commissions within the new Division to deal with, e.g., Space Astronomy (including Astronomy from the Moon), High-Energy Astrophysics (both from space and from the ground), gravitational wave astrophysics, and Non-Photon (particle) Astrophysics. However, these possibilities must wait for the final restructuring decisions at the 2012 GA.

The Division now has a total membership slightly over 1000, which does not reflect the potential of interested Union members whose research includes using space-based instrumentation, or which touches the high-energy physics domain. Obviously, more efforts on the part of the OC to broaden the Division membership are in order.

2. Research developments within the past triennium

In this section we list active and planned space missions, separating them by wavelength (energy). We list first missions dedicated to the observation of deep space, excluding Solar System targets. These are the topic of § 2.2. When discussing different missions, we also include those approved or planned, but not yet launched, to provide a perspective on what is likely to be available in the future. The situation projected from the mission listing by wavelength range indicates a future deficiency in the UV domain, with no major facility available following the eventual termination of the HST mission.

2.1. Deep space research

2.1.1. X and γ-ray

The scientific activities pertaining to the Division's scope of interest include the continuing observational activity of GALEX, Swift, AGILE, CHANDRA, SUZAKU, XMM-Newton, RXTE, INTEGRAL and FERMI at the energy range higher than that of the optical, HST in the UV-optical-near IR domain, Spitzer, WISE (operations terminated in February 2011), and Herschel in the IR, and ACE in the astroparticle field.

Since its launch in 1995, RXTE (the Rossi X-ray Timing Explorer Mission) has allowed the detailed study of X-ray variability on timescales from microseconds to months in the energy range from 2 to 250 keV for compact objects, most notably galactic black holes and neutron stars. LOFT (the Large Observatory for X-ray Timing) is one of four candidate

ESA M class missions now undergoing assessment and if selected for flight, LOFT will provide a very significant extention of the science capabilities of RXTE.

After 12 successful years in space, CHANDRA is still going strong, as demonstrated by the strong response to the 13th call for observing time. The 659 proposals oversubscribed the available observing time by a factor of 5.4. As one example of recent science results, measurement of the growth of structure through Chandra observations of nearby and distant clusters, confirmed the existence of dark energy, originally discovered through optical observations of distant supernovae.

XMM-NEWTON also is completing its twelveth successful year and continues normal operations. Among its accomplishments are the serendipitous detections of X-ray luminous galaxy clusters to z~1.6. XMM has significantly increased the number of discovered distant systems in the first half of cosmic time ($z \geqslant 0.8$) to more than three dozen, which allow now a systematic look at the earliest formation history of the most massive objects in the Universe. Deep X-ray and optical observations of the COSMOS field showed that most AGN are not triggered by galaxy mergers.

Since its launch in 2004, SWIFT has discovered more than 600 Gamma-ray bursts. Their luminous afterglows can be detected at the highest redshifts. Of note is GRB090429B, whose photometric redshift is 9.4. Swift's fast scheduling ability and multiwavelength capability have allowed rapid X-ray and optical follow-up of new supernovae, variable stars, and AGN outbursts. Swift also has detected more than 1000 hard X-ray sources with its BAT instrument, most of which are AGN. Combining observations from different missions often leads to exciting results, such as the recent Swift, Chandra and XMM observations that show evidence for a cooling neutron star crust in the binary EXO 0748-676.

SUZAKU, Japan's fifth X-ray astronomy mission, was launched in 2005. Although the X-ray Spectrometer (calorimeter) lost its cryogen shortly after launch, the X-ray Imaging Spectometer and Hard X-ray Detector continue to provide new observations. Recent scientific results include the detection of possible clumping in the baryons in the outskirts of the Perseus cluster. Suzaku, along with Astro-H, the sixth Japanese X-ray mission which is scheduled for launch in 2014, also can measure spins and study the hard X-ray emission from black holes in our Galaxy.

The Spectrum-X-Gamma (SXG) mission is currently under construction as a Russian-German X-ray astrophysical observatory. SXG will carry two powerful telescopes with imaging instruments, the eROSITA (extended ROentgen Survey with an Imaging Telescope Array)and ART-XC. Launch into an L2 orbit will be in late 2013. The first four years of the mission will be devoted to surveys of the entire sky, while the next 3.5 years will be used for pointed observations. eROSITA and ART-XC will perform deep surveys of the entire sky. eROSITA's sensitivity will be more more than thirty times that of ROSAT in the 0.5-2 keV band. Its best imaging, on-axis, will be 15", while the average blurring in the survey will be 28". ART-XC energy band (6 to 30 keV) will provide the first high energy map of the full sky. Due to the scan pattern, the highest sensitivities will be at the ecliptic poles.

eROSITA's sensitivity will result in a wealth of X-ray information on newly discovered clusters of galaxies, groups, galaxies, AGN, stars and compact objects. All massive clusters of galaxies in the Universe will be detected in the all-sky survey. Approximately 1100 clusters will be at $z \geqslant 1$. Measurements of the approximately 100,000 clusters will determine the growth of clusters over cosmic time and be used to place tight constraints on cosmological parameters, including σ_8 and the Dark Energy equation of state. In addition, tens of thousands of groups within 1 Gpc will be detected and used to trace large-scale filaments and to determine the outburst frequency and power of supermassive black holes. Three million AGN will be discovered, allowing a full census of radio and

Seyfert galaxies, quasars and blazars. In addition the X-ray emission from several 10^6 stars, including stellar flares, will be detected and quantified. Finally these observations will form the basis for the most complete X-ray luminosity function for Galactic X-ray sources. These very large samples of clusters, AGN, stars and galactic sources will allow the most interesting and rare objects to be discovered.

The hard X-ray sensitivity of ART-XC telescope and instrument will lead to the detection of several thousand AGN, including the heavily absorbed Compton-thick AGN. The ART-XC also will allow detailed studies of Galactic sources up to 30 keV and images of bright clusters, including the detection of merger shocks.

INTEGRAL, which was launched in 2002 is now in its extended science operations phase, was oversubscribed by a factor of 2.9 during the last allocation (AO-9). Its ability to carry out spectroscopy from 15 keV to 10 MeV, along with simultaneous X-ray and optical monitoring has led to many new results includine a greater understanding of compact objects, from neutron stars to supermassive black holes, including the recent discovery from combined INTEGRAL, Chandra, XMM-Newton, HST and Swift observations of giant "bullets" of gas from the MKN509 black hole.

The Italian AGILE gamma-ray mission was launched in 2007 to survey the sky in the 30 MeV to 50 GeV band and the 18 to 60 keV hard X-ray band. In addition to imaging, AGILE provides sub-millisecond timing for the study of transient phenomena. Important scientific results include the discovery of variability in the Crab's intensity (also see FERMI results) and the detection of gamma-ray emission from Cygnus X-3.

The FERMI Gamma-ray Space Telescope was launched in 2008 to observe the γ-ray sky from 30 MeV to 300 GeV (the LAT instrument maps with a field of view of about 20% of the sky) and from 150 keV to 30 MeV (the GBM instrument monitors the sky for γ-ray bursts). FERMI is operating nominally and, among its many scientific results, reported that the Crab Nebula is not a stable X-ray and gamma-ray reference source, showing intense flares and month-scale variations at hard X-ray energies (also see AGILE results). Another major breakthrough from Fermi is the discovery of dozens of radio quiet neutron stars.

The Nuclear Array Spectrometer (NuSTAR) satellite will be the first focusing high energy X-ray mission, allowing a sensitive survey of the hard X-ray sky. The Pegasus launch is scheduled for February 3, 2012, to begin NuSTAR's two-year mission. This mission will use a dense nested-foil concentrator with 133 shells to observe in the spectral region from 10 to 80 keV, providing a much higher effective area at these energies than either Chandra or XMM-Newton and reaching a sensitivity of 0.8 μCrab in 10^6 seconds for the 10-40 keV band. NuSTAR will have an imaging capability of 10" (FWHM; 50" half-power) for a field of view of 12'.5x12'.5, and a spectral resolution of 0.6 keV at 6 keV, going down to 1 keV at 60 keV.

We note also that the International X-ray Observatory (IXO) is under intense review and redefinition at ESA and NASA. Both are carrying out studies to determine future missions.

2.1.2. *Ultraviolet and optical*

During the past three years, The Galaxy Evolution Explorer (GALEX) has continued its mission to study the basic structures of the Universe. Launched in April 2003, it has now completed more than 8 years in-orbit and its original 29 month mission has been extended several times. A recent scientific highlight combines GALEX data with observations made by the Anglo-Australian telescope to help confirm that dark energy is driving the Universe apart at accelerating speeds (results from the WiggleZ Dark Energy Survey; Drinkwater *et al.* 2010). The GALEX mission lost its FUV channel in

May 2009. Based on the 2010 NASA Senior Review Panel's recommendations, NASA announced that all GALEX observations will terminate at the end of September 2012. A final data release is planned. Given the mission status and future possible missions under consideration when this report was written, it is not likely that the GALEX capabilities will be significantly expanded within the next decade.

With the end of the FUSE mission on October 2007 and the failure of the electronics of the Space Telescope Imaging Spectrograph (STIS) on HST in August 2004, the availability of spectroscopic observations in the UV has been limited to the low resolution GRISM spectroscopy on GALEX, together with UV monitor telescopes on Swift and XMM-Newton. Therefore, the final HST servicing mission in May 2009 was of particular importance to UV astronomy. Apart from general health servicing of the spacecraft (batteries, gyros etc.), the plan involved the installation of two new scientific instruments, Wide Field Camera 3 and the Cosmic Origins Spectrograph (COS), and repairs to STIS and the Advanced Camera for Surveys. The mission was a tremendous success with all installation and repair goals achieved, making HST more capable than it has ever been.

India's first dedicated astronomy satellite ASTROSAT is planned for launch in April 2012. This mission will cover the spectral domain from hard X-rays to the UV. One of the instruments on board is the ultraviolet imaging telescope (UVIT) which consists of two Ritchey-Chrétien telescopes (37.5-cm diameter) providing simultaneous coverage of a 28' diameter field in three wavelength bands. One of the two telescopes is dedicated to the far-ultraviolet (FUV) channel with a wavelength coverage between 130 and 180 nm. The other telescope splits the incoming light into two channels using a dichroic mirror into an near ultraviolet (NUV) channel with a spectral coverage between 200 and 300 nm and a visible channel with coverage between 320 and 550 nm. The detectors are intensified CMOS detectors with an effective angular resolution of better than 2" in space. A number of broad and narrow band filters may be used to observe targets in different spectral bands while a transmission grating provides slitless spectroscopy in the FUV and NUV channels with a resolution of about 100.

Although most of the UVIT time will be in parallel with the X-ray telescopes to monitor the observed source, the UV team has a fraction of satellite time to observe their UV targets. Time for guest observers on ASTROSAT is planned following a PV phase, with a steadily increasing fraction for national and international proposals.

We note the effective termination of the TAUVEX UV telescope by the Israel Space Agency following its removal from the Indian Space Research Organization GSAT-4 satellite (subsequently lost on launch in May 2010). We also note continued efforts in Russia during this triennium to prepare the World Space Observatory-UV (Spektr-UV) as a spectroscopic and imaging instrument for an eventual launch in 2013-2014.

The COROT satellite, studying transiting exoplanets and stellar astroseismology, has been operating normally in this period, and so has the MOST satellite of the Canadian Space Agency. The Kepler mission, with similar goals to these of MOST and COROT, but having significantly enhanced sensitivity, has been in operation since 2009 and has announced more than 1000 exoplanet candidates. Recently Kepler observations made the first unambiguous detection of a planet orbiting two stars (remember planet Tatooine from Star Wars?).

The GAIA (Global Astrometric Interferometer for Astrophysics) mission, to follow-up and enhance the Hipparcos mission, is being readied for a 2012-2013 launch on a Soyuz rocket. GAIA will operate at the L_2 point and is expected to produce a catalog of approximately 10^9 stars to magnitude 20 with astrometry good to $\sim$20 μas (microarcsecond) at 15 mag, and $\sim$200 μ as at 20 mag.

2.1.3. *Infrared and submillimeter*

The first Japanese infrared satellite AKARI launched in February 2006 and operated successfully, producing an ∼all-sky survey by August 2007. The mission continued despite the boil-off of its liquid Helium coolant, by using active cooling to only 40K and restricting the long-wave IR observations. The survey covered more than 96% of the entire sky with better sensitivity and angular resolution than IRAS in a wider wavelength range. The first point source catalog with 1.3×10^6 sources was released in March 2010. The all-sky images and the faint source catalogs will also be released in several years. Besides the all-sky survey, about 5,000 pointed observations had been made for selected sky areas and individual sources, including planetary sources, young and old stars, galaxies, and the cosmic IR background. More than 12,000 pointed observations were made in the near-infrared. The observations were terminated in May 2011, but the data reduction and archiving activities are continuing.

SPITZER, launched in 2003, has had very broad scientific impacts. For example, with Spitzer, astronomers have probed stellar nurseries in our Galaxy and also detected the red and dead galaxies in very high redshift clusters. Recently Spitzer images of Maffei 2 showed the structure of this nearby galaxy, which is almost completely hidden in visible light by dust clouds in our Milky Way. Spitzer has now run out of coolent and is operating in the "warm" Spitzer phase.

The Wide-Field Infrared Survey Explorer (WISE) mission was a NASA-funded Explorer mission carrying a 40-cm telescope and detectors cooled to 15K by solid hydrogen. It was launched in 2009 and ran out of cryogen on 29 September 2010. During this period WISE conducted a full survey of the sky in four IR bands: 3.1, 4.6, 12 and 22 μm. On 14 April 2011 a preliminary release of WISE data was made public, covering 57% of the sky observed by the spacecraft. The results of the full survey are scheduled to be released by March 2012.

SOFIA, (the Stratospheric Observatory for Infrared Astronomy) a modified Boeing 747 aircraft that carries a 2.5 meter telescope for observations in the mid-infrared to submillimeter range, began science flights in late 2010. SOFIA is a joint program by NASA and German Aerospace Center and is expected to have a 20 year lifetime.

The WMAP mission, which collected invaluable data on the cosmic microwave background for nine years at the L_2 location, was terminated on 28 October 2010. The combination of the WMAP and HST results yielded, for the first time, a Hubble constant good to a few %. Fortunately, the PLANCK satellite also at L_2 took over the CMB measurements from WMAP from July 2009, with enhanced sensitivity and angular resolution. Both the Early Release Compact Source Catalogue with thousands of sources detected by Planck and the Early Release Sunyaev-Zeldovich Catalogue of 189 clusters were released in January 2011.

Together with Planck, ESA launched the HERSCHEL spacecraft carrying the largest astronomical telescope ever launched with a 3.5 m-diameter primary mirror. The key science objectives of Herschel are to study the formation of stars and galaxies, and to investigate the relationship between the two, but other interesting results emerge continuously. For instance, a recent paper in Nature (Hartog *et al.* 2011) reported finding a very similar D/H ratio in outgasing from comet 103P/Hartley 2, strengthening the hypothesis that at least some H_2O on earth is of extraterrestrial origin.

The James Webb Space Telescope (JWST), planned to succeed HST and observe from L_2 in the 0.6 to 28 μm band with a 6.5-m telescope, is under review by the United States Congress after costing significantly more than originally proposed, and launching at least seven years later than planned.

ESA selected in 2011 the EUCLID mission to map the geometry of the dark Universe. The mission will investigate the distance-redshift relationship and the evolution of cosmic structures by measuring shapes and redshifts of galaxies and clusters of galaxies out to redshifts of ~2. EUCLID is optimized for two primary cosmological probes: weak gravitational lensing and Baryonic Acoustic Oscillations (BAO). EUCLID will operate from the L_2 point and will use a modified Korsch telescope with a primary diameter of 1.2-m, and central obscuration 0.4-m. The visible channel will use a 600 Mpixel CCD mosaic that will image a field of view of 0.5 square degrees with 0".2 pixels. The near-IR channel will use a 7.5 Mpixel mosaic to provide photometry from 0.9 to 2.0 μm. EUCLID is now scheduled for a 2019 launch.

The Space Infrared Telescope for Cosmology and Astrophysics (SPICA) is a proposed Japanese-led mission with extensive international collaboration (ESA, Korea, etc.). SPICA is optimized for mid- and far-infrared astronomy using a cryogenically-cooled 3.2-m telescope. Its high spatial resolution and unprecedented sensitivity will enable addressing a number of key problems, ranging from the star-formation history of the Universe to the formation of planets. To reduce the mission mass, SPICA will be launched at ambient temperature and will be cooled down on-orbit using on board mechanical coolers together with an efficient radiative cooling system. This combination allows a 3-m class space telescope cooled to 6K with a moderate total weight (3700 kg). The target launch year of SPICA is around 2020 for a five-year or longer mission.

2.1.4. *Radio Observations from Space*

The space radio interferometry mission RadioAstron (Spekr-R, in Russian) was launched in July 2011 and operates properly in a highly elliptical orbit with an apogee higher than 350,000-km. This is the first of the Russian large space observatories in the Spectrum (Spektr) series to reach orbit. RadioAstron carries a 10-m segmented dish deployed in-orbit, operates at wavelengths from 1.35-cm to 92-cm and, together with ground-based radio observatories, offers an unprecedented angular resolution as fine as 10^{-6} arcsec. First light was on 27 September 2011, when Cassiopeia A was observed at 92 and 18-cm.

2.1.5. *Gravitational wave astrophysics*

The planned Laser Interferometer Space Antenna (LISA) was de-emphasized in April 2011 from the NASA mission list due to budgetary constraints. ESA is planning a full revision of the mission's concept within its Cosmic Vision L-class. LISA was thus renamed the Next Gravitational–Wave Observatory (NGO), with selection of the winning mission candidate expected in February 2012.

2.1.6. *High-energy particles*

In the domain of high-energy particle observatories, we note the continuation of the PAMELA (Payload for Antimatter Matter Exploration and Light-nuclei Astrophysics) operating on the Resurs-DK1 satellite. Among the highlighs of this mission is the discovery of antiprotons trapped in the geomagnetic belts circling the Earth (2011 ApJ 737, 129).

ACE (the Advanced Composition Explorer) was launched in 1997 and from its L1 location continues to detect high energy particles accelerated by the Sun, as well as particles accelerated in the heliosphere and galactic regions, with near real time 24/7 coverage of space weather, providing about one hour advance warnings of geomagnetic storms. Studies with ACE include determinations of the elemental and isotopic composition of the the solar corona, the solar wind and the local interstellar medium.

2.1.7. *Ground-based high-energy astrophysics*

In the domain of ground-based high energy observatories, we note the successful operation of the southern Auger observatory in Argentina, but regret the decision by the US not to host the northern part of the observatory. We note the revision of the HECR spectrum above 10^{18} eV presented by the Auger collaboration at the 32nd ICRC, Beijing, China, in 2011 (arXiv:1107.4809). We also note the continued operation and exciting science results from the imaging Cherenkov telescopes at the VERITAS, CANGAROO, H.E.S.S. and MAGIC sites.

2.2. *Investigating our Solar System*

Among the active space missions exploring the Solar System, CASSINI continues to observe Saturn, its rings and moons, from orbit. JUNO was launched in August 2011 to orbit Jupiter 33 times from 2016. The successful exploration of the minor planet Vesta by DAWN will be followed by a similar exploration of Ceres. An analysis of the sample container retrieved from HAYABUSA showed grains originating from the asteroid Itokawa; these are now being actively studied.

Among the inner planets, the exploration of Venus by the VENUS EXPRESS spacecraft is continuing. So is the orbital exploration of Mercury by MESSENGER; imagery from orbit showed the presence of small depressions with bright interiors and halos, often found in clusters. These could be actively forming today. Mercury will be the target of ESA and JAXA's twin BEPI-COLOMBO spacecraft in 2014.

Following a successful impact of its probe on comet Tempel 1, the DEEP IMPACT spacecraft now named EPOXI visited comet Hartley 2 in 2010. The current comet exploration flagship is ROSETTA, on its way to comet 67P/Churyumov-Gerasimenko to arrive in 2014 for orbit exploration and sample return. In the future, in 2020, NASA's OSIRIS-REx will explore the near-Earth asteroid 1999 RQ36 and return samples from its surface.

We note the constellation of spacecraft exploring Mars either from orbit or on the surface. These include the MARS ODYSSEY (launched in 2001), the MARS EXPRESS (launched also in 2001), and the MARS RECONNAISSANCE ORBITER (launched in 2006). On the surface of Mars, the OPPORTUNITY rover is still functional. In the future, the MARS SURFACE LANDER's CURIOSITY rover is expected to launch in late-2011 and arrive on Mars in August 2012. The Mars ATMOSPHERIC and VOLATILE EVOLUTION (MAVEN) will join the Mars exploration effort in 2014. Before that, the Russian PHOBOS-GRUNT sample return and the YINGHUO-1 Mars orbiter will arrive in 2013. In 2016 the ExoMARS international mission of ESA and NASA will study trace gases in the Martian atmosphere and deploy the ESA lander and rover on the surface.

The NEW HORIZONS Pluto and Kuiper Belt exploring probe is on its way to a fly-by through Pluto's system in 2016. This is likely to be followed by the exploration (also via fly-by) of one or more Kuiper belt objects (KBOs) after passing Pluto.

Observations of the Sun are continuously carried out by the Solar and Heliospheric Observatory (SOHO) launched operating at the Sun-Earth L_1 point. SOHO operates together with the Advanced Composition Explorer (ACE) also at L_1 and with the two STEREO spacecraft, one ahead of Earth in its orbit, the other trailing behind, to provide panoramic views of almost the entire solar surface.

The HINODE satellite of JAXA (with collaborating space agencies) continues to study the solar magnetic activity. Its continued operation, past the design lifetime of three years, is important as the activity cycle ramps up to a new maximum.

SDO, the Solar Dynamics Observatory, is the first mission in NASA's Living with a Star Program and was launched in February 2010. SDO is obtaining observations of the solar atmosphere on small angular scales and in several wavelengths simultaneously. Primary goals are to understand how the Sun's magnetic field is generated and structured and to determine how this stored magnetic energy is converted and released by the solar wind, energetic particles and intensity variations.

The Sun and the interaction of the solar wind with magnetospheres, that of the Earth in particular, are of special interest and significant space research effort is spent in this domain. In this field it is worth noting not only specific missions to study the Sun, but also the flexible retargeting of probes not originally intended for this purpose to such studies. One example is ARTEMIS, "Acceleration, Reconnection, Turbulence and Electrodynamics of the Moon's Interaction with the Sun", composed of two probes originally members of the THEMIS mission in Earth orbit studying Earth's aurora, but which were redirected to the Moon to save the two probes from losing power in Earth's shade. This new mission will study the environment of Earth-Moon Lagrange points, the solar wind, the Moon plasma wake and how the Earth magnetotail and the Moon's magnetism interact with the solar wind.

Following the depletion of cryogen in WISE, the NASA Planetary division provided funding for a short mission extension in warm condition, called NEOWISE, to search for small Solar System bodies close to Earth's orbit. It is likely that the NEOWISE survey will catalog about 300,000 main-belt asteroids, of which approximately 100,000 will be new, and about 700 near-Earth objects including about 300 newly discovered.

The examples of NEOWISE and of ARTEMIS are ones to be applauded. In this era of diminishing resources available for space research, when long-planned missions are descoped because of lack of funding, flexible thinking is important to maximize the science returns. One possibility that should be seriously considered by designers of future space missions is to use the cruise phase of their missions for additional science. This was very successfully done with the Voyager UVS instruments to study astronomical sources and the UV background, and could be replicated with small additional funding to the basic cost of the mission for other deep space probes.

3. Scientific meetings

The vitality of a topic can be measured by a number of metrics. One is the number of publications appearing during a certain period; another is the number and frequency of scientific meeting dedicated to this topic. We note that during the triennium period referred to in this report, numerous scientific meetings with components of Space or High-energy Astrophysics took place throughout the world. In particular several IAU Special Sessions, Joint Discussions, and Symposia at the Rio General Assembly included high energy and space topics. Meetings and workshops included those focused on science results from particular missions, techniques or energies (e.g. Chandra, XMM-Newton, Suzaku, eROSITA, Akari, far IR interferometry, astroparticle and underground physics, cosmic rays), as well as those focused on particular types of objects (e.g. "Thirty Years of Magnetars," "High Energy Phenomena in Massive Stars," "PANDA symposium on Stellar Outflows," "Supersoft X-ray Sources," "GRB Physics," "Black Hole Physics," "Feedback in Galaxies, Groups, and Clusters," "Accretion and Ejecta in AGN," "Physics of Relativistic Flows"). The many meetings, along with the vigorous space programs conducted by the world nations, indicate that the space astrophysics remains a high-interest subject.

Of particular note, we mention the success of the HST refurbishments which were celebrated with the HST III conference, held in Venice in October 2010, which presented results from the new and refurbished instruments. Having passed its 21st birthday in space on 24 April 2011, several more years of observations are promised by the Hubble Space Telescope. However, the actual length of time provided will depend on the telescope health and on funding decisions for orbital boosts and future operations. This, combined with delays in launching ASTROSAT and the effective termination of the TAUVEX project by the Israel Space Agency, imply that the future access to observations in the UV is becoming a critical issue. This has been addressed in a number of meetings. For example, the Network of UV Astronomers (NUVA) in Europe has organized a series of workshops. The latest, held in St Petersburg in May 2010, reviewed current science results and considered the near future in the form of the Russian led World Space Observatory (WSO), which will deliver high resolution UV spectroscopy and imaging with an efficiency several times that of HST.

While the WSO and ASTROSAT will continue UV astronomical capability beyond HST, plans for a true next generation UV/optical facility with an order of magnitude improvement in capability remain uncertain. Also, apart from a handful of sounding rocket observations, there are no firm plans for any new facility in the important EUV waveband, which is yet to be explored with high effective area and spectral resolution. A workshop on "Beyond JWST: The Next Steps in UV-Optical-IR Space Astronomy", was hosted by the Space Telescope Science Institute in March 2009. Several ideas emerged from this meeting and now appear to have merged into a project called ATLAS-T. Research in support of future UV/optical telescope technology was a priority in the recently published US Decadal survey.

On the other hand, the descriptions given above of the high-energy astrophysics field and of the IR missions indicate that these domains are relatively healthy and will provide more years of actively collecting astronomical information.

4. Closing remarks

These are exciting times to be astrophysicists. We are extremely fortunate to have operating space missions and ground based observatories that provide fantastic data across many wavelengths and allow us to address profound questions concerning the origin and evolution of the Universe and all types of celestial bodies within it. Most space missions and many ground based observatories provide rich archives, allowing broad community access to existing observations, often in conjunction with observations at other wavelengths, and often for purposes not imagined by the original observer. These treasure chests of archival data have a led to many new discoveries.

In the past three years, new worlds have been discovered, along with new understandings of our solar system. We have studied the nearest star, our Sun, and the most distant galaxies, AGN and clusters. We have constrained cosmolological parameters, although the nature of both Dark Energy and dark matter still elude us. There is still much to be learned and new observatories are being planned and constructed that will provide answers to many of our questions. However, the new observations, while answering old questions, will likely result in yet more questions that will require yet more observations and theoretical work to answer them. The excitement of astrophysical discoveries continues.

Christine Jones and Noah Brosch
president and vice-president of the Division

Transactions IAU, Volume XXVIIIA
Reports on Astronomy 2009–2012　　　　　© International Astronomical Union 2012
Ian F. Corbett, ed.　　　　　　　　　　doi:10.1017/S1743921312003043

DIVISION XII　　　UNION-WIDE ACTIVITIES
ACTIVITES D'INTERET GENERAL DE L'UAI

PRESIDENT	Françoise Genova
VICE-PRESIDENT	Raymond Norris
PAST PRESIDENT	Malcolm Smith
OC members	Dennis Crabtree
	Olga B. Dluzhnevskaya
	Masatoshi Ohishi
	Rosa M. Ros
	Clive L.N. Ruggles
	Nicolay N. Samus
	Virginia L. Trimble
	Wim van Driel
	Glenn M. Wahlgren

DIVISION XII COMMISSIONS and WORKING GROUPS

Comm. 5 Documentation and Astronomical Data
Comm. 5 / WG Astronomical Data
Comm. 5 / WG Libraries
Comm. 5 / WG FITS Data Format
Comm. 5 / WG Virtual Observatories, Data Centers & Networks
Comm. 5 / TF Preservation & Digitization of
Photographic Plates
Comm. 6 Astronomical Telegrams
Comm. 6 /Service Central Bureau for Astronomical
Telegrams (CBAT)
Comm. 14 Atomic and Molecular Data
Comm. 14 / WG Atomic Data
Comm. 14 / WG Molecular Data
Comm. 14 / WG Collision Processes
Comm. 14 / WG Solids and their Surfaces
Comm. 41 History of Astronomy
Comm. 41 / WG Archives
Comm. 41 / WG Historical Instruments
Comm. 41 / WG Transits of Venus
Comm. 41 / WG Astronomy and World Heritage
Comm. 46 Astronomy Education and Development
Comm. 46 / PG International Schools for Young Astronomers
Comm. 46 / PG World-Wide Development of Astronomy
Comm. 46 / PG Teaching for Astronomy Development

Comm. 46 / PG Collaborative Programs
Comm. 46 / PG National Liaisons on Astronomy
Education & Newsletter
Comm. 46 / PG Public Understanding at the Times of
Solar Eclipses & Transits
Comm. 46 / PG Collaborative Programs; cosponsored activities
Comm. 46 / PG Network for Astronomy Education NASE
Comm. 50 Protection of Existing & Potential Observatory Sites
Comm. 50 / WG Controling Light Pollution
Comm. 55 Communicating Astronomy with the Public
Comm. 55 / WG Washington Charter
Comm. 55 / WG Virtual Astronomy Multimedia Project
Comm. 55 / WG Communicating Astronomy Journal
Comm. 55 / WG CAP Conferences
Comm. 55/ WG Johannes Kepler
Comm. 55/ WG New Media
Div. X-XII / WG Historic Radio Astronomy - See Division X

Transactions IAU, Volume XXVIIIA
Reports on Astronomy 2009–2012   © International Astronomical Union 2012
Ian Corbett, ed.   doi:10.1017/S1743921312003055

COMMISSION 5

DOCUMENTATION AND ASTRONOMICAL DATA

DOCUMENTATION ET DONNÉES ASTRONOMIQUES

PRESIDENT	Masatoshi Ohishi
VICE-PRESIDENT	Robert J. Hanisch
PAST PRESIDENT	Ray P. Norris
ORGANIZING COMMITTEE	Heinz Andernach, Marsha Bishop, Elizabeth Griffin, Ajit Kembhavi, Tara Murphy, Fabio Pasian

COMMISSION 5 WORKING GROUPS AND TASK FORCE

Div. XII / Commission 5 WG	Astronomical Data
Div. XII / Commission 5 WG	Designations
Div. XII / Commission 5 WG	Libraries
Div. XII / Commission 5 WG	FITS
Div. XII / Commission 5 WG	Virtual Observatories, Data Centers and Networks
Div. XII / Commission 5 TF	Preservation and Digitization of Photographic Plates

TRIENNIAL REPORT 2009–2012

1. Introduction

IAU Commission 5 (http://www.nao.ac.jp/IAU/Com5/) deals with data management issues, and its working groups and task group deal specifically with information handling, with data centers and networks, with technical aspects of collection, archiving, storage and dissemination of data, with designations and classification of astronomical objects, with library services, editorial policies, computer communications, ad hoc methodologies, and with various standards, reference frames, etc. FITS (Flexible Image Transport System), the major data exchange format in astronomy, has been standardized, maintained and updated by the FITS working group under Commission 5.

It has been suggested that a new era of astronomical research utilizing large amounts of data, *i.e.*, the 4th paradigm in astronomy, will soon come, and astronomers need to be well-prepared for this new era. Since the data production rate will be 100 to 1000 times larger than that at present, it will be crucial to have a combination of advanced machine learning technologies with immediate access to extant, distributed, multi-wavelength databases. Such an approach is necessary to make these assessments and to construct event notices that will be autonomously distributed to robotic observatories for near-real-time follow-up. Advanced data analyses combined with statistics and data mining will be essential to derive general "rules" and/or "knowledge" on various phenomena in the Universe, as the data volumes will make human inspection and analysis of the data impossible. The most important and exciting astronomical discoveries of the coming decade will rely on research and development in data science disciplines (including data management, access, integration, mining, visualization and analysis algorithms) that enable rapid information extraction, knowledge discovery, and scientific decision support for real-time astronomical research facility operations. Significant scientific results are expected to be obtained from data-intensive astronomical research in the very near future and beyond, and, Commission 5 needs to take a lead towards the new research paradigm.

2. Developments within the Past Triennium

Highlights of the 2009–2012 triennium include

• IAU Symposium 285 "New Horizons in Time Domain Astronomy": The symposium was held between September 19 and 23, 2011, in Oxford, the United Kingdom, and was organized by Robert J. Hanisch and Elizabeth Griffin as the co-chairs, together with some other Commission 5 members as the SOC. There were 240 participants.
• FITS: The third version of the FITS Standard Document was published in Pence *et al.* (2010). The new standard is also available on the FITS Support Office Web site at http://fits.gsfc.nasa.gov.

In addition, Commission 5 handled other issues as follows:

Commission Name: There was a discussion within the organizing committee if the Commission's name, "Documentation and Astronomical Data", should be changed, reflecting increased capacity and importance of astronomical data. There were a few proposals, such as "Astroinformatics", "Astronomical Data and Data Management", "Data Management and Information", "Astronomical Data and Information Management", and "Astronomical Data and Information". However, no consensus was achieved.

Proposal on a new IAU Resolution: Commission 5 received a proposal towards a new IAU Resolution regarding better writing styles. The draft Resolution was extensively discussed electronically in the organizing committee, and it was agreed to simplify the draft, and to further discuss the draft Resolution during the Commission's planned business meeting in Beijing in 2012. The finalized draft Resolution is expected to be submitted to the Executive Committee of the IAU for its approval.

CBAT issue: On May 18th, 2010, Ian Corbett (IAU General Secretary) requested Commissions 5 and 6 to provide their views and perspectives on the possibility of modifying the CBAT activity by means of the VOEvent technologies. Those Commissions were requested to submit their reports by the end of July, 2010. The OC members of Commission 5 and Rob Seaman (USA, chairman of VOEvent WG of the International Virtual Observatory Alliance) developed the following view from Commission 5.

Commission 5 is of the view that introduction of information technologies for the Astronomical Telegram systems would be in line with the primary purpose of the Commission, and welcomes adoption of the VO Technologies to help observers to report newly discovered transient phenomena and to disseminate such information to astronomy communities in the world. Thus Commission 5 would like to provide as much technological support as possible to the IAU EC. Commission 5 believes that the adoption of the VO technologies by the IAU would further accelerate of the astronomical research.

However, Commission 5 would like to draw EC's attention to the following issues that need to be resolved:

1) Although Commission 5 could provide technological support, the Commission itself does not own any resources (e.g., budget and manpower for software development, hardware for servers) for a new system like the CBAT. Therefore it is suggested that the EC consider how the necessary resources might be provided on a long-term basis.

2) It would be a task for Commission 6, not Commission 5, to cross-check newly reported "discoveries" and to "authorize" them as appropriate. Thus both Commissions need to coordinate their roles when designing and developing a new Astronomical Telegram dissemination system. Commission 5 is of the opinion that such coordination would be done by the new Task Force working towards the new system that will be established under Division XII.

Finally, Commission 5 recommends that the chairman of the IVOA VOEvent WG and at least one of the owners/operators of a community-accessible VOEvent broker be members of the above Task Force to ensure that astronomers get the access that they need.

The view above was consolidated with that of Commission 6, and the consolidated report was submitted to the EC.

Science Sessions during the 2012 GA: Commission 5 members submitted proposals towards associated science meetings during the General Assembly in Beijing. One proposal of a special session on "Data Intensive Astronomy" has been accepted; the preparatory process has already been started, chaired by Masatoshi Ohishi (Japan).

2.1. Activity Report of WG Designations

Chair: Marion Schmitz (USA)
Web Sites: http://cdsweb.u-strasbg.fr/cgi-bin/Dic/iau-spec.htx
for IAU Recommendations for Nomenclature,
http://vizier.u-strasbg.fr/viz-bin/DicForm
for Proposal for Registering a new Acronym, and
http://cdsweb.u-strasbg.fr/IAU/starnames.html
for star name history

At the 2009 Rio de Janeiro IAU meeting, Marion Schmitz (Caltech, USA) presided over the Commission 5 Working Group Designations meeting.

The Working Group Designations of IAU Commission 5 clarifies existing astronomical nomenclature and helps astronomers avoid potential problems when designating their sources.

The most important function of WG Designations during the period 2009-2011 was overseeing the IAU REGISTRY FOR ACRONYMS (for newly discovered astronomical sources of radiation; http://cdsweb.u-strasbg.fr/cgi-bin/DicForm) which is sponsored by the WG and operated by the Strasbourg Data Center (CDS). The Clearing House, a subgroup of the WG, screens the submissions for accuracy and conformity to the IAU Recommendations for Nomenclature (http://cdsweb.u-strasbg.fr/iau-spec.html). From its beginning in 1997 through September 2011, there have been 260 submissions and 211 acceptances. Attempts to register asterisms, common star names, and suspected variable stars were rejected.

Assistance was provided for inquiries about naming exo-planets after Roman-Greek mythology and about a systematic method for creating a more scientific convention for exo-planets. Both of these also involved discussions with IAU Commission 53.

2.2. Activity Report of WG Libraries

Chair: Marsha Bishop (USA)
Co-Chair: Robert J. Hanisch (USA)
Web Site: http://www.eso.org/sci/libraries/IAU-WGLib/index.html

The primary activity this triennium for Commission 5 Working Group Libraries has been to broaden the awareness of Working Group Libraries and increase the involvement of astronomy librarians from around the world to strengthen the working relationship between scientific user groups and libraries. In pursuit of this, we worked with Librarians to bring about a confluence of The Library and Information Services in Astronomy (LISA) and Working Group Libraries. While we were unable to schedule a joint meeting for Beijing, we did succeed in expanding the level of interest among astronomy librarians which has led to a more substantial program for Beijing. The program will bring scientists and librarians together to discuss and build on the ability of the library community to provide the information requirements of the scientific community. In addition, shared projects and programs between observatories, operated by librarians, will be reviewed.

2.3. Activity Report of WG FITS

Chair: William D. Pence (USA)
Web Site: http://fits.gsfc.nasa.gov/iaufwg/

The WG-FITS is the international control authority for the FITS (Flexible Image Transport System) data format. It is composed of 23 members from major astronomical institutions distributed around the world. The main activities during this triennium have included:

A major revision to the FITS standard document, which contains the formal definition of the FITS (Flexible Image Transport System) data format, was completed by the WG-FITS in 2008. This large document was subsequently published in the Astronomy & Astrophysics journal in Pence *et al.* (2010).

During the previous IAU triennium, the WG-FITS established a procedure for registering FITS conventions that are in use in the scientific community. This registry provides a central location on the FITS Support Office Web site (also sponsored and maintained by the WG-FITS) for documenting each convention for posterity. This registry continues to grow: 9 new conventions were submitted during the current triennium, bringing the total number of registered FITS conventions to 20. It is anticipated that several new FITS conventions will be submitted to this registry each year.

Major progress has been made on a document that defines a standard World Coordinate System (WCS) convention for specifying date and time coordinates that are to be associated with astronomical FITS data files. This is the next in a series of documents which have previously defined the standard FITS WCS conventions for spatial and for spectral coordinate systems. A complete draft of this new time WCS document was released for public comment in 2011. The final version of the document is expected to be submitted to the WG-FITS for formal approval in 2012.

2.4. Activity Report of WG Virtual Observatories

Chair	Robert J. Hanisch
Vice-Chair	Robert D. Bentley
Board Members	Beatriz Barbuy, Daniel Egret,
	Toshio Fukushima, George Helou,
	Peter Quinn

Web Site: http://cdsweb.u-strasbg.fr/IAU/wgvo.html

The Working Group on Virtual Observatories, Data Centers, and Networks arose from discussions at the IAU General Assembly in Prague (2006), and the first meeting of the WG was held during the 2009 General Assembly in Rio de Janeiro. The International Virtual Observatory is one of the rare truly global endeavors of astronomy. Many projects, each with its own goals, have been set up around the world to develop the IVO. The International Virtual Observatory Alliance (IVOA) is an alliance of the VO projects, with the aims of managing communication between VO projects, defining a common road map, and managing propositions for the definition and evolution of IVO standards through the IVOA Working Groups.

The IAU WG on Virtual Observatories, Data Centers, and Networks is the standard-bearer of the International Virtual Observatory at IAU, and it is the primary point of contact between the IVOA and the IAU. Its primary role is to provide an interface between IVOA activities, in particular IVOA standards and recommendations, and other IAU standards, policies, and recommendations. In particular, it raises VO-related topics (e.g. symposia, GA sessions) that should be handled by the IAU (Commission 5, Division XII and executive level). It helps facilitate take-up of VO standards in the broader community, particularly in liaison with national and international data centers, and provides outreach to VO and data management efforts generally in related fields within the IAU (planetary science, solar astronomy, etc.).

The WG brings to the attention of the IVOA Executive any topics it considers to be important for the IVO. It can be consulted by the IVOA Executive on any topic relevant to the international development of the VO. The WG consists of members of IVO projects together with individuals bringing an external view on the long term vision of the VO and other stakeholders. Participants include the president of Commission 5, the chair of the IVOA, a representative of the WG FITS (Commission 5), a representative of the WG on Astronomical Data (Commission 5), and a representative of the WG on International Solar Data Access (Division II).

The WG met on 4 August 2009 in conjunction with the IAU General Assembly in Rio de Janeiro, with approximately 25 people in attendance, representing at least 10 different countries. The meeting began with a discussion of the general goals for the WG and its relationship to other activities within the IAU (e.g., the WG on FITS, also under Commission 5).

Prof. Albert Bruch gave a presentation on Virtual Observatory activities in Brazil, particularly concerning the formation of BraVO, the Brazilian Virtual Observatory initiative. It was suggested that BraVO might be a catalyst for other VO programs in South America and could help to organize regional VO meetings. (This has, in fact, occurred; BraVO will be hosting the IVOA Interoperability Workshop in October of 2012, and will be running a VO school in conjunction with the 2012 annual meeting of the Brazilian Astronomical Society (SAB).)

Various news items were presented and discussed.

- The US Virtual Astronomical Observatory program was expected to begin very soon, with R. Hanisch as Director. (In fact, US VAO funding began in April 2010.)
- The AstroGrid project in the UK was not successful in attaining funding for ongoing operations.
- France's VO program continues on a strong footing.
- C. Corbally (Vatican Observatory) noted that scanned plates had been published to VO standards.

- Argentina plans to constitute a VO program within one year. (The Argentina Virtual Observatory, "NOVA", was indeed started and joined the IVOA in 2010.)
- Australia plans to renew its VO activities. (In 2010 Australia VO was restored.)
- Japan VO is active and in its operational phase since March 2008. Data from the Subaru telescope is being downloaded through JVO.
- The European VO has begun a program of educational outreach.
- Microsoft's Worldwide Telescope includes built-in access to VO services. Students as young as 10–12 years are using WWT to build tours of the universe.

Discussion continued concerning the take-up of VO tools and infrastructure within the research community. Efforts need to be made to improve professional outreach and provide a simpler introduction to VO standards. (The IVOA has taken action in this regard, with a much improved document describing the VO architecture and the roles and relationships of the various IVOA standards and protocols. In addition, all IVOA standards that have reached the final stage of "recommendation" (REC) are now available through the electronic preprint *arXiv*, http://arxiv.org, and are indexed by ADS, http://adswww.harvard.edu.)

Virtual Observatory standards development and discussions on scientific priorities, science applications, data mining tools and technologies, and education and public outreach are conducted in the semi-annual Interoperability Workshops of the IVOA. During this triennium IVOA "Interop" meetings have been held and are planned as follows:

24-29 May 2009	Strasbourg, France (CDS)
9-13 November 2009	Garching, Germany (ESO)
17-21 May 2010	Victoria, British Columbia, Canada (CADC)
7-11 December 2010	Nara, Japan (JVO)
16-20 May 2011	Naples, Italy (U. Naples/INAF)
17–21 October 2011	Pune, India (IUCAA)
21-25 May 2012	Urbana, Illinois, USA (NCSA)
October/November 2012	São Paolo, Brazil (U. São Paolo)

Further information about IVOA activities can be found at http://www.ivoa.net/.
The WG will meet next during the 2012 General Assembly in Beijing.

2.5. Activity Report of TF Preservation and Digitization of Photographic Plates (TF-PDPP)

Chair: Elizabeth Griffin (Canada)
Web Site: http://www.lhobs.org/PDPP.html

The TF-PDPP is a worldwide organization of (chiefly) professional astronomers who maintain a watchdog surveillance over their own, local, national and other collections of astronomical plates and associated (meta) data. The objectives of the PDPP are necessarily long-term, and cannot – should not – be confined to any short-term goals whose realization could then be said to be "completed". A physical meeting of the members is expensive to arrange, and since many have retired status it is also difficult for the PDPP to hold anything but a token meeting at an IAU General Assembly. However, at IAU S285 held in September 2011, one of the 20 dedicated afternoon workshops was focused on the preservation and digitization of photographic plates, and several PDPP members who were attending the Symposium made contributions.

The existence of the PDPP is becoming more widely known, and queries or messages concerning photographic plates now tend to be (re)-directed to members, which is definitely an improvement compared to just 8 or 9 years ago. Notices of uncatalogued – even unidentified – plates in South Africa and in Australia have thus been referred to the PDPP, as also have notices of impending abandonment of duplicate series such as Sky Survey materials. In the case of the latter we have been able to energize our network to contact any likely recipients. The PDPP's Website, where its occasional newsletter "SCAN-IT" can be accessed, is currently being transferred to another Webmaster and site.

Important strides have been achieved by DASCH (the project to digitize the world's largest collection, at Harvard), and will thus create a tremendous precedent for PDPP's endeavours.

We are pleased to note the following:

1) DASCH has been offered full support by a benefactor to digitize all the Harvard collection. The images will eventually be placed in the public domain.

2) The DAO (Canada) has commenced scanning its major collections of photographic spectra. Each spectrum is calibrated in both intensity and wavelength, and collapsed into a 1-D spectrum in either 50 or 10 mA steps (depending on the original material). A few hundred spectra are already in the public domain, and can be accessed via the DAO Science Archive within the CADC Web-site.

3) The new rapid scanner (DAMIAN), designed and installed at the Royal Observatory of Belgium in Brussels, is now becoming fully operational. Plans to digitize collections from European observatories are under discussion.

4) Projects have been commenced in China, Russia and elsewhere to digitize national collections. Unfortunately the tasks often depend on volunteer labour, or on grant money whose supply is liable to stall, and are thus prone to delays or discontinuity. However, those projects do not the benefit of purpose-built scanning equipment.

5) The Wide Field Plate Archive, installed in Sofia (Bulgaria), continues heroically to collect meta-data about collections of wide-field plates.

6) The Pisgah Astronomical Research Institute (PARI) has been awarded a major NSF grant to upgrade its infrastructure and physical plant, and will thus be able to reconnect the air-conditioning to the rooms where collections of donated plates are stored.

7) PARI has continued to receive collections of unwanted photographic plates.

8) PARI acquired the two large "Gamma" scanning machines from STScI when they were decommissioned in 2008. To date they have been reassembled and tested mechanically but do not yet have full operational driving software.

A Workshop, to follow up the one held at PARI in 2007, has been funded and organized for April 2012, at the AIP in College Park. The prime goal of this second meeting is to write an Action Plan for astronomy's photographic plates in North America.

3. Future Commission 5 activities

Commission 5 plans to hold the following business sessions during the 2012 General Assembly in Beijing: two sessions for the entire commission and one session each for all WG and TF. Some of WG/TF sessions could be held in parallel.

Special session 15 on "Data Intensive Astronomy" will be held between August 28th and 31st, 2012, in Beijing, China Nanjing.

4. Closing remarks

Commission 5, including its Working Groups and Task Force, has made great progress regarding the data and the documentation issues, as is reported here. Such a successful progress owes to the talent of all the Commission members. As the president of the Commission, I would like to acknowledge all the effort made by the Commission members, especially by the Organizing Committee members.

Masatoshi Ohishi (NAOJ, Japan)
president of the Commission

Reference

"Definition of the Flexible Image Transport System (FITS), Version 3.0". W. D. Pence, L. Chiapetti, C. G. Page, R. A. Shaw, & E. Stobie, *A&A.*, 524, A42 (2010).

Transactions IAU, Volume XXVIIIA
Reports on Astronomy 2009–2012   © International Astronomical Union 2012
Ian Corbett, ed.   doi:10.1017/S1743921312003067

COMMISSION 6   ASTRONOMICAL TELEGRAMS

TELEGRAMMES ASTRONOMIQUES

PRESIDENT   N. N. Samus
VICE-PRESIDENT   H. Yamaoka
PAST PRESIDENT   A. C. Gilmore
ORGANIZING COMMITTEE   K. Aksnes, D. W. E. Green,
                          B. G. Marsden, S. Nakano, E. Roemer,
                          J. Ticha

DIRECTOR OF THE BUREAU: D. W. E. Green, Harvard University, 20
Oxford St., Cambridge, MA 02138, USA
(e-mail: dgreen@eps.harvard.edu)
ASSISTANT DIRECTOR OF THE BUREAU: G. V. Williams, Smithsonian
Astrophysical Observatory, 60 Garden Street, Cambridge, MA 02138, USA
(e-mail: gwilliams@cfa.harvard.edu)

TRIENNIAL REPORT 2009–2012

1. Introduction

As earlier, the main activity of the Commission was performed by the Central Bureau for Astronomical Telegrams (CBAT), effectively directed by Dan Green. These three years were a difficult period for the Bureau and thus for the Commission because the Bureau unexpectedly had to move from the Smithsonian Astrophysical Observatory, its home since 1965, to the Harvard University's Department of Earth and Planetary Sciences. This move caused many serious administrative and logistical problems, effectively solved by the CBAT Director, Dan Green, and CBAT Director Emeritus, Brian Marsden. A great shock, not only for our commission but for the whole astronomical community, was Brian's death on November 18, 2010.

The flow of new astronomical discoveries becomes stronger and stronger, making the activity of the CBAT an increasingly difficult job, and funding remains a serious problem. We are most grateful to all institutions, listed in the CBAT report below, who provided funds permitting to continue this work, really needed by the astronomical community. Special thanks are due to the Department of Earth and Planetary Sciences for providing new home to the CBAT. We also gratefully acknowledge the cooperation between the CBAT and the Minor Planet Center.

In my opinion, the *IAU Circulars*, the CBAT Electronic Telegrams (*CBETs*), facilities provided by the CBAT web site remain one of the most important means of informing the astronomical community on discoveries of objects of special interest. Personally, I would like to thank Dan Green for his great effort and energy, which made my work as the Commission President very easy.

N. N. Samus
President of the Commission

2. Report of the Central Bureau for Astronomical Telegrams

There were 264 *IAU Circulars* and 1334 *Central Bureau Electronic Telegrams* (*CBETs*) issued during the triennium 2005-2008:

Dates	Circulars	CBETs
2008 July-Dec.	Nos. 8957-9007	Nos. 1423-1640
2009 Jan.-June	Nos. 9008-9054	Nos. 1641-1864
2009 July-Dec.	Nos. 9055-9103	Nos. 1865-2107
2010 Jan.-June	Nos. 9104-9156	Nos. 2108-2344
2010 July-Dec.	Nos. 9157-9190	Nos. 2345-2612
2011 Jan.-June	Nos. 9191-9220	Nos. 2613-2756

This was the third Triennium in which the Central Bureau issued electronic-only *CBETs* to aid in the rapid dissemination of reports. The *CBETs* have evolved into a full supplemental and complemental publication to the *Circulars* during this past Triennium, reflecting the move toward electronic publication and away from printed publication. It is the intention that all announcements of objects — except for supernovae — requiring designation by the Central Bureau (novae, comets, solar-system satellites) continue to be be noted on the printed *Circulars*; supernovae are now published almost entirely on *CBETs*, due to their great numbers, so publication of most items regarding supernovae on the *Circulars* has ceased due to the unfortunate lack of sufficient financial support from the supernova community (though a few supernova researchers generously continue to be long-time supporters of the CBAT through their paid subscriptions).

Subscribers may receive the *Circulars* in printed and/or electronic form, the latter being available by e-mail or by logging in to the Computer Service, either directly on the Bureau's computers or via the World Wide Web. Since 1997, the *Circulars* have been made freely available at the CBAT website, but following complaints by paying subscribers, the general delay in posting for non-subscribers is now about one year (expanded in late 2004 from the previous 4-6 weeks). The *CBETs* are also posted at the CBAT website, with the earlier *CBETs* also available freely (and, like the *Circulars*, are indexed via the web-based bibliographic "Astrophysics Data System", the ostensible replacement to the now-defunct *Astronomy and Astrophysics Abstracts*).

Funding is currently being sought to allow the CBAT to post all electronic *Circulars* and *CBETs* freely. However, the CBAT still functions primarily based on paid subscriptions. The U.S. National Science Foundation accepted a proposal to fund the CBAT Director's salary at a 50-percent level during February 2008-January 2010. The Director of the Smithsonian Astrophysical Observatory then decided in 2009 that he no longer wished the CBAT to be located at SAO, its home since 1965, and the CBAT Director obtained an invitation from the Department of Earth and Planetary Sciences (under Prof. Stein Jacobsen) at Harvard University to move the CBAT there; this move was accomplished during February-August 2010, with the CBAT transitioning from computers at SAO to new ones at EPS/Harvard during the course of the year 2010.

The CBAT thanks Assistant CBAT Director G. V. Williams for help in the computer transition, as well as Michael Rudenko (SAO) in this regard. Thanks also go to Prof. Jacobsen and the EPS Department at Harvard for generously provided office space and the infrastructure to keep the CBAT running seemlessly during the transition. The CBAT also gratefully acknowledges a couple of grants from the Tamkin Foundation (Los Angeles, California) to help with purchasing new computer equipment, following its long commitment to funding the CBAT with the Minor Planet Center at SAO. Director Emeritus Brian G. Marsden also was instrumental in helping get the CBAT moved from SAO to EPS/Harvard, until his untimely death on 2010 November 18. Great additional effort is being expended by the CBAT Director to seek alternate — and more extensive and long-lasting — sources of income, including from international sources.

The number of subscriptions to the printed *Circulars* was under 100 at the end of June 2008, continuing the slow decline (the number of such subscriptions peaked around 800 in the 1990s); the cost of printing the *Circulars* has continued to pay for itself from paid subscriptions to the printed *IAUCs*, so this printing will be maintained until the point is reached where a loss occurs. The Computer Service has also slowly declined by about 10 percent in the last triennium (to just under 400 in June 2011).

Supernovae and comets have continued to dominate the activities of the Bureau, as related in the annual reports of the Bureau as published in the *IAU Information Bulletins* and made available at the Commission-6 website (URL `http://cfa-www.harvard.edu/iau/Commission6.html`). The pattern continues regarding increasing numbers of comet discoveries from NEO surveys being first reported as objects of asteroidal appearance, where they are often posted on the MPC's "NEO Confirmation" webpage because of their unusual motion – follow-up observations then showing some of the objects to be of cometary appearance. The working link between the CBAT and MPC continues to be highly useful in the joint announcement of such objects. *CBETs* were issued in the past three years by Marsden, Williams, and Rudenko in their generous assisting of the CBAT Director upon his travels.

A huge change in the way that the CBAT deals with extra-solar-system discoveries of novae, supernovae, and other variable/transient objects was planned out in 2010 and implemented at the beginning of 2011: the Bureau's new "Transient Objects Confirmation Page" (TOCP), which was modelled somewhat after the Minor Planet Center's "Near-Earth-Objects Confirmation Page" (NEOCP). The astronomical community had for years been asking for more automation to the manner in which the CBAT handles discovery reports, and the TOCP has successfully answered the problem surrounding manual intervention by CBAT staff, where staff members cannot be available 24 hours a day to issue reports of new discoveries. The TOCP, then, permits registered users to post discovery information immediately at the TOCP website; the main TOCP page (which can be found at website URL `http://www.cbat.eps.harvard.edu/unconf/tocp.html`) lists all of the spectroscopically unconfirmed objects that have been posted at the TOCP, while a second TOCP webpage has objects that have been removed from the TOCP (and noted as having been published on *CBETs/IAUCs* or having some other explanation given). When posted on the TOCP, each object automatically gets a provisional TOCP positional designation in proper IAU format (of the form TCP J12345678+0123456; prefixes PSN and PNV also used) and is given its own webpage, upon which follow-up observations can be contributed by registered users. The TOCP has become very popular among both professional and amateur astronomers, and many amateurs are contributing useful confirming observations in real time to the TOCP website. The CBAT continues to maintain its "Astronomical Headlines" webpage, which contains discovery information of all objects announced by the CBAT; note that, with the change to EPS/Harvard from SAO, the website URL has changed to `http://www.cbat.eps.harvard.edu/Headlines.html`.

The project of working to scan in older printed *IAUCs* continued during the past triennium, as the first few hundred *Circular* cards (including two series of *Circulars* issued prior to the present series from Copenhagen and from Belgium) were scanned in and posted on the CBAT website. This was done initially using the Harvard College Observatory Library's copy, which has many damaged cards due to the fact that they were mailed in the first several decades with no protecting envelopes. Librarians at the U.S. Naval Observatory and the Copenhagen Observatory Library sent copies of hundreds of cards where cleaner copies were needed for the scanning, and their efforts are gratefully acknowledged. Due to the transition from SAO to EPS/Harvard in 2010, this project was temporarily shut down (save for numerous requests that came from researchers who contacted the Director for information that they needed on specific objects, leading to the scanning and web-posting specific old *Circulars*), but the scanning of old *IAUCs* will be hopefully completed in the coming triennium.

The CBAT Director interacts with members of numerous other Commissions at the triennial IAU General Assemblies and during the course of the triennium by e-mail in efforts to increase the value of the CBAT to all astronomers, and all scientists are encouraged to dialogue with the CBAT Director regarding how the work of the CBAT can be more useful to their own work. For example, the CBAT works closely with Commission 22 to announce reports regarding both old and new meteor-shower activity and with the GCVS team in Moscow to issue formal new permanent variable-star designations to new Galactic novae.

The *IAU Circulars* have continued to serve as the official announcement medium for the annual Edgar Wilson Award for amateur discoveries of comets; the CBAT is attempting to move the Awards administration from SAO to EPS/Harvard, as the CBAT is officially responsible for the announcement of comet discoveries and names.

The CBAT, the IAU, and the astronomical community lost a giant with the death of Brian Marsden in late 2010. His selfless work to get astronomical discoveries out via the Central Bureau from its move to Cambridge in 1965 (and the Minor Planet Center from its move to Cambridge in 1978) until his death cannot be understated. A year later, Brian's absence is keenly felt particularly in the area of comets. He ably served as Director Emeritus to the CBAT from his retirement as Director in 2000, in many respects (including working together to issue new comet discoveries with the MPC, issuing *CBETs* while the current Director was travelling, and refereeing many items published by the CBAT).

Since the *IAU Circulars* are in fact refereed (to a greater extent than many contributors realize), the CBAT benefits from consultation with members of Commission 6 in their various areas of astronomical expertise, as well as referees from the general astronomical community.

D. W. E. Green
Director of the Bureau

Transactions IAU, Volume XXVIIIA
Reports on Astronomy 2009–2012
Ian Corbett, ed.

© International Astronomical Union 2012
doi:10.1017/S1743921312003079

COMMISSION 14

ATOMIC AND
MOLECULAR DATA

DONNEES ATOMIQUE ET MOLECULAIRES

PRESIDENT	Glenn M. Wahlgren
VICE-PRESIDENT	Ewine F. van Dishoeck
PAST PRESIDENT	Steven R. Federman
ORGANIZING COMMITTEE	Peter Beiersdorfer, Milan S. Dimitrijevic, Alain Jorissen, Lyudmila I. Mashonkina, Hampus Nilsson, Farid Salama, Jonathan Tennyson

COMMISSION 14 WORKING GROUPS

Div. XII / Commission 14 WG	Atomic Data
Div. XII / Commission 14 WG	Collision Processes
Div. XII / Commission 14 WG	Molecular Data
Div. XII / Commission 14 WG	Solids and Their Surfaces

TRIENNIAL REPORT 2009–2012

1. Introduction

The main purpose of Commission 14 is to foster interactions between the astronomical community and those conducting research to provide data vital to reducing and analysing astronomical observations and conducting theoretical investigations. One way that the Commission accomplishes this goal is through triennial compilations on recent relevant research in astronomy, atomic, molecular and solid state physics, and related fields of chemical analysis. The most recent compilations appear in the accompanying set of Commission 14 WG Triennial Reports, which were produced by members of the Working Groups and the Organizing Committee of Commission 14.

During the most recent triennial period, members of Commission 14 have also been active in organizing and participating in meetings of various types. In part, it is through these meetings that the astronomical community can communicate their needs to data producers, while data producers provide the results of their studies. Input from the astronomical community is critical to maintaining the vitality of the data producing community, fostering collaboration on proposals and projects that can lead to funding opportunities for data producers. The WG reports provide a record of these meetings and other activities relevant to the Commission.

2. Looking forward to the next triennial period

The importance of continued efforts to improve and expand on the current state of atomic and molecular data comes from recent developments in ground-based and orbital astronomical observatories.

Investigations in atomic spectroscopy remain a critical undertaking for the iron-group elements at all wavelengths. These elements provide a large number of spectral lines, which are of interest in their own right as well as contributing to line blending. Data for characterizing spectral lines at ultraviolet wavelengths of post-iron-group elements still remains undetermined for basic abundance analyses. The near-infrared region has come of age in the past few years with the commissioning of high spectral resolution instruments, such as the ESO VLT CRIRES, and requires accurate wavelengths and transition probabilities for most elements. At longer wavelengths, atomic spectra are virtually unchartered territory except for the strongest features. High precision radial velocity determinations and wavelength calibration and standards at infrared wavelengths both require continued laboratory analysis.

With the flow of data at far-infrared and sub-millimeter wavelengths becoming available from the Herschel Space Observatory, ALMA (Atacama Large Millimeter / Submillimeter Array), and SOFIA (Stratospheric Observatory for Infrared Astronomy), molecular data are needed for line identification and chemical abundance studies. Such data include transitions of simple hydrides and complex molecules, infrared spectra of hot bands from molecules that probe brown dwarfs and exoplanet atmospheres, and ultraviolet / optical data of radiative processes associated with photochemistry. These areas continue to draw much activity and the Commission is expected to highlight results in this area for the astronomical community during the next triennial period.

In the last few years the study of particulate matter has experienced an astonishing increase in the number of studies of the formation of molecules on surfaces of dust grain analogs. The long bibliographic list in this area, provided in the WG report, is a subset of a much longer list of publications on this subject. The interest in chemical/physical processes occurring on the surfaces of dust grains has been spurred by the desire to understand observations coming from infrared and sub-millimeter space telescopes.

Collisional cross-sections and processes are necessary for accurate astrophysical plasma modelling and interpretation, as for example modelling atmospheres of exoplanets and developing strategies in the search of biomarkers. Such collisional processes include electron impact excitation and ionization, charge transfer collisions, and neutral hydrogen atom impact excitation and ionization. However, for the elements beyond Ca (with few exceptions, e.g. Fe II), it still appears preferable to use an effective collision strength of unity for forbidden transitions and the van Regemorter formalism for allowed transitions. Thus, experimental and theoretical studies of collision processes remains important for astrophysical research. The WG report on collisional processes provides ample literature appropriate for analysing stellar spectral features. However, developing methods of incorporating these data into spectrum analysis remains fertile ground for future studies.

Concerning line profiles, such data remain of critical importance for analysis, interpretation and synthesis of astrophysical spectra of high resolution, when taking into account plans for Extremely Large Telescopes and space telescopes.

Finally, a continuing effort of importance to the astronomical community is the availability of data and bibliographic information through data bases, along with the critical assessment of these data. The databases most often do not focus on data of solely astrophysical importance, but rather on data of general importance to a wide variety of fields. On-going efforts to standardize formats for spectral line data bases are likely to produce fruitful results within the next few years. The WG Triennial Reports summarize databases of current interest to astrophysics.

Glenn M. Wahlgren
President of the Commission

Transactions IAU, Volume XXVIIIA
Reports on Astronomy 2009–2012
Ian Corbett, ed.

© International Astronomical Union 2012
doi:10.1017/S1743921312003080

ATOMIC DATA

DIVISION XII / COMMISSION 14 / WORKING GROUP
ATOMIC DATA

CHAIR   Gillian Nave
VICE-CHAIR   Glenn M. Wahlgren, Jeffrey R. Fuhr

TRIENNIAL REPORT 2009–2012

This report summarizes laboratory measurements of atomic wavelengths, energy levels, hyperfine and isotope structure, energy level lifetimes, and oscillator strengths. Theoretical calculations of lifetimes and oscillator strengths are also included. The bibliography is limited to species of astrophysical interest. Compilations of atomic data and internet databases are also included. Papers are listed in the bibliography in alphabetical order, with a reference number in the text.

1. Energy levels, wavelengths, line classifications, and line structure

Major analyses of wavelengths, energy levels and line classifications have been published for **V I** 257, **Fe II** [52], [51], **Cr I** [262], **Te II** [246], and **Ho II** [102] in the past three years. Wavelengths and energy levels have also been measured in the symbiotic nova RR Telescopii for the following species: Al VI, Ar III-V, C II, Ca V-VII, Cl IV, K IV-VI, Mg V-VI, N II-III, Na IV-VI, Ne III-V, O IV, P IV, S IV-V, and Si II-IV. Line identifications based on solar flares and active regions were published for Ar XI,XIV, Ca XIV-XV, Fe XII, Fe XIII, Fe XVII, and Ni XV [240].

Additional publications of wavelengths, energy levels and line classifications include:

Al IV-XI [96], **Ar II** [200], **Ba I** [298], **Ca I** [15, 60] IS [229], **Ca XI** [119], **Ce II** [162], **Cl I** [44], **Co II** HFS [38], **Cr I** IS [262], **Cr III** [243], **D I** [213], **Dy I** HFS, IS [166], **Er I** [125], HFS, IS [123], **Er II** [272], **Eu I** [274], **Fe I** HFS, IS [148], **Fe II** [50, 51, 52, 126], **Fe VI-XIV** [170], **Fe VII-IX** [284], **Fe VII** [286], **Fe VIII** [287, 159], **Fe IX-XVI** [173], **Fe IX** [282, 158], **Fe X** [134], **Fe XI** [291, 288, 135], **Fe XII-XXII** [278], **Fe XIII** [260], **Fe XVII** [289, 94], **Gd I** IS [19, 125], **Ge I-II** [200], **H I** IS [213, 21], **Hg I** IS [231], **Ho II** HFS [102], **K I** IS, HFS [151, 37], **La I** HFS [32], **La II** HFS [86, 87, 88], **Li I** IS [230], **Mg I** IS [111, 244], **Mg II** IS [34], **Mg III** [46], **Mg VII** [173], **Na I-XI** [233], **Na I** [147], **Nb I** HFS [149], **Nb II** HFS [203], **O I** [121], **Os I** HFS [33], **Pb I-II** IS [266, 267], **Pr I-II** HFS [89], **S IX-XIII** [279], **S V** [186], **Si I** IS, HFS [165], **Si VII-IX** [173], **Sm I** [97], **Sn I** [299], **Ta I** HFS [95], **Ta II** HFS [261], **Tc I,II** [190], **Te II** [246], **Ti I** IS, HFS [124], **Ti II** IS, HFS [204], **Ti XIII** [119], **Tm I** HFS [13], **V I** [257], **V XXIII** [93], **Xe II** [285], **Zn II** [14].

The references for elements heavier than Ni (Z>28) are limited to the first three or four spectra only, these data being of most interest for astronomical spectroscopy.

Current analyses of neutral through doubly-ionized spectra are underway for iron-group spectra at the National Institute of Standards and Technology (NIST) and the University of Wisconsin, USA and Imperial College, London, UK. Work on rare-earth elements is being performed at the University of Wisconsin, USA; Laboratoire Aimé

Cotton, Orsay, France; Observatoire de Paris-Meudon, France; and the Institute of Spectroscopy, Troitsk, Russia. Studies of more highly-ionized elements are being done using electron beam ion traps at NIST, USA, Lawrence Livermore National Laboratory, USA, and Heidelberg, Germany, and with an accelerator in Beijing, China.

2. Wavelength standards

Much of the work on wavelength standards during the period of this report has focused on standards required for calibration of astronomical spectrographs and for detecting possible changes in the fine-structure constant during the history of the Universe. Wavelengths emitted by a uranium-neon hollow cathode lamp suitable for astronomical spectrograph calibration have been measured using Fourier transform spectroscopy (FTS) [219]. Updated wavelength standards suitable for detecting changes in the fine-structure constant have been measured using FTS for **Mg I, Mg II, Ti II, Cr II, Mn II, Fe II** and **Zn II** [14, 200]. The most accurate wavelength standards are now made using laser spectroscopy with a laser frequency comb for calibration. Frequency standards with uncertainties of below 1 MHz have been published using this method for **Li I** [230], **Mg II** [34, 112], **Ca I** [229], and **Ca II** [271]. Additional laser spectroscopy measurements include the 546 nm line of ^{198}Hg [231], which was a widely used wavelength standard for earlier studies. These laser spectroscopy measurements have been used to validate the scale of FTS measurements [200], putting all of these measurements on the same wavelength scale. Ritz wavelengths based on re-optimized energy levels of ^{198}Hg are found in [145].

3. Transition probabilities

The transition-probability data in the references in section 7 were obtained by both theoretical and experimental methods. The references for elements heavier than Ni (Z>28) are limited to the first three or four spectra only.

4. Compilations, Reviews, Conferences

Major compilations of wavelengths, energy levels or transition probabilities have been published for the following elements: **Al** [136], **Ar** [228], **B** [84, 143, 144], **Be** [84], **Cs** [234], **H I; D I; T I** [146, 269], **He** [269], **K** [232], **Li** [269], **Na** [233], **S** [214], **Si** [137], and **Sr I** [236]. Additional data can be found in *NIST Atomic Transition Probabilities,* section of the Handbook of Chemistry and Physics [85].

Papers on atomic spectroscopic data are included in the proceedings of the 10th International Conference on Atomic Spectra and Oscillator Strengths [2], the 7th International Conference on Atomic and Molecular data and their Applications [1], and the 2010 NASA Laboratory Astrophysics workshop [3]. Additional conferences including papers on atomic data include Atomic Processes in Plasmas, the Congress of the European Group on Atomic Systems; and the meeting of the Division of Atomic, Molecular and Optical Physics of the American Physical Society.

5. Databases

The following databases of atomic spectra at NIST have received significant updates since the last triennial report:

NIST Atomic Spectra Database:
http://www.nist.gov:/pml/data/asd.cfm contains critically compiled data on wavelengths, energy levels and oscillators strengths.

Ground Levels and Ionization Energies for the Neutral Atoms:
http://www.nist.gov/pml/data/ion_energy.cfm

NIST Atomic Spectra Bibliographic Databases:
http://www.nist.gov/pml/data/asbib/index.cfm
Consists of three databases of publications on atomic transition probabilities, atomic energy levels and spectra, and atomic spectral line broadening.

Additional on-line databases including significant quantities of atomic data include:

The MCHF/MCDHF Collection on the Web (C.Froese Fischer *et al.*) at http://nlte.nist.gov/MCHF/index.html contains results of multi-configuration Hartree-Fock (MCHF) or multi-configuration Dirac-Hartree-Fock (MCDHF) calculations for hydrogen and Li-like through Ar-like ions, mainly for $Z \leqslant 30$. Data for fine-structure transitions are included.

The TOPbase and Opacity Projects include transition probability and oscillator strength data for astrophysically abundant ions ($Z \leqslant 26$). A database is available at http://cdsweb.u-strasbg.fr/topbase/topbase.html

CHIANTI, an atomic database for spectroscopic diagnostics of astrophysical plasmas at http://www.chianti.rl.ac.uk/ contains atomic data and programs for computing spectra from astrophysical plasmas, with the emphasis on highly-ionized atoms.

The Vienna Atomic Line Database (VALD) web site (http://ams.astro.univie.ac.at/ vald/) is a collection of atomic line parameters of astronomical interest, with tools for selecting subsets of astrophysical interest.

The bibl database is a comprehensive bibliographic database of experimental and theoretical papers on atomic spectroscopy, with an emphasis on papers published since 1983. It is available at http://das101.isan.troitsk.ru/bibl.htm).

6. Notes for References

The references are identified by a running number. This refers to the general reference list at the end of this report, where the literature is ordered alphabetically according to the first author. Each reference contains one or more code letters indicating the method applied by the authors, defined as follows:

THEORETICAL METHODS:

Q: quantum mechanical calculations. **QF:** Calculations of forbidden lines.

EXPERIMENTAL METHODS:

CL: New classifications **EL:** Energy levels. **WL:** Wavelengths.
HFS: Hyperfine structure. **IS:** Isotope structure. **L:** Lifetimes.
TE: Experimental transition probabilities.

OTHER:

CP: Data compilations. **R:** Relative values only. **F:** Forbidden lines.

Gillian Nave
chair of Working Group

7. References on lifetimes and transition probabilities

Al II: 56
Al III: 172
Al IV: 168
Al IX: 104
Al V-XII: 212
Al XI: 117

Ar V: 255
Ar VIII: 172
Ar IX: 168
Ar XVI: 117

Au III: 281

B I: 292
B IV: 6

Ba I: 298
Ba II: 120, 224

Be III: 6

C I: 29
C II: 127, 247
C V: 6

Ca I: 15
Ca II: 92, 224, 226
Ca VIII: 132
Ca X: 172
Ca XI: 168
Ca XVII: 154
Ca XVIII: 117

Cd II: 224

Ce I: 59, 108
Ce II: 110, 162

Cl I: 44, 206
Cl VII: 172
Cl VIII: 168
Cl XV: 117

Co XII: : 259
Co XIII: 259
Co XIV: 259
Co XV: 98, 259
Co XVII: 172
Co XVIII: 168
Co XXIII: 104

Cr II: 35, 101
Cr VIII: 5
Cr XIV: 172
Cr XV: 168
Cr XXII: 264

Cu II: 48, 76

Er I: 109, 164
Er II: 163, 272

Eu I: 294
Eu III: 273

F VII: 12
F VIII: 10
F IX: 11

Fe II: 51, 52, 66, 100, 191, 218
Fe III: 63, 64, 65
Fe IV: 61, 67, 80, 113
Fe VI: 31
Fe VII: 270, 286
Fe VIII: 159, 287
Fe IX: 158
Fe X: 43
Fe XI: 259, 288, 291
Fe XII: 250, 252, 258, 259
Fe XIII: 245, 259
Fe XIV: 169, 248, 250, 252, 259
Fe XV: 152, 194
Fe XVI: 172, 198
Fe XVII: 168, 177
Fe XVII-XXV: 114
Fe XIX: 49, 142, 197
Fe XX: 142
Fe XXII: 104, 193, 195
Fe XXIII: 277
Fe XXIV: 178, 263
Fe XXVI: 4, 55

Ga I: 280

Gd I: L 107, 160

Ge IV: 74

He I: 16, 192

Hf I,III: 184

Hg II: 224

In I: 225

K I: 20, 54, 223
K II: 254
K IX: 172
K X: 168
K XVII: 117

Kr II: 131

La I: 77, 128, 129
La II: 150

Li I-II: 22
Li II: 6

Mg I: 56, 122, 256
Mg II: 69, 172
Mg III: 168
Mg IX: 118, 290
Mg X: 117

Mn I: 42
Mn XV: 172
Mn XVI: 168
Mn XXI: 104

Mo II: 180

N I-VII: 90
N I: 27, 29, 45
N II: 29, 239, 251
N III: 127
N V: 12
N VI: 10
N VII: 11

References

[1] 7 th International Conference on Atomic and Molecular Data and their Applications: 2010, AIP Conference Proceedings Volume 1344

[2] 10th International Colloquium on Atomic Spectra and Oscillator Strengths for Astrophysical and Laboratory Plasmas: 2011 Canadian J. Physics, 89

[3] The 2010 NASA Laboratory Astrophysics Workshop: 2011, ed. D.R. Schultz, Oak Ridge National Laboratory http://www-cfadc.phy.ornl.gov/nasa_law/proceedings.html

[4] Aggarwal, K. M., Hamada, K., Igarashi, A., Jonauskas, V., Keenan, F. P., & Nakazaki, S.: 2008, Astron. Astrophys. 484, 879, **Q, QF**

[5] Aggarwal, K. M., Kato, T., Keenan, F. P., & Murakami, I.: 2009a, Astron. Astrophys. 506, 1501, **Q, QF**

[6] Aggarwal, K. M., Kato, T., Keenan, F. P., & Murakami, I.: 2011, Phys. Scr. 83, 015302, **Q, QF**

[7] Aggarwal, K. M. & Keenan, F. P.: 2008a, Astron. Astrophys. 489, 1377, **Q, QF**

[8] Aggarwal, K. M. & Keenan, F. P.: 2008b, Astron. Astrophys. 486 , 1053, **Q, QF**

[9] Aggarwal, K. M. & Keenan, F. P.: 2010, Phys. Scr. 82, 065302, **Q, QF**

[10] Aggarwal, K. M., Keenan, F. P., & Heeter, R. F.: 2009b, Phys. Scr. 80, 045301, **Q, QF**

[11] Aggarwal, K. M., Keenan, F. P., & Heeter, R. F.: 2010a, Phys. Scr. 82, 015006, **Q, QF**

[12] Aggarwal, K. M., Keenan, F. P., & Heeter, R. F.: 2010b, Phys. Scr. 81, 015303, **Q, QF**

[13] Akimov, A. V., Chebakov, K. Y., Tolstikhina, I. Y., Sokolov, A. V., Rodionov, P. B., Kanorsky, S. I., Sorokin, V. N., & Kolachevsky, N. N.: 2008, Quantum Electron. 38, 961, **HFS**

[14] Aldenius, M.: 2009, Phys. Scr. T134, 014008, **WL, CL**

[15] Aldenius, M., Lundberg, H., & Blackwell-Whitehead, R.: 2009a, Astron. Astrophys. 502, 989, **WL, CL, TE, EL**

[16] Alexander, S. A. & Coldwell, R. L.: 2008, Int. J. Quantum Chem. 108 , 2813, **Q, QF**

[17] Alonso-Medina, A.: 2010, Spectrochim. Acta, Part B 65, 158, **TE**

[18] Alonso-Medina, A., Colón, C., & Zanón, A.: 2009, Mon. Not. R. Astron. Soc. 395, 567, **Q**

[19] Ankush, B. K. & Deo, M. N.: 2010, Phys. Scr. 81, 055301, **IS, EL**

[20] J. M. P., Serrao, J. M. P. S.: 2008, J. Quant. Spectrosc. Radiat. Transfer 109, 453, **Q**

[21] Arnoult, O., Nez, F., Julien, L., & Biraben, F.: 2010, Eur. Phys. J. D 60, 243, **WL, EL**

[22] Ateş, Ş. & Çelik, G.: 2009, Acta Phys. Pol. A 116(2), 169, **Q**

[23] Ateş, Ş., Tekeli, G., Çelik, G., Akin, E., & Taşer, M.: 2009, Eur. Phys. J. D 54, 21, **Q**

[24] Bacławski, A.: 2008a, J. Quant. Spectrosc. Radiat. Transfer 109, 1986, **TE, R**

[25] Bacławski, A.: 2008b, J. Phys. B 41, 225701, **TE**

[26] Bacławski, A.: 2011, Eur. Phys. J. D 61, 327, **TE, R**

[27] Bacławski, A. & Musielok, J.: 2008a, J. Quant. Spectrosc. Radiat. Transfer 109, 2537, **TE, R**

[28] Bacławski, A. & Musielok, J.: 2008b, Spectrochim. Acta, Part B 63, 1315, **TE**

[29] Bacławski, A. & Musielok, J.: 2009, Acta Phys. Pol. A 116(2), 176, **CP**

[30] Bacławski, A. & Musielok, J.: 2011, J. Phys. B 44, 135002, **TE, R**

[31] Ballance, C. P. & Griffin, D. C.: 2008, J. Phys. B 41, 195205, **Q**

[32] Başar, G., Başar, G., & Kröger, S.: 2009, Opt. Commun. 282, 562, **HFS, EL**

[33] Başar, G., Başar, G., Kröger, S., & Guthöhrlein, G. H.: 2010, J. Phys. B 43, 074008, **HFS**

[34] Batteiger, V., Knünz, S., Herrmann, M., Saathoff, G., Schüssler, H. A., Bernhardt, B., Wilken, T., Holzwarth, R., Hänsch, T. W., & Udem, T.: 2009, Phys. Rev. A 80, 022503, **IS, WL**

[35] Bautista, M. A., Ballance, C., Gull, T. R., Hartman, H., Lodders, K., Martínez, M., & Meléndez, M.: 2009a, Mon. Not. R. Astron. Soc. 393, 1503, **QF**

[36] Bautista, M. A., Quinet, P., Palmeri, P., Badnell, N. R., Dunn, J., & Arav, N.: 2009b, Astron. Astrophys. 508, 1527, **Q**

[37] Behrle, A., Koschorreck, M., & Köhl, M.: 2011, Phys. Rev. A 83, 052507, **IS, HFS**

[38] Bergemann, M., Pickering, J. C., & Gehren, T.: 2010, Mon. Not. R. Astron. Soc. 401, 1334, **HFS**

[39] Bhatia, A. K. & Landi, E.: 2011a, At. Data Nucl. Data Tables 97, 50, **Q, QF**

[40] Bhatia, A. K. & Landi, E.: 2011b, At. Data Nucl. Data Tables 97, 189, **Q, QF**

[41] Biémont, E., Blagoev, K., Engström, L., Hartman, H., Lundberg, H., Malcheva, G., Nilsson, H., Whitehead, R. B., Palmeri, P., & Quinet, P.: 2011, Mon. Not. R. Astron. Soc. 414, 3350, **L, Q**

[42] Blackwell-Whitehead, R., Pavlenko, Y. V., Nave, G., Pickering, J. C., Jones, H. R. A., Lyubchik, Y., & Nilsson, H.: 2011, Astron. Astrophys. 525, p. A44, **TE**

[43] Brenner, G., López-Urrutia, J. R. C., Bernitt, S., Fischer, D., Ginzel, R., Kubiček, K., Mäckel, V., Mokler, P. H., Simon, M. C., & Ullrich, J.: 2009, Astrophys. J. 703, 68, **LF**

[44] Bridges, J. M. & Wiese, W. L.: 2008a, Phys. Rev. A 78, 062508, **TE**

[45] Bridges, J. M. & Wiese, W. L.: 2010, Phys. Rev. A 82, 024502, **TE**

[46] Brown, C. M., Kramida, A. E., Feldman, U., & Reader, J.: 2009a, Phys. Scr. 80, 065302, **EL, CL, WL**

[47] Brown, J. B., Brown, M. S., Cheng, S., Curtis, L. J., Ellis, D. G., Federman, S. R., & Irving, R. E.: 2011, Can. J. Phys. 89, 413, **Q, R**

[48] Brown, M. S., Federman, S. R., Irving, R. E., Cheng, S., & Curtis, L. J.: 2009b, Astrophys. J. 702, 880, **L**

[49] Butler, K. & Badnell, N. R.: 2008, Astron. Astrophys. 489, 1369, **Q**

[50] Castelli, F., Johansson, S., & Hubrig, S.: 2008, J. Phys.: Conf. Ser. 130, 012003, **CL, WL**

[51] Castelli, F. & Kurucz, R. L.: 2010a, Astron. Astrophys. 520, p. A57, **EL, CL, Q**

[52] Castelli, F., Kurucz, R. L., & Hubrig, S.: 2009a, Astron. Astrophys. 508, 401, **CL, WL, EL, Q**

[53] Çelik, G. & Ateş, Ş.: 2008a, Acta Phys. Pol. A 113(6), 1619, **Q**

[54] Çelik, G. & Ateş, Ş.: 2008b, Can. J. Phys. 86, 487, **Q**

[55] Chen, C.-Y., Wang, K., Huang, M., Wang, Y.-S., & Zou, Y.-M.: 2010, J. Quant. Spectrosc. Radiat. Transfer 111, 843, **Q, QF**

[56] Cheng, C., Gao, X., Qing, B., Zhang, X.-L., & Li, J.-M.: 2011, Chin. Phys. B 20, 033103, **Q**

[57] Chwalla, M., Benhelm, J., Kim, K., Kirchmair, G., Monz, T., Riebe, M., Schindler, P., Villar, A. S., Hänsel, W., Roos, C. F., Blatt, R., Abgrall, M., Santarelli, G., Rovera, G. D., & Laurent, P.: 2009, Phys. Rev. Lett. 102, 023002, **WL**

[58] Colón, C. & Alonso-Medina, A.: 2010, J. Phys. B 43, 165001, **Q**

[59] Curry, J. J.: 2009, J. Phys. D 42, 135205, **TE**

[60] Dammalapati, U., Norris, I., Burrows, C., & Riis, E.: 2011, Phys. Rev. A 83, 062513, **CL, WL**

[61] Deb, N. C. & Hibbert, A.: 2008a, J. Phys. B 41, 081007, **Q**

[62] Deb, N. C. & Hibbert, A.: 2008b, At. Data Nucl. Data Tables 94, 561, **Q**

[63] Deb, N. C. & Hibbert, A.: 2008c, J. Phys.: Conf. Ser. 130, 012006, **Q, QF**

[64] Deb, N. C. & Hibbert, A.: 2009a, J. Phys. B 42, 065003, **QF**

[65] Deb, N. C. & Hibbert, A.: 2009b, At. Data Nucl. Data Tables 95, 184, **Q**

[66] Deb, N. C. & Hibbert, A.: 2010a, Astrophys. J. 711, L104, **QF**

[67] Deb, N. C. & Hibbert, A.: 2010b, At. Data Nucl. Data Tables 96, 358, **Q**

[68] Deb, N. C., Hibbert, A., Felfli, Z., & Msezane, A. Z.: 2009, J. Phys. B 42, 015701, **QF**

[69] Destree, J. D., Williamson, K. E., & Snow, T. P.: 2010a, Astrophys. J. 712, L48, **CL, WL, TE**

[70] Dixit, G., Nataraj, H. S., Sahoo, B. K., Chaudhuri, R. K., & Majumder, S.: 2008, J. Phys. B 41, 025001, **Q, QF**

[71] Dixit, G., Sahoo, B. K., Chaudhuri, R. K., & Majumder, S.: 2009, J. Phys. B 42, 165702, **Q, QF**

[72] Djeniže, S., Srećković, A., & Bukvić, S.: 2010, Spectrochim. Acta, Part B 65, 61, **TE, R**

[73] Duan, B., Bari, M. A., Zhong, J. Y., Yan, J., Li, Y. M., & Zhang, J.: 2008, Astron. Astrophys. 488, 1155, **Q, QF**

[74] Dutta, N. N. & Majumder, S.: 2011, Astrophys. J. 737, 25, **Q, QF**

[75] Fan, Q., Liao, Z. J., Yang, J. H., & Zhang, J. P.: 2009, Phys. Scr. 79, 015301, **Q**

[76] Federman, S. R., Curtis, L. J., Brown, M., Cheng, S., Irving, R. E., Torok, S., & Schectman, R. M.: 2008, J. Phys.: Conf. Ser. 130, 012007, **TE, R, L**

[77] Feng, Y.-Y., Zhang, W., Kuang, B., Ning, L.-L., Jiang, Z.-K., & Dai, Z.-W.: 2011, J. Opt. Soc. Am. B 28, 543, **L**

[78] Feng, Y.-Y., Zhang, W., Ning, L.-L., Kuang, B., Sun, G.-J., & Dai, Z.-W.: 2010, J. Phys. B 43, 225001, **L**

[79] Fischer, C. F. & Ralchenko, Y.: 2008, Int. J. Mass Spectrom. 271, 85, **Q**

[80] Fischer, C. F., Rubin, R. H., & Rodríguez, M.: 2008, Mon. Not. R. Astron. Soc. 391 , 1828, **QF**

[81] Fischer, C. F., Tachiev, G., Rubin, R. H., & Rodríguez, M.: 2009, Astrophys. J. 703, 500, **Q, QF**

[82] Fivet, V., Biémont, E., Engström, L., Lundberg, H., Nilsson, H., Palmeri, P., & Quinet, P.: 2008, J. Phys. B 41, 015702, **L, Q**

[83] Fivet, V., Quinet, P., Palmeri, P., Biémont, E., Asplund, M., Grevesse, N., Sauval, A. J., Engström, L., Lundberg, H., Hartman, H., & Nilsson, H.: 2009, Mon. Not. R. Astron. Soc. 396, 2124, **L, Q**

[84] Fuhr, J. R. & Wiese, W. L.: 2010, J. Phys. Chem. Ref. Data 39, 013101, **CP**

[85] Fuhr, J. R. & Wiese, W. L.: 2011, CRC Handbook of Chemistry and Physics, 92nd Edition, (Ed. D. R. Lide) CRC Press, Boca Raton, FL,

[86] Furmann, B., Elantkowska, M., Stefańska, D., Ruczkowski, J., & Dembczyński, J.: 2008a, J. Phys. B 41, 235002, **HFS**

[87] Furmann, B., Ruczkowski, J., Stefańska, D., Elantkowska, M., & Dembczyński, J.: 2008b, J. Phys. B 41, 215004, **HFS**

[88] Furmann, B., Stefańska, D., & Dembczyński, J.: 2010, J. Phys. B 43, 015001, **EL,CL, WL, HFS**

[89] Gamper, B., Uddin, Z., Jahangir, M., Allard, O., Knöckel, H., Tiemann, E., & Windholz, L.: 2011, J. Phys. B 44, 045003, **CL, WL, EL,HFS**

[90] García, J., Kallman, T. R., Witthoeft, M., Behar, E., Mendoza, C., Palmeri, P., Quinet, P., Bautista, M. A., & Klapisch, M.: 2009, Astrophys. J., Suppl. Ser. 185, 477, **Q**

[91] Gattinger, R. L., Lloyd, N. D., Bourassa, A. E., Degenstein, D. A., McDade, I. C., & Llewellyn, E. J.: 2009, Can. J. Phys. 87, 1133, **TE-R-F**

[92] Gerritsma, R., Kirchmair, G., Zähringer, F., Benhelm, J., Blatt, R., & Roos, C. F.: 2008, Eur. Phys. J. D 50, 13, **TE**

[93] Gillaspy, J. D., Chantler, C. T., Paterson, D., Hudson, L. T., Serpa, F. G., & Takács, E.: 2010, J. Phys. B 43, 074021, **CL, EL,WL**

[94] Gillaspy, J. D., Lin, T., Tedesco, L., Tan, J. N., Pomeroy, J. M., Laming, J. M., Brickhouse, N., Chen, G.-X., & Silver, E.: 2011, Astrophys. J. 728, 132, **WL**

[95] Głowacki, P., Uddin, Z., Guthöhrlein, G. H., Windholz, L., & Dembczyński, J.: 2009, Phys. Scr. 80, 025301, **CL, WL, EL, HFS**

[96] Gu, M. F., Beiersdorfer, P., & Lepson, J. K.: 2011, Astrophys. J. 732, 91, **CL, WL**

[97] Guan, F., Dai, C.-J., & Zhao, H.-Y.: 2008, Chin. Phys. B 17, 3655, **EL, CL, WL**

[98] Gupta, G. P. & Msezane, A. Z.: 2008, Eur. Phys. J. D 49, 157, **Q**

[99] Gupta, G. P. & Msezane, A. Z.: 2010, Phys. Scr. 81, 045302, **Q**

[100] Gurell, J., Hartman, H., Blackwell-Whitehead, R., Nilsson, H., Bäckström, E., Norlin, L. O., Royen, P., & Mannervik, S.: 2009a, Astron. Astrophys. 508, 525, **L, F**

[101] Gurell, J., Nilsson, H., Engström, L., Lundberg, H., Blackwell-Whitehead, R., Nielsen, K. E., & Mannervik, S.: 2010, Astron. Astrophys. 511, p. A68, **L, TE**

[102] Gurell, J., Wahlgren, G. M., Nave, G., & Wyart, J.-F.: 2009b, Phys. Scr. 79, 035306, **CL, WL, HFS, EL**

[103] Hamdi, R., Nessib, N. B., Milovanović, N., Popović, L. Č., Dimitrijević, M. S., & Sahal-Bréchot, S.: 2008, Mon. Not. R. Astron. Soc. 387, 871, **Q**

[104] Hao, L.-H. & Jiang, G.: 2011, Phys. Rev. A 83, 012511, **Q**

[105] Hartman, H., Gurell, J., Lundin, P., Schef, P., Hibbert, A., Lundberg, H., Mannervik, S., Norlin, L.-O., & Royen, P.: 2008, Astron. Astrophys. 480, 575, **LF, QF**

[106] Hartman, H., Nilsson, H., Engström, L., Lundberg, H., Palmeri, P., Quinet, P., & Biémont, E.: 2010, Phys. Rev. A 82, 052512, **L, Q**

[107] Hartog, E. A. D., Bilty, K. A., & Lawler, J. E.: 2011, J. Phys. B 44, 055001, **L**

[108] Hartog, E. A. D., Buettner, K. P., & Lawler, J. E.: 2009, J. Phys. B 42, 085006, **L**

[109] Hartog, E. A. D., Chisholm, J. P., & Lawler, J. E.: 2010, J. Phys. B 43, 155004, **L**

[110] Hartog, E. A. D. & Lawler, J. E.: 2008, J. Phys. B 41, 045701, **L**

[111] He, M., Therkildsen, K. T., Jensen, B. B., Brusch, A., Thomsen, J. W., & Porsev, S. G.: 2009, Phys. Rev. A 80, 024501, **IS, HFS**

[112] Herrmann, Batteiger, M. V., Knünz S., Saathoff G., Udem Th., and Hänsch T. W., 2009, Phys. Rev. Lett. 102, 013006 **WL**

[113] Hibbert, A. & Deb, N. C.: 2008, J. Phys.: Conf. Ser. 130, 012012, **Q**

[114] Hou, H.-J., Jiang, G., Hu, F., & Hao, L.-H.: 2009, At. Data Nucl. Data Tables 95, 125, **Q**

[115] Hu, F., Jiang, G., Hong, W., & Hao, L. H.: 2008, Eur. Phys. J. D 49, 293, **Q**

[116] Hu, F., Jiang, G., Yang, J. M., Zhang, J. Y., & Zhao, X. F.: 2011, Acta Phys. Pol. A 120(3), 429, **Q**

[117] Hu, M.-H. & Wang, Z.-W.: 2009, Chin. Phys. B 18, 2244, **Q**

[118] Hudson, C. E.: 2009, Astron. Astrophys. 493, 697, **Q**

[119] Ishikawa, Y., Encarnación, J. M. L., & Träbert, E.: 2009, Phys. Scr. 79, 025301, **Q**

[120] Iskrenova-Tchoukova, E. & Safronova, M. S.: 2008, Phys. Rev. A 78, 012508, **Q, QF**

[121] Ivanov, T. I., Salumbides, E. J., Vieitez, M. O., Cacciani, P. C., de Lange, C. A., & Ubachs, W.: 2008, Mon. Not. R. Astron. Soc. 389, L4, **WL, CL**

[122] Jensen, B. B., Ming, H., Westergaard, P. G., Gunnarsson, K., Madsen, M. H., Brusch, A., Hald, J., & Thomsen, J. W.: 2011, Phys. Rev. Lett. 107, 113001, **LF**

[123] Jin, W.-G., Nakai, H., Kawamura, M., & Minowa, T.: 2009a, J. Phys. Soc. Jpn. 78, 015001, **HFS, IS**

[124] Jin, W.-G., Nemoto, Y., & Minowa, T.: 2009b, J. Phys. Soc. Jpn. 78, 094301, **IS, HFS**

[125] Jin, W.-G., Nemoto, Y., Nakai, H., Kawamura, M., & Minowa, T.: 2008, J. Phys. Soc. Jpn. 77, 124301, **IS**

[126] Johansson, S.: 2009, Phys. Scr. T134, 014013, **EL, CL**

[127] Jönsson, P., Li, J.-G., Gaigalas, G., & Dong, C.-Z.: 2010, At. Data Nucl. Data Tables 96, 271, **Q**

[128] Karaçoban, B. & Özdemir, L.: 2008a, J. Quant. Spectrosc. Radiat. Transfer 109, 1968, **Q**

[129] Karaçoban, B. & Özdemir, L.: 2008b, Acta Phys. Pol. A 113(6), 1609, **Q**

[130] Karaçoban, B. & Özdemir, L.: 2011, J. Kor. Phys. Soc. 58, 417, **Q**

[131] Karmakar, S. & Das, M. B.: 2010, Eur. Phys. J. D 59, 361, **L**

[132] Karpuškienė, R. & Bogdanovich, P.: 2009, At. Data Nucl. Data Tables 95, 533, **Q**

[133] Kedzierski, D., Kusz, J., & Muzolf, J.: 2010, Spectrochim. Acta, Part B 65, 248, **TE**

[134] Keenan, F. P., Jess, D. B., Aggarwal, K. M., Thomas, R. J., Brosius, J. W., & Davila, J. M.: 2008a, Mon. Not. R. Astron. Soc. 389, 939, **CL, Q, QF**

[135] Keenan, F. P., Milligan, R. O., Jess, D. B., Aggarwal, K. M., Mathioudakis, M., Thomas, R. J., Brosius, J. W., & Davila, J. M.: 2010, Mon. Not. R. Astron. Soc. 404, 1617, **CL, WL**

[136] Kelleher, D. E. & Podobedova, L. I.: 2008a, J. Phys. Chem. Ref. Data 37, 709, **CP**

[137] Kelleher, D. E. & Podobedova, L. I.: 2008b, J. Phys. Chem. Ref. Data 37, 1285, **CP**

[138] Kingston, A. E. & Hibbert, A.: 2008, J. Phys. B 41, 155001, **Q**

[139] Kingston, A. E. & Hibbert, A.: 2009, J. Phys. B 42, 185004, **Q**

[140] Kingston, A. E. & Hibbert, A.: 2010, J. Phys. B 43, 165003, **Q**

[141] Kisielius, R., Storey, P. J., Ferland, G. J., & Keenan, F. P.: 2009, Mon. Not. R. Astron. Soc. 397, 903, **Q**

[142] Kotochigova, S., Linnik, M., Kirby, K. P., & Brickhouse, N. S.: 2010, Astrophys. J., Suppl. Ser. 186, 85, **Q**

[143] Kramida, A., Ryabtsev, A. N., & Ekberg, J. O., Kink, I., Mannervik, S., Martinson, I.: 2008, Phys. Scr. 78, 025301 **CP**

[144] Kramida, A., Ryabtsev, A. N., Ekberg, J. O., Kink, I., Mannervik, S., Martinson, I.: 2008 Phys. Scr. 78, 025302 **CP**

[145] Kramida, A.: 2011a, J. Res. Natl. Inst. Stand. Technol. 116(2), 599, **EL, CL, WL**

[146] Kramida, A. E.: 2010a, At. Data Nucl. Data Tables 96, 586, **E, WL, CL, HFS, QF**

[147] Kramida, A. E.: 2010b, J. Phys. B 43, 205001, **EL,WL, CL**

[148] Krins, S., Oppel, S., Huet, N., von Zanthier, J., & Bastin, T.: 2009, Phys. Rev. A 80, 062508, **IS, HFS**

[149] Kröger, S., Er, A., Öztürk, I. K., Başar, G., Jarmola, A., Ferber, R., Tamanis, M., & Začs, L.: 2010, Astron. Astrophys. 516, p. A70, **HFS**

[150] Kułaga-Egger, D. & Migdałek, J.: 2009, J. Phys. B 42, 185002, **Q**

[151] Kumar, P. V. K. & Suryanarayana, M. V.: 2011, J. Phys. B 44, 055003, **IS, HFS, WL**

[152] Landi, E.: 2011, At. Data Nucl. Data Tables 97, 587, **Q**

[153] Landi, E. & Bhatia, A. K.: 2008, At. Data Nucl. Data Tables 94, 1, **Q, QF**

[154] Landi, E. & Bhatia, A. K.: 2009a, At. Data Nucl. Data Tables 95, 155, **Q, QF**

[155] Landi, E. & Bhatia, A. K.: 2009b, At. Data Nucl. Data Tables 95, 547, **Q, QF**

[156] Landi, E. & Bhatia, A. K.: 2010, At. Data Nucl. Data Tables 96, 52, **Q, QF**

[157] Landi, E. & Young, P. R.: 2009a, Astrophys. J. 706, 1, **DB**

[158] Landi, E. & Young, P. R.: 2009b, Astrophys. J. 707, 1191, **CL, WL, EL,Q**

[159] Landi, E. & Young, P. R.: 2010a, Astrophys. J. 713, 205, **CL, WL, EL,Q**

[160] Lawler, J. E., Bilty, K. A., & Hartog, E. A. D.: 2011, J. Phys. B 44, 095001, **TE**

[161] Lawler, J. E., Chisholm, J., Nitz, D. E., Wood, M. P., Sobeck, J., & Hartog, E. A. D.: 2010a, J. Phys. B 43, 085701, **TE**

[162] Lawler, J. E., Sneden, C., Cowan, J. J., Ivans, I. I., & Hartog, E. A. D.: 2009a, Astrophys. J., Suppl. Ser. 182, 51, **L, TE**

[163] Lawler, J. E., Sneden, C., Cowan, J. J., Wyart, J.-F., Ivans, I. I., Sobeck, J. S., Stockett, M. H., & Hartog, E. A. D.: 2008b, Astrophys. J., Suppl. Ser. 178, 71, **TE**

[164] Lawler, J. E., Wyart, J.-F., & Hartog, E. A. D.: 2010b, J. Phys. B 43, 235001, **TE**

[165] Lee, S. A. and Jr., W. M. F.: 2010, Phys. Rev. A 82, 042515, **IS, HFS**

[166] Leefer, N., Cingöz, A., & Budker, D.: 2009, Opt. Lett. 34, 2548, **IS, HFS, EL**

[167] Li, H.-L., Li, P., Cheng, Z., & Ma, H.-R.: 2008, Commun. Theor. Phys. 49, 217, **Q**

[168] Liang, G. Y. & Badnell, N. R.: 2010, Astron. Astrophys. 518, p. A64, **Q**

[169] Liang, G. Y., Badnell, N. R., López-Urrutia, J. R. C., Baumann, T. M., Zanna, G. D., Storey, P. J., Tawara, H., & Ullrich, J.: 2010, Astrophys. J., Suppl. Ser. 190, 322, **Q**

[170] Liang, G. Y., Baumann, T. M., López-Urrutia, J. R. C., Epp, S. W., Tawara, H., Gonchar, A., Mokler, P. H., Zhao, G., & Ullrich, J.: 2009a, Astrophys. J. 696, 2275, **CL**

[171] Liang, G. Y., Whiteford, A. D., & Badnell, N. R.: 2009b, Astron. Astrophys. 499, 943, **Q**

[172] Liang, G. Y., Whiteford, A. D., & Badnell, N. R.: 2009c, Astron. Astrophys. 500, 1263, **Q**

[173] Liang, G. Y. & Zhao, G.: 2010, Mon. Not. R. Astron. Soc. 405 , 1987, **WL, CL**

[174] Liang, L., Jiang, W. X., Zhou, C., & Zhang, L.: 2008a, Opt. Commun. 281, 2107, **Q**

[175] Liang, L. & Zhou, C.: 2008, J. Quant. Spectrosc. Radiat. Transfer 109 , 1995, **Q**

[176] Liang, L., Zhou, C., & Zhang, L.: 2008c, Chin. Opt. Lett. 6, 804, **Q**

[177] López-Urrutia, J. R. C. & Beiersdorfer, P.: 2010, Astrophys. J. 721, 576, **EF, LF**

[178] Louzon, E., Feigel, A., Frank, Y., Raicher, E., Klapisch, M., Mandelbaum, P., Levy, I., Hurvitz, G., Ehrlich, Y., Frankel, M., Maman, S., & Henis, Z.: 2011, High En. Dens. Phys. 7, 124, **Q**

[179] Luna, F. R. T., Mania, A. J., & Hernandes, J. A.: 2009, J. Appl. Spectrosc. 76, 447, **Q**

[180] Lundberg, H., Engström, L., Hartman, H., Nilsson, H., Palmeri, P., Quinet, P., & Biémont, E.: 2010, J. Phys. B 43, 085004, **L, TE, Q**

[181] Lundin, P., Gurell, J., Mannervik, S., Royen, P., Norlin, L.-O., Hartman, H., & Hibbert, A.: 2008, Phys. Scr. 78, 015301, **LF, QF**

[182] Malcheva, G., Mayo, R., Ortiz, M., Ruiz, P., Engström, L., Lundberg, H., Nilsson, H., Quinet, P., Biémont, E., & Blagoev, K.: 2009a, Mon. Not. R. Astron. Soc. 395, 1523, **L, Q**

[183] Malcheva, G., Nilsson, H., Engström, L., Lundberg, H., Biémont, E., Palmeri, P., Quinet, P., & Blagoev, K.: 2011, Mon. Not. R. Astron. Soc. 412 , 1823, **L, Q**

[184] Malcheva, G., Yoca, S. E., Mayo, R., Ortiz, M., Engström, L., Lundberg, H., Nilsson, H., Biémont, E., & Blagoev, K.: 2009b, Mon. Not. R. Astron. Soc. 396, 2289, **L, Q**

[185] Mandal, S., Dixit, G., Sahoo, B. K., Chaudhuri, R. K., & Majumder, S.: 2008, J. Phys. B 41, 055701, **Q, QF**

[186] Mania, A. J., Luna, F. R. T., Borges, F. O., & Cavalcanti, G. H.: 2009a, J. Quant. Spectrosc. Radiat. Transfer 110, 2162, **EL, CL, WL, Q**

[187] Mania, A. J., Luna, F. R. T., & Hernandes, J. A.: 2009c, J. Quant. Spectrosc. Radiat. Transfer 110, 82, **Q**

[188] Mania, A. J., Luna, F. R. T., & Mania, E.: 2011, J. Appl. Spectrosc. 77, 758, **Q**

[189] Mashonkina, L., Ryabchikova, T., Ryabtsev, A., & Kildiyarova, R.: 2009, Astron. Astrophys. 495, 297, **Q**

[190] Mattolat, C., Gottwald, T., Raeder, S., Rothe, S., Schwellnus, F., Wendt, K., Thörle-Pospiech, P., & Trautmann, N.: 2010, Phys. Rev. A 81, 052513, **EL**

[191] Meléndez, J. & Barbuy, B.: 2009, Astron. Astrophys. 497, 611, **TE, Q**

[192] Morton, D. C. & Drake, G. W. F.: 2011, Phys. Rev. A 83, 042503, **Q, QF**

[193] Nahar, S. N.: 2008, J. Quant. Spectrosc. Radiat. Transfer 109 , 2731, **Q**

[194] Nahar, S. N.: 2009, At. Data Nucl. Data Tables 95, 577, **Q, QF**

[195] Nahar, S. N.: 2010a, At. Data Nucl. Data Tables 96, 26, **Q, QF**

[196] Nahar, S. N.: 2010b, At. Data Nucl. Data Tables 96, 863, **Q**

[197] Nahar, S. N.: 2011, At. Data Nucl. Data Tables 97, 403, **Q, QF**

[198] Nahar, S. N., Eissner, W., Sur, C., & Pradhan, A. K.: 2009, Phys. Scr. 79, 035401, **Q, QF**

[199] Nakhate, S. G., Mukund, S., & Bhattacharyya, S.: 2010, J. Quant. Spectrosc. Radiat. Transfer 111, 394, **L**

[200] Nave, G. & Sansonetti, C. J.: 2011, J. Opt. Soc. Am. B 28, 737, **WL, EL**

[201] Nilsson, H., Engström, L., Lundberg, H., Palmeri, P., Fivet, V., Quinet, P., & Biémont, E.: 2008, Eur. Phys. J. D 49, 13, **L, Q**

[202] Nilsson, H., Hartman, H., Engström, L., Lundberg, H., Sneden, C., Fivet, V., Palmeri, P., Quinet, P., & Biémont, E.: 2010, Astron. Astrophys. 511, p. A16, **L, TE, Q, T**

[203] Nilsson, H. & Ivarsson, S.: 2008a, Astron. Astrophys. 492, 609, **HFS, TE**

[204] Nouri, Z., Rosner, S. D., Li, R., Scholl, T. J., & Holt, R. A.: 2010, Phys. Scr. 81, 065301, **IS, HFS**

[205] Oliver, P. & Hibbert, A.: 2008a, J. Phys. B 41, 165003, **Q**

[206] Oliver, P. & Hibbert, A.: 2008b, J. Phys.: Conf. Ser. 130, 012016, **Q**

[207] Oliver, P. & Hibbert, A.: 2010, J. Phys. B 43, 074013, **Q**

[208] Özdemir, L., Ürer, G., & Karaçoban, B.: 2008, J. Quant. Spectrosc. Radiat. Transfer 109 , 1886, **Q**

[209] Palmeri, P., Quinet, P., Biémont, E., Gurell, J., Lundin, P., Norlin, L.-O., Royen, P., Blagoev, K., & Mannervik, S.: 2008, J. Phys. B 41, 125703, **LF, QF**

[210] Palmeri, P., Quinet, P., Fivet, V., Biémont, E., Cowley, C. R., Engström, L., Lundberg, H., Hartman, H., & Nilsson, H.: 2009, J. Phys. B 42, 165005, **L, Q**

[211] Palmeri, P., Quinet, P., Fivet, V., Biémont, E., Nilsson, H., Engström, L., & Lundberg, H.: 2008, Phys. Scr. 78, 015304, **L, Q**

[212] Palmeri, P., Quinet, P., Mendoza, C., Bautista, M. A., García, J., Witthoeft, M. C., & Kallman, T. R.: 2011, Astron. Astrophys. 525, p. A59, **Q**

[213] Parthey, C. G., Matveev, A., Alnis, J., Pohl, R., Udem, T., Jentschura, U. D., Kolachevsky, N., & Hänsch, T. W.: 2010, Phys. Rev. Lett. 104, p. 233001, **IS, EL, WL**

[214] Podobedova, L. I., Kelleher, D. E., & Wiese, W. L.: 2009, J. Phys. Chem. Ref. Data 38, 171, **CP**

[215] Quinet, P., Biémont, E., Palmeri, P., Engström, L., Hartman, H., Lundberg, H., & Nilsson, H.: 2011, J. Electron Spectrosc. Relat. Phenom. 184, 174, **Q**

[216] Quinet, P., Fivet, V., Palmeri, P., Biémont, E., Engström, L., Lundberg, H., & Nilsson, H.: 2009, Astron. Astrophys. 493, 711, **Q, L**

[217] Quinet, P., Palmeri, P., Fivet, V., Biémont, E., Nilsson, H., Engström, L., & Lundberg, H.: 2008, Phys. Rev. A 77, 022501, **L, Q**

[218] Ramsbottom, C. A.: 2009, At. Data Nucl. Data Tables 95, 910, **Q**

[219] Redman, S. L., Lawler, J. E., Nave, G., Ramsey, L. W., & Mahadevan, S.: 2011, Astrophys. J., Suppl. Ser. 195, 24, **CL, WL**

[220] Rehse, S. J. & Ryder, C. A.: 2009, Spectrochim. Acta, Part B 64, 974, **TE**

[221] Safronova, M. S. & Safronova, U. I.: 2011a, Phys. Rev. A 83, 052508, **Q, QF**

[222] Safronova, U. I. & Mancini, R.: 2009, At. Data Nucl. Data Tables 95, 54, **Q**

[223] Safronova, U. I. & Safronova, M. S.: 2008, Phys. Rev. A 78, 052504, **Q**

[224] Safronova, U. I. & Safronova, M. S.: 2011b, Can. J. Phys. 89, 465, **Q, QF**

[225] Sahoo, B. K. & Das, B. P.: 2011, Phys. Rev. A 84, 012501, **Q, QF**

[226] Sahoo, B. K., Das, B. P., & Mukherjee, D.: 2009, Phys. Rev. A 79, 052511, **Q**

[227] Sahoo, B. K., Nataraj, H. S., Das, B. P., Chaudhuri, R. K., & Mukherjee, D.: 2008, J. Phys. B 41, 055702, **QF**

[228] Saloman, E. B.: 2010, J. Phys. Chem. Ref. Data 39, 033101, **CP**

[229] Salumbides, E. J. , Maslinskas, V., Dildar, U. M., Wolf, A. L., van Duijn, E.-J., Eikema, K. S. E., Ubachs, W.: 2011 Phys. Rev. A 83, 012502 **WL**

[230] Sansonetti, C. J., Simien, C. E., Gillaspy, J. D., Tan, J. N., Brewer, S. M., Brown, R. C., Wu, S.-J., & Porto, J. V.: 2011, Phys. Rev. Lett. 107, 023001, **WL, HFS, IS, EL**

[231 Sansonetti, C. J. & Veza, D.: 2010, J. Phys. B 43, 205003, **WL, IS, HFS**

[232] Sansonetti, J. E.: 2008a, J. Phys. Chem. Ref. Data 37, 7, **CP**

[233] Sansonetti, J. E.: 2008b, J. Phys. Chem. Ref. Data 37, 1659, **CP**

[234] Sansonetti, J. E.: 2009a, J. Phys. Chem. Ref. Data 38, 761, **CP**

[235] Sansonetti, J. E.: 2009b, J. Phys. Chem. Ref. Data 38, 761, **CP**

[236 Sansonetti, J. E. & Nave, G.: 2010a, J. Phys. Chem. Ref. Data 39, 033103, **CP**

[237] Santos, J. P., Costa, A. M., Madruga, C., Parente, F., & Indelicato, P.: 2011, Eur. Phys. J. D 63, 89, **Q**

[238] Shah, M. L., Pulhani, A. K., Gupta, G. P., & Suri, B. M.: 2010, J. Opt. Soc. Am. B 27, 423, **L, TE**

[239] Shen, X.-Z., Yuan, P., & Liu, J.: 2010, Chin. Phys. B 19, 053101, **Q, QF**

[240] Shestov, S. V., Bozhenkov, S. A., Zhitnik, I. A., Kuzin, S. V., Urnov, A. M., Beigman, I. L., Goryaev, F. F., & Tolstikhina, I. Y.: 2008, Astron. Lett. 34, 33, **CL, WL**

[241] Singh, J., Jha, A. K. S., & Mohan, M.: 2010a, J. Phys. B 43, 115005, **Q**

[242] Singh, J., Jha, A. K. S., Verma, N., & Mohan, M.: 2010b, At. Data Nucl. Data Tables 96, 759, **Q**

[243] Smillie, D. G., Pickering, J. C., & Smith, P. L.: 2008, Mon. Not. R. Astron. Soc. 390, 733, **WL, CL**

[244] Steenstrup, M. P., Brusch, A., Jensen, B. B., Hald, J., & Thomsen, J. W.: 2010, Phys. Rev. A 82, 054501, **IS**

[245] Storey, P. J. & Zeippen, C. J.: 2010, Astron. Astrophys. 511, p. A78, **Q**

[246] Tauheed, A., Joshi, Y. N., & Steinitz, M.: 2009, Can. J. Phys. 87, 1255, **WL, CL, EL**

[247] Tayal, S. S.: 2008a, Astron. Astrophys. 486, 629, **Q**

[248] Tayal, S. S.: 2008b, Astrophys. J., Suppl. Ser. 178, 359, **Q**

[249] Tayal, S. S.: 2009a, Phys. Scr. 79, 015303, **Q**

[250] Tayal, S. S.: 2009b, Phys. Rev. A 80, 032512, **QF**

[251] Tayal, S. S.: 2011a, Phys. Rev. A 83, 012515, **Q, QF**

[252] Tayal, S. S.: 2011b, At. Data Nucl. Data Tables 97, 481, **Q**

[253] Tayal, S. S. & Zatsarinny, O.: 2010a, Astrophys. J., Suppl. Ser. 188, 32, **Q, QF**

[254] Tayal, S. S. & Zatsarinny, O.: 2010b, Astron. Astrophys. 510, p. A79, **Q**

[255] Tayal, V., Gupta, G. P., & Tripathi, A. N.: 2009, Indian J. Phys. 83, 1271, **Q**

[256] Therkildsen, K. T., Jensen, B. B., Ryder, C. P., Malossi, N., & Thomsen, J. W.: 2009, Phys. Rev. A 79, 034501, **TE**

[257] Thorne, A. P., Pickering, J. C., & Semeniuk, J.: 2011, Astrophys. J., Suppl. Ser. 192, 11, **CL, WL, EL**

[258] Träbert, E., Hoffman, J., Reinhardt, S., Wolf, A., & Zanna, G. D.: 2008, J. Phys.: Conf. Ser. 130, 012018, **LF**

[259] Träbert, E., Hoffmann, J., Krantz, C., Wolf, A., Ishikawa, Y., & Santana, J. A.: 2009, J. Phys. B 42, 025002, **LF**

[260] Träbert, E., Ishikawa, Y., Santana, J. A., & Zanna, G. D.: 2011, Can. J. Phys. 89, 403, **EL, CL, TE**

[261] Uddin, Z. and Windholz, L.: 2009, Chin. J. Phys. 47(4), 454, **HFS, CL, WL**

[262] Wallace, L. & Hinkle, K.: 2009, Astrophys. J. 700, 720, **CL, WL, EL**

[263] Wang, Z.-W., Li, X.-R., Hu, M.-H., Liu, Y., & Wang, Y.-N.: 2008a, Chin. Phys. Lett. 25, 2004, **Q**

[264] Wang, Z.-W., Liu, Y., Hu, M.-H., Li, X.-R., & Wang, Y.-N.: 2008b, Chin. Phys. B 17, 2909, **Q**

[265] Wang, Z.-W., Wang, Y.-N., Hu, M.-H., Li, X.-R., & Liu, Y.: 2008c, Sci. China, Ser. G 51, 1633, **Q**

[266] Wasowicz, T. J.: 2009, Eur. Phys. J. D 53, 263, **IS**

[267] Wasowicz, T. J., Werbowy, S., Kwela, J., & Drozdowski, R.: 2010, J. Opt. Soc. Am. B 27, 2628, **IS**

[268] Wei, H. G., Shi, J. R., Zhao, G., & Liang, Z. T.: 2010, Astron. Astrophys. 522, p. A103, **Q**

[269] Wiese, W. L. & Fuhr, J. R.: 2009, J. Phys. Chem. Ref. Data 38, 565, **CP**

[270] Witthoeft, M. C. & Badnell, N. R.: 2008, Astron. Astrophys. 481, 543, **QF**

[271] Wolf, A. L., van den Berg, S. A., Gohle, C., Salumbides, E. J., Ubachs, W., & Eikema, K. S. E.: 2008, Phys. Rev. A 78, 032511, **WL, EL**

[272] Wyart, J.-F. & Lawler, J. E.: 2009a, Phys. Scr. 79, 045301, **EL, CL, W, Q**

[273] Wyart, J.-F., Tchang-Brillet, W.-U. L., Churilov, S. S., & Ryabtsev, A. N.: 2008, Astron. Astrophys. 483, 339, **Q**

[274] Xie, J., Dai, C.-J., & Li, M.: 2011, J. Phys. B 44, 015002, **EL**

[275] Xu, J.-X., Feng, Y.-Y., Sun, G.-J., & Dai, Z.-W.: 2009, Chin. Phys. B 18, 3828, **L**

[276] y. Zhang, T. & w. Zheng, N.: 2009, Chin. J. Chem. Phys. 22, 246, **Q**

[277] Yang, J.-H., Li, P., Zhang, J.-P., & Li, H.-L.: 2008, Commun. Theor. Phys. 50, 468, **Q**

[278] Yang, Z. H., Du, S. B., Chang, H. W., Zhang, Y. P., Zhang, B. L., Xu, Q. M., Yu, D. Y., & Cai, X. H.: 2010, J. Quant. Spectrosc. Radiat. Transfer 111 , 2007, **WL, CL**

[279] Yang, Z. H., Du, S. B., Zeng, X. T., Chang, H. W., Zhang, B. L., Wang, W., Yu, D. Y., & Cai, X. H.: 2009, Astron. J. 137, 4020, **CL, WL**

[280] Yildiz, M., Çelik, G., & Kiliç, H. Ş.: 2009, Acta Phys. Pol. A 115, 641, **Q**

[281] Yoca, S. E., Biémont, E., Delahaye, F., Quinet, P., & Zeippen, C. J.: 2008, Phys. Scr. 78, 025303, **Q**

[282] Young, P. R.: 2009, Astrophys. J. 691, L77, **CL, WL, EL**

[283] Young, P. R., Feldman, U., & Lobel, A.: 2011, Astrophys. J., Suppl. Ser. 196, 23, **CL, WL, EL**

[284] Young, P. R. & Landi, E.: 2009, Astrophys. J. 707, 173, **CL, WL, EL**

[285] Yüce, K., Castelli, F., & Hubrig, S.: 2011a, Astron. Astrophys. 528, p. A37, **CL, WL, EL, TE**

[286] Zanna, G. D.: 2009a, Astron. Astrophys. 508, 501, **CL, WL, EL, Q**

[287] Zanna, G. D.: 2009c, Astron. Astrophys. 508, 513, **CL, WL, EL, Q**

[288] Zanna, G. D.: 2010a, Astron. Astrophys. 514, p. A41, **EL, CL, WL, Q, QF**

[289] Zanna, G. D. & Ishikawa, Y.: 2009, Astron. Astrophys. 508, 1517, **EL, CL, WL**

[290] Zanna, G. D., Rozum, I., & Badnell, N. R.: 2008, Astron. Astrophys. 487, 1203, **Q, QF**

[291] Zanna, G. D., Storey, P. J., & Mason, H. E.: 2010a, Astron. Astrophys. 514, p. A40, **EL, CL, WL, Q**

[292] Zhang, T.-Y. & Zheng, N.-W.: 2009, Acta Phys. Pol. A 116(2), 141, **Q**

[293] Zhang, T.-Y., Zheng, N.-W., & Ma, D.-X.: 2009, Int. J. Quantum Chem. 109, 145, **Q**

[294] Zhang, W., Du, S., Feng, Y.-Y., Jiang, L.-Y., Jiang, Z.-K., & Dai, Z.-W.: 2011a, Mon. Not. R. Astron. Soc. 413 , 1803, **L**

[295] Zhang, W., Feng, Y.-Y., & Dai, Z.-W.: 2010a, J. Opt. Soc. Am. B 27, 2255, **L**

[296] Zhang, W., Feng, Y.-Y., Sun, G.-J., & Dai, Z.-W.: 2010b, J. Phys. B 43, 235005, **L**

[297] Zhang, W., Feng, Y.-Y., Xu, J.-X., Palmeri, P., Quinet, P., Biémont, E., & Dai, Z.-W.: 2010c, J. Phys. B 43, 205005, **L**

[298] Zhang, W., Palmeri, P., Quinet, P., Biémont, E., Du, S., & Dai, Z.-W.: 2010d, Phys. Rev. A 82, 042507, **EL, CL, L**

[299] Zhang, Y., Xu, J.-X., Zhang, W., You, S., Ma, Z.-G., Han, L.-L., Li, P.-F., Sun, G.-J., Jiang, Z.-K., Yoca, S. E., Quinet, P., Biémont, E., & Dai, Z.-W.: 2008a, Phys. Rev. A 78, 022505, **L**

Transactions IAU, Volume XXVIIIA
Reports on Astronomy 2009–2012   © International Astronomical Union 2012
Ian Corbett, ed.   doi:10.1017/S1743921312003092

DIVISION XII/COMMISSION 14/WORKING GROUP ON MOLECULAR DATA

CHAIR   Steven R. Federman
VICE-CHAIR   Peter F. Bernath
VICE-CHAIR   Holger S. P. Müller

TRIENNIAL REPORT 2009-2012

1. Introduction

The current report covers the period from the second half of 2003 to the first half of 2011, bringing the Working Group's efforts up to date, and is divided into three main sections covering rotational, vibrational, and electronic spectroscopy. Rather than being exhaustive, space limitations only allow us to highlight a representative sample of work on molecular spectra. Related research on collisions, reactions on grain surfaces, and astrochemistry appear in the report by another Working Group. These also recount recent conferences and workshops on molecular astrophysics.

2. Rotational Spectra

A large number of reviews have appeared dealing with rotational spectra of molecules potentially relevant to radio-astronomical observations. Therefore, emphasis is placed on investigations dealing with molecular species already observed in space. Related molecules are included to a large extent also. The following groups have been created – hydride species, anions, molecules which may occur in circumstellar envelopes of late type stars, complex molecules, weed species, and other molecules – and are discussed in turn.

Several databases provide rotational spectra of molecular species of astrophysical and astrochemical relevance. The two most important sources for predictions generated from experimental data by employing appropriate Hamiltonian models are the Cologne Database for Molecular Spectroscopy, CDMS (http://www.astro.uni-koeln.de/cdms/) with its catalog (http://www.astro.uni-koeln.de/cdms/catalog). An updated description appeared in Müller *et al.* (2005) and the JPL catalog (http://spec.jpl.nasa.gov/). Both also provide primary information, i.e. laboratory data with uncertainties, mostly in special archive sections. Additional primary data are available in the Toyama Microwave Atlas (http://www.sci.u-toyama.ac.jp/phys/4ken/atlas/). A useful resource on the detection of certain molecular transitions in space is the NIST Recommended Rest Frequencies for Observed Interstellar Molecular Microwave Transitions, which has been updated and described by Lovas (2004).

The European FP7 project Virtual Atomic and Molecular Data Centre, VAMDC, (http://www.vamdc.org/) aims at combining several spectroscopic, collisional, and kinetic databases. The CDMS is the rotational spectroscopy database taking part; several infrared databases are also involved. The project has been described by Dubernet *et al.* (2010). Other tertiary sources combining data from various databases are, e.g., Cassis (http://cassis.cesr.fr/), which provides tools to analyze astronomical spectra, or splatalogue (http://www.splatalogue.net/).

2.1. *Hydrides*

Hydrides here are all molecules consisting of one non-metal atom and one or more H atoms. They may be neutral or positively charged. Metal hydrides will be dealt with in subsection 2.3. Rotational spectra of some anionic hydride molecules have been obtained also, but such molecules have not yet been detected in space, see also subsection 2.2. Hydrides are usually difficult to observe from the ground as their fundamental transitions occur in the upper millimeter region for heavier species to the terahertz region for lighter ones. Some hydrides have recently been detected from the ground, such as $^{13}CH^+$ with the CSO, or SH^+ or OH^+ with the APEX telescope. The launch of the *Herschel* satellite in May 2009 has created a wealth of opportunities to investigate hydrides, in particular with the high resolution instrument HIFI. H_2O^+, H_2Cl^+, ND, and possibly HCl^+ have been deteceted, and the fundamental transitions of CH^+ and of HF have been observed for the first time.

The very light hydrides H_2D^+ and HD_2^+ have attracted considerable attention (Amano & Hirao 2005; Asvany *et al.* 2008; Yonezu *et al.* 2009). Particularly noteworthy is the measurement of the $1_{0,1} - 0_{0,0}$ transition of H_2D^+ more than 60 MHz away from the previously accepted value by Asvany *et al.* (2008). A similarly large deviation occurred between the initial transition frequency of the $J = 1 - 0$ transition of CH^+ (Pearson & Drouin 2006) and the one measured by Amano (2010a), who also recorded the same transition for $^{13}CH^+$ and CD^+. Müller (2010) used the latter transition frequencies together with data from electronic spectra to derive extensive predictions for rotational and rovibrational transitions of several CH^+ isotopologues. Isotopic CH (Halfen *et al.* 2008), CH_2 (Brünken *et al.* 2005), CHD (Ozeki *et al.* 2011), CH_2D^+ (Amano 2010b), and CH_3D (Drouin *et al.* 2009) have been other carbon containing hydrides studied.

No accurate rotational transition frequencies are available for CH_2^+ or NH_2^+. Their fundamental transitions are beyond the region which can be studied with HIFI.

The nitrogen hydrides NH (Flores-Mijangos *et al.* 2004), NH^+ (Hübers *et al.* 2009), NH_3 (Yu *et al.* 2010), as well as the isotopologues NHD_2 (Endres *et al.* 2006), $^{15}NH_2D$ and $^{15}NHD_2$ (Elkeurti *et al.* 2008) have been investigated recently. Transition frequencies have been determined for the oxygen hydrides ^{17}OH (Polehampton *et al.* 2003), OH^+ (Müller *et al.* 2005), H_2O and $H_2^{18}O$ (Golubiatnikov *et al.* 2006), $H_2^{17}O$ (Puzzarini *et al.* 2009), D_2O (Brünken *et al.* 2007a; Cazzoli *et al.* 2010a), H_3O^+ (Yu *et al.* 2009), and H_2DO^+ (Müller *et al.* 2010).

Halogen containing molecules have attracted increased attention in recent years. Among the hydride species investigated recently are DF (Cazzoli *et al.* 2006a), HF^+ (Allen *et al.* 2004), H_2F^+ (Fujimori *et al.* 2011), and H_2Cl^+ (Araki *et al.* 2001).

Other investigations of heavier hydrides include Lamb-dip measurements of low-lying rotational states of PH_3 (Cazzoli & Puzzarini 2006) and a combined analysis of spectroscopic data of SH^+ (Brown & Müller 2009).

2.2. *Anions*

There had been speculation about the possibility of molecular anions in space for quite a while. However, accurate transition frequencies were only known for OH^- and SH^-; for a few other species approximate values were known from infrared spectroscopy. However, all these species were deemed to be too fragile with respect to photolysis. Moreover, all these species were hydride species with fundamental transitions not easily accessible from the ground.

Interestingly, a molecular line survey of the circumstellar envelope of the carbon-rich star CW Leo, also known as IRC +10216, between 28 and 50 GHz revealed several series of unidentified lines (Kawaguchi *et al.* 1995). One of these turned out to be caused by the anion C_6H^- (McCarthy *et al.* 2006). This finding sparked a flurry of investigations into laboratory rotational spectra of related molecular anions and the search for these species in space. C_2H^- (Brünken *et al.* 2007b), as well as C_4H^- and C_8H^- (Gupta *et al.* 2007), were characterized and, except for the smallest member, found in space. The isoelectronic molecules CN^- (Gottlieb *et al.* 2007) and C_3N^- (Thaddeus *et al.* 2008) were found in the lab as well as in the circumstellar envelope of CW Leo. The identification of C_5N^- also in that source seemed to be certain enough even in the absence of laboratory spectral data. Furthermore, submillimeter transitions have been recorded for CN^-, C_2H^-, and C_4H^- (Amano 2008) as well as for C_3N^- (Amano 2010c). In addition, the laboratory rotational spectrum of NCO^- was recently reported (Lattanzi *et al.* 2010).

2.3. *Circumstellar molecules*

The chemistry of circumstellar envelopes of late-type stars is special as a number of molecules have only been detected or are particularly abundant in such sources. Probably the most fascinating object in this regard is CW Leo. Whereas many astronomers have thought this source has a rather special chemistry, it turned out that it is a rather common source. The reason why molecules are particularly easily detected in this source is its proximity. For instance, several metal containing molecules have been detected first in the envelope of CW Leo, but in recent years MgNC, NaCN, and NaCl have been detected also in the envelopes of other AGB stars, the latter even in those of O-rich AGB stars such as IK Tau and VY CMa. Interestingly, recently the metal-containing molecules AlO and AlOH as well as PO have been detected first in the envelope of VY CMa. In addition, KCN and FeCN were detected toward CW Leo.

For a number of metal-containing molecules rotational spectra have been studied. Hydride species include AlH (Halfen & Ziurys 2004), CrH (Harrison *et al.* 2006), MnH (Halfen & Ziurys 2008), and FeH (Brown *et al.* 2006). Among the nitride and oxide species studied are VN and VO (Flory & Ziurys 2008), CoO (McLamarrah *et al.* 2005), ZnO (Zack *et al.* 2009), and even TiO_2 (Kania *et al.* 2011). Several halogenides have been investigated, such as NaCl and KCl (Caris *et al.* 2004), TiF (Sheridan *et al.* 2003), TiCl (Maeda *et al.* 2001), VCl (Halfen *et al.* 2009b), CoF (Harrison *et al.* 2007), CoCl (Flory *et al.* 2004), ZnF (Flory *et al.* 2006), and ZnCl (Tenenbaum *et al.* 2007). Also recorded were rotational spectra of CoCN and NiCN (Sheridan *et al.* 2004; Sheridan & Ziurys 2003).

The study of the rotational spectrum of C_2P (Halfen *et al.* 2009a) laid the ground work for its detection in the circumstellar envelope of CW Leo. This detection, in turn, suggests that molecules such as C_2F, C_2Cl, C_3F, and C_3Cl may be detectable as well. Rotational spectra have been detected for all of these species except for the first one (Sumiyoshi *et al.* 2003; Yoshikawa *et al.* 2009b,a).

Other molecules investigated, including predominantly or potentially circumstellar ones, are C_2H and C_6H in excited vibrational states (Killian *et al.* 2007; Gottlieb *et al.* 2010) and CH_3CP (Bizzocchi *et al.* 2003).

2.4. *Complex molecules*

Some of the smaller complex molecules with many emission or absorption lines observable by radio-astronomical means are presented in subsection 2.5.

Among the larger saturated or almost saturated molecules, propenal, propanal, and acetamide have been detected with the GBT in the microwave region. Interestingly, propylene has been detected with the IRAM 30 m telescope toward TMC-1. Three complex molecules have been detected with the same instrument in the course of a molecular line survey of Sagittarius B2(N) at 3 mm. These are aminoacetonitrile, n-propyl cyanide, and ethyl formate. The detection articles also feature critical evaluations of the spectroscopic parameters of the former two molecules (Belloche et $al.$ 2008, 2009). The rotational spectrum of ethyl formate has also been revisited (Medvedev et $al.$ 2009). Other studies include $cyclo$-propyl cyanide (Bizzocchi et $al.$ 2008) and iso-propyl cyanide (Müller et $al.$ 2011), two conformers of ethylene glycol (Christen & Müller 2003; Müller & Christen 2004), microwave spectra of several conformers of 1,2- and 1,3-propanediol (Lovas et $al.$ 2009; Plusquellic et $al.$ 2009), propane (Drouin et $al.$ 2006), millimeter wave spectra of the amino acids glycine and alanine (Ilyushin et $al.$ 2005; Hirata et $al.$ 2008), acetamide (Ilyushin et $al.$ 2004), methylamine (Ilyushin & Lovas 2007), acetic acid (Ilyushin et $al.$ 2008), n-propanol (Kisiel et $al.$ 2010), and diethyl ether (Walters et $al.$ 2009). Other investigations may turn out to also be relevant once telescope arrays such as EVLA, NOEMA, or ALMA conduct large scale line surveys or dedicated seaches for particular molecules.

2.5. *Weed species*

The term "weed species" has been coined for molecules that have very many emission or absorption lines in various sources, but considered mostly for star-forming regions. It should be emphasized that the plethora of lines is not only a nuisance, but there is also considerable information about, for example, temperature or density in these lines. The molecule with particularly many rather strong lines is methanol. It has been extensively studied up to the second torsional state (Xu et $al.$ 2008); even higher states have been studied to some extent (Pearson et $al.$ 2009), but these are difficult to model at present. CH_3OH has also been proposed as a particularly well suited molecule to investigate the possibility of temporal or spatial variations of fundamental constants (Jansen et $al.$ 2011; Levshakov et $al.$ 2011). One requirement is the knowledge of very highly accurate transition frequencies. Some of these have been summarized as well as newly reported ones by Müller et $al.$ (2004). Other methanol isotopologues studied recently include CH_2DOH (Mukhopadhyay et $al.$ 2002; Lauvergnat et $al.$ 2009), CH_3OD (Duan et $al.$ 2003), and $CH_3{}^{18}OH$ (Fisher et $al.$ 2007).

Other weed molecules, for which several isotopic species have been investigated, are methyl cyanide (Müller et $al.$ 2009), ethyl cyanide (Brauer et $al.$ 2009; Demyk et $al.$ 2007; Margulès et $al.$ 2009b), vinyl cyanide (Kisiel et $al.$ 2009), methyl formate (Ilyushin et $al.$ 2009; Carvajal et $al.$ 2009, 2010; Margulès et $al.$ 2009a, 2010), formamide (Kryvda et $al.$ 2009), and acetone (Groner et $al.$ 2008; Lovas & Groner 2006). Moreover, dimethyl ether (Endres et $al.$ 2009), its isomer ethanol (Pearson et $al.$ 2008), and the only inorganic weed species SO_2 (Müller & Brünken 2005) have been studied.

One should keep in mind that weed species are not only seen in the ground vibrational state, but often in several excited states, and this is true for minor isotopic species as well. While deriving an appropriate Hamiltonian model is usually the most compact form to represent the rotational spectrum of a molecule in a given vibrational state, this task can be formidable and worse in cases of strong vibration-rotation interaction that may take much more than months to be tackled. As an alternative, the De Lucia group proposed recording spectra at different temperatures. While this procedure yields huge amounts of data, and extrapolation in frequency or to higher temperatures is impossible and extrapolation to lower temperatures only to some extent, it seems to be still a pragmatic approach for some molecules. Several reports have been published on ethyl cyanide (Fortman et $al.$ 2010) and one on vinyl cyanide (Fortman et $al.$ 2011).

2.6. Other molecules

Other molecules, for which rotational spectra have been (re-) investigated, include cations such as $HCNH^+$ and CH_3CNH^+ (Amano et al. 2006), CS^+ (Bailleux et al. 2008), CF^+ (Cazzoli et al. 2010b), ^{15}N isotopologues of N_2H^+ (Dore et al. 2009), HCS^+ (Margulès et al. 2003), and HCO^+ (Tinti et al. 2007).

Other short-lived molecules include DNC (Brünken et al. 2006), C_3H and ^{13}C isotopologues of C_nH with $n = 3, 5 - 7$ (Caris et al. 2009; McCarthy & Thaddeus 2005), HOCN, HONC, and HSCN (Brünken et al. 2009a; Mladenović et al. 2009; Brünken et al. 2009b) as well as SiN and PN (Bizzocchi et al. 2006; Cazzoli et al. 2006b).

In addition, several stable or fairly stable molecules have been studied, including isotopologues of CO in their ground vibrational states (Puzzarini et al. 2003) and the main isotopologue in excited vibrational states (Gendriesch et al. 2009), various isotopologues in the ground and excited states of CS and SiS (Müller et al. 2005, 2007), isotopologues of HCN in their ground vibrational states (Cazzoli & Puzzarini 2005) and the main isotopologue in excited vibrational states (Zelinger et al. 2003), H_2CO (Brünken et al. 2003), the isoelectronic H_2CNH (Dore et al. 2010), H_2CS (Maeda et al. 2008), isotopologues of HCOOH (Lattanzi et al. 2008), and CH_3C_2H and CH_3C_4H (Cazzoli & Puzzarini 2008).

3. Vibrational Spectra

Here we describe vibration-rotation spectra of gaseous molecules of astronomical or potential astronomical interest. It is based in part on a review article entitled "Molecular astronomy of cool stars and sub-stellar objects" aimed at physical chemists and laboratory spectroscopists (Bernath 2009).

In addition to the references to particular molecules given below there are a number of spectral database compilations that are useful. Perhaps the most helpful is the HITRAN database that contains vibration-rotation line parameters for a large number of species such as H_2O, CO_2, CO, etc. found in the Earth's atmosphere (Rothman et al. 2009). HITRAN is widely used for astronomical applications although it is not always suitable because of missing lines and bands, particularly in the near infrared region. For example, "cool" astronomical objects such as brown dwarfs can have surface temperatures in excess of 1000 K, and HITRAN is designed for temperatures near 300 K. In this regard, there is a HITEMP database (Rothman et al. 2010) for H_2O, CO_2, CO, NO, and OH that is more suitable for high temperature sources such as stellar atmospheres.

For larger molecules, individual vibration-rotation lines are no longer clearly resolved and it becomes necessary to replace line-by-line calculations by absorption cross sections. The main drawback to using cross sections is that a considerable number of laboratory measurements are needed to match the temperature and pressure conditions of the objects under observation. HITRAN also includes a number of high resolution infrared absorption cross sections for organic molecules such as ethane and acetone, but the broadening gas in HITRAN is air rather than H_2, N_2, or CO_2. While the GEISA database has significant overlap with HITRAN, it contains additional molecules of interest for studies of planetary atmospheres (Jacquinet-Husson et al. 2011). A very useful set of infrared absorption cross sections for several hundred molecules have been measured at the Pacific Northwest National Laboratory (PNNL) for the 600-6500 cm^{-1} (1.54–16.7 μm) range (Sharpe et al. 2004). The PNNL IR database, however, may not be completely suitable for astronomical applications, for example for planetary atmospheres. All PNNL spectra are recorded at relatively low resolution (0.112 cm^{-1}) as mixtures with pure nitrogen gas at pressures of 760 Torr and temperatures of 278, 293, or 323 K. Nevertheless, in the absence of spectra recorded under more appropriate experimental conditions, they can be very useful.

Other interesting general sources for infrared data are various high resolution spectral atlases of the Sun (Livingston & Wallace 2003; Hase *et al.* 2010) and sunspots (Wallace *et al.* 2001, 2002) because they include molecular (and atomic) line assignments. The web site spectrafactory (Cami *et al.* 2010a) is also useful for calculating infrared spectra of astronomical interest, although not all of the input line parameters are the most recent or recommended ones.

3.1. *Diatomic Molecules*

Diatomic molecules, particularly diatomic hydrides of the more abundant elements, are often observed through their infrared spectra. Molecular hydrogen itself is difficult to observe in the infrared because all transitions are electric-dipole forbidden; however, HD has a small dipole moment and very recently the 2–0 band was measured near 1.4 μm by cavity ring down spectroscopy (Kassi & Campargue 2011). In the case of OH, NH, and CH, the new infrared solar atlas (Hase *et al.* 2010) measured by the Atmospheric Chemistry Experiment (ACE) infrared Fourier transform spectrometer from orbit (Bernath *et al.* 2005) has led to an improvement in the spectroscopy. The infrared Fraunhofer lines were combined with laboratory measurements to extend the measured line positions to higher vibrational and rotational quantum numbers for OH (Bernath & Colin 2009), NH (Ram & Bernath 2010), and CH (Colin & Bernath 2010).

Laboratory vibration-rotation spectra for a considerable number of metal hydrides are known with recent measurements of, for example, BeH (Shayesteh *et al.* 2003a), MgH (Shayesteh *et al.* 2004a), and CaH (Shayesteh *et al.* 2004b); even the metal dihydrides BeH_2 (Shayesteh *et al.* 2003b) and MgH_2 (Shayesteh *et al.* 2003c) have been detected in the laboratory. Although metal hydrides such as MgH and CaH are detected in stellar atmospheres by their electronic transitions, there have been no infrared astronomical observations. Other diatomics such as CO, SO, SiO, HF, HCl, SH, CS, and SiS are commonly seen by infrared observations of cool stellar atmospheres (Bernath 2009); their spectroscopy however has not been improved much in recent years. An exception to this is some work on the line intensities of SiS (Cami *et al.* 2009) and HCl (Li *et al.* 2011).

3.2. *Small Polyatomics*

The spectra of "cold" water, ammonia, and methane as given in the HITRAN database are generally satisfactory for astronomical purposes, except for overtone and combination bands of NH_3 and CH_4 in the near infrared and visible regions. For spectra of hot samples as in brown dwarfs and exoplanets, the situation is much less sanguine, except in the case of water for which there has been extensive experimental (Zobov *et al.* 2008) and theoretical work (Barber *et al.* 2006). Much work is continuing on highly-excited water levels near dissociation and on computing line intensities ab initio, but the BT2 line list (Barber *et al.* 2006) is generally suitable for computing water opacities. For hot ammonia there has been rapid recent progress with new laboratory spectra recorded (Hargreaves *et al.* 2011) and with at least two groups providing rather good calculated spectra (Huang *et al.* 2011; Yurchenko *et al.* 2011). Ammonia is seen in brown dwarfs and is thought to be the key molecule in defining a new class of ultracool Y-type dwarfs (Cushing *et al.* 2011). Methane is the laggard because of the difficulty of the problem, and the existing experimental data (Nassar & Bernath 2003) and calculations are not very satisfactory for simulating brown dwarf and exoplanet spectra. For CO_2, HITRAN for cold molecules and HITEMP or CDSD-4000 databases (Tashkun & Perevalov 2011) for hot molecules are recommended.

3.3. *Large Molecules*

The discovery of C_{60} and C_{70} in the young planetary nebula Tc1 was reported by Cami *et al.* (2010b) using the Spitzer Space Telescope. The discovery was rapidly confirmed with detections in other planetary nebulae; some of these sources also have spectral features generally attributed to polycyclic aromatic hydrocarbons (PAHs). Infrared emission spectra of gaseous C_{60} and C_{70} were recorded as a function of temperature by Nemes *et al.* (1994) and more recently the integrated molar absorptivity of the infrared bands of the solid has been measured as a function of temperature (Iglesias-Groth *et al.* 2011). There have also been extensive laboratory measurements and quantum chemical calculations of PAH spectra, some of which is summarized in the NASA Ames database (Bauschlicher *et al.* 2010). Traditionally infrared spectra of neutral and ionized PAHs have been recorded by matrix isolation spectroscopy, but now gas-phase measurements are possible (Ricks *et al.* 2009; Galué *et al.* 2011). Computational studies of PAHs and related species have also advanced, and the quality and quantity of calculations available is remarkable (Hudgins *et al.* 2005; Bauschlicher *et al.* 2010).

4. Electronic Spectra

In this section we describe recent work on electronic spectra that includes line identification, energy levels, as well as data needed for photochemical models. The data come in a variety of forms, such as absorption cross sections (or equivalently oscillator strengths), predissociation widths, and analyses of line anomalies resulting from perturbations between energy levels. Here we can only present a representative sampling of work on electronic spectra. The number of experiments on electron excitation/scattering is great, and we provide a few illustrative examples of research in this area. This section is divided into three topics: interstellar matter, which includes diffuse molecular clouds and disks around newly formed stars as well as comets whose chemistry is similar, metal hydrides and oxides in the spectra of late-type stars, and the atmospheres of planets and their satellites. Recent attempts to identify the diffuse interstellar bands is not included, but instead we refer the reader to a review on laboratory astrophysics (Savin *et al.* 2011) that includes a discussion of this topic. It is worthwhile noting, however, that the study of the diffuse interstellar bands has led to a renaissance in the gas phase spectroscopy of complex carbon molecules that begun in 2003. A typical example is the case of PAH molecules. Traditionally UV-Visible spectra of neutral and ionized PAHs could only be recorded by matrix isolation spectroscopy, but now gas-phase measurements are possible (see reviews in Joblin & Tielens (2011)).

4.1. *Interstellar matter*

There is continued interest in spectroscopic studies of CO and much of the recent efforts are needed for improved photochemical modeling. A comprehensive analysis of Rydberg states has been published (Eidelsberg *et al.* 2004a), as has a compilation of triplet and singlet transitions (Eidelsberg & Rostas 2003) that included oscillator strengths. Improved wavelengths have appeared for the $A - X$ system of bands in $C^{17}O$ and $C^{18}O$ (Steinmann *et al.* 2003; du Plessis *et al.* 2006, 2007), for the $E - X$ (0,0) band in CO, ^{13}CO, and $^{13}C^{18}O$ (Cacciani & Ubachs 2004), and for triplet-singlet bands involving the e, d, and a' states in several isotopologues (du Plessis *et al.* 2007; Yang *et al.* 2008; Dickenson *et al.* 2010). Several quantum mechanical calculations of potential energy curves (Chakrabarti & Tennyson 2006; Vázquez *et al.* 2009; Lefebvre-Brion *et al.* 2010) have appeared. Empirical determinations of oscillator strengths for Rydberg transitions have been published, including one based on interstellar spectra acquired with the *Far Ultraviolet Spectroscopic*

Explorer (Sheffer *et al.* 2003), two using synchrotron radiation (Eidelsberg *et al.* 2004b; Eidelsberg *et al.* 2006), and one employing electron scattering techniques (Kawahara *et al.* 2008). Gilijamse *et al.* (2007) have obtained the lifetime for the $a\,^3\Pi\,v=0$ level. In their study, Eidelsberg *et al.* (2006) have also reported predissociation rates for B and W states; the large rates found for the $B\,^1\Sigma^+\,v=6$ level result from interactions with the D' state. A number of other experimental and theoretical efforts describing the $B - D'$ interaction have been published (Andric *et al.* 2004; Grozdanov *et al.* 2004; Baker 2005a; Bitencourt *et al.* 2007). The triplet k and c states have been the focus of studies (Baker 2005b; Baker & Launay 2005a,b,c) on perturbations caused with other states, while Ben *et al.* (2007) have described their experiment on perturbations between the a and d levels. Finally, a review (Lefebvre-Brion & Lewis 2007) discussing perturbations in the isoelectronic molecules, CO and N_2, has also appeared. More spectroscopic work on N_2 is provided below.

New spectroscopic studies on CH, CH^+, NH have appeared since 2003. Extending the study of Watson (2001) on Rydberg transitions in CH, Sheffer & Federman (2007) have inferred oscillator strengths and predissociation rates for the $3d - X$, $4d$ - X, $F - X$, and $D - X$ bands seen in interstellar spectra acquired with the *Hubble Space Telescope*. Theoretical efforts on CH include structure calculations of the $3d$ complex (Vázquez *et al.* 2007) and oscillator strengths for Rydberg transitions (Lavín *et al.* 2009); the latter study also determined photoionization cross sections. A new analysis of the $A\,^1\Pi - X\,^1\Sigma^+$ system in CH^+ has been completed (Hakalla *et al.* 2006), and Weselak *et al.* (2009) have obtained oscillator strengths for the (1,0), (2,0), (3,0), and (4,0) bands of this system from ground-based interstellar spectra. CH^+ photodissociation cross sections have been a focus of recent theoretical work (Barinovs & van Hemert 2004; Bouakline *et al.* 2005). Revised term values have been determined for the $X\,^3\Sigma^-$ and $A\,^3\Pi$ states of NH by Ram & Bernath (2010), who analyzed spectra from the ACE and ATMOS instruments.

Other molecules of interest to interstellar and cometary studies, such as C_2, CN, and C_3, have been studied recently. Theoretical efforts have computed oscillator strengths and radiative lifetimes for singlet (Phillips) and triplet (Swan, Ballik-Ramsay, and $d - c$) systems in C_2 (Kokkin *et al.* 2007; Schmidt & Bacskay 2007); of particular note is that a self-consistent set of oscillator strengths for the Phillips ($A - X$) system of bands is emerging. Experimental studies involving triplet states have been performed with a variety of techniques (Joester *et al.* 2007; Tanabashi *et al.* 2007; Nakajima *et al.* 2009; Bornhauser *et al.* 2010, 2011), many times focusing on the presence of perturbations. Toffoli & Lucchese (2004) have calculated near threshold photoionization cross sections for C_2. Spectroscopic studies of the violet and red band systems for isotopologues of CN (Ram *et al.* 2006, 2010b,c) have improved the precision of the molecular constants for this molecule. This work will also be of interest to stellar astronomy. Shi *et al.* (2010) have obtained spectroscopic parameters in the isotopologues of CN through *ab initio* calculations. Moreover, experimental work on bands in the $A\,^1\Pi_u - X\,^1\Sigma_g^+$ system of C_3 has appeared (McCall *et al.* 2003; Tanabashi *et al.* 2005; Zhang *et al.* 2005; Chen *et al.* 2010, 2011). Perturbations have been seen in many of these bands. Zhang *et al.* (2005) have measured lifetimes as well.

Spectroscopic studies on H_2O and HCl have also been conducted. Fillion *et al.* (2004) used synchrotron radiation to examine the intense Rydberg nd series in H_2O for energies up to 12 eV. Another study with synchrotron radiation found several new vibrational progressions around 10 eV and also reported abolute cross sections (Mota *et al.* 2005). Additional experimental efforts obtained absolute absorption cross sections with a synchrotron source (Cheng *et al.* 2004) and oscillator strengths via electron impact-excitation (Thorn *et al.* 2007); Cheng *et al.* (2004) compared their results with additional theoretical computations. Borges (2006b) extended his earlier calculations (Borges 2006a) in a study of oscillator strengths for the $\tilde{A}\,^1B_1 - \tilde{X}\,^1A_1$ transition in H_2O. Another electron scattering experiment (Li *et al.* 2006) derived oscillator strengths for valence-shell excitations in HCl.

4.2. *Late-type stars*

The transition from metal oxides to metal hydrides is a signature of the latest stellar spectral types. During the reporting period papers on line lists, line strengths, and opacities have appeared. For metal hydrides, these studies include high-resolution spectra acquired with a Fourier transform spectrometer of bands in the $A\,^2\Pi - X\,^2\Sigma^+$ and $B\,^2\Sigma^+ - X\,^2\Sigma^+$ systems of ^{24}MgH (Shayesteh *et al.* 2007) and of the $E\,^2\Pi - X\,^2\Sigma^+$ transition in CaH and CaD (Ram *et al.* 2011b). The work on MgH provided the data on the highest lying vibrational level in the ground electronic state. A depertubation analysis of the MgH spectra was recently completed (Shayesteh & Bernath 2011). Using laser-induced fluorescence, Chowdhury *et al.* (2006) examined the (1,0) band of the $A\,^6\Sigma^+ - X\,^6\Sigma^+$ transition in CrH. New theoretical calculations have led to improved line lists and opacities for transitions in TiH (Burrows *et al.* 2005) and the $F\,^4\Delta_i - X\,^4\Delta_i$ transitions in FeH (Dulick *et al.* 2003). Magnetic properties of FeH involving the F and X states, including a study of the Zeeman effect, have been determined (Harrison *et al.* 2008; Harrison & Brown 2008).

4.3. *Planetary atmospheres*

Nitrogen and sulfur dioxide have received a considerable amount of attention recently. Spectroscopic work on N_2 (Sprengers *et al.* 2003, 2005b; Lewis *et al.* 2008a; Vieitez *et al.* 2008) has focused on extreme UV transitions, where perturbations and predissociation play a significant role; the study by Sprengers *et al.* (2003) considered the $^{15}N_2$ and $^{14}N^{15}N$ isotopologues. A theoretical calculation (Lewis *et al.* 2008b) has studied the perturbations involving these high-lying Rydberg states. Oscillator strengths and line widths have been obtained from a number of experimental (Stark *et al.* 2005, 2008; Heays *et al.* 2009; Huber *et al.* 2009) and theoretical (Jungen *et al.* 2003; Haverd *et al.* 2005; Lavín & Velasco 2011) efforts. Lifetimes of Rydberg states have been determined through a combination of experiments and calculations (Sprengers *et al.* 2004; Lewis *et al.* 2005a,b; Sprengers *et al.* 2005a; Sprengers & Ubachs 2006). As for SO_2, the focus has been on absorption cross sections at UV wavelengths (Rufus *et al.* 2003, 2009; Danielache *et al.* 2008; Hermans *et al.* 2009; Blackie *et al.* 2011), where Danielache *et al.* (2008) studied isotolopogues for sulfur. Furthermore, absorption cross sections for ammonia and its isotopologues (Cheng *et al.* 2006; Wu *et al.* 2007) and carbon dioxide (Stark *et al.* 2007) have been determined experimentally.

Steven R. Federman
chair of Working Group

•

References

Allen, M. D., Evenson, K. M., & Brown, J. M. 2004, *J. Mol. Spectrosc.*, 227, 13
Amano, T. & Hirao, T. 2005, *J. Mol. Spectrosc.*, 233, 7
Amano, T., Hashimoto, K., & Hirao, T. 2006, *J. Mol. Struct.*, 795, 190
Amano, T. 2008, *J. Chem. Phys.*, 129, 244305
Amano, T. 2010a, *ApJ*, 716, L1
Amano, T. 2010b, *A&A*, 516, L4
Amano, T. 2010c, *J. Mol. Spectrosc.*, 259, 16

Andric, L., Bouakline, F., Grozdanov, T. P., & McCarroll, R. 2004, *A&A*, 421, 381

Araki, M., Furuya, T., & Saito, S. 2001, *J. Mol. Spectrosc.*, 210, 132

Asvany, O., Ricken, O., Müller, H. S. P., Wiedner, M. C., Giesen, T. F., & Schlemmer, S. 2008, *Phys. Rev. Lett.*, 100, 233004

Bailleux, S., Walters, A., Grigorova, E., & Margulès, L. 2008, *ApJ*, 679, 920

Baker, J., 2005a, *Chem. Phys. Lett.*, 408, 312

Baker, J., 2005b, *J. Mol. Spectrosc.*, 234, 75

Baker, J. & Launay, F. 2005a, *Chem. Phys. Lett.*, 404, 49

Baker, J. & Launay, F. 2005b, *Chem. Phys. Lett.*, 415, 296

Baker, J. & Launay, F. 2005c, *J. Chem. Phys.*, 123, 234302

Barber, R. J., Tennyson, J., Harris, G. J., & Tolchenov, R. N. 2006, *MNRAS*, 368, 1087

Barinovs, G. & van Hemert, M. C. 2004, *Chem. Phys. Lett.*, 399, 406

Bauschlicher, C. W., et al. 2010, *ApJS*, 189, 341. See http://www.astrochem.org/pahdb/

Belloche, A., Menten, K. M., Comito, C., Müller, H. S. P., Schilke, P., Ott, J., Thorwirth, S., & Hieret, C. 2008, *A&A*, 482, 179

Belloche, A., Garrod, R. T., Müller, H. S. P., Menten, K. M., Comito, C., & Schilke, P. 2009, *A&A*, 499, 215

Ben, J., Li, L., Zheng, L., Chen., Y., & Yang, X. 2007, *Chem. Phys.*, 335, 109

Bernath, P. F., et al. 2005, Geophys. Res. Lett., 32, L15S01. See http://www.ace.uwaterloo.ca/

Bernath, P. F. 2009, *Int. Rev. Phys. Chem.*, 28, 681

Bernath, P. F. & Colin, R. 2009, *J. Mol. Spectrosc.*, 257, 20

Bitencourt, A. C. P., Prudente, F. V., & Vianna, J. D. M. 2007, *J. Phys. B.*, 40, 2075

Bizzocchi, L., Cludi, L., & Degli Esposti, C. 2003, *J. Mol. Spectrosc.*, 218, 53

Bizzocchi, L., Degli Esposti, C., & Dore, L. 2006, *A&A*, 455, 1161

Bizzocchi, L., Degli Esposti, C., Dore, L., & Kisiel, Z. 2008, *J. Mol. Spectrosc.*, 251, 138

Blackie, D., Blackwell-Whitehead, R., Stark, G., Pickering, J. C., Smith, P. L., Rufus, J., & Thorne, A. P. 2011, *JGRE*, 116, 03006

Borges, I. 2006a, *J. Phys. B*, 39, 641

Borges, I. 2006b, *Chem. Phys.*, 328, 284

Bornhauser, P., Knopp, G., Gerber, T., & Radi, P. P. 2010, *J. Mol. Spectrosc.*, 262, 69

Bornhauser, P., Sych, Y., Knopp, G., Gerber, T., & Radi, P. P. 2011, *J. Chem. Phys.*, 134, 044302

Bouakline, F., Grozdanov, T. P., Andric, L., & McCarroll, R. 2005, *J. Chem. Phys.*, 122, 044108

Brauer, C. S., Pearson, J. C., Drouin, B. J., & Yu, S. 2009, *ApJS*, 184, 133

Brown, J. M., Körsgen, H., Beaton, S. P., & Evenson, K. M. 2006, *J. Chem. Phys.*, 124, 234309

Brown, J. M. & Müller, H. S. P. 2009, *J. Mol. Spectrosc.*, 255, 68

Brünken, S., Müller, H. S. P., Lewen, F., & Winnewisser, G. 2003, Phys. Chem. *Chem. Phys.*, 5, 1515

Brünken, S., Müller, H. S. P., Lewen, F., & Giesen, T. F. 2005, *J. Chem. Phys.*, 123, 164315

Brünken, S., Müller, H. S. P., Thorwirth, S., Lewen, F., & Winnewisser, G. 2006, *J. Mol. Struct.*, 780, 3

Brünken, S., Müller, H. S. P., Endres, C., Lewen, F., Giesen, T., Drouin, B., Pearson, J. C., & Mäder, H. 2007a, Phys. Chem. *Chem. Phys.*, 9, 2103

Brünken, S., Gottlieb, C. A., Gupta, H., McCarthy, M. C., & Thaddeus, P. 2007b, *A&A*, 464, L33

Brünken, S., Gottlieb, C. A., McCarthy, M. C., & Thaddeus, P. 2009a, *ApJ*, 697, 880

Brünken, S., Yu, Z., Gottlieb, C. A., McCarthy, M. C., & Thaddeus, P. 2009b, *ApJ*, 706, 1588

Burrows, A., Dulick, M., Bauschlicher, C. W., Bernath, P. F., Ram, R. S., Sharp, C. M., & Milsom, J. A. 2005, *ApJ*, 624, 988

Cacciani, P. & Ubachs, W. 2004, *J. Mol. Spectrosc.*, 225, 62

Cami, J., et al. 2009, *ApJ*, 690, L122

Cami, J., van Malderen, R., & Markwick, A. J. 2010a, *ApJS*, 187, 409. See http://www.spectrafactory.net/

Cami, J., Bernard-Salas, J., Peeters, E., & Malek, S. E. 2010b, Science, 329, 1180

Caris, M., Lewen, F., Müller, H. S. P., & Winnewisser, G. 2004, *J. Mol. Struct.*, 695–696, 243

Caris, M., Giesen, T. F., Duan, C., Müller, H. S. P., Schlemmer, S., & Yamada, K. M. T. 2009, J. Mol. Spectrosc., 253, 99

Carvajal, M., et al. 2009, A&A, 500, 1109

Carvajal, M., Kleiner, I., & Demaison, J. 2010, ApJS, 190, 315

Cazzoli, G. & Puzzarini, C. 2005, J. Mol. Spectrosc., 233, 280

Cazzoli, G., Puzzarini, C., Tamassia, F., Borri, S., & Bartalini, S. 2006a, J. Mol. Spectrosc., 235, 265

Cazzoli, G., Cludi, L., & Puzzarini, C. 2006b, J. Mol. Struct., 780, 260

Cazzoli, G. & Puzzarini, C. 2006, J. Mol. Spectrosc., 239, 64

Cazzoli, G. & Puzzarini, C. 2008, A&A, 487, 1197

Cazzoli, G., Dore, L., Puzzarini, C., & Gauss, J. 2010a, Mol. Phys., 108, 2335

Cazzoli, G., Cludi, L., Puzzarini, C., & Gauss, J. 2010b, A&A, 509, A1

Chakrabarti, K. & Tennyson, J. 2006, J. Phys. B., 39, 1485

Chen, C.-W., Merer, A. J., Chao, J.-M., & Hsu, Y.-C. 2010, J. Mol. Spectrosc., 263, 56

Chen, K.-S., Zhang, G., Merer, A. J., Hsu, J.-C., & Chen, W.-J. 2011, J. Mol. Spectrosc., 267, 169

Cheng, B.-M., Chung, C.-Y., Bahou. M., Lee, Y.-P., Lee, L. C., van Harrevelt, R., & van Hemert, M. C. 2004, J. Chem. Phys., 120, 224

Cheng, B.-M., et al. 2006, ApJ, 647, 1535

Chowdhury, P. K., Merer, A. J., Rixon, S. J., Bernath, P. F., & Ram, R. S. 2006, Phys. Chem. Chem. Phys., 8, 822

Christen, D. & Müller, H. S. P. 2003, Phys. Chem. Chem. Phys., 5, 3600

Colin, R. & Bernath, P. F. 2010, J. Mol. Spectrosc., 263, 120

Cushing, M. C., et al. 2011, ApJ, accepted, arXiv:1108.4678

Danielache, S. O., Eskebjerg, C., Johnson, M. S., Ueno, Y., & Yoshida, N. 2008, JGRD, 113, 17314

Demyk, K., et al. 2007, A&A, 466, 255

Dickenson, G. D., Nortje, A. C., Steenkamp, C. M., Rohwer, E. G., & du Plessis, A. 2010, ApJ, 714, L268

Dore, L., Bizzocchi, L., Degli Esposti, C., & Tinti, F. 2009, A&A, 496, 275

Dore, L., Bizzocchi, L., Degli Esposti, C., & Gauss, J. 2010, J. Mol. Spectrosc., 263, 44

Drouin, B. J., Pearson, J. C., Walters, A., & Lattanzi, V. 2006, J. Mol. Spectrosc., 240, 227

Drouin, B. J., Yu, S., Pearson, J. C., & Müller, H. S. P. 2009, JQSRT, 110, 2077

du Plessis, A., Rohwer, E. G., & Steenkamp, C. M. 2006, ApJS, 165, 432

du Plessis, A., Rohwer, E. G., & Steenkamp, C. M. 2007, J. Mol. Spectrosc., 243, 124

Duan, Y.-B., Ozier, I., Tsunekawa, S., & Takagi, K. 2003, J. Mol. Spectrosc., 218, 95

Dubernet, M. L., Boudon, V., Culhane, J. L., et al. 2010, JQSRT, 111, 2151

Dulick, M., Bauschlicher, C. W., Burrows, A., Sharp, C. M., Sharp, R. S., Ram, R. S., & Bernath, P. F. 2003, ApJ, 594, 651

Eidelsberg, M. & Rostas, F. 2003, ApJS, 145, 89

Eidelsberg, M., Launay, F., Ito, K., Matsui, T., Hinnen, P. C., Reinhold, E., Ubachs, W., & Huber, K. P. 2004a, J. Chem. Phys., 121, 292

Eidelsberg, M. Lemaire, J. L., Fillion, J. H., Rostas, F., Federman, S. R., & Sheffer, Y. 2004b, A&A, 424, 355

Eidelsberg, M., Sheffer, Y., Federman, S. R., Lemaire, J. L., Fillion, J. H., Rostas, F., & Ruiz, J. 2006, ApJ, 647, 1543

Elkeurti, M., Coudert, L. H., Orphal, J., Wlodarczak, G., Fellows, C. E., & Toumi, S. 2008, J. Mol. Spectrosc., 251, 90

Endres, C. P., Müller, H. S. P., Brünken, S., Paveliev, D. G., Giesen, T. F., Schlemmer, S., & Lewen, F. 2006, J. Mol. Struct., 795, 242

Endres, C. P., Drouin, B. J., Pearson, J. C., Müller, H. S. P., Lewen, F., Schlemmer, S., & Giesen, T. F. 2009, A&A, 504, 635

Fillion, J.-H., Ruiz, J., Yang, X.-F., Castillejo, M., Rostas, F., & Lemaire, J.-L. 2004, J. Chem. Phys., 120, 6531

Fisher, J., Paciga, G., Xu, L.-H., Zhao, S. B., Moruzzi, G., & Lees, R. M. 2007, *J. Mol. Spectrosc.*, 245, 7

Flores-Mijangos, J., Brown, J. M., Matsushima, F., Odashima, H., Takagi, K., Zink, L. R., & Evenson, K. M. 2004, *J. Mol. Spectrosc.*, 225, 189

Flory, M. A., Halfen, D. T., & Ziurys, L M. 2004, *J. Chem. Phys.*, 121, 8385

Flory, M. A., McLamarrah, S. K., & Ziurys, L. M. 2006, *J. Chem. Phys.*, 125, 194304

Flory, M. A. & Ziurys, L. M. 2008, *J. Mol. Spectrosc.*, 247, 76

Fortman, S. M., Medvedev, I. R., Neese, C. F., & De Lucia, F. C. 2010, *ApJ*, 725, 1682

Fortman, S. M., Medvedev, I. R., Neese, C. F., & De Lucia, F. C. 2011, *ApJ*, 737, 20

Fujimori, R., Kawaguchi, K., & Amano, T. 2011, *ApJ*, 729, L2

Galué, H. A., Rice, C. A., Steill, J. D., & Oomens, J. 2011, *J. Chem. Phys.*, 134, 054310

Gendriesch, R., Lewen, F., Klapper, G., Menten, K. M., Winnewisser, G., Coxon, J. A., & Müller, H. S. P. 2009, *A&A*, 497, 927

Gilijamse, J. J., Hoekstra, S., Meek, S. A., Metsälä, M., van de Meerakker, S. Y. T., Meijer, G., & Groenenboom, G. C. 2007, *J. Chem. Phys.*, 127, 221102

Golubiatnikov, G. Y., Markov, V. N., Guarnieri, A., & Knöchel, R. 2006, *J. Mol. Spectrosc.*, 240, 251

Gottlieb, C. A., Brünken, S., McCarthy, M. C., & Thaddeus, P. 2007, *J. Chem. Phys.*, 126, 191101

Gottlieb, C. A., McCarthy, M. C., & Thaddeus, P. 2010, *ApJS*, 189, 261

Groner, P., Medvedev, I. R., De Lucia, F. C., & Drouin, B. J. 2008, *J. Mol. Spectrosc.*, 251, 180

Grozdanov, T. P., Bouakline, F., Andric, L., & McCarroll, R. 2004, *J. Phys. B.*, 37, 1737

Gupta, H., Brünken, S., Tamassia, F., Gottlieb, C. A., McCarthy, M. C., & Thaddeus, P. 2007, *ApJ*, 655, L57

Hakalla, R., Kepa, R., Szajna, W., & Zachwieja, M. 2006, Eur. Phys. J. D, 38, 481

Halfen, D. T. & Ziurys, L. M. 2004, *ApJ*, 607, L63

Halfen, D. T. & Ziurys, L. M. 2008, *ApJ*, 672, L77

Halfen, D. T., Ziurys, L. M., Pearson, J. C., & Drouin, B. J. 2008, *ApJ*, 687, 731

Halfen, D. T., Sun, M., Clouthier, D. J., & Ziurys, L. M. 2009a, *J. Chem. Phys.*, 130, 014305

Halfen, D. T., Ziurys, L. M., & Brown, J. M. 2009b, *J. Chem. Phys.*, 130, 164301

Hargreaves, R. J., Li, G., & Bernath, P. F. 2011, *ApJ*, 735, 111

Harrison, J. J. & Brown, J. M. 2008, *ApJ*, 686, 1426

Harrison, J. J., Brown, J. M., Halfen, D. T., & Ziurys, L. M. 2006, *ApJ*, 637, 1143

Harrison, J. J., Brown, J. M., Flory, M. A., Sheridan, P. M., McLamarrah, S. K., & Ziurys, L. M. 2007, *J. Chem. Phys.*, 127, 194308

Harrison, J. J., Brown, J. M., Chen., J., Steimle, T. C., & Sears, T. J. 2008, *ApJ*, 679, 854

Hase, F., Wallace, L., McLeod, S. D., Harrison, J. J., & Bernath, P. F. 2010, *JQSRT*, 111, 521. See http://www.ace.uwaterloo.ca/solaratlas.html/

Haverd, V. E., Lewis, B. R., Gibson, S. T., & Stark, G. 2005, *J. Chem. Phys.*, 123, 214304

Heays, A. N., Lewis, B. R., Stark, G., Yoshino, K., Smith, P. L., Huber, K. P., & Ito, K. 2009, *J. Chem. Phys.*, 131, 194308

Hermans, C., Vandaele, A. C., & Fally, S. 2009, *JQSRT*, 110, 756

Hirata, Y., Kubota, S., Watanabe, S., Momose, T., & Kawaguchi, K. 2008, *J. Mol. Spectrosc.*, 251, 314

Huang, X., Schwenke, D. W., & Lee, T. J. 2011, *J. Chem. Phys.*, 134, 044320 and 044321

Huber, K. P., Chan, M.-C., Stark, G., Ito, K., & Matsui, T. 2009, *J. Chem. Phys.*, 131, 084301

Hübers, H.-W., Evenson, K. M., Hill, C., & Brown, J. M. 2009, *J. Chem. Phys.*, 131, 034311

Hudgins, D. M., Bauschlicher, C. W., & Allamandola, L. J. 2005, *ApJ*, 632, 316

Iglesias-Groth, S., Cataldo, F., & Manchado, A. 2011, *MNRAS*, 413, 213

Ilyushin, V. V., Alekseev, E. A., Dyubko, S. F., Kleiner, I., & Hougen, J. T. 2004, *J. Mol. Spectrosc.*, 227, 115

Ilyushin, V. V., Alekseev, E. A., Dyubko, S. F., Motiyenko, R. A., & Lovas, F. J. 2005, *J. Mol. Spectrosc.*, 231, 15

Ilyushin, V. & Lovas, F. J. 2007, J. Phys. Chem. Ref. Data, 36, 1141

Ilyushin, V., Kleiner, I., & Lovas, F. J. 2008, J. Phys. Chem. Ref. Data, 37, 97

Ilyushin, V., Kryvda, A., & Alekseev, E. 2009, *J. Mol. Spectrosc.*, 255, 32

Jacquinet-Husson, N., Crepeau, L., Armante, R., *et al.* 2011, *JQSRT*, 112, 2395

Jansen, P., Xu, L.-H., Kleiner, I., Ubachs, W., & Bethlem, H. L. 2011, *Phys. Rev. Lett.*, 106, 100801

Joblin, C. & Tielens, A. G. G. M. (Eds.) 2011, EAS Publications Series, Vol. 46

Joester, J. A., Nakajima, M., Reilly, N. J., Kokkin, D. L., Nauta, K., Kable, S. H., & Schmidt, T. W. 2007, *J. Chem. Phys.*, 127, 214303

Jungen, Ch., Huber, K. P., Jungen, M., & Stark, G. 2003, *J. Chem. Phys.*, 118, 4517

Kania, P., Hermanns, M., Brünken, S., Müller, H. S. P., & Giesen, T. F. 2011, *J. Mol. Spectrosc.*, 268, 173

Kassi, S. & Campargue, A. 2011, *J. Mol. Spectrosc.*, 267, 36

Kawaguchi, K., Kasai, Y., Ishikawa, S.-I., & Kaifu, N. 1995, PASJ, 47, 853

Kawahara, H., Kato, H., Hoshino, M., Tanaka, H., & Brunger, M. J. 2008, Phys. Rev., A77, 012713

Killian, T. C., Gottlieb, C. A., & Thaddeus, P. 2007, *J. Chem. Phys.*, 127, 114320

Kisiel, Z., Pszczółkowski, L., Drouin, B. J., Brauer, C. S., Yu, S., & Pearson, J. C. 2009, *J. Mol. Spectrosc.*, 258, 26

Kisiel, Z., *et al.* 2010, Phys. Chem. *Chem. Phys.*, 12, 8329

Kokkin, D. L., Bacskay, G. B., & Schmidt, T. W. 2007, *J. Chem. Phys.*, 126, 084302

Kryvda, A. V., Gerasimov, V. G., Dyubko, S. F., Alekseev, E. A., & Motiyenko, R. A. 2009, *J. Mol. Spectrosc.*, 254, 28

Lattanzi, V., Walters, A., Drouin, B. J., & Pearson, J. C. 2008, *ApJS*, 176, 536

Lattanzi, V., Gottlieb, C. A., Thaddeus, P., Thorwirth, S., & McCarthy, M. C. 2010, *ApJ*, 720, 1717

Lauvergnat, D., Coudert, L. H., Klee, S., & Smirnov, M. 2009, *J. Mol. Spectrosc.*, 256, 204

Lavín, C., Velasco, A. M., & Martín, I. 2009, *ApJ*, 692, 1354

Lavín, C. & Velasco, A. M. 2011, *ApJ*, 739, 16

Lefebvre-Brion, H. & Lewis, B. R., 2007, Mol. Phys., 105, 1625

Lefebvre-Brion, H., Liebermann, H. P., & Vázquez, G. J. 2010, *J. Chem. Phys.*, 132, 024311

Levshakov, S. A., Kozlov, M. G., & Reimers, D. 2011, *ApJ*, 738, 26

Lewis, B. R., Gibson, S. T., Zhang, W., Lefebvre-Brion, H., & Robbe, J.-M. 2005a, *J. Chem. Phys.*, 122, 144302

Lewis, B. R., Gibson, S. T., Sprengers, J. P., Ubachs, W., Johansson, A., & Wahlström, C.-G. 2005b, *J. Chem. Phys.*, 123, 236101

Lewis, B. R., Baldwin, K. G. H., Sprengers, J. P., Ubachs, W., Stark, G., & Yoshino, K. 2008a, *J. Chem. Phys.*, 129, 164305

Lewis, B. R., Heays, A. N., Gibson, S. T., Lefebvre-Brion, H., & Lefebvre, R. 2008b, *J. Chem. Phys.*, 129, 164306

Li, G., Gordon, I. E., Bernath, P. F., & Rothman, L. S. 2011, *JQSRT*, 112, 1543

Li, W.-B., Zhu, L.-F., Yuan, Z.-S., Liu, X.-J., & Xu, K.-Z. 2006, *J. Chem. Phys.*, 125, 154310

Livingston, W. & Wallace, L. 2003, *An Atlas of the Solar Spectrum in the Infrared from* 1850 − 9000 cm^{-1} (1.1 to 5.4 μm), *Revised*, N.S.O. Technical Report # 03-001, National Solar Observatory, Tucson. See ftp://nsokp.nso.edu/pub/atlas/photatl/

Lovas, F. J. 2004, J. Phys. Chem. Ref. Data, 33, 177

Lovas, F. J. & Groner, P. 2006, *J. Mol. Spectrosc.*, 236, 173

Lovas, F. J., Plusquellic, D. F., Pate, B. H., Neill, J. L., Muckle, M. T., & Remijan, A. J. 2009, *J. Mol. Spectrosc.*, 257, 82

Maeda, A., Hirao, T., Bernath, P. F., & Amano, T. 2001, *J. Mol. Spectrosc.*, 210, 250

Maeda, A., *et al.* 2008, *ApJS*, 176, 543

Margulès, L., Lewen, F., Winnewisser, G., Botschwina, P., & Müller, H. S. P. 2003, Phys. Chem. *Chem. Phys.*, 5, 2770

Margulès, L., Coudert, L. H., Møllendal, H., Guillemin, J.-C., Huet, T. R., & Janečková, R. 2009a, *J. Mol. Spectrosc.*, 254, 55

Margulès, L., *et al.* 2009b, *A&A*, 493, 565

Margulès, L., *et al.* 2010, *ApJ*, 714, 1120

McCall, B. J., Casaes, R. N., Ádámkovics, M., & Saykally, R. J. 2003, *Chem. Phys. Lett.*, 374, 583

McCarthy, M. C. & Thaddeus, P. 2005, *J. Chem. Phys.*, 122, 174308

McCarthy, M. C., Gottlieb, C. A., Gupta, H., & Thaddeus, P. 2006, *ApJ*, 652, L141

McLamarrah, S. K., Sheridan, P. M., & Ziurys, L. M. 2005, *Chem. Phys. Lett.*, 414, 301

Medvedev, I. R., De Lucia, F. C., & Herbst, E. 2009, *ApJS*, 181, 433

Mladenović, M., Lewerenz, M., McCarthy, M. C., & Thaddeus, P. 2009, *J. Chem. Phys.*, 131, 174308

Mota, R., et al. 2005, *Chem. Phys. Lett.*, 416, 152

Mukhopadhyay, I., Perry, D. S., Duan, Y.-B., Pearson, J. C., Albert, S., Butler, R. A. H., Herbst, E., & De Lucia, F. C. 2002, *J. Chem. Phys.*, 116, 3710

Müller, H. S. P. & Christen, D. 2004, *J. Mol. Spectrosc.*, 228, 298

Müller, H. S. P., Menten, K. M., & Mäder, H. 2004, *A&A*, 428, 1019

Müller, H. S. P. & Brünken, S. 2005, *J. Mol. Spectrosc.*, 232, 213

Müller, H. S. P., Schlöder, F., Stutzki, J., & Winnewisser, G. 2005, *J. Mol. Struct.*, 742, 215

Müller, H. S. P., et al. 2007, *Phys. Chem. Chem. Phys.*, 9, 1579

Müller, H. S. P., Drouin, B. J., & Pearson, J. C. 2009, *A&A*, 506, 1487

Müller, H. S. P., Dong, F., Nesbitt, D. J., Furuya, T., & Saito, S. 2010, *Phys. Chem. Chem. Phys.*, 12, 8362

Müller, H. S. P. 2010, *A&A*, 514, L6

Müller, H. S. P., Coutens, A., Walters, A., Grabow, J.-U., & Schlemmer, S. 2011, *J. Mol. Spectrosc.*, 267, 100

Nakajima, M., Joester, J. A., Page, N. I., Reilly, N. J., Bacskay, G. B., Schmidt, T. W., & Kable, S. H. 2009, *J. Chem. Phys.*, 131, 044301

Nassar, R. & Bernath, P. F. 2003, *JQSRT*, 82, 279

Nemes, L. et al. 1994, *Chem. Phys. Lett.*, 218, 295

Ozeki, H., Bailleux, S., & Wlodarczak, G. 2011, *A&A*, 527, A64

Pearson, J. C. & Drouin, B. J. 2006, *ApJ*, 647, L83

Pearson, J. C., Brauer, C. S., & Drouin, B. J. 2008, *J. Mol. Spectrosc.*, 251, 394

Pearson, J. C., Brauer, C. S., Drouin, B. J., & Xu, L.-H. 2009, *Can. J. Phys.*, 87, 449

Plusquellic, D. F., Lovas, F. J., Pate, B. H., Neill, J. L., Muckle, M. T., & Remijan, A. J. 2009, *J. Phys. Chem. A*, 1131, 12911

Polehampton, E. T., Brown, J. M., Swinyard, B. M., & Baluteau, J.-P. 2003, *A&A*, 406, L47

Puzzarini, C., Dore, L., & Cazzoli, G. 2003, *J. Mol. Spectrosc.*, 217, 19

Puzzarini, C., Cazzoli, G., Harding, M. E., Vázquez, J., & Gauss, J. 2009, *J. Chem. Phys.*, 131, 234304

Ram, R. S., Davis, S. P., Wallace, L., Engleman, R., Appadoo, D. R. T., & Bernath, P. F. 2006, *J. Mol. Spectrosc.*, 237, 225

Ram, R. S. & Bernath, P. F. 2010, *J. Mol. Spectrosc.*, 260, 115

Ram, R. S., Wallace, L., & Bernath, P. F. 2010b, *J. Mol. Spectrosc.*, 263, 82

Ram, R. S., Wallace, L., Hinkle, K., & Bernath, P. F. 2010c, *ApJS*, 188, 500

Ram, R. S. & Bernath, P. F. 2011a, *ApJS*, 194, 34

Ram, R. S., Tereszchuk, K., Gordon, I. E., Walker, K. A., & Bernath, P. F. 2011b, *J. Mol. Spectrosc.*, 266, 86

Ricks, A. M., Douberly, G. E., & Duncan, M. A. 2009, *ApJ*, 702, 301

Rothman, L. S., et al. 2009, *JQSRT*, 110, 523. See http://www.cfa.harvard.edu/HITRAN/

Rothman, L. S., et al. 2010, *JQSRT*, 111, 2139

Rufus, J., Stark, G., Smith, P. L., Pickering, J. C., & Thorne, A. P. 2003, *JGRE*, 108, 5011

Rufus, J., Stark, G., Thorne, A. P., Pickering, J. C., Blackwell-Whitehead, R. J., Blackie, D., & Smith, P. L. 2009, *JGRE*, 114, 06003

Savin, D. W., et al. 2011, *Prog. Phys.*, in press

Schmidt, T. W. & Bacskay, G. B. 2007, *J. Chem. Phys.*, 127, 234310

Sharpe, S. W., Johnson, T. J., Sams, R. L., Chu, P. M., Rhoderick, G. C., & Johnson, P. A. 2004, *Appl. Spectrosc.*, 58, 1452. See http://nwir.pnl.gov/

Shayesteh, A. & Bernath, P. F. 2011, *J. Chem. Phys.*, 135, 094308

Shayesteh, A., Tereszchuk, K., Bernath, P. F., & Colin, R. 2003a, *J. Chem. Phys.*, 118, 1158

Shayesteh, A., Tereszchuk, K., Bernath, P. F., & Colin, R. 2003b, *J. Chem. Phys.*, 118, 3622

Shayesteh, A., Appadoo, D. R. T., Gordon, I., & Bernath, P. F. 2003c, *J. Chem. Phys.*, 119, 7785

Shayesteh, A., Appadoo, D. R. T., Gordon, I., LeRoy, R. J., & Bernath, P. F. 2004a, *J. Chem. Phys.*, 120, 10002

Shayesteh, A., Walker, K. A., Gordon, I., Appadoo, D. R. T., & Bernath, P. F. 2004b, *J. Mol. Struct.*, 695-696, 23

Shayesteh, A., Henderson, R. D. E., Le Roy, R. J., & Bernath, P. F. 2007, J. Phys. Chem. A, 111, 12495

Sheffer, Y., Federman, S. R., & Andersson, B.-G. 2003, *ApJ*, 597, L29

Sheffer, Y. & Federman, S. R. 2007, *ApJ*, 659, 1352

Sheridan, P. M. & Ziurys, L. M. 2003, *J. Chem. Phys.*, 118, 6370

Sheridan, P. M., McLamarrah, S. K., & Ziurys, L. M. 2003, *J. Chem. Phys.*, 119, 9496

Sheridan, P. M., Flory, M. A., & Ziurys, L. M. 2004, *J. Chem. Phys.*, 121, 8360

Shi, D., Liu, H., Zhang, X., Sun, J., Zhu, Z., & Liu, Y. 2010, *J. Mol. Struct.*, 956, 10

Sprengers, J. P., Ubachs, W., Baldwin, K. G. H., Lewis, B. R., & Tchang-Brillet, W-ÜL. 2003, *J. Chem. Phys.*, 119, 3160

Sprengers, J. P., Johansson, A., L'Huillier, A., Wahlström, C.-G., Lewis, B. R., & Ubachs, W. 2004, *Chem. Phys. Lett.*, 389, 348

Sprengers, J. P., Ubachs, W., & Baldwin, K. G. H. 2005a, *J. Chem. Phys.*, 122, 144301

Sprengers, J. P., Reinhold, E., Ubachs, W., Baldwin, K. G. H., & Lewis, B. R. 2005b, *J. Chem. Phys.*, 123, 144315

Sprengers, J. P. & Ubachs, W. 2009, *J. Mol. Spectrosc.*, 235, 176

Stark, G., Huber, K. P., Yoshino, K., Smith, P. L., & Ito, K. 2005, *J. Chem. Phys.*, 123, 214303

Stark, G., Yoshino, K., Smith, P. L., & Ito, K. 2007, *JQSRT*, 103, 67

Stark, G., Lewis, B. R., Heays, A. N., Yoshino, K., Smith, P. L., & Ito, K. 2008, *J. Chem. Phys.*, 128, 114302

Steinmann, C. M., Rohwer, E. G., & Stafast, H. 2003, *ApJ*, 590, L123; errat. *ApJ*, 591, L167

Sumiyoshi, Y., Ueno, T., & Endo, Y. 2003, *J. Chem. Phys.*, 119, 1426

Tanabashi, A., Hirao, T., Amano, T., & Bernath, P. F. 2005, *ApJ*, 624, 1116

Tanabashi, A., Hirao, T., Amano, T., & Bernath, P. F. 2007, *ApJS*, 169, 472

Tashkun, S. A. & Perevalov, V. I. 2011, *JQSRT*, 112, 1403

Tenenbaum, E. D., Flory, M. A., Pulliam, R. L., & Ziurys, L. M. 2007, *J. Mol. Spectrosc.*, 244, 153

Thaddeus, P., Gottlieb, C. A., Gupta, H., Brünken, S., McCarthy, M. C., Agúndez, M., Guélin, M., & Cernicharo, J. 2008, *ApJ*, 677, 1132

Thorn, P. A., *et al.* 2007, *J. Chem. Phys.*, 126, 064306

Tinti, F., Bizzocchi, L., Degli Esposti, C., & Dore, L. 2007, *ApJ*, 669, L113

Toffoli, D. & Lucchese, R. R. 2004, *J. Chem. Phys.*, 120, 6010

Vázquez, G. J., Amero, J. M., Liebermann, H. P., Buenker, R. J., & Lefebvre-Brion, H. 2007, *J. Chem. Phys.*, 126, 164302

Vázquez, G. J., Amero, J. M., Liebermann, H. P., & Lefebvre-Brion, H. 2009, J. Phys. Chem. A, 113, 13395

Vieitez, M. O., Ivanov, T. I., de Lange, C. A., Ubachs, W., Heays, A. N., Lewis, B. R., & Stark, G. 2008, *J. Chem. Phys.*, 128, 134313

Wallace, L., Hinkle, K., & Livingston, W. C. 2001, *Sunspot Umbral Spectra in the Region* 4000 to 8640 cm^{-1} (1.16 to 2.50 μm), N.S.O. Technical Report # 01-001, National Solar Observatory, Tucson. See ftp://nsokp.nso.edu/pub/atlas/spot5atl/

Wallace, L., Hinkle, K., & Livingston, W. C. 2002, *Sunspot Umbral Spectra in the Regions* 1925 to 2226 and 2392 to 3480 cm^{-1} (2.87 to 4.18 and 4.48 to 5.35 μm), N.S.O. Technical Report # 02-001, National Solar Observatory, Tucson. See ftp://nsokp.nso.edu/pub/atlas/spot6atl/

Walters, A., Müller, H. S. P., Lewen, F., & Schlemmer, S. 2009, *J. Mol. Spectrosc.*, 257, 24

Watson, J. K. G. 2001, *ApJ*, 555, 472

Weselak, T., Galazutdinov, G. A., Musaev, F. A., Beletsky, Y., & Krelowski, J. 2009, *A&A*, 495, 189

Wu, Y.-J., Lu, H.-C., Chen, H.-K., Cheng, B.-M., Lee, Y.-P., & Lee, L. C. 2007, *J. Chem. Phys.*, 127, 154311

Xu, L.-H., *et al.* 2008, *J. Mol. Spectrosc.*, 251, 305

Yang, X., Ben, J., Li, L., & Chen, Y. 2008, *JQSRT*, 109, 468

Yonezu, T., Matsushima, F., Moriwaki, Y., Takagi, K., & Amano, T. 2009, *J. Mol. Spectrosc.*, 256, 238

Yoshikawa, T., Sumiyoshi, Y., & Endo, Y. 2009a, *J. Chem. Phys.*, 130, 094302

Yoshikawa, T., Sumiyoshi, Y., & Endo, Y. 2009b, *J. Chem. Phys.*, 130, 164303

Yu, S., Drouin, B. J., Pearson, J. C., & Pickett, H. M. 2009, *ApJS*, 180, 119

Yu, S., *et al.* 2010, *J. Chem. Phys.*, 133, 174317

Yurchenko, S. N., Barber, R. J., & Tennyson, J. 2011, *MNRAS* 413, 1828

Zack, L. N., Pulliam, R. L., & Ziurys, L. M. 2009, *J. Mol. Spectrosc.*, 256, 186

Zelinger, Z., Amano, T., Ahrens, V., Brünken, S., Lewen, F., Müller, H. S. P., & Winnewisser, G. 2003, *J. Mol. Spectrosc.*, 220, 223

Zhang, G., Chen, K.-S., Merer, A. J., Hsu, Y.-C., Chen, W.-J., Shaji, S., & Liao, Y.-A. 2005, *J. Chem. Phys.*, 122, 244308

Zobov, N. F., *et al.* 2008, *MNRAS*, 387, 1093

Transactions IAU, Volume XXVIIIA
Reports on Astronomy 2009–2012   © International Astronomical Union 2012
Ian Corbett, ed.   doi:10.1017/S1743921312003109

DIVISION XII / COMMISSION 14 / WORKING GROUP COLLISION PROCESSES

CO-CHAIRS   Gillian Peach
                   Milan S. Dimitrijevic

TRIENNIAL REPORT 2009–2012

1. Introduction

Research in atomic and molecular collision processes and spectral line broadening has been very active since our last report, Peach, Dimitrijević & Stancil 2009. Given the large volume of the published literature and the limited space available, we have attempted to identify work most relevant to astrophysics. Since our report can not be comprehensive, additional publications can be found in the databases at the web addresses listed in the final section. Elastic and inelastic collisions among electrons, atoms, ions, and molecules are included and charge transfer can be very important in collisions between heavy particles.

Numerous meetings on collision processes and line broadening have been held throughout the report period. Important international meetings that provide additional sources of data through their proceedings are: the 19th *International Conference on Spectral Line Shapes (ICSLS)* (Gigosos & González 2008), the 7th *Serbian Conference on Spectral Line Shapes in Astrophysics* (SCSLSA) (Popović & Dimitrijević 2009), the XXVI *International Conference on Photonic, Electronic, and Atomic Collisions* (ICPEAC) (Orel *et al.* 2009), the 20th *ICSLS* (Lewis & Predoi-Cross 2010), the 22nd *International Conference on Atomic Physics* (ICAP) (Bachor, Drummond & Hannaford 2011), and the 7th *International Conference on Atomic and Molecular Data and their Applications* (Bernotas, Karazija & Rudzikas 2011). The 8th SCSLSA and the XXVIIth ICPEAC took place in June and July 2011 and their proceedings will be published in *Baltic Astronomy* and *Journal of Physics: Conference Series*, respectively.

2. Electron collisions with atoms and molecules

Collisions of electrons with atoms, molecules and atomic and molecular ions are the major excitation mechanism for a wide range of astrophysical environments. In addition, electron collisions play an important role in ionization and recombination, contribute to cooling and heating of the gas, and may contribute to molecular fragmentation and formation. In the following sections we summarize recent work on collisions for astrophysically relevant species, including elastic scattering, excitation, ionization, dissociation, recombination and electron attachment and detachment.

A review has been published of the atomic data necessary for the non-LTE analysis of stellar spectra (Mashonkina 2009). Other references are listed below for scattering by the atoms and molecules specified.

2.1. *Electron scattering by neutral atoms*

Elastic scattering: H (screened Coulomb interactions) (Zhang *et al.* 2010), Mg (Zatsarinny *et al.* 2009), Ar (Gargioni & Grosswendt 2008), I (Zatsarinny *et al.* 2011), Mn, Cu, Zn,

Ni, Ag, Cd (Felfli *et al.* 2011), Rb, Cs, Fr (Gangwar *et al.* 2010).

Excitation: H (screened Coulomb interactions) (Zhang *et al.* 2010, Zhang *et al.* 2011), $He(2^{1,3}S)$ (Wang *et al.* 2009, Wang *et al.* 2010), He, Ne (Kretinin *et al.* 2008), Mg (Zatsarinny *et al.* 2009), Ar (Gargioni & Grosswendt 2008).

Ionization: He (Bray *et al.* 2010, Ren *et al.* 2011), $He(2^{1}S)$ (Wang *et al.* 2010), Ar (Gargioni & Grosswendt 2008).

Total cross section: Na (Jiao *et al.* 2010).

2.2. *Electron scattering by atomic ions*

Elastic scattering: Mg^{+}, Ca^{+} (Mitroy & Zhang 2008).

Excitation: Hydrogen isoelectronic sequence $Cr^{23+}-Ni^{27+}$ (Malespin *et al.* 2011), Lithium isoelectronic sequence $Be^{+}-Kr^{33+}$ (Liang & Badnell 2011), Ne^{3+}, Ne^{6+} (Ludlow *et al.* 2011), Neon isoelectronic sequence $Na^{+}-Kr^{26+}$ (Liang & Badnell 2010), Sodium isoelectronic sequence $Mg^{+}-Kr^{25+}$ (Liang *et al.* 2009b), Mg^{4+} (Hudson *et al.* 2009), Mg^{8+} (Del Zanna *et al.* 2008), Si^{9+} (Liang *et al.* 2009a), $S^{8+}-S^{11+}$ (Liang *et al.* 2011), K^{+} (Tayal & Zatsarinny 2010), Fe^{10+} (Del Zanna *et al.* 2010), Fe^{12+} (Storey & Zeippen 2010), Fe^{18+} (Butler & Badnell 2008), Ni^{10+} (Aggarwal & Keenan 2008a), Ni^{18+} (Aggarwal & Keenan 2008c).

Recombination: H^{+} (Chluba *et al.* 2010), N^{+} (Fang *et al.* 2011), Aluminium isoelectronic sequence $Si^{+}-Zn^{17+}$ (Abdel-Naby *et al.* 2011), Argon isoelectronic sequence $K^{+}-Zn^{12+}$ (Nikolić *et al.* 2010), $Fe^{7}+$, Fe^{8+} (Schmidt *et al.* 2008), Selenium ions Se^{q+}, $q = 1 - 6$ (Sterling & Witthoeft 2011).

Energy levels, radiative and excitation rates: O^{3+} (Aggarwal & Keenan 2008b, Keenan *et al.* 2009), O^{6+} (Aggarwal & Keenan 2008d), Si^{+} (Bautista *et al.* 2009), Ar^{17+} (Aggarwal *et al.* 2008), Selenium ions Se^{q+}, $q = 1 - 6$ (Sterling & Witthoeft 2011).

X-ray line emission: Na^{9+} (Phillips *et al.* 2010).
Radiative and Auger decay: Aluminium ions Al^{q+}, $q = 0 - 11$ (Palmeri *et al.* 2011).

2.3. *Electron scattering by molecules*

Elastic scattering: H_2O (Liu & Zhou 2010), $(H_2O)_2$ (Bouchiha *et al.* 2008), Li_2 (Tarana & Tennyson 2008), CO (Allan 2010), NH (Rajvanshi & Baluja 2010), NO_2 (Munjal *et al.* 2009), S_2 (Rajvanshi & Baluja 2011), SO_2 (Machado *et al.* 2011), SOS (Kaur *et al.* 2010), S_3 (Kaur *et al.* 2011).
Electron exchange: O_2, NO, NO_2 (Holtkötter & Hanne 2009).

Excitation: H_2 (Kretinin *et al.* 2008), Li_2 (Tarana & Tennyson 2008), CO (Allan 2010), N_2 (Kato *et al.* 2010, Johnson *et al.* 2010, Mavadat *et al.* 2011), NH (Rajvanshi & Baluja 2010), NO_2 (Munjal *et al.* 2009), S_2 (Rajvanshi & Baluja 2011), SOS (Kaur *et al.* 2010), S_3 (Kaur *et al.* 2011), C_2H_4 (da Costa *et al.* 2008).

Ionization: N_2 (Gochitashvili *et al.* 2010), NH (Rajvanshi & Baluja 2010), SiCl, $SiCl_2$, $SiCl_3$, $SiCl_4$ (Kothari *et al.* 2011), SO, SO_2 (Vinodkumar *et al.* 2008), SOS (Kaur *et al.* 2010), S_2 (Rajvanshi & Baluja 2011), S_3 (Kaur *et al.* 2011).

Total cross sections: SO, SO_2, SO_2Cl_2, SO_2ClF, SO_2F_2 (Joshipura & Gangopadhyay 2008), SO_2 (Machado *et al.* 2011), C_2H_4O (Szmytkowski *et al.* 2008), CH_3OH, CH_3NH_2 (Vinodkumar *et al.* 2008).

Dissociative processes: H_2 (Celiberto *et al.* 2009, Bellm *et al.* 2010, Celiberto *et al.* 2011), HCl (Fedor *et al.* 2010), HCl, DCl, HBr, DBr (Fedor *et al.* 2008), C_2H_2 (Chourou & Orel 2008).

Attachment: C_2H, C_2N (Harrison & Tennyson 2011).

2.4. *Electron scattering by molecular ions*

Detachment: C_2^- (Halmová *et al.* 2008).

Excitation: C_2^- (Halmová *et al.* 2008), CO^+ (Stäuber & Bruderer 2009).

Rotational cooling: HD^+ (Shafir *et al.* 2009).

Dissociative and recombination processes: HD^+ (Fifirig & Stroe 2008, Takagi *et al.* 2009, Stroe & Fifirig 2011), H_3^+ (Glosík *et al.* 2009), HF^+ (Roos *et al.* 2008, Roos *et al.* 2009), CH_3^+, CD_3^+ (Bahati *et al.* 2009), $CD_3OCD_2^+$, $(CD_3)_2OD^+$ (Hamberg *et al.* 2010a), CD_3CDOD^+, $CH_3CH_2OH_2^+$ (Hamberg *et al.* 2010b).

3. Collisions between heavy particles

A review entitled 'Energetic ion, atom and molecule reactions and excitation in H_2 discharges' has recently been published (Phelps 2009). Other references are listed below for the atomic and molecular processes specified.

3.1. *Collisions between neutral atoms and atomic ions*

Inelastic scattering: H + H (Barklem *et al.* 2011), Na + H (Barklem *et al.* 2010).

Excitation: $H + H^+$ (Winter 2009).

Charge transfer processes: H, D, $T + He^{2+}$ (Stolterfoht *et al.* 2010), $H^- + H^+$, mutual neutralization (Stenrup *et al.* 2009), $H + H^+$ (Winter 2009), $H + Li^{2+}$ (Mančev 2009), $H + B^{5+}$, C^{4+} (Barragán *et al.* 2010), $H + C^{3+}$, O^{3+}, Si^{3+} (Guevara *et al.* 2011), $H^+ + Li^+$, Be^{2+}, B^{3+}, C^{4+} (Samanta & Purkait 2011), $He + H^+$ (Guzmán *et al.* 2009, Harris *et al.* 2010, Fischer *et al.* 2010), $He + H^+$, He^{2+} (Zapukhlyak & Kirchner 2009), $He + H^+$, He^+, He^{2+} (Schöffler *et al.* 2009), $He + {}^3He^{2+}$ (Alessi *et al.* 2011), $He + He^{2+}$, Li^+, Li^{2+}, Li^{3+}, C^{6+}, O^{8+} (Samanta *et al.* 2011), $He + C^{3+}$ (Wu *et al.* 2009b), $He + N^{3+}$ (Liu *et al.* 2011b), $He + O^{3+}$ (Kamber *et al.* 2008, Wu *et al.* 2009a), He, Ne, Ar, Kr, Xe $+ C^{3+}$ (Santos *et al.* 2010), $He^+ + He^+$ (Mančev 2009), Li + H, H^+ (Cabrera-Trujillo *et al.* 2008), $Li + H^+$ (Liu *et al.* 2011a), $Ne + C^{2+}$, C^{3+}, O^{2+}, O^{3+} (Ding *et al.* 2008), Na + H ion-pair production (Barklem *et al.* 2010), $Mg + H^+$, He^{2+} (Kumari *et al.* 2011), $Mg + Cs^+$ (Sabido *et al.* 2008), $Ar + H^+$ (Cabrera-Trujillo *et al.* 2009), $Ar + O^{3+}$ (Kamber *et al.* 2008).

Ionization: $H + H^+$ (Winter 2009), $He + H^+$ (Guzmán *et al.* 2009), $He + He^{2+}$ (Ogurtsov *et al.* 2011), He, Ne, Ar, Kr, $Xe + H^+$ (Miraglia & Gravielle 2008), He, Ne, Ar, Kr, Xe $+ C^{3+}$ (Santos *et al.* 2010), Li^+, Na^+, K^+, $Rb^+ + H^+$ (Miraglia & Gravielle 2008).

Detachment: C^-, O^-, F^-, Na^-, Si^-, S^-, Cl^-, Ge^- + He, Ne, Ar (Jalbert *et al.* 2008), F^-, Cl^-, Br^-, $I^- + H^+$ (Miraglia & Gravielle 2008).

Energy loss and stopping cross sections: Li + H, H^+ (Cabrera-Trujillo *et al.* 2008), H, D, $T + He^{2+}$ (Cabrera-Trujillo *et al.* 2011).

3.2. *Collisions between atoms and molecules*

Dissociative processes: $H_2O + H^+$ (Monce *et al.* 2009), $H_2O + He^+$ (Garcia *et al.* 2008).

Radiative association: $H_2 + H^-$ (Ayouz *et al.* 2011).

Excitation and/or fragmentation: $CN^- + H_2$ (Agúndez *et al.* 2010), $CO^+ + H$ (Andersson *et al.* 2008), $CO_2^+ + He$ (González-Magaña *et al.* 2008), $CO^+ + H$, H_2 (Stäuber & Bruderer 2009).

Charge transfer processes: D_2, O_2, H_2O, $CO_2 + O^{3+}$ (Kamber *et al.* 2008), $H_2 + He^{2+}$ (Khoma *et al.* 2009), $CH_3 + H^+$ (Nagao *et al.* 2008), $CO_2 + He^+$ (Lin & Mayer 2010), $HCl + C^{2+}$ (Rozsályi *et al.* 2011).

Ionization and/or capture: $H_2O + H^+$, He^{2+}, C^{6+} (Illescas *et al.* 2011), $C_2H_4 + H^+$ (Getahun *et al.* 2010), $N_2 + H^+$ (Gochitashvili *et al.* 2010).

4. Stark broadening

Knowledge of Stark broadening parameters (line widths and shifts) for a large number of atomic transitions is very important for the analysis, interpretation and modelling of stellar spectra, circumstellar conditions and H II regions. For hot dense stars such as white dwarfs this is often the most important broadening mechanism.

4.1. *Developments in line broadening theory*

Rosato *et al.* (2009) have reexamined the Stark broadening of hydrogen lines in the presence of a magnetic field and developed an impact theory for ions, valid for low electron densities ($N_e \leqslant 10^{14}$ cm^{-3}), which takes into account the Zeeman splitting of the atomic energy levels. Rosato *et al.* (2010) have also studied numerically the role of time ordering in such plasmas, by using a simulation code that accounts for the evolution of the microscopic electric field generated by the charged particles moving close to the atom. Calisti *et al.* (2010) have developed a very fast method to account for the dynamical effects of charged particles on the spectral line shape emitted by plasmas, based on a formulation of the frequency fluctuation model.

Ab initio calculations of Stark broadening parameters, i.e. calculations where the required atomic energy levels and oscillator strengths are determined during the calculation and are not taken from other sources, have been considered and reviewed by Ben Nessib (2009). A book has recently been published (Gordon & Sorochenko 2009) that gives a detailed account of the surprising discovery in the 1960's of the radio recombination lines and their subsequent analysis. Even now some features have still not been satisfactorily explained.

4.2. *Isolated lines*

For isolated lines Stark broadening is dominated by collisions with plasma electrons. Broadening parameters have been determined theoretically for:

One line from the 3s-3p transition array for each of the spectra Si XI, Ti XI, Cr XIII, Cr XIV, Fe XV, Fe XVI, Ni XVIII and Fe XXIII and two lines from the array for K VIII, Ca IX, Sc X and Ti XI (Elabidi & Sahal-Bréchot 2011); two lines for 3s-3p transitions for ions C IV, N V, O VI, F VII, Na IX, Mg X, Al XI, Si XII and P XIII and one line for Ne VIII (Elabidi *et al.* 2009); the 2s-2p resonance doublets of C IV, N V, O VI, F VII and Ne VIII ions (Elabidi *et al.* 2011). These calculations all use a quantum mechanical approach.

For five lines of Cu I (Zmerli *et al.* 2010) and the lines Ne I 837.8 nm (Christova *et al.* 2010b) and Ar I 737.2 nm (Christova *et al.* 2010a), new Stark broadening parameters are obtained using a semiclassical perturbation approach. A semi-empirical approach, which uses a set of wave functions obtained from Hartree-Fock relativistic calculations and includes core polarization effects, has been applied to 58 lines of Pb IV (Alonso-Medina *et al.* 2010) and 171 lines of Sn III (Alonso-Medina & Colón 2011).

Broadening parameters have been obtained experimentally for the following numbers of lines:

34 Pb I (Alonso-Medina 2008), 25 Pb III (Alonso-Medina 2011), 34 Pb IV and 4 Pb V (Bukvić *et al.* 2011), 28 Cd III (Djeniže *et al.* 2009, Bukvić *et al.* 2009b), 13 Si I, 15 Si II, 28 Si III and 9 Si IV (Bukvić *et al.* 2009a), 29 (Bukvić *et al.* 2008) and 19 (Djurović *et al.* 2011) Ar III, 30 Kr III (Ćirišan *et al.* 2011), 12 Ne II, 8 Kr II and 5 Xe II (Peláez *et al.* 2010b), 38 Xe II (Peláez *et al.* 2009a, Peláez *et al.* 2009b), 10 Xe III (Peláez *et al.* 2009b), 5 Au I and 26 Au II (Djeniže 2009), 9 Sb III (Djeniže 2008), 15 Mn I and 10 Fe I (Zielinska *et al.* 2010), 21 Fe II (Aragón *et al.* 2011) and C I 833.5 nm (Bartecka *et al.* 2011).

The regularities and systematic trends of Stark broadening parameters and reasons for deviations have been investigated within the multiplets (Peláez *et al.* 2010b, Peláez *et al.* 2009a), along the homologous sequence of singly-ionized noble gases (Peláez *et al.* 2010a), within the spectral series (Christova *et al.* 2010a) and along isoelectronic sequences (Elabidi *et al.* 2009, Elabidi *et al.* 2011). Also the dependence of electron- and proton-impact Stark widths on the upper-level ionization potential within different series of spectral lines of neutral magnesium (Tapalaga *et al.* 2011) and as a function of charge on the atomic core (Elabidi & Sahal-Bréchot 2011) have been evaluated and discussed. This kind of trend and regularity analysis can be useful for the prediction of Stark broadening parameters and therefore for the spectroscopic diagnostic of astrophysical plasmas.

4.3. *Transitions in hydrogenic and helium-like systems*

Stark-broadened line profiles of the hydrogen Brackett series have been computed within the Model Microfield Method for the conditions of stellar atmospheres and circumstellar envelopes (Stéhlé & Fouquet 2010), and Tremblay & Bergeron (2009) have performed improved calculations for the Stark broadening of hydrogen lines in dense plasmas typical of white-dwarf atmospheres. The central asymmetry of the Hβ line has been measured and analysed (Djurović *et al.* 2009) and new experimental results for Hα and Hγ have been published (Mijatović *et al.* 2010a, Mijatović *et al.* 2010b).

Omar (2010, 2011) published new calculations for the Stark broadening of the He I lines at 504.8 nm, 388.9 nm, 318.8 nm, 667.8 nm and 501.6 nm formed in a dense plasma. Tables of Stark broadening for the He I 447.1 nm line have been generated using computer simulations (Gigosos & González 2009). This line and its forbidden component have also been studied theoretically (González *et al.* 2011) and experimentally (Ivković *et al.* 2010, González *et al.* 2011). Gao *et al.* (2008) have carried out experiments for the He I 388.9 nm and 706.5 nm lines.

5. Broadening by neutral atoms and molecules

The analysis of experimental molecular spectra in order to extract line shape parameters is often very difficult. Line shapes can be affected by collisional narrowing and the dependence of collisional broadening and shifting on molecular speed. When these effects are sufficiently important, fitting Voigt profiles to experimental spectra produces systematic errors in the parameters retrieved. Here the experimental and theoretical results selected have been confined to the basic atomic and molecular data required for a description of the pressure broadening and shift of lines and molecular bands.

Since the last report an important book has been published (Hartman *et al.* 2008) that gives a comprehensive review of experimental and theoretical work on collisional effects in molecular spectra. In the following sections the items are labelled by 'E' and 'T' to indicate experimental work and theoretical analysis, respectively.

5.1. *Broadening and shift of atomic lines*

New research has been published in the period 2008-2011 and the transitions studied together with the perturbing atoms or molecules are listed below. The work is theoretical except where indicated by 'E'.

H: line wings of Lyα broadened by H and He (Allard *et al.* 2009a, Allard & Christova 2009); line wings of Lyγ by H$^+$ (Allard *et al.* 2009b) and Hα by H (Allard *et al.* 2008).

He: self broadening of line $3s^3S-2p^3P$ (Allard *et al.* 2009c, Allard *et al.* 2011).

Li: self broadening of resonance line (Reggami *et al.* 2009); resonance line broadened by He (Peach & Whittingham 2009, Peach 2010a, Peach 2010b); 2s-3d transition broadened by Ne and Ar (Rosenberry & Stewart 2011).

Na: resonance line broadened by H (Peach 2010b); lines 3s-3p and 3p-3d broadened by He (Peach & Whittingham 2009, Peach 2010a, Peach 2010b).

K: self broadening of resonance line (Reggami *et al.* 2009) and line wings (Talbi *et al.* 2008).

K, Rb and Cs: self broadening of principal series (E) (Vadla *et al.* 2009).

Rb: 5s-5p D2 line, broadening by He, CH_4, C_2H_6, C_3H_8, n-C_4H_{10} (E) (Zameroski *et al.* 2011).

Cs: 6s-6p D2 line, broadening by ^{3}He, H_2, HD, D_2, N_2, CH_4, C_2H_6, CF_4 (E) (Pitz *et al.* 2010).

5.2. *Broadening and shift of molecular lines*

Much new data have been published since the last report was prepared. The molecules are listed below with their perturbing atomic or molecular species and are labelled by 'E' and 'T' to indicate experimental work and theoretical analysis, respectively.

H_2-Ar: collision-induced absorption (T) (Tran *et al.* 2011b).

D_2-Kr: collision-induced absorption (E) (Abu-Kharma *et al.* 2010).

HI: lines broadened by N_2 (E) (Domanskaya *et al.* 2011).

HBr: self broadening (E) (Domanskaya *et al.* 2009).

HI and HBr: lines broadened by rare gases (E) (Domanskaya *et al.* 2009).

HDO: lines broadened by CO_2 (T) (Gamache *et al.* 2011).

HCl: lines broadened and shifted by N_2, He, Ar and Xe (E) (Hurtmans *et al.* 2009).

HCN: lines broadened by N_2, O_2 and air (E) (Yang *et al.* 2008).

H_2CO: lines broadened by H_2CO and N_2 (E+T) (Jacquemart *et al.* 2010).

HNO_3: lines broadened by N_2 (T) (Laraia *et al.* 2009).

HO_2: lines broadened by N_2 (E) (Miyano & Tonokura 2011).

H_2O_2: lines broadened by N_2, O_2 and air (E) (Sato *et al.* 2010).

H_2O: lines broadened by H_2 (E) (Krupnov 2010), (T) (Wiesenfeld & Faure 2010); by H_2 and He (E) (Dick *et al.* 2010); by N_2 (E) (Lavrentieva *et al.* 2010; by N_2 and O_2 (T) (Gamache & Laraia 2009); by O_2 (E) (Petrova *et al.* 2011; by H_2O (E) (Lisak *et al.* 2009, Ptashnik & Smith 2010); by H_2O, N_2, O_2 (E+T) (Cazzoli *et al.* 2008, Cazzoli *et al.* 2009, Koshelev 2011); by air (T) (Voronin *et al.* 2010); by CO_2 (T) (Sagawa *et al.* 2009); by H_2, He, N_2, O_2 and CO_2 (E) (Dick *et al.* 2009b); by rare gases (E+T) (Fiadzomor *et al.* 2008).

CH_4: lines broadened by N_2 (T) (Gabard & Boudon 2010); by N_2 and O_2 (E) (Lyulin *et al.* 2009); by CH_4 (E) (Smith *et al.* 2010, Lyulin *et al.* 2011); by CH_4 and N_2 (E) (McRaven *et al.* 2011); by O_2 and air (E) (Martin & Lepère 2009); by air (E) (Smith *et al.* 2009, Smith *et al.* 2011).

C_2H_2: broadened by H_2 (T) (Thibault *et al.* 2011a); by H_2 and D_2 (E+T) (Thibault *et al.* 2009); by N_2 (E) (Dhyne *et al.* 2009, Fissiaux *et al.* 2009, Dhyne *et al.* 2010); by C_2H_2 (E) (Li *et al.* 2010, Povey *et al.* 2011, Dhyne *et al.* 2011); by He and Ar (T) (Ivanov & Buzykin 2010); by Ne and Kr (E) (Nguyen *et al.* 2009a).

C_2H_4: lines broadened by C_2H_4 (E) (Flaud *et al.* 2011); by Ar (E+T) (Nguyen *et al.* 2009b).

C_2H_6: lines broadened by N_2 (E) (Blanquet *et al.* 2009); by C_2H_6 and N_2 (E) (Devi *et al.* 2010b, Devi *et al.* 2010c); by O_2 and air (E) (Fissiaux *et al.* 2010).

CH_3Br: lines broadened by N_2 (T) (Boussetta *et al.* 2011); by CH_3Br, N_2 and O_2 (E) (Hoffman & Davies 2009); by CH_3Br (T) (Goméz *et al.* 2010).

CH_3F: lines broadened by CH_3F and He (E) (Koubek *et al.* 2011).

CO: lines broadened by CO, N_2 and O_2 (E) (Koshelev & Markov 2009); by H_2, N_2, O_2, CO, CO_2 and He (E) (Dick *et al.* 2009a).

CO_2: by O_2 (E) (Devi *et al.* 2010a); by CO_2 (E+T) (Predoi-Cross *et al.* 2010, Tran *et al.* 2011a); by CO_2, N_2 and O_2 (E) (Li *et al.* 2008); by air (T) (Hartmann 2009), (E) (Gulidova *et al.* 2010), (E+T) (Lamouroux *et al.* 2010); by He (E) (Deng *et al.* 2009); by air and Ar (E) (Farooq *et al.* 2010).

Cs_2: lines broadened by Cs_2 (E) (Misago *et al.* 2009).

N_2: lines broadened by H_2 (T) (Goméz *et al.* 2011); by N_2 (E+T) (Thibault *et al.* 2011b).

NH_3: lines broadened by H_2 and He (E) (Hanley *et al.* 2009); by He (T) (Dhib 2010; by NH_3 (E) (Aroui *et al.* 2009, Guinet *et al.* 2011); by NH_3 and O_2 (E+T) (Nouri *et al.* 2009).

O_2: lines broadened by O_2 (E) (Lisak *et al.* 2010, Wójtewicz *et al.* 2011); by O_2 and OO isotopologues (E) (Long *et al.* 2011); by O_2 and air (E) (Long *et al.* 2010).

O_2-CO_2: collision-induced absorption (E) (Vangvichith *et al.* 2009).

O_3: lines broadened by air (Drouin & Gamache 2008); by N_2 and air (E+T) (Tran *et al.* 2011c).

OH: lines broadened by N_2, H_2O and Ar (E) (Hwang *et al.* 2008).

OCS lines broadened by N_2, O_2 and OCS (E) (Koshelev & Tretyakov 2009); by N_2 and O_2 (E) (Galalou *et al.* 2011).

I_2: lines broadened by Ar (E) (Phillips & Perram 2008).

6. Databases

Some useful databases are:

Vienna Atomic Line Database (VALD) of atomic data for analysis of radiation from astrophysical objects, containing central wavelengths, energy levels, statistical weights, transition probabilities and line broadening parameters for all chemical elements of astronomical importance. It can be found at http://vald.astro.univie.ac.at/ (Kupka *et al.* 1999).

The database of Robert L. Kurucz comprises atomic line parameters, including line broadening. An update to this database is discussed by Kurucz 2011. (http://kurucz.harvard.edu)

CHIANTI database (Dere *et al.* 2009) contains a critically evaluated set of up-to-date atomic data for the analysis of optically thin collisionally ionized astrophysical plasmas. It lists experimental and calculated wavelengths, radiative data and rates for electron and proton collisions, see websites http://sohowww.nascom.nasa.gov/solarsoft and http://www.damtp.cam.ac.uk/user/astro/chianti/.

CDMS – Cologne Database for Molecular Spectroscopy, see website http://www.ph1.uni-koeln.de/vorhersagen/, provides recommendations for spectroscopic transition frequencies and intensities for atoms and molecules of astronomical interest in the frequency range 0-10 THz, i.e. 0-340 cm^{-1} (Müller *et al.* 2005).

BASECOL database (http://basecol.obspm.fr) contains excitation rate coefficients for ro-vibrational excitation of molecules by electrons, He and H_2 and it is mainly used for the study of interstellar, circumstellar and cometary atmospheres.

TIPTOPbase (http://cdsweb.u-strasbg.fr/topbase/home.html) contains:
(i) TOPbase, that lists atomic data computed in the Opacity Project; namely LS-coupling energy levels, gf-values and photoionization cross sections for light elements (Z $\leqslant$ 26) of astrophysical interest and
(ii) TIPbase that lists intermediate-coupling energy levels, transition probabilities and electron impact excitation cross sections and rates for astrophysical applications (Z $\leqslant$ 28), computed by the IRON Project.

HITRAN – (HIgh-resolution TRANsmission molecular absorption database) is at http://www.cfa. harvard.edu/hitran/ (Rothman *et al.* 2009). It lists individual line parameters for molecules in the gas phase (microwave through to the UV), photoabsorption cross-sections for many molecules, and refractive indices of several atmospheric aerosols. A high temperature extension to HITRAN is HITEMP (To access the HITEMP data: ftp to cfa-ftp.harvard.edu; user = anonymous; password = e-mail address). It contains data for water, CO_2, CO, NO and OH (Rothman *et al.* 2010).

GEISA – (Gestion et Etude des Informations Spectroscopiques Atmosphériques) is a computer-accessible spectroscopic database, designed to facilitate accurate forward radiative transfer calculations using a line-by-line and layer-by-layer approach. It can be found at http://ether.ipsl.jussieu.fr/etherTypo/?id=950 (Jacquinet-Husson *et al.* 2008).

NIST – The National Institute of Standards and Technology hosts a number of useful databases for Atomic and Molecular Physics. A list can be found at http://www.nist.gov/srd/atomic.cfm. Among them are: An atomic spectra database and three bibliographic databases providing references on atomic energy levels and spectra, transition probabilities and spectral line shapes and line broadening.

STARK-B database (http://stark-b.obspm.fr) contains theoretical widths and shifts of isolated lines of atoms and ions due to collisions with charged perturbers, obtained using the impact approximation (Sahal-Bréchot 2010).

The European FP7 project will finish at the end of 2012. The virtual Atomic and Molecular Data Centre (VAMDC - http://www.vamdc.eu/) is being created with the aim of building an accessible and interoperable e-infrastructure for atomic and molecular data that will upgrade and integrate European (and other) A&M database services (Dubernet *et al.* 2011, Rixon *et al.* 2011).

References

Abdel-Naby, Sh. A., Nikolić, D., Gorczyca, T. W., *et al.* 2011, *A&A*, in press, DOI: http://dx.doi.org/10.1051/0004-6361/201117544
Abu-Kharma, M., Omari, H. Y., Shawaqfeh, N., & Stamp, C. 2010, *JMoSp*, 259, 111
Aggarwal, K. M., Hamada, K., Igarashi, A., *et al.* 2008, *A&A*, 487, 383
Aggarwal, K. M. & Keenan, F. P. 2008a, *EuPhJD*, 46, 205
Aggarwal, K. M. & Keenan, F. P. 2008b, *A&A*, 486, 1053
Aggarwal, K. M. & Keenan, F. P. 2008c, *A&A*, 488, 365
Aggarwal, K. M. & Keenan, F. P. 2008d, *A&A*, 489, 1377
Agúndez, M., Cernicharo, J., Guélin, M., *et al.* 2010, *A&A*, 517, L2
Alessi M., Otranto, S. & Focke, P. 2011, *PhRvA*, 83, 014701
Allan, M. 2010, *PhRvA*, 81, 042706

Allard, N. F., Kielkopf, J. F., Cayrel, R., & Van't Veer-memmerat, C. 2008, A&A, 480, 581

Allard, N. F. & Christova, M. 2009, New Astron. Rev., 53, 252

Allard, N. F. Kielkopf, J. F. 2009a, A&A, 493, 1155

Allard, N. F., Noselidze, I., & Kruk, J. W. 2009b, A&A, 506, 993

Allard, N. F., Deguilhem, B., Gadea, F. X., et al. 2009c, EPL, 88, 53002

Allard, N. F., Bonifaci, N., & Denat, A. 2011, EuPhJD, 61, 365

Alonso-Medina, A. 2008, Spectrochim. Acta B, 63, 598

Alonso-Medina, A., Colón, C., Montero, J. L., & Nation, L. 2010, MNRAS, 401, 1080

Alonso-Medina, A. 2011, Spectrochim. Acta B, 66, 439

Alonso-Medina, A. & Colón, C. 2011, MNRAS, 414, 713

Andersson, S., Barinovs, G., & Nyman, G. 2008, ApJ, 678, 1042

Aragón, C., Vega, P., & Aguilera, J. A. 2011, JPhB: At. Mol. Opt. Phys., 44, 055002

Aroui, H., Laribi, H., Orphai, J., & Chelin, P. 2009, JQSRT, 110, 2037

Ayouz, M., Lopes, R., Raoult, M., et al. 2011, PhRvA, 83, 052712

Bachor, H., Drummond, P., & Hannaford, P., eds. 2011, Proc. 22nd International Conference on Atomic Physics, JP-CS, 264

Bahati, E. M., Fogle, M., Vane, C. R., et al. 2009, PhRvA, 79, 052703

Barklem, P. S., Belyaev, A. K., Dickinson, A. S., & Gadéa, F. X. 2010, A&A, 519, A20

Barklem, P. S., Belyaev, A. K., Guitou, M., et al. 2011, A&A, 530, A94

Barragán, P., Errea, L. F., Guzmán, F., et al. 2010, PhRvA, 81, 062712

Bartecka, A., Baclawski, A., & Musielok, J. 2011, Cent. Eur. J. Phys., 9, 131

Bautista, M. A., Quinet, P., Palmeri, P., et al. 2009, A&A, 508, 1527

Bellm, S., Lower, J., Weigold, E., & Mueller, D. W. 2010, PRL, 104, 023202

Ben Nessib, N. 2009, New Astron. Rev., 53, 255

Bernotas, A., Karazija, R., & Rudzikas, Z., eds. 2011, Proc. 7th International Conference on Atomic and Molecular Data and their Applications, AIP-CP, 1344

Blanquet, G., Auwera, J. V., & Lepère, M. 2009, JMoSp, 255, 72

Bouchiha, D., Caron, L. G., Gorfinkiel, J. D., & Sanche, L. 2008, JPhB: At. Mol. Opt. Phys., 41, 045204

Boussetta, Z., Aroui, H., Jacquemart, D., et al. 2011, JQSRT, 112, 769

Bray, I., Fursa, D. V., Kadyrov, A. S., & Stelbovics, A. T. 2010, PhRvA, 81, 062704

Bukvić, S., Žigman, V., Srećković, A. & Djeniže, S. 2008, JQSRT, 109, 2869

Bukvić, S., Djeniže, S., & Srećković, A. 2009a, A&A, 508, 491

Bukvić, S., Djeniže, S., Srećković, A., & Nikolić, Z. 2009b, Phys. Lett. A, 373, 2750

Bukvić, S., Djeniže, S., Nikolić, Z., & Srećković, A. 2011, A&A, 529, A83

Butler, K. & Badnell, N. R. 2008, A&A, 489, 1369

Cabrera-Trujillo, R., Sabin, J. R., Deumens, E., & Öhrn, Y. 2008, PhRvA, 78, 012707

Cabrera-Trujillo, R., Amaya-Tapia, A., & Antillón, A. 2009, PhRvA, 79, 012712

Cabrera-Trujillo, R., Sabin, J. R., Öhrn, Y., et al. 2011, PhRvA, 83, 012715

Calisti, A., Mossé, C., Ferri, S., et al. 2010, PhRvE, 81, 016406

Cazzoli, G., Puzzarini, C., Buffa, G., & Tarrini, O. 2008, JQSRT, 109, 2820

Cazzoli, G., Puzzarini, C., Buffa, G., & Tarrini, O. 2009, JQSRT, 110, 609

Celiberto, R., Janev, R. K., Wadehra, J. M., & Laricchiuta, A. 2009, PhRvA, 80, 012712

Celiberto, R., Janev, R. K., Wadehra, J. M., & Laricchiuta, A. 2011, PhRvA, 84, 012707

Chluba, J., Vasil, G. M., & Dursi, L. J. 2010, MNRAS, 407, 599

Chourou, S. T. & Orel, A. E. 2008, PhRvA, 77, 042709

Christova, M., Dimitrijević, M. S., & Kovačević, A. 2010a, JP-CS, 207, 2024

Christova, M., Dimitrijević, M. S., Simić, S., & Sahal-Bréchot, S. 2010b, JP-CS, 207, 2025

Ćirišan, M., Peláez, R. J., Djurović, S., et al. 2011, PhRvA, 83, 012513

da Costa, R. F., Bettega, M. H. F., & Lima, M. A. P. 2008, PhRvA, 77, 042723

Del Zanna, G., Rozum, I. & Badnell, N. R. 2008 A&A, 487, 1203

Del Zanna, G., Storey, P. J. & Mason, H. E. 2010 A&A, 514, A40

Deng, W., Mondelain, D., Thibault, F., et al. 2009, JMoSp, 256, 102

Dere, K. P., Landi, E., Young, P. R., et al. 2009, A&A, 498, 915

Devi, V. M., Benner, D. C., Miller, C. E., & Predoi-Cross, A. 2010a, *JQSRT*, 111, 2355
Devi, V. M., Benner, D. C., Rinsland, C. P., *et al.* 2010b, *JQSRT*, 111, 2481
Devi, V. M., Rinsland, C. P., Benner, D. C., *et al.* 2010c, *JQSRT*, 111, 1234
Dhib, M. 2010, *JMoSp*, 259, 80
Dhyne, M., Fissiaux, L., Populaire, J.-C., & Lepère, M. 2009, *JQSRT*, 110, 358
Dhyne, M., Joubert, P., Populaire, J.-C., & Lepère, M. 2010, *JQSRT*, 111, 973
Dhyne, M., Joubert, P., Populaire, J.-C., & Lepère, M. 2011, *JQSRT*, 112, 969
Dick, M. J., Drouin, B. J., Crawford, T. J., & Pearson, J. C. 2009a, *JQSRT*, 110, 628
Dick, M. J., Drouin, B. J., & Pearson, J. C. 2009b, *JQSRT*, 110, 619
Dick, M. J., Drouin, B. J., & Pearson, J. C. 2010, *PhRvA*, 81, 022706
Ding, B. W., Chen, X. M., Yu, D. Y., *et al.* 2008, *PhRvA*, 78, 062718
Djeniže, S. 2008, *Phys. Lett. A*, 372, 6658
Djeniže, S. 2009, *Spectrochim. Acta B*, 64, 242
Djeniže, S., Srećković, A., & Bukvić, S. 2009, *EuPhJD*, 62, 185
Djurović, S., Ćirišan, M., Demura, A. V., *et al.* 2009, *PhRvE*, 79, 046402
Djurović, S., Mar, S., Peláez, R. J., & Aparicio, J. A. 2011, *MNRAS*, 414, 1389
Domanskaya, A. V., Bulanin, M. O., Kerl, K., & Maul, C. 2009, *JMoSp*, 253, 20
Domanskaya, A. V., Asfin, R. E., Maul, C., *et al.* 2011, *JMoSp*, 265, 69
Drouin, B. J. & Gamache, R. R. 2008, *JMoSp*, 251, 194
Dubernet, M. L., Boudon, V., Culhane, J. L., *et al.* 2011, *JQSRT*, 111, 2151
Elabidi, H. & Sahal-Bréchot, S. 2011, *EuPhJD*, 61, 285
Elabidi, H., Sahal-Bréchot, S., & Ben Nessib, N. 2009, *EuPhJD*, 54, 51
Elabidi, H., Sahal-Bréchot, S., Dimitrijević, M. S., & Ben Nessib, N. 2011, *MNRAS*, 417, 2624
Fang, X., Storey, P. J. & Liu, X.-W. 2011 *A&A*, 530, A18
Farooq, A., Jeffries, J. B., & Hanson, R. K. 2010, *JQSRT*, 111, 949
Fedor, J., May, O., & Allan, M. 2008, *PhRvA*, 78, 032701
Fedor, J., Winstead, C., McKoy, V., *et al.* 2010, *PhRvA*, 81, 042702
Felfi, Z., Msezane, A. Z., & Sokolovski, D. 2011, *PhRvA*, 83, 052705
Fiadzomor, P. A. Y., Keen, A. M., Grant, R. B., & Orr-Ewing, A. J. 2008, *Chem. Phys. Lett.*, 462, 188
Fifirig, M. & Stroe, M. 2008, *Phys. Scr.*, 78, 065302
Fischer, D., Gudmundsson, M., Berényi, Z., *et al.* 2010, *PhRvA*, 81, 012714
Fissiaux, L., Dhyne, M., & Lepère, M. 2009, *JMoSp*, 254, 10
Fissiaux, L., Blanquet, G., & Lepère, M. 2010, *JQSRT*, 111, 2037
Flaud, J.-M., Lafferty, W. J., Devi, V. M., *et al.* 2011, *JMoSp*, 267, 3
Gabard, T. & Boudon, V. 2010, *JQSRT*, 111, 1328
Galalou, S., Ben Mabrouk, K., Aroui, H., *et al.* 2011, *JQSRT*, 112, 2750
Gamache, R. R. & Laraia, A. L. 2009 *JMoSp*, 257, 116
Gamache, R. R., Laraia, A. L. & Lamouroux, J. 2011 *Icarus*, 213, 720
Gangwar, R. K., Tripathi, A. N., Sharma, L., & Srivastava, R. 2010, *JPhB: At. Mol. Opt. Phys.*, 43, 085205
Gao, H. M., Ma, S. I., Xu, C. M., & Vu, L. 2008, *EuPhJD*, 47, 191
Garcia, P. M. Y., Sigaud, G. M., Luna, H., *et al.* 2008, *PhRvA*, 77, 052708
Gargioni, E. & Grosswendt, B. 2008, *Rev. Mod. Phys.*, 80, 451
Getahun, H., Errea, L. F., Illescas, C., *et al.* 2010, *EuPhJD*, 60, 45
Gigosos, M. A. & González, M. Á. (eds.) 2008, Proc. 19th *International Conference on Spectral Line Shapes*, AIP-CP, 1058
Gigosos, M. A. & González, M. Á. 2009, *A&A*, 503, 293
Glosík, I., Plašil, R., Korolov, I., *et al.* 2009, *PhRvA*, 79, 052707
Gochitashvili, M. R., Kezerashvili, R. Ya., & Lomsadze, R. A. 2010, *PhRvA*, 82, 022702
Goméz, L., Jacquemart, D., Bouanich, J.-P., *et al.* 2010, *JQSRT*, 111, 1252
Goméz, L., Ivanov, S. V., Buzykin, O. G., & Thibault, F. 2011, *JQSRT*, 112, 1942
González-Magaña, O., Cabrera-Trujillo, R. & Hinojosa, G. 2008, *PhRvA*, 78, 052712
González, M. Á., Ivković, M., Gigosos, M. A., *et al.* 2011, *A&A*, 503, 293

Gordon, M. A. & Sorochenko, R. L. 2009, *Radio Recombination Lines* (New York:Springer Science+Business Media)

Guevara, N. L., Teixeira, E., Hall, B., *et al.* 2011, *PhRvA*, 83, 052709

Guinet, M., Jeseck, P., Mondelain, D., *et al.* 2011, *JQSRT*, 112, 1950

Gulidova, O. S., Asfin, R. E., Grigoriev, I. M., & Filippov, N. N. 2010, *JQSRT*, 111, 2315

Guzmán, F., Errea, L. F., & Pons, B. 2009, *PhRvA*, 80, 042708

Halmová, G., Gorfinkiel, J. D., & Tennyson, J. 2008, *JPhB: At. Mol. Opt. Phys.*, 41, 155201

Hamberg, M., Österdahl, F., Thomas, R. D., *et al.* 2010a, *A&A*, 514, A83

Hamberg, M., Zhaunerchyk, V., Vigren, E., *et al.* 2010b, *A&A*, 522, A90

Hanley, T. R., Steffes, P. G., & Karpowicz, B. M. 2009, *Icarus*, 202, 316

Harris, A. L., Peacher, J. L., Schulz, M., & Madison, D. H. 2010, *JP-CS*, 212, 012031

Harrison, S. & Tennyson, J. 2011, *JPhB: At. Mol. Opt. Phys.*, 44, 045206

Hartmann, J.-M. 2009, *JQSRT*, 110, 2019

Hartmann, J.-M., Boulet, C., & Robert, D. 2008, *Collisional Effects on Molecular Spectra* (Amsterdam:Elsevier)

Hoffman, K. J. & Davies, P. B. 2009, *JMoSp*, 254, 69

Holtkötter, I. & Hanne, G. F. 2009, *PhRvA*, 80, 022709

Hudson, C. E., Ramsbottom, C. A., Norrington, P. H. & Scott, M. P. 2009, *A&A*, 494, 729

Hurtmans, D., Henry, A., Valentin, A., & Boulet, C. 2009, *JMoSp*, 254, 126

Hwang, S. M., Kojima, J. N., Nguyen, Q.-V., & Rabinowitz, M. J. 2008, *JQSRT*, 109, 2715

Illescas, C., Errea, L. F., & Méndez, L., *et al.* 2011, *PhRvA*, 83, 052704

Ivanov, S. V. & Buzykin, O. G. 2010, *JQSRT*, 111, 2341

Ivković, M., González, M. Á., & Jovićević, S., *et al.* 2010, *Publ. Astron. Obs. Belgrade*, 89, 201

Jacquemart, D., Laraia, A. L., & Tchana, F. K., *et al.* 2010, *JQSRT*, 111, 1209

Jacquinet-Husson, N., Scott, N. A., & Chedin, A., *et al.* 2008, *JQSRT*, 109, 1043

Jalbert, G., Wolff, W., Magalhães, S. D., & de Castro Faria, N. V. 2008, *PhRvA*, 77, 012722

Jiao, L., Zhou, Y., & Wang, Y. 2010, *PhRvA*, 81, 042713

Johnson P. V., Young, J. A. & Malone, C. P., *et al.* 2010, *JP-CS*, 204, 012003

Joshipura, K. N. & Gangopadhyay, S. 2008, *JPhB: At. Mol. Opt. Phys.*, 41, 215205

Kamber, E. Y., Abu-Haija, O., & Wardwell, J. A. 2008, *PhRvA*, 77, 012701

Kato, H., Suzuki, D., & Ohkawa, M., *et al.* 2010, *PhRvA*, 81, 042717

Kaur, S. & Baluja, K. L. 2010, *PhRvA*, 82, 022717

Kaur, S., Bharadvaja, A., & Baluja, K. L. 2011, *PhRvA*, 83, 062707

Keenan, F. P. Crockett, P. J., & Aggarwal, K. M., *et al.* 2009, *A&A*, 495, 359

Khoma, M. V., Lazur, V. Yu., & Janev, R. K. 2009, *PhRvA*, 80, 032706

Koshelev M. A. & Tretyakov, M. Yu. 2009, *JQSRT*, 110, 118

Koshelev M. A. & Markov, V. N. 2009, *JQSRT*, 110, 526

Koshelev M. A. 2011, *JQSRT* 112, 550

Kothari, H. N., Pandya, S. H., & Joshipura, K. N. 2011, *JPhB: At. Mol. Opt. Phys.*, 44, 125202

Koubek, J. & Boulet, C., Perrin A., *et al.* 2011, *JMoSp*, 266, 12

Kretinin, I. Yu, Krisilov, A. V. & Zon, B. A. 2008 *JPhB: At. Mol. Opt. Phys.*, 41, 215206

Krupnov, A. F. 2010, *PhRvA*, 82, 036703

Kumari, S., Chatterjee, S. N., Jha, L. K., & Roy, B. N. 2011, *EuPhJD*, 61, 355

Kupka, F., Piskunov, N., & Ryabchikova, T. A., *et al.* 1999, *A&AS*, 138, 119

Kurucz, R. L. 2011, *CanJPhys*, 89, 417

Lamouroux, J., Tran, H., & Laraia, A. L., *et al.* 2010, *JQSRT*, 111, 2321

Laraia, A. L., Gamache, R. R., & Hartmann, J.-M., *et al.* 2009, *JQSRT*, 110, 687

Lavrentieva, N. N., Petrova, T. M., Solodov, A. M., & Solodov, A. A. 2010, *JQSRT*, 111, 2291

Lewis, J. K. C. & Predoi-Cross, A. (eds.) 2010, Proc. 20th *International Conference on Spectral Line Shapes*, AIP-CP, 1290

Li, J. S., Liu, K., & Zhang, W. J., *et al.* 2008, *JMoSp*, 252, 9

Li, J. S., Durry, G., & Cousin, J., *et al.* 2010, *JQSRT*, 111, 2332

Liang, G. Y., Whiteford, A. D., & Badnell, N. R. 2009a, *A&A*, 499, 943

Liang, G. Y., Whiteford, A. D., & Badnell, N. R. 2009b, *A&A*, 500, 1263

Liang, G. Y. & Badnell, N. R. 2010, *A&A*, 518, A64

Liang, G. Y. & Badnell, N. R. 2011, *A&A*, 528, A69

Liang, G. Y., Badnell, N. R., & Zhao, G., *et al.* 2011, *A&A*, 533, A87

Lin, Y. & Mayer, P. M. 2010, *Chem. Phys.*, 378, 103

Lisak, D., Havey, D. K., & Hodges, J. T. 2009, *PhRvA*, 79, 052507

Lisak, D., Masłowski, P., & Cygan, A., *et al.* 2010, *PhRvA*, 81, 042504

Liu, J.-B. & Zhou, Ya.-J. 2010, *Chin. Phys. B*, 19, 093403

Liu, L., Liu, C. H., Wang, J. G., & Janev, R. K. 2011, *PhRvA*, 84, 032710

Liu, X. J., Wang, J. G., Qu, Y. Z., & Buenker, R. J. 2011, *PhRvA*, 84, 042706

Long, D. A., Havey, D. K., & Okumura, M., *et al.* 2010, *JQSRT*, 111, 2021

Long, D. A., Havey, D. K., & Yu, S. S., *et al.* 2011, *JQSRT*, 112, 2527

Ludlow, J. A., Lee, T. G., & Ballance, C. P., *et al.* 2011, *PhRvA*, 84, 022701

Lyulin, O. M., Nikitin, A. V., & Perevalov, V. I., *et al.* 2009, *JQSRT*, 110, 654

Lyulin, O. M., Perevalov, V. I., & Morino, I., *et al.* 2011, *JQSRT*, 112, 531

Machado, L. E., Sugohara, R. T., & dos Santos, A. S., *et al.* 2011, *PhRvA*, 84, 032709

McRaven, C. P., Cich, M. J., & Lopez, G. V., *et al.* 2011, *JMoSp*, 266, 43

Malespin, C., Ballance, C. P., & Pindzola, M. S., *et al.* 2011, *A&A*, 526, A115

Mančev, I. 2009, *EuPhJD*, 51, 213

Martin, B. & Lepère, M. 2009, *JMoSp*, 255, 6

Mashonkina, L. 2009, *Phys. Scr.*, T134, 014004

Mavadat, M., Ricard, A., Sarra-Bournet, C., & Laroche, G. 2011, *JPhD: Appl. Phys.*, 44, 155207

Mijatović, Z., Nikolić, D., Kobilarov, R. & Ivković, M. 2010a, *Publ. Astron. Obs. Belgrade*, 89, 217

Mijatović, Z., Nikolić, D., Kobilarov, R. & Ivković, M. 2010b, *JQSRT*, 111, 990

Miraglia, J. E. & Gravielle, M. S. 2008, *PhRvA*, 78, 052705

Misago, F., Lepèere, M., Bouanich, J.-P., & Blanquet, G. 2009, *JMoSp*, 254, 16

Mitroy, J. & Zhang, J. Y. 2008, *EuPhJD*, 46, 415

Miyano, S. & Tonokura, K. 2011, *JMoSp*, 265, 47

Monce, M. N., Pan, S., Radeva, N. L., & Pepper, J. L. 2009, *PhRvA*, 79, 012704

Müller, H. S. P., Schlöder, F., Stutzki, J., & Winnewisser, G. 2005, *JMoStr*, 742, 215

Munjal, H., Baluja, K. L., & Tennyson, J. 2009, *PhRvA*, 79, 032712

Nagao, M., Hida, K., & Kimura, M., *et al.* 2008, *PhRvA*, 78, 012708

Nguyen, L., Blanquet, G., Dhyne, M., & Lepèere, M. 2009a, *JMoSp*, 254, 94

Nguyen, L., Blanquet, G., Populaire, J.-C., & Lepèere, M. 2009b, *JQSRT*, 110, 367

Nikolić, D., Gorczyca, T. W., Korista, K. T., & Badnell, N. R. 2010, *A&A*, 516, A97

Nouri, S., Ben Mabrouk, K., & Chelin, P., *et al.* 2008, *JMoSp*, 258, 75

Ogurtsov, G. N., Ovchinnikov, S. Yu., Macek, J. H. & Mikoushkin, V. M. 2011, *PhRvA*, 84, 032706

Omar, B. 2010, *Int. J. Spectrosc.*, 506346

Omar, B. 2011, *J. At. Mol. Opt. Phys*, ID 850807

Orel, A. E., Starace, A. F., & Nikolić, D., *et al.* (eds.) 2009, Proc. XXVI *International Conference on Photonic, Electronic & Atomic Collisions, JP-CS*, 194

Palmeri, P., Quinet, P., & Mendoza, C., *et al.* 2011, *A&A*, 525, A59

Peach, G., Dimitrijević, M. S., & Stancil, P. C. 2009, in Karel A. van der Hucht (ed.), *Reports on Astronomy 2006-2009, IAU Transactions XXVIIA* (Cambridge: CUP), 385

Peach, G. & Whittingham, I. B. 2009, *New Astron. Rev.*, 53, 227

Peach, G. 2010a, in: *Mem. S.A.It. Suppl.*, 15, 68

Peach, G. 2010b, in: *Spectral Line Shapes*, Proc. 20th Intern. Conf., AIP-CP, 1290, 14

Peláez, R. J., Djurović, S., & Ćirišan, M., *et al.* 2009a, *JPhB: At. Mol. Opt. Phys.*, 42, 125002

Peláez, R. J., Djurović, S., & Ćirišan, M., *et al.* 2009b, *A&A*, 507, 1697

Peláez, R. J., Djurović, S., & Ćirišan, M., *et al.* 2010a, *A&A*, 518, A60

Peláez, R. J., Djurović, S., Ćirišan, M., Aparicio, J. A., & Mar, S. 2010b, *JP-CS*, 257, 2021

Petrova, T. M., Solodov, A. M., & Solodov, A. A., *et al.* 2011, *JQSRT*, 112, 2741

Phelps, A. V. 2009, *PhRvE*, 79, 066401

Phillips, G. T. & Perram, G. P. 2008, *JQSRT*, 109, 1875

Phillips, K. J. H., Aggarwal, K. M., Landi, E., & Keenan, F. P. 2010, *A&A*, 518, A41

Pitz, G. A., Fox, C. D., & Perram, G. P. 2010, *PhRvA*, 82, 042502

Popović, L. Č. & Dimitrijević, M. S. (eds.) 2009, Proc. 7th Serbian Conf. *Spectral Line Shapes in Astrophysics, New Astron. Rev.*, 53, 107

Povey, C., Predoi-Cross, A., & Hurtmans, D. R. 2011, *JMoSp*, 268, 177

Predoi-Cross, A., Liu, W., & Murphy, R., *et al.* 2010, *JQSRT*, 111, 1065

Ptashnik, I. V. & Smith, K. M. 2010, *JQSRT*, 111, 1317

Rajvanshi, J. S. & Baluja, K. L. 2010, *PhRvA*, 82, 062710

Rajvanshi, J. S. & Baluja, K. L. 2011, *PhRvA*, 84, 042711

Reggami, L., Bouledroua, M., Allouche, A. R., & Aubert-Frécon, M. 2009, *JQSRT*, 110, 72

Ren, X., Bray, I., & Fursa, D. V., *et al.* 2011, *PhRvA*, 83, 052711

Rixon, G., Dubernet, M. L., & Piskunov, N., *et al.* 2011, *AIP-CP*, 1344, 107

Roos, J. B., Larson, Å. A., & Orel, A. E. 2008, *PhRvA*, 78, 022508

Roos, J. B., Orel, A. E., & Larson, Å. A. 2009, *PhRvA*, 79, 062510

Rosato, J., Marandet, Y., & Capes, H., *et al.* 2009, *PhRvE*, 79, 046408

Rosato, J., Boland, D., & Difallah, M., *et al.* 2010, *Int. J. Spectrosc.*, 374372

Rosenberry, M. A. & Stewart B. 2011, *JPhB: At. Mol. Opt. Phys.*, 44, 055207

Rothman, L. S., Gordon, I. E., & Barbe, A., *et al.* 2009, *JQSRT*, 110, 533

Rothman, L. S., Gordon, I. E., & Barbe, A., *et al.* 2010, *JQSRT*, 111, 2139

Rozsályi, E., Bene, E., & Halász, G. J., *et al.* 2011, *PhRvA*, 83, 052713

Sabido, M, de Andrés, J. & Sogas, J., *et al.* 2008, *EuPhJD*, 47, 63

Sagawa, H., Mendrok, J., & Seta, T., *et al.* 2009, *JQSRT*, 110, 2027

Sahal-Bréchot, S. 2010, *JP-CS*, 257, 2037

Samanta, R. & Purkait, M. 2011, *EuPhJD*, 64, 311

Samanta, R., Purkait, M., & Mandal, C. R. 2011, *PhRvA*, 83, 032706

Santos, A. C. F., Sigaud, G. M., & Melo, W. S., *et al.* 2010, *PhRvA*, 82, 012704

Sato, T. O., Mizoguchi, A., & Mendrok, J., *et al.* 2010, *JQSRT*, 111, 821

Schmidt, E. W., Schippers, S., & Bernhardt, D., *et al.* 2008, *A&A*, 492, 265

Schöffler, M. S., Titze, J., & Schmidt, Ph. H., *et al.* 2009, *PhRvA*, 79, 064701

Shafir, D., Novotny, S., & Buhr, H., *et al.* 2009, *PRL*, 102, 223202

Smith, M. A. H., Benner, D. C., Predoi-Cross, A., & Devi, V. M. 2009, *JQSRT*, 110, 639

Smith, M. A. H., Benner, D. C., Predoi-Cross, A., & Devi, V. M. 2010, *JQSRT*, 111, 1152

Smith, M. A. H., Benner, D. C., Predoi-Cross, A., & Devi, V. M. 2011, *JQSRT*, 112, 952

Stäuber, P. & Bruderer, S. 2009, *A&A*, 505, 195

Stéhle, C. & Fouquet, S. 2010, *Int. J. Spectrosc.*, 506346

Stenrup, M., Larson, Å. A., & Elander, N. 2009, *PhRvA*, 79, 012713

Sterling, N. C. & Witthoeft, M. C. 2011, *A&A*, 529, A147

Stolterfoht, N., Cabrera-Trujillo, R., & Krstić, P. S., *et al.* 2010, *PhRvA*, 81, 052704

Storey, P. J. & Zeippen, C. J. 2010, *A&A*, 511, A78

Stroe, M. & Fifirig, M. 2011, *EuPhJD*, 61, 63

Szmytkowski, C., Domaracka, A. Możejko, P., & Ptasińska-Denga, E. 2008, *JPhB: At. Mol. Opt. Phys.*, 41, 065204

Takagi, H., Hara, S., & Sato, H. 2009, *PhRvA*, 79, 012715

Talbi, F., Bouledroua, M., & Alioua, K. 2008, *EuPhJD*, 50, 141

Tapalaga, I., Dojčinović, I. P., & Purić, J. 2011, *MNRAS*, 415, 503

Tarana, M. & Tennyson, J. 2008, *JPhB: At. Mol. Opt. Phys.*, 41, 205204

Tayal, S. S. & Zatsarinny, O. 2010, *A&A*, 510, A79

Thibault, F., Fuller, E. P., & Grabow, K. A., *et al.* 2009, *JMoSp*, 256, 17

Thibault, F., Ivanov, S. V., & Buzykin, O. G., *et al.* 2011a, *JQSRT*, 112, 1429

Thibault, F., Martinez, R. Z., Bermejo, D., & Goméz, L. 2011b, *JQSRT*, 112, 2542

Tran, H., Boulet, C., & Stefani, S., *et al.* 2011a, *JQSRT*, 112, 925

Tran, H., Thibault, F., & Hartmann, J.-M. 2011b, *JQSRT*, 112, 1035

Tran, H., Picquet-Varrault, B., & Boursier, C., *et al.* 2011c, *JQSRT*, 112, 2287

Tremblay, P.-E. & Bergeron, P. 2009, *ApJ*, 696, 1755
Vadla, C., Horvatic, V., & Niemax, K. 2009, *PhRvA*, 80, 052506
Vangvichith, M., Tran, H., & Hartmann, J.-M. 2009, *JQSRT*, 110, 2212
Vinodkumar, M., Limbachiya, C., & Joshipura, K. N., *et al.* 2008, *JP-CS*, 115, 012013
Voronin, B. A., Lavrentieva, N. N., & Mishina, T. P., *et al.* 2010, *JQSRT*, 111, 2308
Wang, Y. C., Zhou, Y., Cheng, Y., & Ma, J. 2009, *Chin. Phys. Lett.*, 26, 083401
Wang, Y. C., Zhou, Y., & Cheng, Y., *et al.* 2010, *JPhB: At. Mol. Opt. Phys.*, 43, 045201
Wiesenfeld, L. & Faure, A. 2010, *PhRvA*, 82, 040702(R)
Winter, T. G. 2009, *PhRvA*, 80, 032701
Wójtewicz, S., Lisak, D., & Cygan, A., *et al.* 2011, *PhRvA*, 84, 032511
Wu, Y., Qi, Y. Y., & Zou, S. Y., *et al.* 2009, *PhRvA*, 79, 062711
Wu, Y., Qi, Y. Y., & Yan, J., *et al.* 2009, *PhRvA*, 80, 022715
Yang, C., Buldyreva, J., Gordon, I. E., *et al.* 2008 *JQSRT*, 109, 2857
Zameroski, N. D., Hager, G. D., & Rudolph, W., *et al.* 2011, *JQSRT*, 112, 59
Zapukhlyak, M. & Kirchner, T. 2009, *PhRvA*, 80, 062705
Zatsarinny, O., Bartschat, K., & Gedeon, S., *et al.* 2009, *PhRvA*, 79, 052709
Zatsarinny, O., Bartschat, K., & Garcia, G., *et al.* 2011, *PhRvA*, 83, 042702
Zhang, S. B., Wang, J. G., & Janev, R. K. 2010, *PhRvA*, 81, 032707
Zhang, S. B., Wang, J. G., Janev, R. K., & Chen, X. J. 2011, *PhRvA*, 83, 032724
Zielinska, S., Pellerin, S., & Dzierzega, K., *et al.* 2010, *JPhD: Appl. Phys.*, 43, 434005
Zmerli, B., Ben Nessib, N., Dimitrijević, M. S., & Sahal-Bréchot, S. 2010, *Phys. Scr.*, 82, 055301

Transactions IAU, Volume XXVIIIA
Reports on Astronomy 2009–2012   © International Astronomical Union 2012
Ian Corbett, ed.   doi:10.1017/S1743921312003110

DIVISION XII / COMMISSION 14 / WORKING GROUP SOLIDS AND THEIR SURFACES

CHAIR   Gianfranco Vidali

TRIENNIAL REPORT 2009–2011

1. Introduction

The *ISO* and *Spitzer* space observatories yielded a treasure trove of data on dust and ices covering dust grains. Now *Herschel*, and soon *SOFIA* and *ALMA*, will provide unprecedented views of the molecular world of the interstellar medium (ISM). It is on dust grains that key ISM molecules, such as hydrogen, formaldehyde, methanol, and water are formed. As a result of these new observations, there is a great need to know more about the interaction processes of atoms and molecules with dust grains. (The Proceedings of the 2010 NASA Laboratory Astrophysics Workshop (www-cfadc.phy.ornl.gov/nasa_law/) give a good view of recent accomplishments in the study of atom/molecule - solid interactions as well as other aspects of laboratory astrophysics.)

In the last decade and a half there has been a tremendous increase of interest in laboratory studies of ISM processes occurring on interstellar dust grains. This has prompted the entrance into this field of a number of laboratories with a tradition in surface science. Besides the standard probes that have been used in the past, techniques are now available that can give precise information at the atomic/molecular level about the formation of molecules on dust, including: Thermal Programmed Desorption (TPD), Reflection Absorption Infrared Spectrometry (RAIRS), Resonant Enhanced Multiphoton Ionization (REMPI), and Atom Force Microscopy (AFM). These techniques yield information about the kinetics and energetics of atomic/molecular diffusion on and desorption from surfaces, the products of reactions, the ro-vibrational state of ejected products, and the morphology of the solid surfaces, respectively. The success of research in atom/molecule/charged particle/photon-dust interaction has produced a surge of publications. Studies of interest to astrochemistry are now regularly published in chemical physics/surface science journals. A representative sample of such literature is listed below. It can be of use to astronomers and astrochemists in understanding the crucial steps of reactions on dust analogues.

In theoretical research, there are two developments of note: the use of new stochastic tools to predict the molecule formation process on grains of different sizes, and the study of reaction mechanisms (Langmuir-Hinshelwood, Ealy-Rideal, and hot-atom) on surfaces of materials of astrophysical interest. Most of these studies pertain to hydrogen interaction with graphite/graphene/PAHs and appear in the chemical-physics literature.

Several research groups that are currently working in this area are listed here, each with its group leader and main research focus:
- Catania Observatory, E. Palumbo (ions in ices)
- Heriot-Watt University Edinburgh, M. R. S. McCoustra (desorption of mixed ices)

- Hokkaido University, N. Watanabe / A. Kouchi (ice formation, photon-ice interaction)
- Jet Propulsion Laboratory, A. Chutjian (ions on ices)
- Leiden University, H. Linnartz / E. van Dishoeck (photodesorption from ices, water formation on ices)
- Max Planck Institute for Astronomy, T. Henning (solids)
- NASA Ames Research Center, L. Allamandola (UV on ices, PAHs)
- NASA Ames Research Center, F. Salama (dust exposure, dust formation, PAHs)
- NASA - Goddard Space Flight Center, M. Moore (ions in ices)
- Syracuse University USA, G. Vidali (formation of H_2 and water on dust grain analogues)
- University College London, W. A. Brown (desorption of mixed ices)
- University College London, S. Price (H_2 formation on graphite)
- University of Cergy-Pontoise, J.-L. Lemaire (D_2 on ices, silicates)
- University of Hawai'i, R. Kaiser (keV electron in ices)
- University of Missouri, A. Speck (dust)
- University of Virginia, R. Baragiola (ions in ices)

2. Meetings

During the reporting period, a number of meetings containing sessions about atomic and molecular interaction with surfaces have been held. They are often featured at regularly scheduled COSPAR, American Astronomical Society and Lunar and Planetary Institute meetings. Unfortunately, meeting Web sites may no longer be accessible.

Important meetings (listed in inverse chronological order):
- European Conference on Laboratory Astrophysics, Paris, France, 2011
- The Molecular Universe, IAU Symposium 280, Toledo, Spain 2011
- Fifth Workshop on Titan Chemistry, Kauai, Hawai'i, 2011
(http://www.chem.hawaii.edu/Bil301/Titan2011.html)
- Herschel and the Characteristics of Dust in Galaxies, Lorentz Center, Leiden, Netherlands, 2011
- Pacifichem, Honolulu, Hawai'i, USA, 2010
- NASA Laboratory Astrophysics Workshop, Gatlinburg, TN, USA, 2010
- Stormy Cosmos: the Evolving ISM from Spitzer to Herschel and Beyond, Pasadena, CA, USA, 2010
- WittFest: Origin and Evolution of Dust, Toledo, OH, USA, 2010
- Molecules in Galaxies, Oxford Physics Conference Series, Oxford, United Kingdom, 2010
- Dust and Ice: Their Roles in Astrophysical Environments, Univ. of Georgia, Athens, GA, USA, 2010
- Recent Advances in Experimental and Observational Astrochemistry, Amer.Chem.Soc. Symposium, San Francisco, CA, USA, 2010
- Infrared Emission, ISM and Star Formation, MPI, Heidelberg, Germany, 2010
- International Conference on Laboratory Astrophysics, Dunhuang, Gansu, China, 2009
- Bridging Laboratory and Astrophysics: Molecules, Dust and Ices in Regions of Stellar and Planetary Formation, AAS 214th, Pasadena, CA, 2009
- The Chemical Enrichment of the Intergalactic Medium, Lorentz Center, Leiden, the Netherlands, 2009

• Interstellar Surfaces: from Laboratory to Models, Lorentz Center, Leiden, the Netherlands, 2009

3. Notable publications

Most of the works cited below regard the laboratory experiments and theories of photon and particle interaction with solid surfaces that are relevant to understanding similar processes occurring in space. Included in this selection are papers about PAHs (Polycyclic Aromatic Hydrocarbons) that are relevant to atom/surface interactions. Additional information on PAHs can be obtained in the report of the Commission 14 Working Group on Molecular Data. Key observations that are related to dust are included.

Gianfranco Vidali
Chair of Working Group
Solids and Their Surfaces

References

3.1. *2011*

Accolla, M., Congiu, E., Dulieu, F., Manico', G., Chaabouni, H., Matar, E., Mokrane, H., Lemaire, J.-L., & Pironello, V. 2011 Changes in the morphology of interstellar ice analogues after hydrogen atom exposure *PhysChemChemPhys* 13, 8037.

Acharyya, K., Hassel, G. E., & Herbst, E. 2011 The Effects of Grain Size and Grain Growth on the Chemical Evolution of Cold Dense Clouds *ApJ* 732, 73.

Andersson, S., Arasa, C., Yabushita, A., Yokoyama, M., Hama, T., Kawasaki, M., Western, C. M.,, & Ashfold, M. N. R.A theoretical and experimental study on translational and internal energies of H_2O and OH from the 157 nm irradiation of amorphous solid water at 90 K *PhysChemChemPhys*13, 15810.

Bennett, C. J., Hama, T., Kim, Y. S., Kawasaki, M., & Kaiser, R. I. 2011 Laboratory Studies on the Formation of Formic Acid (HCOOH) in Interstellar and Cometary Ices *ApJ* 727, 27.

Das, A. & Chakrabarti, S. K. 2011 Composition and evolution of interstellar grain mantle under the effects of photodissociation *MNRAS* in press.

Dukes, C. A., Chang, W-Y. Fam, M., & Baragiola, R. A. 2011 Laboratory studies on the sputtering contribution to the sodium atmospheres of Mercury and the Moon *Icarus* 212, 436.

Ennis, C., Bennett, C. J.; Jones, B. M., & Kaiser, R. I. 2011 Formation of D_2-water and D_2-carbonic Acid in Oxygen-rich Solar System Ices via D_2^+ Irradiation *ApJ* 733, 79.

Fayolle, E. C., Bertin, M., Romanzin, C., Michaut, X., Oberg, K. I., Linnartz, H., & Fillion, J.-H. 2011 CO Ice Photodesorption: A Wavelength-dependent Study *ApJ* 739, 36.

Fayolle, E. C., Oberg, K. I., Cuppen, H. M., Visser, R., & Linnartz, H. 2011 Laboratory H_2O:CO_2 ice desorption data: entrapment dependencies and its parameterization with an extended three-phase model *A& A* 529, 74.

Garozzo, M., La Rosa, L., Kanuchova, Z., Ioppolo, S., Baratta, G. A., Palumbo, M. E, & ; Strazzulla, G. 2011 The influence of temperature on the synthesis of molecules on icy grain mantles in dense molecular clouds *A&A* 528, 118.

Godard, M., Fraud, G., Chabot, M., Carpentier, Y., Pino, T., Brunetto, R., Duprat, J., Engrand, C., Brchignac, P., D'Hendecourt, L., & Dartois, E. 2011 Ion irradiation of carbonaceous interstellar analogues. Effects of cosmic rays on the 3.4 μm interstellar absorption band *A&A* 529. 146.

Goumans, T. P. M. 2011 Isotope effects for formaldehyde plus hydrogen addition and abstraction reactions: rate calculations including tunnelling *MNRAS* 413, 261.

Hama, T., Watanabe, N., Kouchi, A., & Yokoyama, M. 2011 Spin Temperature of Water Molecules Desorbed from the Surfaces of Amorphous Solid Water, Vapor-deposited and Produced from Photolysis of a CH_4/O_2 Solid Mixture *ApJ* 738, 15.

Hama, T., Yokoyama, M., Yabushita, A., Kawasaki, M., & Watanabe, N. 2011 Translational and rotational energy measurements of desorbed water molecules in their vibrational ground state following 157 nm irradiation of amorphous solid water *NuclInstrMethB* 269, 1011.

He, J., Frank, P., & Vidali, G. 2011 Interaction of hydrogen with surfaces of silicates: single crystal vs. amorphous *PhysChemChemPhys* 13, 15803.

Ioppolo, S., van Boheemen, Y., Cuppen, H. M., van Dishoeck, E. F., & Linnartz, H. 2011 Surface formation of CO_2 ice at low temperatures. *MNRAS* 413, 2281.

Ioppolo, S., Cuppen, H M., van Dishoeck, E. F., & Linnartz, H. 2011 Surface formation of HCOOH at low temperature *MNRAS* 410, 1089.

Jing, D., He, J., Brucato, J., De Sio, A., Tozzetti, L., & Vidali, G. 2011 On water formation in the interstellar medium: laboratory study of the O+D reaction on surfaces *ApJ* 741, L9.

Kinugawa, T., Yabushita, A., Kawasaki, M., Hama, T., & Watanabe, N. 2011 Surface abundance change in vacuum ultraviolet photodissociation of CO2 and H_2O mixture ices *PhysChemChemPhys* 13, 15785.

Kristensen, L. E., Amiaud, L., Fillion, J.-H., Dulieu, F., & Lemaire, J.-L. 2011 H_2, HD, and D_2 abundances on ice-covered dust grains in dark clouds *A&A* 527, 44.

Laas, J. C., Garrod, R. T., Herbst, E., & Widicus Weaver, S. L. 2011 Contributions from Grain Surface and Gas Phase Chemistry to the Formation of Methyl Formate and Its Structural Isomers *ApJ* 728, 71.

Lattelais, M., Bertin, M., Mokrane, H., Romanzin, C., Michaut, X., Jeseck, P., Fillion, J.-H., Chaabouni, H., Congiu, E.& Dulieu, F. 2011 Differential adsorption of complex organic molecules isomers at interstellar ice surfaces *A&A* 532, 12L.

Lepetit, B., Lemoine, D., Medina, Z., & Jackson, B. 2011 Sticking and desorption of hydrogen on graphite: A comparative study of different models *JChemPhys* 134, 114705

Oba, Y., Watanabe, N., Kouchi, A., Hama, T., & Pirronello, V. 2011 Experimental studies of surface reactions among OH radicals that yield H_2O and CO_2 at 40-60 K *PhysChemChemPhys* 13, 15792.

Oberg, K., Boogert, A., Pontoppidan, K. M. *et al.* 2011 Ices in starless and star forming cores *The Molecular Universe Proceedings IAU Symposium No. 280* J. Cernicharo, & R. Bachiller, Eds.

Ormel, C. W., Min, M.,Tielens, A. G. G. M., Dominik, C., & Paszun, D. 2011 Dust coagulation and fragmentation in molecular clouds. II. The opacity of the dust aggregate size distribution *A&A* 532, 43.

Roman-Duval, J., Israel, F. P., Bolatto, A., Hughes, A., Leroy, A., Meixner, M., Gordon, K., Madden, S. C., Paradis, & D. Kawamura, A. 2010 Dust/gas correlations from Herschel observations *A&A* 518, 74L.

Romanzin, C.; Ioppolo, S.; Cuppen, H. M.; van Dishoeck, E. F.; Linnartz, H. 2011 Water formation by surface O_3 hydrogenation *JChemPhys* 134, 084504.

Shi, J., Raut, U., Kim, J.-H., Loeffler, M., & Baragiola, R. A. 2011 Ultraviolet Photon-induced Synthesis and Trapping of H_2O_2 and O_3 in Porous Water Ice Films in the Presence of Ambient O_2: Implications for Extraterrestrial Ice *ApJ* 738, L3.

Tachibana, S., Nagahara, H., Ozawa, K., Ikeda, Y., Nomura, R., Tatsumi, K., & Joh, Y. 2011 Kinetic Condensation and Evaporation of Metallic Iron and Implications for Metallic Iron Dust Formation *ApJ* 736, 16.

Talbi, D., Wakelam, V., the Kida team 2011 KIDA : The new kinetic database for astrochemistry *JPhysConfSeries* 300, 012019.

Vidali, G. 2011 Molecule formation on interstellar grains *Proc.2010NASA LabAstWorkshop*I-19.

3.2. *2010*

Anderson, L. D., Zavagno, A., Rodn, and 36 coauthors 2010 The physical properties of the dust in the RCW 120 H II region as seen by Herschel *A&A* 518, 99.

Bennett, C. J., Jamieson, C. S., & Kaiser, R. I. 2010 Mechanistical studies on the formation and destruction of carbon monoxide (CO), carbon dioxide (CO2), and carbon trioxide (CO_3) in interstellar ice analog samples *PhysChemChemPhys* 12, 4032.

Burke, D. J., & Brown, W. A. 2010 Ice in space: surface science investigations of the thermal desorption of model interstellar ices on dust grain analogue surfaces *PhysChemChemPhys* 12, 5947.

Cazaux, S., Cobut, V., Marseille, M., Spaans, M., & Caselli, P. 2010 Water formation on bare grains: When the chemistry on dust impacts interstellar gas *A&A* 522, 74.

Cooper, P. D., Moore, M. H., & Hudson, R. L. 2010 O atom production in water ice: Implications for O_2 formation on icy satellites *JGRE* 11510013

Cuppen, H. M., Ioppolo, S., Romanzin, C., & Linnartz, H. 2011 Water formation at low temperatures by surface O_2 hydrogenation II: the reaction network *PhysChemChemPhys* 12, 12077

Seperuelo D. E., Domaracka, A., Boduch, P., Rothard, H., Dartois, E., & da Silveira, E. F. 2010 Laboratory simulation of heavy-ion cosmic-ray interaction with condensed CO *A&A* 512, 71.

Dulieu, F., Amiaud, L., Congiu, E., Fillion, J.-H., Matar, E., Momeni, A., Pirronello, V., & Lemaire, J. L. 2010 Experimental evidence for water formation on interstellar dust grains by hydrogen and oxygen atoms *A&A* 512, 30.

Fam, M., Loeffler, M. J., Raut, U., & Baragiola, R. A. 2010 Radiation-induced amorphization of crystalline ice *Icarus* 207, 314.

Ferullo, R. M., Domancich, N. F., & Castellani, N. J. 2010 On the performance of van der Waals corrected-density functional theory in describing the atomic hydrogen physisorption on graphite *ChemPhysLett* 500, 283.

Fleming, B., France, K., Lupu, R. E., & McCandliss, S. R. 2010 Spitzer Mapping of Polycyclic Aromatic Hydrocarbon and H_2 Features in Photodissociation Regions *ApJ* 725, 195.

Godard, M.;& Dartois, E. 2010 Photoluminescence of hydrogenated amorphous carbons. Wavelength-dependent yield and implications for the extended red emission *A&A* 529, 39.

Hama, T., Yokoyama, M., Yabushita, A., & Kawasaki, M. 2010 Role of OH radicals in the formation of oxygen molecules following vacuum ultraviolet photodissociation of amorphous solid water *JChemPhys* 133, 104504.

Hama, T., Yokoyama, M., Yabushita, A., Kawasaki, M., Andersson, S., Western, C. M., Ashfold, M. N. R., Dixon, R. N., & Watanabe, N. 2010 A desorption mechanism of water following vacuum-ultraviolet irradiation on amorphous solid water at 90 K *JChemPhys* 132, 164508.

Hama, T.,Yokoyama, M.,Yabushita, A., Kawasaki, M., Wickramasinghe, P., Guo, W., Loock, H.-P., Ashfold, M. N. R., & Western, C. M. 2009 Translational and internal energy distributions of methyl and hydroxyl radicals produced by 157 nm photodissociation of amorphous solid methanol *JChemPhys* 131, 224512

He, J., Gao, K., Vidali, G., Bennett, C. J., & Kaiser, R. I. 2010 Formation of Molecular Hydrogen from Methane Ice *ApJ* 721, 1656

Henning, T. 2010 Laboratory Astrophysics of Cosmic Dust Analogues *LectNotesPhys* 815, 313.

Henning, T. 2010 Cosmic Silicates *ARA&A* 48, 21.

Henning, T. & Mutschke, H. 2010 Optical properties of cosmic dust analogs: a review *JNanophotonics* 4, 041580.

Ioppolo, S., Cuppen, H. M., Romanzin, C., van Dishoeck, E. F. & Linnartz, H. 2010 Water formation at low temperatures by surface O_2 hydrogenation I: characterization of ice penetration *PhysChemChemPhys* 12, 12065.

Islam, F., Cecchi-Pestellini, C., Viti, S., & Casu, S. 2010 Formation Pumping of Molecular Hydrogen in Dark Clouds *ApJ* 725, 1111.

Juhsz, A., Bouwman, J., Henning, Th., Acke, B., van den Ancker, M. E., Meeus, G., Dominik, C., Min, M., Tielens, A. G. G. M. & Waters, L. B. F. M. 2010 Dust Evolution in Protoplanetary Disks Around Herbig Ae/Be Stars the Spitzer View *ApJ* 721, 431.

Kalvans, J. & Shmeld, I. 2010 Subsurface chemistry of mantles of interstellar dust grains in dark molecular cores *A&A* 521, 37.

Lemaire, J.-L., Vidali, G., Baouche, S., Chehrouri, M., Chaabouni, H., & Mokrane, H. 2010 Competing Mechanisms of Molecular Hydrogen Formation in Conditions Relevant to the Interstellar Medium *ApJ* 725, 156L.

Matar, E., Bergeron, H., Dulieu, F., Chaabouni, H., Accolla, M., & Lemaire, J. L. 2010 Gas temperature dependent sticking of hydrogen on cold amorphous water ice surfaces of interstellar interest *JChemPhys* 133, 104507.

Oba, Y., Watanabe, N., Kouchi, A., Hama, T.,, & Pirronello, V. 2010 Experimental Study of CO_2 Formation by Surface Reactions of Non-energetic OH Radicals with CO Molecules *ApJ* 712, L174.

Oba, Y., Watanabe, N., Kouchi, A., Hama, T., & Pirronello, V. 2011 Formation of Carbonic Acid (H_2CO_3) by Surface Reactions of Non-energetic OH Radicals with CO Molecules at Low Temperatures *ApJ* 722, 1598.

Oberg, K. I., van Dishoeck, E. F., Linnartz, H., & Andersson, S. 2010 The Effect of H_2O on Ice Photochemistry *ApJ* 718, 832.

Pagani, L., Steinacker, J., Bacmann, A., Stutz, A., & Henning, T. 2010 The Ubiquity of Micrometer-Sized Dust Grains in the Dense Interstellar Medium *Science* 329, 1622.

Palumbo, M. E., Baratta, G. A., Leto, G., & Strazzulla, G. 2010 H bonds in astrophysical ices *JMolecStruct* 972, 64.

Quan, D., Herbst, E., Osamura, Y., & Roueff, E. 2010 Gas-grain Modeling of Isocyanic Acid (HNCO), Cyanic Acid (HOCN), Fulminic Acid (HCNO), and Isofulminic Acid (HONC) in Assorted Interstellar Environments *ApJ* 725, 2101.

Sandstrom, K. M., Bolatto, A. D., Draine, B. T., Bot, C., & Stanimirovi, S. 2010 The Spitzer Survey of the Small Magellanic Cloud (S3MC): Insights into the Life Cycle of Polycyclic Aromatic Hydrocarbons *ApJ* 715, 701.

Semenov, D., Hersant, F., Wakelam, V., Dutrey, A., Chapillon, E., Guilloteau, St.. Henning, Th., Launhardt, R., Pitu, V., & Schreyer, K. 2010 Chemistry in disks. IV. Benchmarking gas-grain chemical models with surface reactions *A&A* 522, 42.

Siebenmorgen, R., & Krugel, E. 2010 The destruction and survival of polycyclic aromatic hydrocarbons in the disks of T Tauri stars *A&A* 511, 6.

Sturm, B., Bouwman, J., Henning, Th., Evans, N. J., Acke, B., Mulders, G. D., Waters, L. B. F. M., van Dishoeck, E. F., Meeus, G., Green, J. D *et al.* 2010 First results of the Herschel key program "Dust, Ice and Gas In Time" (DIGIT): Dust and gas spectroscopy of HD 100546 *A&A* 518, 129.

Tsvetkov, A. G. & Shematovich, V. I. 2010 Kinetic Monte Carlo method for simulating astrochemical kinetics: Hydrogen chemistry in diffuse clouds *SolarSystemRes* 44, 177.

Yang, Z., Eichelberger, B., Carpenter, M. Y., Martinez, O., Jr., Snow, T. P., &Bierbaum, V. M. 2010 Experimental and Theoretical Studies of Reactions Between H Atoms and Carbanions of Interstellar Relevance *ApJ* 723, 1325.

van Hoof, P. A. M., van de Steene, G. C., Barlow, M. J., Exter, K. M., Sibthorpe, B., Ueta, T., Peris, V., Groenewegen, M. A. T., Blommaert, J. A. D. L., Cohen, M., *et al.* 2010 Herschel images of NGC 6720: H_2 formation on dust grains *A&A* 518, L137.

Vasyunin, A. I., Wiebe, D. S., Birnstiel, T., Zhukovska, S., Henning, T., & Dullemond, C. P. 2011 Impact of Grain Evolution on the Chemical Structure of Protoplanetary Disks *ApJ* 727, 76.

Vidali, G.; Li, L. 2010 Molecular hydrogen desorption from amorphous surfaces at low temperature *JPhysCondMat* 22, 304012.

Voshchinnikov, N. V. & Henning, Th. 2010 From interstellar abundances to grain composition: the major dust constituents Mg, Si, and Fe *A&A* 517, 45.

Wakelam, V. , Smith, I. W. M., Herbst, E., Troe, J., Geppert, W., Linnartz, H., Oberg, K., Roueff, E., Agndez, M., P. Pernot, P. *et al.* 2010 Reaction Networks for Interstellar Chemical Modelling: Improvements and Challenges *SSRv* 156, 13.

Watanabe, N., Kimura, Y., Kouchi, A., Chigai, T., Hama, T., & Pirronello, V. 2010 Direct Measurements of Hydrogen Atom Diffusion and the Spin Temperature of Nascent H_2 Molecule on Amorphous Solid Water *ApJ* 714, 233.

3.3. *2009*

Arakawa, M., Kagi, H., & Fukazawa, H. 2009 Laboratory Measurements of Infrared Absorption Spectra of Hydrogen-ordered Ice: A Step to the Exploration of Ice XI in Space *ApJS* 184, 361

Bachellerie, D., Sizun, M., Aguillon, F., Teillet-Billy, D., Rougeau, N., & Sidis, V. 2009 Unrestricted study of the Eley-Rideal formation of H_2 on graphene using a new multidimensional graphene-H-H potential: role of the substrate *PhysChemChemPhys* 11, 2715

Bennett, C. J., Jamieson, C. S., & Kaiser, R. I. 2009 An Experimental Investigation of the Decomposition of Carbon Monoxide and Formation Routes to Carbon Dioxide in Interstellar Ices *ApJS* 182,1.

Bennett, C. J., Jamieson, C. S., & Kaiser, R. I. 2009 Mechanistical studies on the formation of carbon dioxide in extraterrestrial carbon monoxide ice analog samples *PhysChemChemPhys* 11, 4210.

Brickhouse, N., Cowan, J., Drake, P., Federman, S., Ferland, G., Frank, A., Haxton, W., Herbst, E., Olive, K., Salama, F., and 2 coauthors 2009 Laboratory Astrophysics and the State of Astronomy and Astrophysics *Astro2010: The Astronomy and Astrophysics Decadal Survey, Position Papers, no. 68*

Brunetto, R., Pino, T., Dartois, E., Cao, A.-T., D'Hendecourt, L., Strazzulla, G., & Brchignac, Ph. 2009 Comparison of the Raman spectra of ion irradiated soot and collected extraterrestrial carbon *Icarus* 200, 323.

Casolo, S., Martinazzo, R., Bonfanti, M., &Tantardini, G. F. 2009 Quantum Dynamics of the Eley-Rideal Hydrogen Formation Reaction on Graphite at Typical Interstellar Cloud Conditions *JPhysChemA* 113, 14545

Congiu, E., Matar, E., Kristensen, L. E., Dulieu, F., & Lemaire, J. L. 2009 Laboratory evidence for the non-detection of excited nascent H_2 in dark clouds *MNRAS* 397, 96.

Cuppen, H. M., van Dishoeck, E. F., Herbst, E., & Tielens, A. G. G. M. 2009 Microscopic simulation of methanol and formaldehyde ice formation in cold dense cores *A&A* 508, 275.

Seperuelo, D. E., Boduch, P., Rothard, H., Been, T., Dartois, E., Farenzena, L. S., & da Silveira, E. F. 2009 Heavy ion irradiation of condensed CO_2: sputtering and molecule formation *A&A* 502, 599.

Fuchs, G. W., Cuppen, H. M., Ioppolo, S., Romanzin, C., Bisschop, S. E., Andersson, S., van Dishoeck, E. F., & Linnartz, H. 2011 Hydrogenation reactions in interstellar CO ice analogues. A combined experimental/theoretical approach *A&A* 505, 629.

Garrod, R. T., Vasyunin, A. I., Semenov, D. A., Wiebe, D. S., & Henning, Th. 2009 A New Modified-Rate Approach For Gas-Grain Chemistry: Comparison with a Unified Large-Scale Monte Carlo Simulation *ApJ* 700 L43

Glauser, A. M., Gdel, M., Watson, D. M., Henning, T., Schegerer, A. A., Wolf, S., Audard, M., & Baldovin-Saavedra, C. 2009 Dust amorphization in protoplanetary disks *A&A* 508, 247.

Gontareva, N. B., Kuzicheva, E. A., & Shelegedin, Vladimir N. 2009 Synthesis and characterization of peptides after high-energy impact on the icy matrix: Preliminary step for further UV-induced formation *P&SS* 57, 441.

Goumans, T. P. M., Richard, C., Catlow, A., & Brown, W. A. 2009 Formation of H_2 on an olivine surface: a computational study *MNRAS* 393, 1403.

Goumans, T. P. M., Catlow, C. R. A., Brown, W. A., Kstner, J., & Sherwood, P. 2009 An embedded cluster study of the formation of water on interstellar dust grains *PhysChemChemPhys* 11, 5431.

Green, S.,D., Bolina, A. S., Chen, R., Collings, M. P., Brown, W. A., & McCoustra, M. R. S. 2009 Applying laboratory thermal desorption data in an interstellar context: sublimation of methanol thin films *MNRAS* 398, 357.

Hama, T., Yabushita, A., Yokoyama, M., Kawasaki, M., & Watanabe, Naoki 2009 Formation mechanisms of oxygen atoms in the O(3PJ) state from the 157 nm photoirradiation of amorphous water ice at 90 K *JChemPhys* 131, 054511.

Hama, T., Yabushita, A.,Yokoyama, M., Kawasaki, M., & Watanabe, Naoki 2009 Formation mechanisms of oxygen atoms in the O(1D2) state from the 157 nm photoirradiation of amorphous water ice at 90 K *JChemPhys* 131, 054510.

Hama, T., Yabushita, A., Yokoyama, M., Kawasaki, M., & Andersson, S. 2009 Desorption of hydroxyl radicals in the vacuum ultraviolet photolysis of amorphous solid water at 90 K *JChemPhys* 131, 054508.

Herbst, E. & van Dishoeck, E. F. 2009 Complex Organic Interstellar Molecules*ARA&A* 47, 427.

Hersant, F., Wakelam, V., Dutrey, A., Guilloteau, S., & Herbst, E. 2009 Cold CO in circumstellar disks. On the effects of photodesorption and vertical mixing *A&A* 493, 49.

Hollenbach, D., Kaufman, M. J., Bergin, E. A., & Melnick, G. J. 2009 Water, O_2, and Ice in Molecular Clouds *ApJ* 690, 1497.

Jger, C., Huisken, F., Mutschke, H., Jansa, I. L., & Henning, Th. 2009 Formation of Polycyclic Aromatic Hydrocarbons and Carbonaceous Solids in Gas-Phase Condensation Experiments *ApJ* 696, 706

Juhsz, A., Henning, Th., Bouwman, J., Dullemond, C. P., Pascucci, I., & Apai, D. 2009 Do We Really Know the Dust? Systematics and Uncertainties of the Mid-Infrared Spectral Analysis Methods *ApJ* 695, 1024

Kozasa, T., Nozawa, T., Tominaga, N., Umeda, H., Maeda, K., & Nomoto, K. 2009 Dust in Supernovae: Formation and Evolution *Cosmic Dust - Near and Far, ASP Conf. Series, Th. Henning, E. Grun, J. Steinacker, eds.* 414, 43.

Le Page, V., Snow, T. P., & Bierbaum, V. M. 2009 Molecular Hydrogen Formation Catalyzed by Polycyclic Aromatic Hydrocarbons in the Interstellar Medium *ApJ* 704, 274

Le Petit, F., Barzel, B., Biham, O., Roueff, E., & Le Bourlot, J. 2009 Incorporation of stochastic chemistry on dust grains in the Meudon PDR code using moment equations. I. Application to the formation of H_2 and HD *A&A* 505, 1153.

Loeffler, M. J. & Baragiola, R. A. 2009 Physical and chemical effects on crystalline H_2O_2 induced by 20 keV protons *JChemPhys* 130, 114504

Lohmar, I., Krug, J.& Biham, O. 2009 Accurate rate coefficients for models of interstellar gas-grain chemistry *A&A* 504, 5L.

Mastrapa, R. M., Sandford, S. A., Roush, T. L., Cruikshank, D. P.,& Dalle Ore, C. M. 2009 Optical Constants of Amorphous and Crystalline H2O-ice: 2.5-22 μm (4000-455 cm^{-1}) Optical Constants of H_2O-ice *ApJ* 701, 1347

Madzunkov, S. M., MacAskill, J. A., Chutjian, A., Ehrenfreund, P., Darrach, M. R., Vidali, G., & Shortt, B. J. 2009 Formation of Formaldehyde and Carbon Dioxide on an Icy Grain Analog Using Fast Hydrogen Atoms *ApJ* 697, 801.

Mokrane, H., Chaabouni, H., Accolla, M., Congiu, E., Dulieu, F., Chehrouri, M., & Lemaire, J. L. 2009 Experimental Evidence for Water Formation Via Ozone Hydrogenation on Dust Grains at 10 K *ApJ* 705, 195.

Muoz Caro, G. M., & Dartois, E. 2009 A tracer of organic matter of prebiotic interest in space, made from UV and thermal processing of ice mantles *A&A* 494, 109.

Nuevo, M., Milam, S. N., Sandford, S. A., Elsila, J. E., & Dworkin, J. P. 2009 Formation of Uracil from the Ultraviolet Photo-Irradiation of Pyrimidine in Pure H_2O Ices *AsBio* 9, 683.

Oba, Y., Miyauchi, N., Hidaka, H., Chigai, T., Watanabe, N., & Kouchi, A. 2009 Formation of Compact Amorphous H_2O Ice by Codeposition of Hydrogen Atoms with Oxygen Molecules on Grain Surfaces *ApJ* 701, 464.

Oberg, K. I., Garrod, R. T., van Dishoeck, E. F., & Linnartz, H. 2009 Formation rates of complex organics in UV irradiated CH_3OH-rich ices. I. Experiments *A&A* 504, 891.

Oberg, K. I., Linnartz, H., Visser, R., & van Dishoeck, E. F. 2009 Photodesorption of Ices. II. H_2O and D_2O *ApJ* 693, 1209.

Oberg, K. I., van Dishoeck, E. F., & Linnartz, H. 2009 Photodesorption of ices I: CO, N_2, and CO_2 *A&A* 496, 281.

Oberg, K. I., Fayolle, E. C., Cuppen, H. M., van Dishoeck, E. F.& Linnartz, H. 2009 Quantification of segregation dynamics in ice mixtures *A&A* 505, 183.

Ormel, C. W., Paszun, D., Dominik, C.; &Tielens, A. G. G. M. 2009 Dust coagulation and fragmentation in molecular clouds. I. How collisions between dust aggregates alter the dust size distribution *A&A* 502, 845.

Ratajczak, A., Quirico, E., Faure, A., Schmitt, B., & Ceccarelli, C. 2009 Hydrogen/deuterium exchange in interstellar ice analogs *A&A* 496, L21.

Shi, J., Teolis, B. D., & Baragiola, R. A. 2009 Irradiation-enhanced adsorption and trapping of O_2 on nanoporous water ice *PhysRevB* 79, 235422.

Shiraiwa, M., Garland, R M., & Poschl, U. 2009 Kinetic double-layer model of aerosol surface chemistry and gas-particle interactions (K2-SURF): degradation of polycyclic aromatic hydrocarbons exposed to O_3, NO_2, H_2O, OH and NO_3 *AtmosChemPhys* 9, 9571

Schou, J. & Hilleret, N. 2009 Sputtering of cryogenic films of hydrogen by keV ions: Thickness dependence and surface morphology *NuclInstrMethodsB* 267, 2748.

Teolis, B. D., Shi, J., & Baragiola, R. A. 2009 Formation, trapping, and ejection of radiolytic O_2 from ion-irradiated water ice studied by sputter depth profiling *JChemPhys* 130, 134704.

Tsvetkov, A. G. & Shematovich, V. I. 2010 Kinetic Monte Carlo method for simulating astro-chemical kinetics: Test calculations of molecular hydrogen formation on interstellar dust particles *SolarSystemRes* 43, 301.

Thrower, J. D., Collings, M. P., Rutten, F. J. M., & McCoustra, M. R. S. 2009 Laboratory investigations of the interaction between benzene and bare silicate grain surfaces *MNRAS* 394, 1510.

Vasyunin, A. I., Semenov, D. A., Wiebe, D. S., & Henning, Th. 2009 A Unified Monte Carlo Treatment of Gas-Grain Chemistry for Large Reaction Networks. I. Testing Validity of Rate Equations in Molecular Clouds *ApJ* 691, 1459

Verhoelst, T., van der Zypen, N., Hony, S., Decin, L., Cami, J., & Eriksson, K. 2009 The dust condensation sequence in red supergiant stars *A&A* 498, 127.

Vidali, G., Li, L., Roser, J., E., & Badman, R. 2009 Catalytic activity of interstellar grains: Formation of molecular hydrogen on amorphous silicates *AdvSpaceRes* 43, 1291.

Vidali, G. 2009 Division XII / Commission 14 / Working Group: Solids and their Surfaces *TransIAUv.4*, p.400.

Visser, R., van Dishoeck, E. F., Doty, S. D., & Dullemond, C. P. 2009 The chemical history of molecules in circumstellar disks. I. Ices *A&A* 495, 881.

Yabushita, A., Hama, T., Yokoyama, M., Kawasaki, M., Andersson, S., Dixon, R. N., Ashfold, Michael N. R., & Watanabe, N. 2009 Translational and Rotational Energy Measurements of Photodesorbed Water Molecules in their Vibrational Ground State from Amorphous Solid Water *ApJ* 699, 80.

Transactions IAU, Volume XXVIIIA
Reports on Astronomy 2009–2012
Ian Corbett, ed.
© International Astronomical Union 2012
doi:10.1017/S1743921312003122

COMMISSION 41

HISTORY OF ASTRONOMY
HISTOIRE DE L'ASTRONOMIE

PRESIDENT	Clive Ruggles
VICE-PRESIDENT	Rajesh Kochhar
PAST PRESIDENT	Nha Il-Seong
ORGANIZING COMMITTEE	Juan Belmonte, Brenda Corbin,
	Teije de Jong, Ray Norris,
	Luisa Pigatto, Mitsuru Soma,
	Chris Sterken, Sun Xiaochun

COMMISSION 41 WORKING GROUPS

Div. XII / Commission 41 WG	Archives
Div. XII / Commission 41 WG	Astronomy and World Heritage
Div. XII / Commission 41 WG	Historic Radio Astronomy
Div. XII / Commission 41 WG	Historical Instruments
Div. XII / Commission 41 WG	Johannes Kepler
Div. XII / Commission 41 WG	Transits of Venus

TRIENNIAL REPORT 2009–2012

1. Introduction

Commission 41 was created at the VIIth IAU General Assembly in Zürich in 1948. From an inauspicious start—Otto Neugebauer was appointed the first President in his absence, but proceeded to express his conviction that "an international organization in the history of astronomy has no positive function... my only activity during my term of service consisted in iterated attempts to resign"—the Commission quickly assumed a key role in the international development of the history of astronomy as an academic discipline.

In 1970, the commission proposed the idea of providing a major synthesis of the field on a quite unprecedented scale. This project, the General History of Astronomy, proceeded under the general editorship of Michael Hoskin, with three volumes being published by Cambridge University Press between 1984 and 1995.

The year 2001 saw the creation of the Inter-Union Commission on the History of Astronomy (ICHA), a joint Commission of the IAU and the International Union of the History and Philosophy of Science, Division of Science and Technology (IUHPS/DHST). The ICHA was founded in order to include scholars who, being primarily historians rather than astronomers, are not members of the IAU. All C41 members are, *ipso facto*, members of the ICHA.

In recent years C41/ICHA has expanded its remit to include the research fields of archaeoastronomy and ethnoastronomy, as well as issues such as the recognition and preservation of astronomical heritage.

The Commission maintains a website at http://www.historyofastronomy.org/.

2. Developments within the past triennium

The year 2009 marked the final completion of the General History of Astronomy. While this hugely ambitious project was never realized in the form originally envisaged— seven books were to be published by Cambridge University Press, but only three ever appeared—it has been brought to a satisfactory conclusion, after nearly 40 years, through the very considerable efforts of Michael Hoskin, including the publication of various key articles in the *Journal for the History of Astronomy*. Taken as a whole, the GHA forms by far the most detailed synthesis available of academic history of astronomy and will endure as an archive of immense value representing the state of the discipline in the late twentieth century. The complete contents list is available on the Commission's website.

3. Conferences

Since the Rio GA, the Commission has been directly involved in the organization of the following major international meetings:

• An international conference on "Astronomy and its Instruments before and after Galileo", held in San Servolo Island, Venice, Italy, on Sep 29–Oct 3, 2009. This joint symposium of the IAU and the INAF Astronomical Observatory of Padova was co-chaired by Luisa Pigatto and Clive Ruggles and included sessions on "Galileo and his Time: the Venetian Cultural Environment"; "Structures through the Ages, from Stone Monuments to Modern Observatories"; "Mathematical and Mechanical Instruments for Astronomy"; and "Space Observatories as Astronomical Instruments". At a round table on astronomical heritage, the meeting also formulated recommendations for urgent actions to achieve some of the objectives of the UNESCO–IAU Astronomy and World Heritage Initiative. The Proceedings have been published as Pigatto & Zanini (2010).

• "The Seventh International Conference on Oriental Astronomy" (ICOA–7), held at the National Astronomical Observatory of Japan in Mitaka, Tokyo, on 6–10 September 2010. The primary purpose of the ICOA conferences is to inspire studies on the history of astronomy in Asian countries and to encourage mutual communication and collaboration between researchers in an international context. ICOA–7 covered topics ranging from archaeoastronomy through to modern astrophysics. Thus in addition to numerous reports on modern astronomy and astrophysics up to post-World War II times, including contributions from young researchers and graduate students in Korea, Iran and Indonesia, there were several reports on megalithic and stone-circle astronomy from prehistoric India, Korea, Japan and Europe and, for the first time in an ICOA conference, papers on astronomy relating to culture, ethnology and climatology. The Proceedings have been published as Nakamura *et al.* (2011).

• IAU Symposium S278, entitled "Archeoastronomy and Ethnoastronomy: Building Bridges between Cultures", held in Lima, Peru, on January 5–14, 2011. This was also an 'Oxford' Symposium on Archeoastronomy, the ninth in what is unarguably the foremost series of international conferences on this highly interdisciplinary topic, one of interest not only to astronomers but also to archaeologists, historians, anthropologists, architects, art historians, historians of religions and others. Meetings in the series have been held at roughly four-yearly intervals since the time when Michael Hoskin, the then-Commission 41 President, organised the first one in Oxford itself. The conference attracted 108 participants from 25 countries.

One of the most important aspects of the conference was encapsulated in the "Building Bridges" subtitle and the strong link with the IAU's new decadal strategic plan, "Astronomy for the Developing World". Archaeoastronomy and ethnoastronomy (often

referred to together as 'cultural astronomy') concern themselves with beliefs and practices relating to the sky across human cultures and across time, from prehistory through
to the indigenous present. The focus thus reaches far beyond just the history of modern
scientific astronomy, reflecting instead a much broader interest in the many different ways
in which human communities perceive the cosmos they are situated within. Understanding engenders respect and exchange, so that opening people's minds to the wonders of
modern astronomy is often achieved far better if the 'Western' astronomers doing the
disseminating have a better understanding of (and respect for) the context of indigenous knowledge and beliefs and practices within which they are working. Inspiration is
a two-way process.

The Proceedings have been published as Ruggles (2011).

4. Other Major Publications

The Proceedings of the wide-ranging meeting on "Accelerating the rate of astronomical
discovery", organized by Ray Norris on behalf of C5 and C41 as Special Session 5 at the
Rio GA, have been published as Norris (2010). Nussbaumer & Bieri (2009) have described
the beginnings of modern cosmology from 1917 through to Einstein's conversion to the
expanding universe in 1931, while Sullivan (2009) has presented a detailed account of the
history of radio astronomy up to 1954 from an intellectual, technical, and social point of
view. Other major publications by Commission members during the triennium include
Curir (2009), Gaulke & Hamel (2010), Hamel (2011), Hearnshaw (2009), Véron (2009),
Whiting (2010), and Wali (2011).

Finally, there have been two major publications on the subject of astronomical heritage:
Wolfschmidt (2009) and Ruggles & Cotte (2010).

5. Closing remarks

Further information on the Commission's activities can be found in its Newsletters,
which are published at roughly six-monthly intervals and are available for download from
the Commission's website.

Clive Ruggles
President of the Commission

References

Curir, A. (ed.) 2009, *Osservar le Stelle: 250 Anni di Astronomia a Torino*, Silvana Editoriale,
    Milano.
Gaulke, K. & Hamel, J. (eds) 2010, *Kepler, Galilei, das Fernrohr und die Folgen*, Acta Historica
    Astronomiae, Vol. 40, Verlag Harri Deutsch, Frankfurt-am-Main.
Hamel, J. (ed.) 2011, *Gottfried Kirch (1639–1710) und die Berliner Astronomie im 18. Jahrhun-
    dert*, Acta Historica Astronomiae, Vol. 41, Verlag Harri Deutsch, Frankfurt-am-Main.
Hearnshaw, J. 2009, *Astronomical Spectrographs and Their History*, Cambridge University Press,
    Cambridge.
Nakamura, T., Orchiston, W., Sôma, M., & Strom, R. (eds) 2011, *Proceedings of the Seventh
    International Conference on Oriental Astronomy (ICOA-7)*, NAOJ, Toyko.
Norris, R. P. (ed.) 2010, *Accelerating the Rate of Astronomical Discovery*, SpS5, August 11–14
    2009, Rio de Janeiro, Brazil. *Proceedings of Science*, PoS(sps5)029. http://pos.sissa.it/
    archive/conferences/099/029/sps5_029.pdf.
Nussbaumer & Bieri 2009, *Discovering the Expanding Universe*, Cambridge University Press,
    Cambridge.

Pigatto L. & Zanini V. (eds) 2010, *Astronomy and its Instruments before and after Galileo*, CLEUP, Padova.

Ruggles, C. L. N. (ed.) 2011, *Archaeoastronomy and Ethnoastronomy: Building Bridges between Cultures*, Proc. IAU Symp. No. 278, Cambridge University Press, Cambridge.

Ruggles, C. L. N. & Cotte, M. (eds) 2010, *Heritage Sites of Astronomy and Archaeoastronomy in the Context of the UNESCO World Heritage Convention: a Thematic Study*, ICOMOS–IAU, Paris.

Sullivan, W. T. III 2009, *Cosmic Noise: A History of Early Radio Astronomy*, Cambridge University Press, Cambridge.

Véron, P. (ed.) 2009, *Verres d'optique et lunettes astronomiques*, Fondazione Ronchi, Firenze.

Wali, K. C. 2011, *A Scientific Autobiography: S. Chandrasekhar*, World Scientific Publishing Company.

Whiting, A. 2010, *Hindsight and Popular Astronomy*, World Scientific Publishing Company.

Wolfschmidt, G. (ed.) 2009, *Cultural Heritage of Astronomical Observatories: from Classical Astronomy to Modern Astrophysics*, ICOMOS–Bässler-Verlag.

Transactions IAU, Volume XXVIIIA
Reports on Astronomy 2009–2012   © International Astronomical Union 2012
Ian Corbett, ed.   doi:10.1017/S1743921312003134

DIVISION IX/COMMISSION 41/WORKING GROUP ARCHIVES

CHAIR	Ileana Chinnici
VICE-CHAIR	Oscar T. Matsuura
MEMBERS	Brenda Corbin
	Suzanne Débarbat
	Daniel Green
	Tsuko Nakamura
	Adam Perkins
	Irakli Simonia

TRIENNIAL REPORT 2009–2012

1. Introduction

In the triennium 2006–2009, the WG Archives promoted an action of recognition of the archival documents concerning the foundation of IAU. The pretext was given by the fact that the year 2009 was the 90th anniversary of the establishment of IAU and the WG considered worth to start checking the conditions of preservation and conservation of such documents, in view of the approaching centenary.

As suggested by S. Débarbat, a Working Plan, entitled "IAU Archives" was proposed by I. Chinnici and approved by the WG members (see: Chinnici *et al.* 2009). The plan was mainly aimed at collecting information about the IAU archival documents kept in the countries first members (class B documents).

Following the plan guidelines, the search of such documents started in some countries, and the first results were mentioned in the triennial report 2006–2009.

The WG meeting held at the past IAU General Assembly (2009). During the IAU General Assembly held in Rio de Janeiro in 2009, the WG Archives has discussed the results and confirmed the value of the Working Plan and, consequently, the importance to continue the efforts for surveying the archival documents related to IAU in the first adhering countries.

It was remarked that many difficulties were met in contacting people in some countries and obtaining collaboration, mainly because of the voluntary nature of this kind of research. The creation of sub-committees was therefore discussed, in order to confer to the delegates a formal authority which could make easier their access to the resources. Moreover, the involvement of historians, librarians and archivists of each country in the WG tasks was strongly recommended.

2. Difficulties and ongoing actions

Unfortunately, the advancement of the work has slowed down and it has been halted in the past triennium. It has been difficult to establish the sub-committees and, in spite of the WG auspices, the collaboration of people involved in archives has been poor. This is probably due to the above-mentioned voluntary work, to the spreading of such documents in several archives, which renders the search very laborious, and to an insufficient circulation of the WG plan in some countries.

However, Latin America is by far the area which has carried out the most advanced recognition, mostly under the impulse given by O. T. Matsuura (see Triennial Report 2006–2009). Recent news from Matsuura concern the elaboration of a model file card for the archival documents, proposed by C. Barboza (see: Matsuura 2010). The model file card is in preparation and it could be suggested as a standard for the IAU document to be inventoried and catalogued in other countries, when possible.

Finally, an inquiry on the conservation of the documents concerning the adhesion of Italy to IAU and the first IAU General Assembly (Rome, 1922) is in progress and a report will be published by I. Chinnici in spring 2012.

3. Next steps

The WG insists on the importance of continuing the actions and tasks of the Working Plan "IAU Archives". At the same time, the WG is aware that it is necessary to re-launch it. Having identified the main problems in the establishment of sub-commissions on a voluntary base, in the spreading of the materials and in the insufficient circulation of the WG Plan, the WG recognizes that this last point can be the new starting point. The WG plan circulation can actually be improved by far, trying to involve other colleagues in the creation of a WG network for sharing and exchanging information. Before the next GA, the WG will ask the C41 Secretary to send the WG plan to all the C41 members, launching a Call for Volunteer Collaborators in all countries.

The aim is that of publishing a report as extensive as possible on the conservation of the archival documents concerning the establishment of IAU in all IAU participant countries, a work hopefully to be achieved before the centenary of IAU in 2019.

Ileana Chinnici
Chair of Working Group

References

Chinnici, I. 2009, IAU Transactions, vol. XXVIIA, Division XII, Commission 41, Working Group Archives, Triennial Report 2006–2009.

Matsuura, O. T. 2010, "Report on the business meeting of the IAU Commission 41 Working Group – Archives in the IAU XXVII General Assembly in Rio de Janeiro, Brazil", ICHA Newsletter, No. 9, pp. 6–8.

Transactions IAU, Volume XXVIIIA
Reports on Astronomy 2009–2012   © International Astronomical Union 2012
Ian Corbett, ed.   doi:10.1017/S1743921312003146

DIVISION IX / COMMISSION 41 / WORKING GROUP
HISTORICAL INSTRUMENTS

CHAIR   Sara J. Schechner
VICE-CHAIR   Tsuko Nakamura
PAST CHAIR   Luisa Pigatto
BOARD MEMBERS   Juergen Hamel
                         Kevin Johnson
                         Rajesh Kochhar
                         Nha Il-Seong
                         Wayne Orchiston
                         Bjørn Ragnvaid Pettersen
                         Shi Yunli

TRIENNIAL REPORT 2009–2012

1. Introduction

The Working Group on Historical Instruments (WG-HI) was founded by the members of Commission 41 at the 2000 Manchester IAU General Assembly with two main objectives: to assemble a bibliography of existing publications relating to historical instruments, and to encourage colleagues to carry out research and publish their results. Since then the concerns of the Working Group have expanded to include efforts to preserve and protect old astronomical instruments, observatories, and related sites as world cultural heritage and material evidence of the development of astronomy in different parts of the globe.

The WG-HI maintains liaisons with sister organizations through the involvement of its officers and board members in them. These include the Scientific Instrument Commission of the International Union for the History and Philosophy of Science / Division of History of Science and Technology (IUHPS-Scientific Instrument Commission); and the American Astronomical Society (AAS) Working Group for the Preservation of Astronomical Heritage.

2. International Year of Astronomy (IYA) 2009 Initiative

In 2007, the WG-HI of IAU Commission 41 (History of Astronomy) began to organize an interdisciplinary conference – "Astronomy and Its Instruments before and after Galileo"- to be held in Venice in 2009 on the 400th anniversary of Galileo's first observations with a telescope. The goals were expressed as follows:

"The conference aims to highlight mankind's path towards an improved knowledge of the sky using mathematical and mechanical tools as well as monuments and buildings, giving rise, in so doing, to scientific astronomy. It will analyze similarities and differences among cultures and countries in exploiting the shared resource that the sky represents, and will examine the historical-political and scientific background favoring the progress

of scientific astronomy in different epochs and countries, progress that led to a crucial turning-point for observational astronomy when Galileo turned the telescope to the night sky and initiated the New Astronomy. A major aim of the meeting is to help move forward the process of ensuring the recognition and protection of cultural properties around the globe that bear powerful witness to the development of astronomy in diverse cultural contexts."

The plan was endorsed by Commission 46 (Astronomy Education and Development), Commission 55 (Communicating Astronomy with the Public), and by IAU Division XII (Union-Wide Activities).

Promoted as a joint symposium of the IAU and the INAF-Astronomical Observatory of Padova, "Astronomy and Its Instruments before and after Galileo" was held in Venice-San Servolo Isle on 27 September – 3 October 2009. It was listed as an official event of UNESCO's International Year of Astronomy 2009, and sponsored by the IUHPS Scientific Instrument Commission. Patrons included UNESCO; the Istituto Nazionale di Astrofisica (INAF); the Università degli Studi di Padova, Italy; the Facoltà di Scienze, Matematiche, Fisiche e Naturali, Università di Padova; the Centro Interdipartimentale di Ricerca in Storia e Filosofia delle Scienze (CIRSFIS); the Centro per la Storia dell'Università di Padova; the Accademia Galileiana di Scienze, Lettere ed Arti in Padova; the Arab Union for Astronomy and Space Sciences (AUASS); the Società Astronomica Italiana (SAIt); the Comune di Venezia; the Provincia di Venezia; and the Regione del Veneto.

The conference program and other details can be found on the website: http://web.oapd.inaf.it/venice2009/index.php. The proceedings were published: Pigatto & Zanini (2010).

3. Conferences

In addition to the aforementioned conference, several WG-HI members presented papers on astronomical instruments at the International Conference on Oriental Astronomy (ICOA-7), which was held at Mitaka, Tokyo in 2010. The proceedings are now in print: Nakamura, Orchiston, Sôma, & Strom (2011).

Between 2009 and 2012, other Working Group members have taken part (or will take part) in meetings of the IUHPS Scientific Instrument Commission (Budapest, Florence, Kassel), the Historical Astronomy Division of the American Astronomical Society (Washington-DC, Seattle, Austin), the Antique Telescope Society (Ann Arbor, Charlottesville, Tuscon), and various special symposia featuring the history of the telescope. Many of these papers are in press.

Members of the Working Group are currently planning sessions on instruments for the General Assembly in Beijing in 2012. Among these, one session will focus on field expeditions, covering not only Transits of Venus (of particular interest in 2012), but also eclipses, determination of longitude, and so forth.

4. Projects

Prior to this triennium, the WG-HI had begun preparation of a thesaurus of historical instruments used in astronomy and related disciplines such as geography, geodesy, navigation, meteorology, and chronology. It was to be a list of terms plus variants and synonyms from different countries, etymologies, general definitions, related bibliographic sources, and images. A preliminary list of instruments was circulated among Working Group members. In 2009 during the current triennium, the project was discontinued when the Working Group learned that it duplicated work already done by the IUHPS Scientific Instrument Commission, museums with major holdings of historical scientific instruments, and other learned societies at the intersection of the history of astronomy and early scientific instruments. Moreover, the availability of these resources on the Web made the publication of a thesaurus by the WG-HI redundant.

In anticipation of the 2012 Transit of Venus, the WG-HI is encouraging knowledgeable scholars and museums holding apparatus used for past transits to collaborate by adding additional material to a Transit of Venus website created in 2004 by the IUHPS Scientific Instrument Commission (http://transits.mhs.ox.ac.uk).

Sara J. Schechner
Chair of Working Group

References

Nakamura, T., Orchiston, W., Sôma, M., & Strom, R. (eds.) 2011. *Proceedings of the Seventh International Conference on Oriental Astronomy*. Tokyo, National Astronomical Observatory of Japan.

Pigatto, L., & Zanini, V. (eds.) 2010. *Astronomy and Its Instruments before and after Galileo*. Padua, Cooperativa Libraria Editrice Università di Padova. pp. 512.

Transactions IAU, Volume XXVIIIA
Reports on Astronomy 2009–2012   © International Astronomical Union 2012
Ian Corbett, ed.   doi:10.1017/S1743921312003158

DIVISION XII / COMMISSION 41 / WORKING GROUP TRANSITS OF VENUS

CHAIR   Hilmar W. Duerbeck (Germany)
PAST CHAIR   Steven Dick (USA)
BOARD   Robert van Gent (Netherlands)
                      David Hughes (UK)
                      Willie Koorts (South Africa)
                      Wayne Orchiston (Australia)
                      Luisa Pigatto (Italy)

TRIENNIAL REPORT 2009–2011

1. Introduction

The previous Transits of Venus Working Group report #6, covering the time mid-2006 to mid-2009, was published by the undersigned in the Journal of Astronomical History and Heritage, Vol. 12, p. 254 (2009). The present report #7, covering the time up to mid-2011, has been prepared for the Reports presented at the IAU General Assembly in Beijing, August 2012. It is expected that after a flurry of publications before and after the transit of 2012, the activities in the field will drop dramatically, and it is planned to terminate the activities of the working group after the Beijing General Assembly, and to write a closing report for the subsequent General Assembly.

As already observed in the previous report, activities between the transits of 2004 and 2012 were most of the time at a low level. At the time of the 2012 transit, symposia are planned in Tromsø (Norway) by Peer Pippin Aspaas and in East Asia or Australia by Wayne Orchiston. There will also be a Historical Astronomy Division special meeting at the American Astronomical Society Austin meeting, Sunday, 8 Jan 2012.

2. Publications in the past triennium

A listing of publications that have appeared since 2009, with some older overlooked references, is given in section 4. A web bibliography, mainly on the 17th to 19th century transits, with many links to original sources, is kept by R. van Gent (see section 3).

The list of publications includes two popular books by Bucher (2011) and Lomb (2011), with a forthcoming second edition of Bill Sheehan's and John Westfall's book *The Transits of Venus*, now titled *Eclipses, Transits and Occultations*, is foreseen for early 2012. Historical topics are covered in a thesis on *Solar eclipses and transits of Venus, 1874-1887, and their role in the popularisation of astronomy in the USA* by Cottam (2011), also presented by Cottam, Orchiston and Stephenson (2011a,b), and in studies on the 1769 and 1882 transit observations by William Wales (Metz 2009), José Antonio Alzate (Vaquero, Trigo, Gallego and Moreno-Corral 2007) and Charles Houzeau (Sterken 2009). The observation of a transit 3000 years ago by Australian Aboriginals is questioned (Herrmann 2011).

Finally, besides the fine satellite-imaging of the 2004 transit (Pasachoff, Schneider and Widemann 2011), more observations and analysis are reported by McKim, Blaxall and Heath (2009) and by Popescu and Rusu (2007). Venus transit spectroscopy (2004) as a tool for exoplanet studies is discussed by Hedelt, Alonso, Brown, *et al.* (2011).

3. Websites

Websites dedicated to historical Venus transits are:

Robert van Gent's Transit of Venus bibliography: http://www.phys.uu.nl/ vgent/venus/ venustransitbib.htm; another version of the 17th and 18th century transits is also available at: http://transitofvenus.nl/wp/ past-transits/bibliography-1631-1639/ and past-transits/bibliography-1761-1769/

Steven van Roode's Historical observations of the transit of Venus (with reports, photos and engravings, coordinates, maps of sites of the 17th to 19th century observers, as well as photos of commemorative plaques: http://transitofvenus.nl/wp/past-transits/

Other websites of interest:

French National Node of the VT-2004-2012 project: http://www.imcce.fr/vt2004/en/index.html

Chuck Bueter's page: http://www.transitofvenus.org/

and Jay Pasachoff's page: http://web.williams.edu/astronomy/eclipse/transits/ToV2012/index.htm

Hilmar W. Duerbeck
chair of Working Group

References

Bucher, G. 2011, *Die Spur des Abendsterns. Die abenteuerliche Erforschung des Venustransits* (Darmstadt: Wissenschaftliche Buchgesellschaft), 215 p.

Cottam, S., Orchiston, W., & Stephenson, F. R. 2011, in: W. Orchiston, T. Nakamura, and R. Strom (eds.), *Highlighting the History of Astronomy in the Asia-Pacific Region* (New York: SpringerP, p. 225

Cottam, S., Orchiston, W., & Stephenson, F. R. 2011, *AAS Meeting #217*, #146.07, *BAAS*, 43.

Cottam, S. 2011, *Solar eclipses and transits of Venus, 1874-1887*, Ph.D. Thesis, Centre for Astronomy, James Cook University

Hedelt, P., Alonso, R., & Brown, T., *et al.* 2011, *A&A*, 533, A136, doi: 10.1051/0004-6361/201016237

Herrmann, D. B. 2011, *Acta Historica Astronomiae*, 43, 351

Lomb, N. 2011, *Transit of Venus: 1631 to the present* (Sydney, New South Publishing), 228 pp.

McKim, R. J., Blaxall, K., & Heath, A. 2009, *JBAA*, 117, 65

Metz, D. 2009, *Science and Education*, 18, 581

Pasachoff, J. M., Schneider, G., & Widemann, T. 2011, *AJ*, 141, art. 112; doi: 10.1088/0004-6256/141/4/112

Popescu, A. S. & Rusu, M. 2007, *Romanian Astr. J.*, 17, 77

Sterken, C. 2009, *AN*, 330, 582

Vaquero, J. M., Trigo, R. M., Gallego, M. C., & Moreno-Corral, M. A. 2007, *Solar Physics*, 240, 165

Transactions IAU, Volume XXVIIIA
Reports on Astronomy 2009–2012
Ian Corbett, ed.

DIVISION IX/COMMISSION 41/WORKING GROUP ASTRONOMY AND WORLD HERITAGE

CHAIR	Clive Ruggles
VICE-CHAIR	Gudrun Wolfschmidt

TRIENNIAL REPORT 2009–2012

1. Introduction

The WG was created in 2008 to progress UNESCO's Astronomy and World Heritage Initiative (AWHI) jointly with the World Heritage Centre, following the signing of a formal Memorandum of Understanding between the IAU and UNESCO.

The AWHI is a thematic initiative aiming to identify, safeguard and promote cultural properties connected with astronomy. The places in question do not just include sites (such as observatories) important in the development of modern scientific astronomy, but also much older constructions whose design or location relate to celestial objects and events, reflecting the ways in which ancient cultures attempted to make sense of the world—the cosmos—within which they dwelt. One of its main objectives of the WG is to help to establish guidelines for State Parties to the World Heritage Convention who wish to nominate sites for inscription on the World Heritage List on the grounds of their relationship to astronomy.

The WG is also concerned with the link between the cultural heritage of astronomy and the natural heritage of the dark night sky. For this reason, it collaborates with organisations concerned with protecting and preserving the dark night sky, such as the Starlight Initiative and the Dark Skies Advisory Group.

The WG maintains a website at http://www.astronomicalheritage.org/.

2. The Thematic Study

From the outset, it had been evident that it would be unproductive to try to develop an 'IAU list' of the most important astronomical heritage sites, since this could be interpreted as undermining the process by which State Parties to the World Heritage Convention identify cultural properties that they consider worthy of nomination to the World Heritage List (WHL) and their applications are considered by UNESCO and its advisory bodies.

Instead, the WG quickly identified two flagship projects. The first of these was to work with the International Council on Monuments and Sites (ICOMOS), the Advisory Body to UNESCO responsible for cultural sites, to produce a 'Thematic Study' on the Heritage Sites of Astronomy. The purpose of ICOMOS Thematic Studies is to present an overall vision of some area of cultural heritage, which helps to establish firm criteria by which nominations of sites to the WHL relating to the topic in question can be judged. This in turn encourages State Parties to identify and submit viable nominations. Case studies of particular sites form an integral part of a Thematic Study, but their purpose is simply to raise and illustrate general issues for the broader discussion; the inclusion or otherwise of any particular site as a Case Study is understood to have no bearing on the outcome should it ever be nominated to the WHL.

Thanks to the intense efforts of 41 members of the Working Group, the ICOMOS-IAU Thematic Study on astronomical heritage, Ruggles & Cotte (2010), was published in June 2010 in time to be presented at the 2010 meeting of UNESCO's World Heritage Committee at the end of July, where it was duly approved. It was subsequently circulated officially by the WHC to all UNESCO National Commissions and is also freely available publicly.

The subject matter ranges from early prehistory to modern astrophysics and space heritage, and also prominently includes dark sky issues and modern observatory sites. In view of the existence of Wolfschmidt (2009), a report produced a year earlier following a meeting organised by ICOMOS-Germany and the University of Hamburg, it was not considered necessary to give special emphasis to classical observatories from the renaissance to the mid-twentieth century, which were treated in equal measure to 14 other cultural heritage themes.

The ICOMOS–IAU Thematic Study has been recognised to be of broader importance, being the first in any field of science heritage, and leads the way in tackling a number of broader issues that apply to science heritage in general and have impeded the successful inscription of science heritage sites.

The 2010 edition was published as an e-book; a printed version, Ruggles & Cotte (2011), followed in 2011.

3. The Astronomical Heritage Web Portal

The second flagship project is the development of an 'Astronomical Heritage Portal'. This is in every sense a "follow-up" to the Thematic Study, since it contains all the information currently presented in the Thematic Study together with more detailed site data, further case study sites, and facilities to comment upon and discuss both the sites themselves and broader issues. In other words, this is a dynamic, publicly accessible database, together with a moderated discussion forum and document-repository, concerning astronomical heritage sites throughout the world.

At the time of writing the site is being tested by WG members and Thematic Study authors with a view to launching it publicly within a timescale of weeks. Once launched, it will become the main website for UNESCO's Astronomy and World Heritage Initiative, formally supported and approved by UNESCO and carrying the UNESCO and World Heritage Centre logos, but implemented and maintained independently.

The longer-term aim is to ensure that the portal becomes 'the' place visited by National Commissions seeking further information, guidance and support once they seek to promote astronomical heritage sites (natural or cultural) to their "national tentative lists" with a view, in the longer term, to nominating them for inscription onto the World Heritage List. This means that the information on the portal must be maintained to the highest professional standards, so that constant monitoring and careful moderation of the content will be needed. This will be an ongoing process into which the WG could continue to have a vital input.

4. Extended Case Studies

Following a meeting at the UNESCO World Heritage Centre in September 2010, attended by a number of WG members together with representatives of ICOMOS and the IAU General Secretary, the WG undertook to develop a selection of the case studies included in the Thematic Study in more detail, following the structure of actual nomination dossiers that are presented to UNESCO by State Parties. The principal aim of these 'extended case studies' is to develop components of potential dossiers that relate particularly to astronomical heritage aspects, this being the 'uncharted territory' that the Thematic Study has attempted to start to map out. In doing so, we hope to identify for further discussion and clarification some of the key general issues that can arise in the particular case of astronomical heritage sites.

Each of the eleven sites that have been chosen as extended case studies raises a particular set of issues. For example, Stonehenge is included—despite already being on the WHL—since it raises issues relating to the re-inscription of existing World Heritage Sites to give more explicit recognition of their astronomical significance. Another four of the extended case studies are cultural sites with a wide geographical and temporal span—The Royal Observatory, Cape of Good Hope, South Africa; Bayconur Space Launch Facility, Russian Federation; The seven-stone antas (a group of prehistoric dolmens) of Portugal and Spain; and the 'thirteen towers' of Chankillo, Peru. A further five are dark sky parks and modern observatory sites where the main concern is to preserve the natural heritage of the dark night sky. The last extended case study—the Star Clocks of Oman—brings the cultural heritage and natural heritage issues together in a very direct way. Here, modern indigenous cultural practices of star observation, linked to vital environmental issues such as water management, are threatened by the erosion of dark skies.

We are developing these extended case studies in stages, for completion by Easter 2012. They will then be discussed and finalised at a Working Group Forum to be held in New Zealand during June 2012, and presented in a report to the IAU General Assembly in Beijing in August. By agreement, the extended case studies will not be made public during this period. After that, however, the aim is to release our report publicly within the context of the Astronomy and World Heritage Initiative.

Clive Ruggles
Chair of Working Group

References

Ruggles, C. L. N. & Cotte, M. (eds) 2010, *Heritage Sites of Astronomy and Archaeoastronomy in the Context of the UNESCO World Heritage Convention: a Thematic Study*, Electronic edition, ICOMOS–IAU, Paris.

Ruggles, C. L. N. & Cotte, M. (eds) 2011, *Heritage Sites of Astronomy and Archaeoastronomy in the Context of the UNESCO World Heritage Convention: a Thematic Study*, Printed edition, ICOMOS–IAU, Paris.

Wolfschmidt, G. (ed.) 2009, *Cultural Heritage of Astronomical Observatories: from Classical Astronomy to Modern Astrophysics*, ICOMOS–Bässler-Verlag.

Transactions IAU, Volume XXVIIIA
Reports on Astronomy 2009–2012
Ian Corbett, ed.

© International Astronomical Union 2012
doi:10.1017/S1743921312003171

DIVISION XII / COMMISSION 50
PROTECTION OF EXISTING AND POTENTIAL OBSERVATORY SITES

CHAIR	Wim van Driel
VICE-CHAIR	Richard Green
PAST CHAIR	Richard J. Wainscoat
BOARD	Elizabeth Alvarez del Castillo
	Carlo Blanco
	David L. Crawford
	Margarita Metaxa
	Masatoshi Ohishi
	Woodruff T. Sullivan III
	Anastasios Tzioumis

COMMISSION 50 WORKING GROUP
Div. XII / Commission 50 WG   Controlling Light Pollution

TRIENNIAL REPORT 2009–2012

1. Introduction

The activities of the Commission have continued to focus on controlling unwanted light and radio emissions at observatory sites, monitoring of conditions at observatory sites, and education and outreach. Commission members have been active in securing new legislation in several locations to further the protection of observatory sites as well as in the international regulation of the use of the radio spectrum and the protection of radio astronomical observations.

To kick off the IAU/UNESCO International Year of Astronomy, 2009, IAU Symposium 260 on the Rôle of Astronomy in Society and Culture was held in Paris. Commission 50 members gave presentations on the importance of saving the magnificent night sky from light pollution and on the need to protect radio astronomy from unwanted interference.

Also in 2009, at its XXVII General Assembly in Rio de Janeiro, the IAU passed Resolution B5 in Defence of the Night Sky and the Right to Starlight. The Resolution encourages IAU members to assist in raising public awareness about the contents and objectives of the International Conference in Defence of the Quality of the Night Sky and the Right to Observe Stars [http://www.starlight2007.net/], in particular the importance of preserving access to an unpolluted night sky for all mankind.

2. Controlling light pollution

A new challenge to protection of observatories from light pollution is coming from a widespread switch towards blue-rich artificial light sources, whose light energy emission peaks near 450 nm, away from sodium-based light sources, which peak near 590 nm.

At most observatory sites, Rayleigh scattering is the dominant mechanism that brings artificial light into the telescope, and this scattering is strongly wavelength dependent – blue light scatters much more efficiently than amber (sodium) light. The emission from blue rich light sources peaks at wavelengths where the human eye is not very sensitive, but the natural night sky is very dark and astronomical detectors very sensitive indeed.

Recent advances in metal halide, induction, fluorescent and particularly light emitting diodes (LEDs) have made these blue rich light sources competitive with sodium lights in terms of energy efficiency (lumens per Watt), and some LEDs are now more efficient than sodium lamps. Some municipalities are selecting the most energy efficient LEDs, which are very blue rich, despite the harm that they do to the night sky; blue rich white light is also harmful to many species of animals, and affects the circadian rhythm in humans and animals.

Several approaches can help to minimize the impact of LEDs and other blue-rich white light sources on astronomy. These include 3 methods, of which nos. 2 and 3 result in much less impact on observatories than method 1:

- 1. Use of low correlated color temperature light sources – CCT < 3,000K is strongly recommended. More energy efficient "warm white" LEDs, which use a phosphor to convert some of their blue light to longer wavelengths, have recently become available, and metal halide, induction and fluorescent lamps and LEDs are all readily available with low CCT.
- 2. Filtering out the blue light. On the island of Hawaii, where Mauna Kea Observatory is located, LED light sources have been installed that use a filter that absorbs nearly all light below 500 nm.
- 3. Use of a carefully crafted mix of LEDs. For example, a mix of 3 amber LEDs and one low CCT white LED produces a light with adequate color rendition, low CCT and minimal blue light.

On the positive side, light from LED light sources can be much more efficiently directed, and most lighting tasks can efficiently be performed using fully shielded light sources. So a lighting task can be achieved using many less total lumens using LEDs. Use of fully shielded light sources has been shown to be of paramount importance for the protection of observatories. We have been cooperating with the *International Dark Sky Association* (IDA) in some aspects of this.

Commission 50 members have increased their presence at meetings of the *Commission internationale de l'éclairage* (CIE), the professional organization of lighting engineers, and participated in a number of technical committees. The IAU liaison with the CIE is Elizabeth Alvarez del Castillo, and David Galadí-Enríquez is our representative on the new CIE Lighting Quality and Energy Efficiency Network, of which the IAU is one of the participating organizations. At the CIE's 2009 Interim Conference, Connie Walker explained how light emitted near the horizontal contributes significantly to sky glow (Luginbuhl 2009), while Richard Wainscoat showed why the newer blue-rich white light sources are especially harmful to astronomical research (see the Section above). In addition to technical committee meetings, CIE Divisions 4 and 5 held a special workshop on LEDs, at which Francisco Javier Díaz Castro and Elizabeth Alvarez represented the IAU. In spring 2010, Ferdinando Patat gave an invited talk at a special CIE conference on energy efficient lighting. In fall 2010, we had standard representation at the CIE Divisions 4 and 5 meetings. C50 members David Galadí-Enríquez and Ramotholo Sefako attended the CIE Quadrennial in 2011. Emphasis has been put this past triennium on raising awareness of the 2 main issues on which we gave talks at the 2009 CIE Conference.

Commission 50 has strengthened its links with the *International Dark Sky Association* (IDA). Connie Walker chairs the IDA Education Committee and has been nominated as member of the IDA Board of Directors; she was awarded the 2011 IDA Hoag-Robinson Award.

One way the legacy of the 2009 IAU/UNESCO International Year of Astronomy's Global Cornerstone Project "Dark Skies Awareness" has continued is through the Global Astronomy Month's "Dark Skies Awareness" programs (Connie Walker, chair). Global Astronomy Month (GAM) has taken place in April over the last two years and is continuing this coming year. One of the main "take-away" messages from GAM is why we should preserve our dark night skies. Several dark skies events and activities are being held worldwide on behalf of GAM to promote public awareness on how to save energy and save our night sky. Specific programs include:

• GLOBE at Night: an international citizen-science campaign to raise public awareness of the impact of light pollution by inviting people to measure their night sky brightness and submit their observations to a website. The campaign has run for two weeks each winter/spring for the last six years. People in 115 countries have contributed 66,000 measurements, making GLOBE at Night one of the most successful light pollution awareness campaigns (Connie Walker, director).

• Dark Skies Rangers: an environmental/astronomy-based program that includes student participation in GLOBE at Night. Students learn the importance of dark skies and immerse themselves in, e.g. activities that illustrate proper lighting technology and what light pollution's effects are on wildlife (Connie Walker, director).

• International Dark Sky Week: with activities, events and info on how to light more responsibly (Connie Walker, director), done with the IDA.

• International Earth & Sky Photo Contest: inviting landscape astrophotographers to capture the beauty of the night sky and/or light pollution's effect on it, done with The World At Night and NOAO.

• World Night in Defense of Starlight (April 20): recognizes the importance of a pristinely dark sky as it relates to our culture, done with the Starlight Initiative.

• One Star at a Time: creates accessible public places to observe a dark night sky.

• Dark Skies Awareness: with 10 minute audio podcasts on the "Dark Skies Crusader" and the effects of light pollution on wildlife, energy, health and astronomy, done with NOAO.

• Astropoetry: on dark skies awareness.

Commission 50 and other IAU members participated in the 9th, 10th and 11th European Symposium for the Protection of the Night Sky (Dark Sky Symposium), which were held in 2009, 2010 and 2011, respectively. The aim of the Symposium is to address issues of light pollution, its causes, its negative effects and possible remedies. Some of the specific issues have been measuring methods of light pollution, the quality of dark sky parks, efficient dark sky friendly lighting and environmental impact (e.g. on insects, humans) of modern light sources (e.g. LEDs). We presented the activities of our Commission, biological studies and numerous Dark Skies Awareness Programs.

Commission 50 endorsed a proposal submitted by Richard Green for a special session on light pollution at the upcoming 2012 IAU General Assembly in Beijing, which was later merged with another proposal submitted by Beatriz Garcia of Commission 46 (Astronomy Education and Development), to be ultimately approved as SpS17 – Light Pollution: Protecting Astronomical Sites and Increasing Global Awareness through Education. The progam will cover a combination of education and outreach topics, and more technical ones focused on astronomical site protection and the potential of spectral encroachment of new types of light sources in the blue.

3. Controlling Radio Frequency Interference

The main goal is to represent the interests of the international astronomical community in matters concerning the protection of radio frequencies allocated to the Radio Astronomy Service and minimize interference to our scientific observations and measurements.

The *International Telecommunication Union* (ITU) is an international organization under the United Nations, which establishes and maintains international rules on frequency use. IAU interaction with the ITU is primarily made via the *Scientific Committee on Frequency Allocations for Radio Astronomy and Space Science* (IUCAF). This committee is composed of representatives from the IAU, URSI (*International Union of Radio Science*) and COSPAR (*Committee on Space Research*), and it operates as an inter-disciplinary committee of the *International Council for Science* (ICSU).

Three Commission 50 officers are members of IUCAF: Masatoshi Ohishi, Tasso Tzioumis and Wim van Driel. Masatoshi Ohishi, the IUCAF chair, served as liaison between the IAU and the ITU, and Tasso Tzioumis chaired the Working Group on Radio Frequency Interference of IAU Division X until August 2009.

The interests and activities of IUCAF range from preserving what has been achieved through regulatory measures or mitigation techniques, to looking far into the future of high frequency use, giant radio telescope use and large-scale distributed radio telescopes. Current priorities, which will certainly keep us busy through the next years, include the use of powerful radars and satellite down-links close in frequency to the radio astronomy bands, the coordination of the operation in shared bands of radio observatories and powerful transmissions from downward-looking satellite radars, the possible detrimental effects of ultra-wide band (UWB) transmissions at around 24/79 GHz regions and high-frequency power line communications (HF-PLC) on all passive services, the scientific use of the 275 to 3000 GHz frequency range, and studies on the operational conditions that will allow the successful operation of future giant radio telescopes.

In the period 2009–2011, IAU representatives on IUCAF participated in a total of 14 international meetings, mainly of various ITU Working Parties and Study Groups. At the ITU, the work of interest to the IAU was focused on the relevant agenda items that were adopted in 2007 for the World Radiocommunication Conference in 2012, WRC-12, as well as on the creation and maintenance of various ITU-R Recommendations and ITU-R Reports. They were also active in the organization of the 3rd IUCAF Summer School on Spectrum Management for Radio Astronomy, which was held in 2010, at the National Astronomical Observatory of Japan (NAOJ) in Tokyo.

The WRC-12 agenda item which is most relevant to radio astronomy concerns the use of the radio spectrum between 275 and 3000 GHz ($\lambda\lambda$ 0.1-1 mm), a frequency range used for observations of important spectral lines and continuum bands. Significant infrastructure investments are being made for the use of these bands, such as the Atacama Large Millimeter/submillimeter Array (ALMA), a facility currently under construction in northern Chile.

No frequency allocations for the use of this frequency range will be made at WRC-12, but the radio astronomy community was tasked to identify a list of specific bands of interest. This list was established in close collaboration with the IAU Working Group on

Important Spectral Lines (chaired by Masatoshi Ohishi), and in 2010 a new ITU-R Recommendation RA.1860 was adopted on preferred frequency bands for radio astronomical measurements in the range 1-3 THz. Another recent Report, ITU-R RA.2189, showed that the frequency range above 1000 GHz can be used by both passive (i.e., receive-only – such as radio astronomy) and active (transmitting) radio services with little possibility of interference.

Power Line Communications (PLC) utilizing the 2-30 MHz frequency range is a technology to send electrical signals for communication purposes through power lines that were designed and installed to carry current at 50/60Hz only. There has been serious concern that the electromagnetic field radiated by power lines may cause harmful interference to radio astronomical observations. Radio astronomers submitted several documents containing measurement results of actual harmful interference from PLC and theoretical analyses to the ITU-R, which were included in ITU-R Report SM.2158.

4. Closing remarks

Our Commission continues to recognize that more professional astronomers, especially IAU members, need to become involved in efforts to reduce light pollution and radio interference – see also the 2009 IAU Resolution B5 referred to in the Introduction.

Wim van Driel
president of the Commission

Reference

Luginbuhl, C. B., Walker, C. E., & Wainscoat, R. J. 2009, *Physics Today*, Vol 62, 12, p32

Transactions IAU, Volume XXVIIIA
Reports on Astronomy 2009–2012
Ian Corbett, ed.

© International Astronomical Union 2012
doi:10.1017/S1743921312003183

COMMISSION 55	COMMUNICATING ASTRONOMY WITH THE PUBLIC COMMUNIQUER ASTRONOMIE AVEC LE PUBLIC

PRESIDENT	Dennis R. Crabtree
VICE-PRESIDENT	Lars Lindberg Christensen
PAST PRESIDENT	Ian Robson
ORGANIZING COMMITTEE	Oscar Alvarez Pomare, Augusto Damineli Neto, Richard T. Fienberg, Anne Green, Ajit K. Kembhavi, Kazuhiro Sekiguchi, Patricia Ann Whitelock, Jin Zhu

COMMISSION 5 WORKING GROUPS

Div. XII / Commission 55 WG	CAP Journal
Div. XII / Commission 55 WG	New Ways
Div. XII / Commission 55 WG	CAP Conferences
Div. XII / Commission 55 WG	Best Practices
Div. XII / Commission 55 WG	VAMP
Div. XII / Commission 55 TF	Washington Charter

TRIENNIAL REPORT 2009–20012

1. Introduction

Commission 55 was approved at the IAU General Assembly in Prague following the great success of the Communicating Astronomy Working Group, which had been set up in 2003. It resides within Division XII and the mission statement of the Working Group has been incorporated into the Commission:

• To encourage and enable a much larger fraction of the astronomical community to take an active role in explaining what we do (and why) to our fellow citizens.

• To act as an international, impartial coordinating entity that furthers the recognition of outreach and public communication on all levels in astronomy.

• To encourage international collaborations on outreach and public communication.

• To endorse standards, best practices and requirements for public communication.

2. Commission Activities

Commission 55 submitted a successful proposal for a Special Session at the General Assembly in Beijing in 2012. The Session, *Communicating Astronomy with the Public for Scientists*, has the goal of providing interested astronomers with training and tips on effective communication with the public. This Special Session will be especially valuable to those astronomers from smaller institutes that do not have communication specialists on staff and who are responsible for communicating with the public in their area.

The SOC will identify key speakers knowledgeable in various areas of science communication who will share their best practices via Invited Talks. Contributed Talks will be invited from astronomers who have successfully communicated astronomy with the public so that they can share their experiences with the other participants.

Commission 55 is supporting, and working with, a small of astronomy communication professionals on establishing a professional association for astronomy communicators. The *International Organization of Professional Astronomy Communicators* would bring together the people involved in astronomy communication to support advancing the profession across both old and new media. These communication professionals form a significant aspect of the participants at the CAP meetings and the synergy of the intereaction between them and the astronomers doing astronomy communication is invaluable.

3. Activities of the Working Groups

This section reports the activities of those Working Groups that have been active. Unfortunately some of the Working Groups were dormant during this period.

3.1. *CAP Conferences*

The Communicating Astronomy with the Public (CAP) conference is the brand name of C55, following two highly successful meetings: ESO HQ in 2005 and Athens in 2007.

CAP2010 was held in Capetown, South Africa from March 15th through 19th and focussed on summarizing and building upon the great success of IYA2009. This was a huge success with approximate 150 participants from many countries around the world. The participants learned about many of the successful programs during IYA2009, from the global cornerstone projects guided by the IAU Secretariat to the small one person projects in many of the smaller countries not usually associated with astronomy. The participatns were able to share their lessons learned and discuss ways in which to build upon the momentum of IYA2009. Capetown provided an excellent venue and the SAAO provided excellent support for the meeting.

CAP2010 was held in Beijing, China. This is the first time a CAP meeting has been held in Asia and given the recent financial turmoil and austerity of many goverements, we were somwhat concerned about the meeting. While the attendance was much smaller than recent CAP meetings (80 people), the participants represented 20 countries and the program was simply excellent. There were many presentations on successful IYA2009 projects that had becmae established as ongoing activities in their respective countries. The Beijing Planetarium were excellent hosts and provided a wonderful venue, top-notch logistics, and a memorable expereience for all participants.

3.2. *VAMP*

The Virtual Astronomy Multimedia Project (VAMP) working group has had an ongoing goal to develop and promote metadata standards appropriate for capturing the rich contextual information in various astronomical image assets. The last two years has seen substantial progress in both the development of the metadata standard and its implementation amongst a variety of observatories. A more detailed description is available on the Commission 55 web pages as well as on the VAMP web pages.

3.2.1. *Astronomy Visualization Metadata (AVM) Standard*

The Astronomy Visualization Metadata (AVM) Standard defines the metadata schema that captures the contextual information for astronomy graphics assets, including astronomical images derived from one or more observational datasets, illustrations and artistic renderings, graphs and data plots, and photographs. The 1.1 version of the standard is an official note of the IVOA, and in the last year, small updates have advanced AVM to version 1.2 (currently in draft form at the VAMP website).

3.2.2. *Tools and Resources*

AVM has been implemented in a variety of astronomical software tools to facilitate its use by the community. Aside from plug-in panels allowing tagging in Adobe Photoshop, two sets of Python scripting libraries have been developed. The PinpointWCS application (currently Mac, but in development for Windows and Linux) allows coordinate information to be captured from astrometric FITS files and applied to JPEG/TIF image files. Microsoft's WorldWide Telescope (Windows desktop client, and multi-platform web client) will read AVM-tagged images and will display them in their proper location in the sky.

Two web resources are in beta development at NASA's Infrared Processing and Analysis Center (IPAC). AVM Toolkit (avm.ipac.caltech.edu) allows web uploads and tagging using the most current 1.2 standard. The Astropix Archive (http://astropix.ipac.caltech.edu), currently in an alpha development stage, is a web archive that ingests AVM-tagged image galleries (provided through RSS feeds) and allows users to search for imagery using sophisticated filtering in both a web interface and automated query form.

3.2.3. *AVM Partners and Supported Websites*

A number of key observatories are routinely tagging imagery as part of the release process and are making these images available in their public downloads. Websites that have integrated AVM tagging into the core database systems include the Spitzer Space Telescope and the European Southern Observatory. Other sites that publish AVM-tagged images currently include Hubble (through both the ESA and STScI sites), the Chandra X-Ray Observatory, Wide-field Survey Explorer (WISE), the Galaxy Evolution Explorer (GALEX). Other missions actively interested in AVM support include Thirty Meter Telescope (TMT) and Large Synoptic Survey Telescope (LSST).

3.2.4. *Current Focus and Community Outreach*

VAMP/AVM topics have been promoted consistently at conferences and workshops. Regular presentations are made at AAS, APS, and Communicating Astronomy with the Public (CAP) conferences, and is a regular topic at astronomy visualization workshops. A special AVM Workshop, "Tag, Get, Use", was held at the Spitzer Science Center in October of 2010 to wrap up changes in the 1.2 version of AVM and bring partner projects together to promote best practices in tagging. There are ongoing discussions between the AVM/VAMP working group and the Virtual Observatory (VO) to bring future AVM projects into VO compliance.

4. Closing remarks

Commission 55 is now established as an essential aspect of the internation network of astronomy communicators. This network consists of both professional astronomers and communication professionals that span developed and developing nations. Commission 55's role in the future will be to foster the recognition of the role of professioinal astronomers as stated in the Washington Charter and to foster relationships between professional astronomers and communicators.

Transactions IAU, Volume XXVIIIA
Reports on Astronomy 2009–2012
Ian Corbett, ed.

EXECUTIVE COMMITTEE WORKING GROUP
FUTURE LARGE SCALE FACILITIES

CHAIR	**Roger Davies**
MEMBERS	**I. Corbett – IAU General Secretary**
	R. Ekers
	R. Green
	M. Iye
	R. Kraan-Korteweg
	M. T. Ruiz
	L. Tacconi
	M. Tarenghi
	C. Wilson
	G. Zhao

TRIENNIAL REPORT 2009–2012

1. Introduction

Some 10 years ago the IAU Executive Committee created a Working Group on Future Large Scale Facilities, but its activities have fallen somewhat into abeyance in the last few years. It has been decided to revive the group, and Roger Davies (Oxford) has agreed to chair it.

2. Terms of Reference

The revised Terms of Reference agreed at EC89 are:

1. To review the status of current planned or proposed large scale ground based and space projects in astronomy.

2. In doing so, to encourage contacts and cooperation between projects.

3. To report on progress against 1. above to the Executive Committee meeting EC90 in April 2012.

4. To organize a Special Session at the 2012 IAU General Assembly to hear presentations on selected projects and to develop a strategic overview of planned and required investment in large scale facilities.

5. To consider what further work might be undertaken by the Working Group, and present a proposal to the meeting of the new Executive Committee at the 2012 General Assembly.

Transactions IAU, Volume XXVIIIA
Reports on Astronomy 2009–2012
Ian Corbett, ed.

© International Astronomical Union 2012
doi:10.1017/S1743921312003201

EXECUTIVE COMMITTEE WORKING GROUP WOMEN IN ASTRONOMY

CO-CHAIRS	Sarah Maddison, Francesca Primas
MEMBERS	Conny Aerts
	Geoffery Clayton
	Françoise Combes
	Gloria Dubner
	Luigina Feretti
	Anne Green
	Elizabeth Griffin
	Yanchun Liang
	Yuko Motizuki
	Birgitta Nordström

TRIENNIAL REPORT 2009–2012

1. Formation and aims

The Working Group was created at the 25th IAU General Assembly in Sydney, Australia, in July 2003 by the IAU Executive Council as a Working Group of IAU Executive. The aims of the Working Group are to evaluate the status of women in astronomy through the collection of statistics over all countries where astronomy research is carried out; and to establish strategies and actions that can help women to attain true equality as research astronomers, which will add enormous value to all of astronomy.

The WG membership changed in August 2009 at the 27th IAU General Assembly in Rio de Janeiro, Brazil. As well as new members, IAU Executive Vice President Martha Haynes was also appointed as the liaison between the Working Group and the IAU Executive.

2. Activities in the last triennium

Since the IAU General Assembly in Rio de Janeiro in August 2009, the WG has been working towards a system for the collation of gender statistics on attendance, speakers and the award of grants at IAU-sponsored events. Thanks to the support of Thierry Montmerle and Lars Holm Nielsen, we now have an online system in place where gender statistics from post-meeting reports is collated online and reported in the WG's website. IAU Symposium organisers complete the online form (www.iau.org/science/meetings/past/meeting_report/womeninastronomy/) and the WG add the data annually to their statistics website (astronomy.swin.edu.au/IAU-WIAWG/projects/IAUmeetingstats.html)

As well as statistics on IAU-sponsored meeting, the WG continues to collate global astronomy gender statistics as they come to hand. Without appropriate funding, it is unfortunately not possible for the WG to conduct its own global survey of the status of women in astronomy.

The WG has also been working with the Special Nominating Committee as required to provide the names of potential candidates for IAU Officer positions, to ensure that prominent women, as well as prominent men, be considered for these important roles.

The WG continues to advocate for affordable child-care facilities for IAU-sponsored meetings. The WG notes that different people have different requirements (e.g. on-site child care, funds to cover child care costs back home, funds to support travel for accompanying family members) and that the IAU does not have a reserve of funds. The WG welcomes that the IAU recognises that support for child care should be taken into account when grant money is distributed. This is now reflected in the rules for the "IAU grants for IAU Symposia". (http://www.iau.org/science/meetings/rules/#231) and "Travel grants for GA Joint Discussions and Special Sessions" (http://www.iau.org/science/meetings/rules/#33).

3. Meetings & workshops

The WG hosted a Women in Astronomy lunch meeting at the 11th Asian-Pacific Regional IAU Meeting, in Chiang Mai, Thailand, on Wednesday 27 July 2011. Prof Jocelyn Bell Burnell presented the keynote address on the status of women in astronomy. About 50 participants attended the meeting.

There was also a women in astronomy lunch meeting at the IAU Symposium 268 "Light Elements in the Universe" held in Geneva, Switzerland, in November 2009. The main message to come from the meeting was that we do not need to wait for the IAU GA lunch meeting which is held every three years, but that each symposium should try and hold their own informal women in astronomy lunch meetings to assist with networking.

4. Plans for the future

It has become clear that efforts to establish strategies and actions that are positive to women in their various career stages and positions within astronomy can best be pursued through person-to-person contact and group discussions. The WG is therefore seeking ways to sustain its momentum during the 3 years between its main meetings (at the IAU GA), and is currently considering a mentoring scheme. Details of the scheme (e.g. whether mentor and mentee should come from the same country or region) will be debated next year in Beijing, and a trial programme will then be launched.

The WG will continue hosting Women in Astronomy meetings at the IAU General Assemblies, but will explore expanding the format of these meetings. At the 28th IAU GA in Beijing 2012 the WG will include a mentoring session as well a poster session to allow more participants to be actively involved in the meeting. The WG will also try to assist Regional IAU Meetings to run Women in Astronomy meetings.

The track the topics that are important to women in astronomy the WG will also summarise the main topics of discussion and any outcomes from various the Women in Astronomy meetings. This will allow the WG to see what issues are common to all women in astronomy, what issues are specific to regions, and what progress is made over time. The WG will continue to collate statistics on the status of women in astronomy and work with the IAU Executive to promote women in astronomy.

Sarah Maddison
Chair of Working Group

CAMBRIDGE JOURNALS

International Journal of Astrobiology

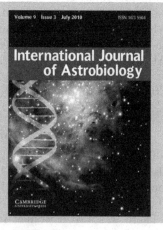

Volume 9 Issue 3 July 2010   ISSN 1473 5504

International Journal of Astrobiology

Managing Editor
Simon Mitton, University of Cambridge, UK

International Journal of Astrobiology is the peer-reviewed forum for practitioners in this exciting interdisciplinary field. Coverage includes cosmic prebiotic chemistry, planetary evolution, the search for planetary systems and habitable zones, extremophile biology and experimental simulation of extraterrestrial environments, Mars as an abode of life, life detection in our solar system and beyond, the search for extraterrestrial intelligence, the history of the science of astrobiology, as well as societal and educational aspects of astrobiology. Occasionally an issue of the journal is devoted to the keynote plenary research papers from an international meeting. A notable feature of the journal is the global distribution of its authors.

International Journal of Astrobiology
is available online at:
http://journals.cambridge.org/ija

**To subscribe contact
Customer Services**

in Cambridge:
Phone +44 (0)1223 326070
Fax +44 (0)1223 325150
Email journals@cambridge.org

in New York:
Phone +1 (845) 353 7500
Fax +1 (845) 353 4141
Email
subscriptions_newyork@cambridge.org

Price information
is available at: **http://journals.cambridge.org/ija**

Free email alerts
Keep up-to-date with new material – sign up at
http://journals.cambridge.org/ija-alerts

For free online content visit:
http://journals.cambridge.org/ija

CAMBRIDGE
UNIVERSITY PRESS

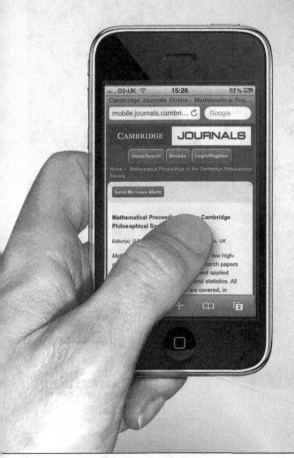

Printed in the United States
by Baker & Taylor Publisher Services

Printed in the United States
by Baker & Taylor Publisher Services